Sustainable Forestry

Sustainable Forestry

Edited by **Aduardo Hapke**

Callisto Reference
New York

Published by Callisto Reference,
106 Park Avenue, Suite 200,
New York, NY 10016, USA
www.callistoreference.com

Sustainable Forestry
Edited by Aduardo Hapke

International Standard Book Number: 978-1-63239-726-3 (Hardback)

Printed in China.

Contents

Preface

This book deals with the topic of sustainable management of forests. It refers to the practice of maintaining and preserving the forest. It works on managing three basic spheres that are ecology, socio-culture and economy. Nurturing the forest will benefit the environment as a whole. This book traces all the developments made in this field. It will give valuable insights into the concepts and innovative techniques used for sustainable forestry. The topics covered in this book will offer the readers a better understanding related to this field. It includes the contributions of experts and scientists which will help the readers keep pace with the latest advances in this field. Thus, this book is a beneficial source of reference for students and experts alike.

Various studies have approached the subject by analyzing it with a single perspective, but the present book provides diverse methodologies and techniques to address this field. This book contains theories and applications needed for understanding the subject from different perspectives. The aim is to keep the readers informed about the progress in the field; therefore, the contributions were carefully examined to compile novel researches by specialists from across the globe.

Indeed, the job of the editor is the most crucial and challenging in compiling all chapters into a single book. In the end, I would extend my sincere thanks to the chapter authors for their profound work. I am also thankful for the support provided by my family and colleagues during the compilation of this book.

Editor

1

Evaluation of sampling strategies to estimate crown biomass

Krishna P Poudel[1], Hailemariam Temesgen[1*] and Andrew N Gray[2]

Abstract

Background: Depending on tree and site characteristics crown biomass accounts for a significant portion of the total aboveground biomass in the tree. Crown biomass estimation is useful for different purposes including evaluating the economic feasibility of crown utilization for energy production or forest products, fuel load assessments and fire management strategies, and wildfire modeling. However, crown biomass is difficult to predict because of the variability within and among species and sites. Thus the allometric equations used for predicting crown biomass should be based on data collected with precise and unbiased sampling strategies. In this study, we evaluate the performance different sampling strategies to estimate crown biomass and to evaluate the effect of sample size in estimating crown biomass.

Methods: Using data collected from 20 destructively sampled trees, we evaluated 11 different sampling strategies using six evaluation statistics: bias, relative bias, root mean square error (RMSE), relative RMSE, amount of biomass sampled, and relative biomass sampled. We also evaluated the performance of the selected sampling strategies when different numbers of branches (3, 6, 9, and 12) are selected from each tree. Tree specific log linear model with branch diameter and branch length as covariates was used to obtain individual branch biomass.

Results: Compared to all other methods stratified sampling with probability proportional to size estimation technique produced better results when three or six branches per tree were sampled. However, the systematic sampling with ratio estimation technique was the best when at least nine branches per tree were sampled. Under the stratified sampling strategy, selecting unequal number of branches per stratum produced approximately similar results to simple random sampling, but it further decreased RMSE when information on branch diameter is used in the design and estimation phases.

Conclusions: Use of auxiliary information in design or estimation phase reduces the RMSE produced by a sampling strategy. However, this is attained by having to sample larger amount of biomass. Based on our finding we would recommend sampling nine branches per tree to be reasonably efficient and limit the amount of fieldwork.

Keywords: Aboveground biomass; Crown; Sampling strategies; Pacific Northwest

* Correspondence: Hailemariam.Temesgen@oregonstate.edu
[1]Department of Forest Engineering, Resources, and Management, College of Forestry, Oregon State University, 280 Peavy Hall, Corvallis, OR 97331, USA
Full list of author information is available at the end of the article

Background

The global issue of climate change and an increasing interest in the reduction of fossil fuel carbon dioxide emissions by using forest biomass for energy production has increased the importance of forest biomass quantification in recent years. Different national and international reports have presented the amount of carbon sequestered by forest ecosystems. For example, the Intergovernmental Panel on Climate Change reports that forests contain about 80% of aboveground and 40% of belowground carbon stock (IPCC 2007). Additionally, it is reported that the amount of carbon stored in dry wood is approximately 50% by weight (Brown 1986; Paladinic et al. 2009; Sedjo and Sohngen 2012).

Biomass, in general, includes both above and below ground living and dead mass of trees, shrubs, vines, and roots. However, most of the researches on biomass estimation have focused on aboveground biomass because of the difficulty in collecting belowground data (Lu 2006). The amount of biomass in a forest is influenced by various site factors such as stand density and site productivity; soil characteristics such as texture and moisture content; and tree characteristics such as species and age. On the other hand, distribution of crown biomass affects the carbon cycle, soil nutrient allocation, fuel accumulation, and wildlife habitat environments in terrestrial ecosystems and it governs the potential of carbon emission due to deforestation (Lu 2005). The major components of aboveground tree biomass are merchantable stem biomass (bole including bark and wood), stump biomass, foliage biomass, and branches/top biomass (Zhou and Hemstrom 2009).

The common biomass estimation approach selects some trees, which are representative of the populations of interest, for destructive sampling and weighs their components. Regression models are then fit to relate some easily measurable attributes, such as diameter at breast height and total tree height, with tree (or component) biomass. The amount of biomass distributed in different components is dependent on species and their geographic location (Pooreter et al. 2012), management practices (Tumwebaze et al. 2013) and tree size and stand density (Jenkins et al. 2003). Ritchie et al. (2013) found that for the given DBH and crown ratio, thinned stands had more foliage biomass but slightly less branch biomass than unthinned stands. Similarly, the contribution of component biomass to the total aboveground biomass varies by tree size (de-Miguel et al. 2014b). Henry et al. (2011) found differences in biomass due to floristic composition, tree species and growth strategies for the tree species within a given climatic zone. Thus, the component biomass estimations, for example branch or crown biomass, bole biomass, and bark biomass, are important to account for the variability within the tree. The common understanding among researchers and practitioners is that an accurate carbon stock estimate requires improved and consistent methods for tree and component biomass estimation (Hansen 2002; Zhou and Hemstrom 2009).

Crown biomass is the oven dry weight of the entire crown, including the leading shoot above the last-formed whorl, excluding the main bole (Hepp and Brister 1982). The components of crown biomass are wood, bark, and foliage weights. Crown biomass accounts for a significant portion of total tree biomass but the amount and its distribution vary by tree and site characteristics. Using the data from two Alaskan *Picea mariana* ecosystems, Barney et al. (1978) reported that foliage comprised 17% to 37% of the total tree mass for the lowland stands and 17% to 50% of the total tree mass in the upland stands. Total bole mass ranged from 11% to 58% in lowland stands and 21% to 61% in the upland stands. In a study to determine the patterns of biomass allocation in dominant and suppressed loblolly pine (*Pinus taeda*), Naidu et al. (1998) found that the dominant trees allocated 24.5% of biomass to the crown (13.2% in branch and 11.3% in needle) and the suppressed trees allocated 12.3% (6.7% in branch and 5.6% in needle). Kuyaha et al. (2013) found that crown biomass formed up to 26% (22% in branch and 4% in needle) of aboveground biomass in farmed eucalyptus species. In assessing the importance of crown dimensions to improve tropical tree biomass estimate, Goodman et al. (2013) found the trees in their study to have nearly half of the total aboveground tree biomass in branches (44% ± 2%).

Estimates of crown biomass for each stand condition is necessary to understand nutrient depletion and for evaluating the economic feasibility of crown utilization for energy production or forest products (Hepp and Brister 1982). Furthermore, estimates of crown biomass aid in fuel load assessments and fire management strategies (He et al. 2013) because it is one of the important input variables in most wildfire models (Saatchi et al. 2007). Much of the focus in estimating crown biomass has been in the form of regression models and in the selection of predictor variables rather than in the methods of sample selection. In addition, comparisons of sampling strategies have been carried out mainly for foliar biomass sampling rather than the total crown (branch wood, bark, and foliage) biomass. Thus, the evaluation of different sampling designs and sample size in estimating crown biomass is an important aspect of aboveground biomass estimation.

Common sampling strategies used in aboveground biomass estimation include simple random sampling, systematic sampling, stratified random sampling, and randomized branch sampling. The suitability of a technique is determined by the availability of funds, required accuracy, structure and composition of vegetation, and desired specificity of estimation (Catchpole and Wheeler 1992). Additionally, the amount of time a particular technique takes to implement in the field is also important. The simple random sampling is generally used as the basis to evaluate the performance of other sampling designs (e.g. Snowdon 1986; Temesgen 2003).

Gregoire et al. (1995) have proposed a number of sampling procedures (randomized branch sampling, importance sampling, control-variate sampling, two-stage and three-stage sampling) that can be used to estimate foliage and other characteristics of individual trees. The randomized branch sampling (RBS) was originally introduced by Jessen (1955) to determine the fruit count on orchard trees. Valentine and Hilton (1977) used this method to obtain estimates of leaf counts, foliar area, and foliar mass of mature *Quercus* species. Good et al. (2001) have employed RBS with importance sampling for estimating tree component biomass. Since the sample is accumulated sequentially along the path, RBS does not require locating and counting the total number of branches beforehand. However, Chiric et al. (2014) posed some doubts on the effectiveness of RBS in sampling big trees or trees with irregular forms. According to Valentine and Hilton (1977), the accuracy of RBS is largely dependent on the probability assignment and the time required to take RBS samples depends on the size of the trees and experience of those taking the samples.

Swank and Schreuder (1974) compared stratified two-phase sampling, two-phase sampling with a regression estimator, and two-phase sampling with a ratio-of-means estimator. They found the stratified two-phase sampling as the most precise and appropriate method for estimating surface area and biomass for a young eastern white pine forest. Temesgen (2003) found that stratified random sampling produced the lowest mean squared error value in comparing five sampling designs to quantify tree leaf area. Stratification in branch biomass sampling can be done in many different ways. Snowdon (1986) showed improved accuracy of estimates by stratification based on crown position compared to those obtained by simple random sampling, especially at low sampling intensities. Their findings suggest that stratification by whorl was slightly but not significantly inferior to stratification based on crown position or branch diameter. Another approach used in selecting branches for estimating crown biomass is to divide the bole into sections and pile up the branches from each section into different size classes and randomly select a number of branches proportional to the total number of branches in each size class (e.g. Harrison et al. 2009, Devine et al. 2013). In an evaluation of ten different sampling strategies, Temesgen et al. (2011) found that systematic sampling with ratio estimation as the most efficient estimate of individual tree foliage biomass. de-Miguel et al. (2014a) developed generalized, calibratable, mixed-effects meta-models for large-scale biomass prediction. One of their objectives was to investigate and demonstrate how the biomass prediction differed when calibration trees were selected using different sampling strategies. They found that a stratified sampling was better compared to the simple random sampling. Thus there is no strong rationale to support one method as being superior to another.

Crown biomass is difficult to predict because of the variability within and among species and various sites. A good allometric equation for predicting aboveground biomass should be based on data collected with an appropriate (precise and unbiased) sampling method. In this context, the objective of this study was to evaluate different sampling strategies to estimate crown biomass. We also evaluated how the performance of different methods was affected when different number of branches (3, 6, 9, and 12) per tree was sampled in estimating crown biomass.

Methods

Study area

This study was conducted in the McDonald-Dunn Forest, an approximately 4,550 ha property, managed by the Oregon State University in the western of the edge of the Willamette Valley in Oregon and on the eastern foothills of the Coast Range (123°15' W, 44°35' N, 120 m elevation). The forest consists predominantly of the Douglas-fir (*Pseudotsuga menziesii* (Mirbel) Franco) and a small Grand fir (*Abies grandis* (Dougl. ex D. Don) Lindl.) and has a wide range of overstory age-class distribution with majority of the stands less than 80 years old and some stands that are 80 to 120 years old. The forest receives approximately 110 cm of annual rainfall and average annual temperature ranges from 6°C to 17°C.

Data

Twenty sample trees (11 Douglas-fir and 9 Grand fir) were subjectively selected from stands of different ages for destructive sampling avoiding the trees with obvious defects and trees close to stand edges. The field work was carried out between the first week of July and third week of September 2012. Trees that were forked below breast height and with damaged tops were not included in sampling. Tree level attributes including total height, height to the base of first live branch, crown width, and main stem diameter at 0.15, 0.76, 1.37, and 2.40 m above ground, and every 1.22 m afterwards were recorded. The branches were divided into four diameter classes (1.3 cm class = 0–2.5 cm, 3.8 cm class = 2.6–5.1 cm, 6.4 cm class = 5.2–7.6 cm, 8.9 cm class = 7.7–10.2 cm). For all first order branches, height to- and diameter- at branch base were measured.

For the first and every third branch, when proceeding from the base, in each diameter class, length and weight of both live and dead branches were recorded. From those selected branches, four branches per diameter class were weighed with and without foliage. The needles were removed in the field to obtain the green weight of foliage and branch wood with bark. Two of these four branches were taken to the lab, keeping branch and foliage in separate paper bags, for drying. The branches were chipped in to small pieces to expedite the drying process and placed in a kiln for drying at 105°C. The oven dry weight was recorded by tracking the weight lost by each sample until no further weight was lost. Table 1 presents the tree and branch level summary of the felled-tree data used in this study.

Individual branch biomass

Kershaw and Maguire (1995) developed a tree specific log linear model (Equation 1) using branch diameter (BD) and depth into the crown (DINC: the distance from tip to the base of the branch) as covariates to estimate branch foliage biomass. Temesgen et al. (2011) successfully used this model in comparing sampling strategies for tree foliage biomass estimation.

$$\ln\left(y_{ij}\right) = \beta_{0i} + \beta_{1i}\ln\left(BD_{ij}\right) + \beta_{2i}\ln\left(DINC_{ij}\right) + \varepsilon_{ij} \qquad (1)$$

This model was modified by replacing DINC with branch length (Equation 2). The modified model provided the best fit (Adj-R^2 = 0.93), therefore was used to predict individual branch biomass within each tree.

$$\ln\left(y_{ij}\right) = \beta_{0i} + \beta_{1i}\ln\left(BD_{ij}\right) + \beta_{2i}\ln\left(BL_{ij}\right) + \varepsilon_{ij} \qquad (2)$$

where y_{ij}, BD_{ij} and BL_{ij} are oven dry weight (kg) of branch (wood, bark, and foliage combined), branch diameter (cm), and branch length (m) of the jth branch on i^{th} tree respectively; β_{ij} 's are regression parameters to be estimated; ln(·) is the natural logarithm; and ε_{ij} 's are the random error. The full model included other variables such as height to the base of the branch, crown width and crown length, but were dropped because they were not statistically significant (p-value > 0.05). Lengths for the 2/3rd branches (not measured in the field) were obtained by fitting the following log linear model (Adj-R^2 = 0.74):

$$\ln\left(BL_{ij}\right) = \beta_{0i} + \beta_{1i}\ln\left(BD_{ij}\right) + \beta_{2i}\ln\left(RBD_{ij}\right) + \varepsilon_{ij} \qquad (3)$$

where BD_{ij}, BL_{ij}, and ε_{ij} are same as defined in Equation 2 and RBD_{ij} is the relative branch depth (relative position of the subject branch from the crown base) of jth branch in i^{th} tree and is computed as follows (Ishii and Wilson 2001):

$$RBD = \frac{\text{total tree height} - \text{height to the base of subject branch}}{\text{total tree height} - \text{height to the base of lowest live branch}}$$

The RBD is 1.0 for the first live branch. The logarithmic regressions are reported to result in a negative bias when data are back transformed to arithmetic scale. The commonly used remedy to this is to multiply the back transformed results by a correction factor $\left[exp\left(\frac{MSE}{2}\right)\right]$, where MSE is the mean squared error obtained by the

Table 1 Summary of felled-tree and branch-level attributes used in this study

Variable	*n*	Mean	SD	Minimum	Maximum
Tree data					
DBH (cm)		62.8	27.0	23.9	114.0
Height (m)		34.6	10.3	18.3	46.5
Crown base height (m)		10.2	6.3	0.8	23.7
Crown length (m)		24.4	6.2	12.2	34.4
Crown width (m)		8.6	2.4	5.2	13.4
Branch data					
Diameter (cm)	3464	3.7	2.1	0.8	10.4
Length (m)	1178	2.8	1.6	0.1	10.2
Total green weight (kg)	1102	4.14	5.61	0.05	46.99
Green weight of branch wood (kg)	326	3.98	5.61	0.05	38.01
Green foliage weight (kg)	264	1.80	1.84	0.05	9.71
Total dry weight (kg)	128	3.44	4.08	0.12	18.40
Dry weight of branch wood (kg)	128	2.68	3.49	0.07	15.37
Dry foliage weight (kg)	128	0.76	0.79	0.04	4.58

least-squares regression. However, there are conflicting remarks about the correction factor itself. For example Beauchamp and Olson (1973) and Flewelling and Pienaar (1981) suggested that this correction factor was still biased because the sample variance is consistent but it is biased for finite sample sizes. We did not use the correction factor in our study. The trend in the relationship between crown biomass and branch diameter and length was similar but the variability in biomass increased with increasing branch length (Figure 1). All statistical procedures were performed using statistical software R (R Core Team 2014).

Methods for crown biomass sampling

We evaluated 11 sampling methods to select branches for estimating crown biomass. The 11 sampling strategies belonged to three main categories: simple random sampling, systematic sampling, and stratified sampling. Methods 1 and 2 are based on simple random sampling (SRS) strategy. In each of these methods, each branch was chosen randomly such that each individual branch has equal probability of selection at any stage of selection. The difference in these methods is in the estimation of total tree biomass: method 1 uses SRS estimator while method 2 (SRS-RAT) uses the ratio estimator with squared branch diameter as auxiliary information. Method 1 is also the basis for comparing the performance of other methods.

Method 3, probability proportional to size (PPS), uses branch size as auxiliary information in sample selection. Total crown biomass in this method was calculated using Horvitz-Thompson estimator (Horvitz and Thompson 1952). Methods 4 (SYS) and 5 (SYS-RAT) are systematic sampling with similar design phase but different estimation phase. Method 4 uses the SRS estimator while method 5 uses the ratio estimator. The fractional interval systematic sample selection procedure was used in the systematic selection of the branches because it ensures the equal probability of selection for all the branches (Temesgen et al. 2011). The interval was determined based on the total number of branches in each tree. In fractional interval systematic sample selection, first a random starting point between 1 and total number of branches was randomly chosen, the interval then is added obtaining exactly n (sample size) branches. Then the numbers are divided by the sample size and rounded to the nearest whole number to get the selected samples.

Methods 6–11 belonged to different stratified sampling strategies. The stratified sampling method divides the population into subpopulations of size n_h, where n_h is the number of elements in stratum h. The total crown length was divided into three sections having equal number of branches as three strata. In methods 6 (STR) and 7 (STR-RAT), $n/3$ branches were randomly selected with equal probability, where n is the sample size. Again, the difference between these two methods lies in the estimation of total crown biomass. STR method uses the SRS estimation technique while STR-RAT method uses the ratio estimation technique to obtain the total crown biomass. Method 8 (STR-PPS), stratified sampling with PPS, selected branches in each stratum with probability proportional to the square of branch diameter. Total crown biomass in this method was obtained by summing the stratum total crown biomass calculated using Horvitz-Thompson unequal probability estimator (Horvitz and Thompson 1952).

Methods 9–11 (stratified, unequal) are based on the idea that the distribution of crown biomass in different strata depends on the relative position of the branches in the tree. Ishii and McDowell (2001) found that mean branch volume increased from upper- to lower-crown. For a given density, biomass (oven dry weight) is the function of volume. Therefore, the stratified sampling method was modified to incorporate the variability of biomass distribution within a tree. Trees were first divided into three sections having equal number of branches. Then, 4, 3, and 2 branches from the lower, middle, and upper section of the trees were selected respectively. This corresponds that the number of branches selected in each section is proportional to the observed biomass in that section of the tree. Because stratification based on crown length resulted in the biased estimation of crown biomass, the balanced stratification method was applied. The total number of

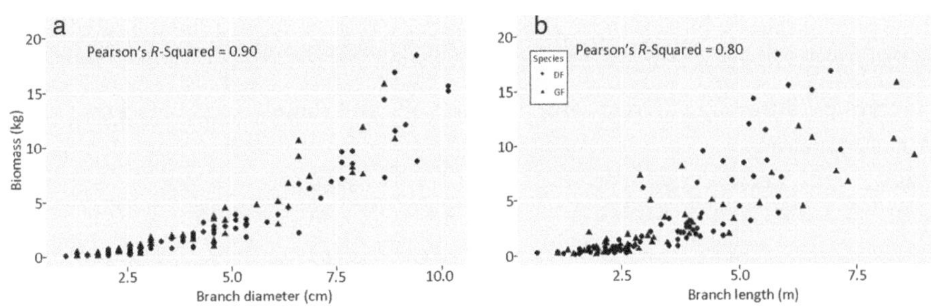

Figure 1 Scatterplot of dry biomass (kg) against branch diameter (a) and branch length (b) by species (DF = Douglas-fir, GF = Grand fir).

Table 2 Summary of methods used for crown biomass estimation in this study

Methods	Equations for total crown biomass	Selection probability	Inclusion probability
Simple random sampling			
SRS	$\hat{\tau}_i = NB_i\left(\frac{1}{n}\sum_{j=1}^{n} y_{ij}\right)$	$\left(\frac{1}{NB_i}\right)^n$	$\frac{n}{NB_i}$
SRS-RAT	$\hat{\tau}_i = \frac{\sum_{j=1}^{n} y_{ij}}{\sum_{j=1}^{n} BD_{ij}^2}\sum_{j=1}^{NB_i} BD_{ij}^2$	$\left(\frac{1}{NB_i}\right)^n$	$\frac{n}{NB_i}$
PPS	$\hat{\tau}_i = \sum_{j\in S}\frac{y_{ij}}{\pi_{ij}^{PPS}}$	$\pi_{ij} = \frac{BD_{ij}^2}{\sum_{j=1}^{NB} BD_{ij}^2}$	$\pi_{ij}^{PPS} = 1-\left(1-\pi_{ij}\right)^n$
Systematic sampling			
SYS	$\hat{\tau}_i = NB_i\left(\frac{1}{n}\sum_{j=1}^{n} y_{ij}\right)$	$\frac{n}{NB_i}$	$\frac{n}{NB_i}$
SYS-RAT	$\hat{\tau}_i = \frac{\sum_{j=1}^{n} y_{ij}}{\sum_{j=1}^{n} BD_{ij}^2}\sum_{j=1}^{NB_i} BD_{ij}^2$	$\frac{n}{NB_i}$	$\frac{n}{NB_i}$
Stratified sampling			
STR	$\hat{\tau}_i = \sum_{h=1}^{H}\sum_{j=1}^{n_h}\frac{N_{ih}}{n_h} y_{ijh}$	$\frac{BD_{ij}^2}{\sum_{j=1}^{NB} BD_{ij}^2}$	$\frac{n_h}{N_{ih}}$
STR-RAT	$\hat{\tau}_i = \frac{\sum_{h=1}^{H}\sum_{j=1}^{n_h}\frac{N_{ih}}{n_h} y_{ijh}}{\sum_{h=1}^{H}\sum_{j=1}^{n_h}\frac{N_{ih}}{n_h} BD_{ijh}^2}$	$\frac{BD_{ij}^2}{\sum_{j=1}^{NB} BD_{ij}^2}$	$\frac{n_h}{N_{ih}}$
STR-PPS	$\hat{\tau}_{i(STR-PPS)} = \sum_{h=1}^{H}\sum_{j\in S_h}\frac{y_{ijh}}{\pi_{ijh}^{(STR-PPS)}}$	$\frac{BD_{ijh}^2}{\sum_{j=1}^{NB_{ih}} BD_{ijh}^2}$	$\pi_{ij}^{STR-PPS} = 1-\left(1-\pi_{ij}\right)^n$

Notation: $\hat{\tau}_i$ is the estimated total crown biomass for i^{th} tree; y_{ij} is the oven dry weight of jth branch on i^{th} tree; NB_i is the number of branches on i^{th} tree; n is number of branches sampled; N_{ih} is the number of branches in h^{th} stratum on i^{th} tree; and n_h is the number of branches sampled in h^{th} stratum.

branches selected in each tree (nine) was determined based on the amount of biomass sampled. Total crown biomass in each stratum was computed using the SRS estimation technique in method 9 (Un-STR), PPS in method 10 (Un-PPS), and ratio estimation in method 11 (Un-STRRAT). Total crown biomass in each tree was computed by summing the crown biomass in each stratum. The unequal branch selection strategy was also evaluated using similar evaluation statistics used for the other eight methods.

Performances of first eight methods were evaluated by selecting four different sample sizes (3, 6, 9, and 12) in each tree. These sample sizes were chosen for the ease of distributing samples into three different strata in stratified sampling with equal number of branches per stratum. Methods 9–11 were based on selecting nine branches in each tree. Table 2 summarizes the inclusion probability, selection probability, and the estimator of the total crown biomass in each of the sampling strategies evaluated in this study.

Evaluation of sampling strategies

We evaluated the performance of 11 sampling strategies to estimate crown biomass using the following six statistics estimated from 5,000 iterations. These measures were successfully used to evaluate the performance of sampling strategies to estimate foliage biomass in Temesgen et al. (2011).

Table 3 Average bias (kg) produced by different sampling methods and sample sizes based on 5,000 simulations

Sample size	Sampling strategies							
	SRS	SRS-RAT	PPS	SYS	SYS-RAT	STR	STR-RAT	STR-PPS
3	0.237	−4.091	−0.158	0.433	−9.320	−0.659	−3.554	−0.104
6	0.215	−1.711	0.010	0.139	−1.922	0.399	−1.387	−0.101
9	−0.081	−0.937	−0.068	0.133	0.166	0.191	−0.857	−0.094
12	−0.030	−0.894	0.158	0.043	−0.078	−0.242	−0.776	−0.073

Table 4 Relative bias (percent) produced by different sampling methods and sample sizes based on 5,000 simulations

Sample size	Sampling strategies							
	SRS	SRS-RAT	PPS	SYS	SYS-RAT	STR	STR-RAT	STR-PPS
3	0.101	−2.437	−0.169	0.107	−6.086	−0.321	−2.107	−0.034
6	−0.001	−1.115	0.024	0.061	−1.061	0.197	−0.900	−0.012
9	0.063	−0.600	−0.027	0.038	0.033	0.148	−0.543	−0.069
12	−0.011	−0.523	0.100	0.037	0.032	−0.035	−0.439	−0.040

1. **Bias:** For each tree the bias (kg) was calculated as the mean difference between observed and predicted total crown biomass for that tree as follows:

$$B_i = \frac{1}{5000}\sum_{s=1}^{5000}(\tau_{is} - \hat{\tau}_{is})$$

where τ_{is} and $\hat{\tau}_{is}$ are the observed and predicted total crown biomasses for i^{th} tree in s^{th} iteration, respectively.

2. **Relative bias:** Relative bias percentage is the ratio of bias to the total observed crown biomass for that tree and computed as follows:

$$RB_i = \frac{1}{5000}\sum_{s=1}^{5000}\frac{(\tau_{is} - \hat{\tau}_{is})}{\tau_{is}}$$

where all the variables are same as defined previously.

3. **Root mean square error (RMSE):**

$$RMSE_i = \sqrt{\frac{1}{5000}\sum_{s=1}^{5000}(\tau_{is} - \hat{\tau}_{is})^2}$$

4. **Relative RMSE:**

$$R\text{-}RMSE_i = \sqrt{\frac{1}{5000}\sum_{s=1}^{5000}\left(\frac{\tau_{is} - \hat{\tau}_{is}}{\tau_{is}}\right)^2}$$

5. **Biomass sampled (BS):** Amount of cost for crown biomass estimation is directly proportional to the amount of crown biomass sampled. Therefore the amount of crown biomass sampled was also used as a criterion for the evaluation of sampling strategies. The amount of crown biomass sampled (sampling intensity) is calculated as follows:

$$BS_i = \frac{1}{5000}\sum_{s=1}^{5000}\sum_{j\in S} y_{ijs}$$

y_{ijs} is the observed total crown biomasses for i^{th} tree, j^{th} sample branch in s^{th} iteration.

6. **Relative biomass sampled (RBS%)** indicates the proportion of crown biomass sampled with respect to the total crown biomass measured and is calculated as follows:

$$RBS_{ij} = \frac{1}{5000}\sum_{s=1}^{5000}\sum_{j\in S}\frac{y_{ijs}}{\tau_{ijs}}$$

Results and discussions

Except for the ratio estimators, the estimators of population totals were unbiased, with biases close to zero for all sample sizes (Tables 3 and 4). The squared bias for these methods ranged from zero to 0.435 kg. Ratio estimators resulted in greater bias than the other methods. The absolute bias of the ratio estimators decreased with increasing sample size as expected.

As expected, the RMSE (and relative RMSE) decreased with increasing sample size (Tables 5 and 6) for all sampling strategies. Based on the RMSE values obtained from 5,000 simulations, the stratified sampling with PPS estimation was the superior method compared to

Table 5 Average RMSE produced by different sampling methods and sample sizes based on 5,000 simulations

Sample size	Sampling strategies							
	SRS	SRS-RAT	PPS	SYS	SYS-RAT	STR	STR-RAT	STR-PPS
3	88.57	42.56	38.04	113.17	48.54	85.05	42.08	34.95
6	60.49	30.01	29.23	57.94	29.98	58.54	29.56	27.99
9	48.60	23.92	25.19	37.88	22.12	46.48	23.55	24.46
12	40.77	20.18	22.78	30.77	17.91	38.89	19.75	22.33

Table 6 Relative RMSE percent produced by different sampling methods and sample sizes based on 5,000 simulations

Sample size	Sampling strategies							
	SRS	SRS-RAT	PPS	SYS	SYS-RAT	STR	STR-RAT	STR-PPS
3	55.70	25.72	23.46	72.02	28.80	53.41	25.61	21.57
6	38.07	17.93	18.23	35.23	17.53	36.63	17.78	17.59
9	30.38	14.19	15.83	23.58	13.29	29.03	14.07	15.55
12	25.42	11.92	14.40	18.37	11.11	24.27	11.76	14.22

all other methods when sample size is 3 or 6 branches per tree. However, while using PPS, stratification of the crown into sections did not reduce the RMSE and relative RMSE significantly. On the other hand, when at least nine branches per tree were sampled, the SYS-RAT was the best and the SRS-RAT was the second best method. Number of branches required to achieve desired precision is another important aspect of estimating crown biomass. On average, the RMSE decreased by 34.3% when the sample size increased from three branches per tree to six branches per tree. The RMSE further decreased by 22.1% and 15.4% when the sample size increased from 6 to 9 and 9 to 12 respectively.

The amount of biomass sampled determines the cost that would be incurred in estimating crown biomass. Biomass sampled and relative biomass sampled in different sampling strategies are presented in Tables 7 and 8. The Strategy-Cost-Accuracy graph (Figure 2) shows the efficiency trade-off across the strategies compared in the study. The SRS and SYS methods resulted in the lowest amount of biomass sampled. On average, the amount of biomass sampled using the PPS method was 1.6, 1.5, 1.4, and 1.4 times higher than the amount of biomass sampled in stratified random sampling when 3, 6, 9, and 12 branches per tree were selected respectively.

On average, selecting 12 instead of 9 branches per tree increased the amount of biomass sampled by 29.2%. Therefore, nine branches in each tree were selected in evaluating the performance of unequal stratified sampling strategy. Results from unequal branch selection are presented in Table 9. This strategy reduced the relative RMSE by 0.6%, 4.5% and 3.5% compared to selecting 9 branches using stratified random sampling, stratified sampling with ratio estimation, and stratified sampling with PPS respectively. This reduction in relative RMSE is obtained by sampling just a little more amount of biomass (1.03 times on average).

Use of allometric equations is inevitable in above-ground biomass estimation because weighing the trees and their components for direct biomass determination is destructive and prohibitively expensive. Choice of biomass sampling strategy determines the quality of data available for fitting such equations. Use of auxiliary information in design and/or estimation phase (ratio estimation and PPS) produced better results in terms of RMSE compared to the methods that do not make use of such information in this study. Previous researches (e.g. Temesgen et al. 2011) have also shown the benefits from using auxiliary information in the design and/or estimation concerning tree biomass.

The model used in estimating branch biomass which is later used as a dependent variable in the test population, was a logarithmic model (Equation 2). There is an inherent negative bias in this method because the dependent variable is transformed prior to estimation (Snowdon 1991). The ratio estimation strategies, SRS-RAT, SYS-RAT, and STR-RAT in this study, were negatively biased. However, in terms of RMSE, these strategies were clearly superior methods compared to the SRS approach. As noted in Temesgen et al. (2011), however, the efficiency of sampling strategies with ratio estimation may be affected by the amount of work and difficulty in implementing these techniques in the field.

The amount of biomass sampled determines the cost that would be incurred in estimating crown biomass. Choice of a sampling strategy determines the amount of biomass and relative biomass sampled. This ultimately determines the amount of time and cost required for a biomass estimation project. The SRS and SYS methods resulted in the lowest amount of biomass sampled. Our

Table 7 Amount of biomass sampled (kg) by different sampling strategies and sample sizes

Sample size	Sampling strategies				
	SRS	PPS	SYS	STR	STR-PPS
3	7.74	12.58	7.73	7.68	12.40
6	15.49	23.79	15.46	15.50	23.29
9	23.17	33.77	23.18	23.19	33.05
12	30.91	42.83	30.92	30.86	41.79

Table 8 Relative amount of biomass sampled (%) by different sampling strategies and sample sizes

Sample size	Sampling strategies				
	SRS	PPS	SYS	STR	STR-PPS
3	5.36	8.67	5.35	5.33	8.45
6	10.71	16.37	10.71	10.72	15.87
9	16.08	23.16	16.06	16.07	22.45
12	21.42	29.27	21.43	21.40	28.33

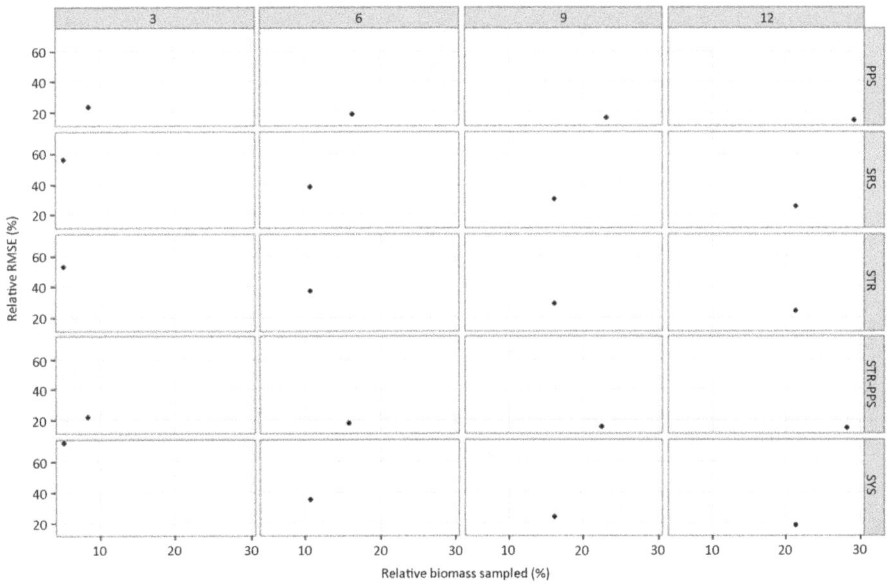

Figure 2 Relative RMSE (%) produced Vs. relative biomass sampled (percent of total crown mass) in different sampling strategies and sample sizes.

results in terms of RMSE values reported and the amount of biomass sampled by each strategy are consistent with the findings of Temesgen et al. (2011) in estimating foliar biomass of Douglas-fir (*Pseudotsuga menziesii* var. *menziesii*) and ponderosa pine (*Pinus ponderosa*).

Conclusions

Crown biomass estimation is a complex process that requires intensive manual field work involving destructive sampling. The amount of fieldwork required and the accuracy of biomass estimation is dependent on the sampling strategy used. Furthermore, the accuracy of the estimation can be improved by adopting appropriate techniques in both the design and estimation phases, beginning with the selection of sample plots and sample trees through model development. In this study, we evaluated 11 different sampling strategies that belonged to three main categories: simple random sampling, systematic sampling and stratified sampling. The SRS, PPS, and ratio estimation techniques were used to obtain the total crown biomass in each tree.

Based on the RMSE values obtained from 5,000 simulations, the stratified sampling with PPS estimation produced better results as compared to all other methods when 3 or 6 branches per tree were sampled. However, the SYS-RAT was the best and the SRS-RAT was the second best method when at least nine branches per tree were sampled. It should also be noted that the lower RMSE values in the PPS estimation techniques are obtained with an increased amount of biomass sampled in each tree. On the other hand, if the auxiliary information on branch size is not used, the systematic sampling provided better results than the SRS or STR method when at least 6 branches per trees were selected. Thus the selection of a specific sampling strategy is dependent on the availability of the time and cost for the given biomass sampling project. Based on our finding we would recommend sampling 9 branches per tree to obtain reasonable efficiency and amount of work involved in the field.

The logic for selecting unequal numbers of branches per stratum within a tree is justified by the fact that the biomass distribution within a tree is not uniform. Selecting equal branches per stratum produced approximately similar results to unequal sampling when the SRS estimation technique was used. However, making use of auxiliary information on branch size in the design and estimation phases further decreased the relative RMSE. Once again, the decreased RMSE by use of auxiliary information is attained by having to sample slightly higher

Table 9 Evaluation statistics produced when selecting 4, 3, and 2 branches from lower, middle, and upper stratum

Method	Bias	Relative bias	RMSE	Relative RMSE	Biomass sampled	Relative biomass sampled
Un-STR	0.100	0.061	46.16	28.87	24.25	16.87
Un-STRRAT	−0.731	−0.491	22.58	13.44		
Un-PPS	0.140	0.063	23.67	15.00	33.85	23.04

amount of biomass. Findings of this study should prove beneficial for the stakeholders working in the field of aboveground biomass and carbon estimation. Additional work using the data from different species and location should be done to further validate the findings in this study.

Competing interests
The authors declare that they have no competing interests.

Authors' contributions
KP designed the sampling experiments, performed the sampling experiments, analyzed the data, and wrote the paper. TH conceived the sampling experiments, critically reviewed the manuscript, contributed to coding and data analysis, and edited the manuscript. AG critically reviewed the manuscript, edited the manuscript and contributed ideas at all phases of the project. All authors read and approved the final manuscript.

Acknowledgements
We thank Professors Lisa Madsen and Glen Murphy (both at Oregon State University) for their insights and comments on an earlier draft, and the Forest Inventory Analysis Unit for funding the data collection and analysis phases of this project.

Author details
[1]Department of Forest Engineering, Resources, and Management, College of Forestry, Oregon State University, 280 Peavy Hall, Corvallis, OR 97331, USA. [2]USDA Forest Service, PNW Research Station, 3200 SW Jefferson Way, Corvallis, OR 97331, USA.

References
Barney RJ, Vancleve K, Schlenter R (1978) Biomass distribution and crown characteristics in two Alaskan Picea mariana ecosystems. Can J For Res 8:36–41

Beauchamp JJ, Olson JS (1973) Corrections for bias in regression estimates after logarithmic transformation. Ecology 54(6):1403–1407

Brown S (1986) Estimating Biomass and Biomass Change of Tropical Forests: A Primer. FAO Forestry Paper 134. Food and Agriculture Organization of the United Nations, Rome

Catchpole WR, Wheeler CJ (1992) Estimating plant biomass: a review of techniques. Aust J Ecol 17:121–131

Chiric G, Puletti N, Salvati R, Arbi F, Zolli C, Corona P (2014) Is randomized branch sampling suitable to assess wood volume of temperate broadleaved old-growth forests? For Ecol Manag 312:225–230

R Core Team (2014) R: A language and environment for statistical computing. R Foundation for Statistical Computing, Vienna, Austria. http://www.R-project.org/

de-Miguel S, Mehtatlo L, Durkaya A (2014a) Developing generalized, calibratable, mixed-effects meta models for large-scale biomass prediction. Can J For Res 44:648–656, a or b

de-Miguel S, Pukkala T, Assaf N, Shater Z (2014b) Intra-specific difference in allometric equations for aboveground biomass of eastern Mediterranean *Pinus brutia*. Ann For Sci 71:101–112, a or b

Devine WD, Footen PW, Harrison RB, Terry TA, Harrington CA, Holub SM, Gould PJ (2013) Estimating Tree Biomass, Carbon, and Nitrogen in two Vegetation Control Treatments in an 11-Year-old Douglas-fir Plantation on a Highly Productive Site. Res. Pap. PNW-RP-591. U.S. Department of Agriculture, Forest Service, Pacific Northwest Research Station, Portland, OR, p 29

Flewelling JW, Pienaar LV (1981) Multiplicative regression with lognormal errors. For Sci 27(2):281–289

Good NM, Paterson M, Brack C, Mengersen K (2001) Estimating tree component biomass using variable probability sampling methods. J Agric Biol Environ Stat 6(2):258–267

Goodman RC, Phillips OL, Baker TR (2013) The importance of crown dimensions to improve tropical tree biomass estimates. Ecol Appl. http://dx.doi.org/10.1890/13-0070.1

Gregoire TG, Valentine HT, Furnival GM (1995) Sampling methods to estimate foliage and other characteristics of individual trees. Ecology 76(4):1181–1194

Hansen M (2002) Volume and biomass estimation in FIA: national consistency vs. regional accuracy. In: McRoberts RE, Reams GA, Van Deusen PC, Moser JW (eds) Proceedings of the third annual Forest Inventory and Analysis symposium. General Technical Report NC-230. U.S. Department of Agriculture, Forest Service, North Central Research Station, St. Paul, MN, pp 109–120

Harrison RB, Terry TA, Licata CW, Flaming BL, Meade R, Guerrini IA, Strahm BD, Xue D, Lolley MR, Sidell AR, Wagoner GL, Briggs D, Turnblom EC (2009) Biomass and stand characteristics of a highly productive mixed Douglas-Fir and Western Hemlock plantation in Coastal Washington. West J Appl For 24(4):180–186

He Q, Chen E, An R, Li Y (2013) Above-ground biomass and biomass components estimation using LiDAR data in a coniferous forests. Forests 4:984–1002

Henry M, Picard N, Trotta C, Manlay RJ, Valentini R, Bernoux M, Saint-André L (2011) Estimating tree biomass of sub-Saharan African forests: a review of available allometric equations. Silva Fenn 45(3B):477–569

Hepp TE, Brister GH (1982) Estimating crown biomass in loblolly pine plantations in the Carolina Flatwoods. For Sci 28(1):115–127

Horvitz DG, Thompson DJ (1952) A generalization of sampling without replacement from a finite universe. J Am Stat Assoc 47:663–685

IPCC (2007) Climate change 2007: synthesis report. In: Core Writing Team, Pachauri RK, Reisinger A (eds) Contribution of working groups I, II and III to the fourth assessment report of the intergovernmental panel on climate change. IPCC, Geneva, Switzerland, p 104

Ishii H, McDowell N (2001) Age-related development of crown structure in coastal Douglas-fir trees. For Ecol Manag 169:257–270

Ishii H, Wilson ME (2001) Crown structure of old-growth Douglas-fir in the western Cascade Range, Washington. Can J For Res 31:1250–1261

Jenkins CJ, Chojnacky DC, Heath LS, Birdsey RA (2003) National-scale biomass estimators for United States tree species. For Sci 49(1):12–35

Jessen RJ (1955) Determining the fruit count on a tree by randomized branch sampling. Biometrics 11(1):99–109

Kershaw JA, Maguire DA (1995) Crown structure in Western hemlock, Douglas-fir, and grand fir in western Washington: trends in branch-level mass and leaf area. Can J For Res 25:1897–1912

Kuyaha S, Dietz J, Muthuri C, Noordwijk MV, Neufeldt H (2013) Allometry and partitioning of above- and below-ground biomass in farmed eucalyptus species dominant in Western Kenyan agricultural landscapes. Biomass Bioenergy 55:276–284

Lu D (2005) Aboveground biomass estimation using Landsat TM data in the Brazilian Amazon. Int J Remote Sens 26(12):509–2525

Lu D (2006) The potential and challenge of remote sensing-based biomass estimation. Int J Remote Sens 7:1297–1328

Naidu SL, DeLucia EH, Thomas RB (1998) Contrasting patterns of biomass allocation in dominant and suppressed loblolly pine. Can J For Res 28:1116–1124

Paladinic E, Vuletic D, Martinic I, Marjanovic H, Indir K, Benko M, Novotny V (2009) Forest biomass and sequestered carbon estimation according to main tree components on the forest stand scale. Period Biol 111(4):459–466

Pooreter H, NIklas KJ, Reich PB, Oleksyn J, Poot P, Mommer L (2012) Biomass allocation to leaves, stems and roots: meta-analyses of interspecific variation and environmental control. New Phytol 193:30–50

Ritchie MW, Zhang J, Hamilton TA (2013) Aboveground tree biomass for *Pinus ponderosa* in northeastern California. Forests 4:179–196

Saatchi S, Halligan K, Despain DG, Crabtree RL (2007) Estimation of forest fuel load from radar remote sensing. IEEE Trans Geosci Remote Sens 45:1726–1740

Sedjo R, Sohngen B (2012) Carbon sequestration in forests and soils. Annu Rev Resour Econ 4:127–144

Snowdon P (1986) Sampling strategies and methods of estimating the biomass of crown components in individual trees of *Pinus radiata* D Don. Aust For Res 16(1):63–72

Snowdon P (1991) A ratio estimator for bias correction in logarithmic regressions. Can J For Res 21:720–724

Swank WT, Schreuder HT (1974) Comparison of three methods of estimating surface area and biomass for a forest of young eastern white pine. For Sci 20:91–100

Temesgen H (2003) Evaluation of sampling alternatives to quantify tree leaf area. Can J For Res 33:82–95

Temesgen H, Monleon V, Weiskittel A, Wilson D (2011) Sampling strategies for efficient estimation of tree foliage biomass. For Sci 57(2):153–163

Tumwebaze SB, Bevilacqua E, Briggs R, Volk T (2013) Allometric biomass equations for tree species used in agroforestry systems in Uganda. Agroforest Syst 87:781–795
Valentine HT, Hilton SJ (1977) Sampling oak foliage by the randomized-branch method. Can J For Res 7:295–298
Zhou X, Hemstrom MA (2009) Estimating aboveground tree biomass on forest land in the Pacific Northwest: a comparison of approaches. Res. Pap. PNW-RP-584. U.S. Department of Agriculture, Forest Service, Pacific Northwest Research Station, Portland, OR, p 18

2

Optimizing continuous cover management of boreal forest when timber prices and tree growth are stochastic

Timo Pukkala

Abstract

Background: Decisions on forest management are made under risk and uncertainty because the stand development cannot be predicted exactly and future timber prices are unknown. Deterministic calculations may lead to biased advice on optimal forest management. The study optimized continuous cover management of boreal forest in a situation where tree growth, regeneration, and timber prices include uncertainty.

Methods: Both anticipatory and adaptive optimization approaches were used. The adaptive approach optimized the reservation price function instead of fixed cutting years. The future prices of different timber assortments were described by cross-correlated auto-regressive models. The high variation around ingrowth model was simulated using a model that describes the cross- and autocorrelations of the regeneration results of different species and years. Tree growth was predicted with individual tree models, the predictions of which were adjusted on the basis of a climate-induced growth trend, which was stochastic. Residuals of the deterministic diameter growth model were also simulated. They consisted of random tree factors and cross- and autocorrelated temporal terms.

Results: Of the analyzed factors, timber price caused most uncertainty in the calculation of the net present value of a certain management schedule. Ingrowth and climate trend were less significant sources of risk and uncertainty than tree growth. Stochastic anticipatory optimization led to more diverse post-cutting stand structures than obtained in deterministic optimization. Cutting interval was shorter when risk and uncertainty were included in the analyses.

Conclusions: Adaptive optimization and management led to 6%–14% higher net present values than obtained in management that was based on anticipatory optimization. Increasing risk aversion of the forest landowner led to earlier cuttings in a mature stand. The effect of risk attitude on optimization results was small.

Keywords: Adaptive optimization; Anticipatory optimization; Stochastic optimization; Risk preferences; Risk; Uncertainty; Reservation price

Background

Maximizing the economic benefits from timber production is equal to maximizing the net present value of future net incomes. Unfortunately, the future net incomes are unknown at the moment when management decision should be made. Future net incomes depend on future timber prices, which show substantial temporal variation (Leskinen and Kangas 1998).

Also the growth and development of trees and stands are poorly known. Deterministic models explain only a part of the growth variation between years, stands and trees. Measurements of past growth show that there are periods of good growth while in other years or during longer periods trees grow less than the long-term average (e.g. Pasanen 1998). In addition to these weather-related seasonal variations in annual growth, there are also between-tree growth differences which cannot be explained by deterministic models. Another factor causing uncertainty in growth prediction is climate change. It is usually assumed that the growth rate will increase in the boreal forests of North Europe (e.g. Pukkala and

Correspondence: timo.pukkala@uef.fi
University of Eastern Finland, PO Box 11180101 Joensuu, Finland

Kellomäki 2012), but the estimated growth trends represent very uncertain knowledge.

Flowering, pollination, seed production and germination are sub-processes of the regeneration process of trees and stands. All these sub-processes are very sensitive to weather conditions such as temperature and rainfall. In addition, the eventual size of the seed crop depends on the fluctuations of seed predators and seed diseases. Since many sub-processes critical to regeneration success depend on weather conditions, it is impossible to predict the exact amount of regeneration in a certain year in the future, even when there are plenty of empirical regeneration data to fit models. The best that can be done is to predict the distribution of regeneration results or the probability of successful regeneration. Mortality is also hard to predict exactly. However, the so-called regular mortality (competition-related mortality) is very low in regularly thinned managed boreal forest. Therefore, if catastrophic events are excluded from the analysis (like in this study) uncertainty in mortality does not add much to the total degree of uncertainty in the prediction of stand development. For an attempt to include catastrophic events see Zhou and Buongiorno (2006).

The above discussion shows that decisions on future forest management must be made under risk and uncertainty. Risk is usually understood to be a situation in which the probabilities of different states of nature are known, which makes it possible to calculate the distribution of outcomes for a certain decision alternative. Uncertainty refers to situations in which the probabilities are unknown. The prevailing situation is uncertainty. However, to make analyses easier, the situation is transformed from uncertainty to risk, by assuming some distributions for the uncertain factors. This allows the analyst to calculate the probabilities of different outcomes of decision alternatives.

Forest landowners have different attitudes toward risk and uncertainty. Most people are risk avoiders, especially in "big" decisions with a major potential impact on their livelihood. A risk-averse person seeks decision alternatives, which are at least reasonable when the states of nature develop in an unfavorable way. Risk avoiders tend to select decision alternatives for which the lower end of the distribution of outcomes is as good as possible (Pukkala and Kangas 1996). They may also minimize the "regret", i.e. the maximum loss compared to the best decision alternative under certain states on nature. On the contrary, risk takers are optimistic and favor decision alternatives that are good under favorable states of nature, even though the probability of such an outcome may be low.

There are two basic approaches to the optimization of stand management in a risk situation: anticipatory and adaptive optimization. Anticipatory optimization seeks a single management schedule, which produces the most favorable distribution of net present values or some other objective function (Valsta 1992). Risk neutral decision makers select management schedules which produce high average net present values. Risk takers often select management schedules for which the best outcomes are good whereas risk avoiders tend to maximize the worst outcomes of alternative management schedules.

Adaptive optimization does not try to find a single management prescription for the stand. Instead, it aims at finding rules that help the landowner to make right decisions in changing environment (Lohmander 2007). A well-known rule is the reservation price function indicating the minimum price that the seller should obtain from timber (Brazee and Mendelsohn 1988; Lohmander 1995; Gong and Yin 2004). A more general approach is the Markov decision process model (Lembersky and Johnson 1975; Kaya and Buongiorno 1987).

It can be assumed that reservation price decreases with increasing financial maturity of the stand: the lower the relative value increment of the stand, the lower is the minimum selling price of a certain timber assortment. Since the relative value increment decreases with increasing stand density and mean tree size, it can be assumed that reservation price is negatively correlated with stand basal area and mean tree diameter (Lohmander 1995; Gong 1998; Lu and Gong 2003).

The aim of this study was to describe a system for stochastic optimization of the management of boreal forests in a situation where future timber prices, tree growth and regeneration are not known exactly. The developed simulation–optimization system was used to compare deterministic and stochastic optima, as well as the results of anticipatory and adaptive optimization approaches. Pukkala and Kellomäki (2012) compared anticipatory and adaptive management in even-aged forestry and Zhou et al. (2008) compared adaptive and anticipatory policies in uneven-aged forests. In this study, continuous cover management of both even-and uneven-aged initial stands was optimized. Continuous cover management refers to any sequence of cuttings that keep a minimum post-cutting residual stand basal area. Regeneration by planting or sowing is not used.

Based on previous studies, it was hypothesized that in a risk situation it is optimal to grow more diverse stands than under certainty (Rollin et al. 2005). Risk avoiders were assumed to maintain more diverse stand structures than risk seekers (Pukkala and Kellomäki 2012). The third hypothesis was that adaptive optimization and management results in higher average net present value than anticipatory optimization (Gong 1998; Lu and Gong 2003).

Methods

Growth and yield model

The set of models that was used to simulate stand development (Pukkala et al. 2013) consists of individual-tree model for diameter increment, individual-tree survival function, and an ingrowth model (Vanclay 1994). To calculate the assortment volumes of removed trees, the height model of Pukkala et al. (2009) and the taper models of Laasasenaho (1982) were used. The article of Pukkala et al. (2013) reports also methods for simulating the residual variation around the diameter increment and ingrowth models. The deviation of diameter increment from deterministic model prediction was modelled as follows (Miina 1993):

$$dev_{it} = a_i + v_{it} \quad (1)$$

$$v_{it} = \rho v_{it-1} + e_{it} \quad (2)$$

where dev_{it} is the deviation from model prediction for tree i and 5-year period t, a_i is normally distributed random tree factor for tree i, v_{it} is random autocorrelated residual for tree i and period t, ρ is correlation coefficient between the residuals of consecutive 5-year periods and e_{it} is normally distributed random number, $\text{var}[e_i] = \text{var}[v_{it}](1 - \rho_i^2)$. It was assumed that 1/3 of dev is accounted for by the tree factors (a_i) and the rest is accounted for by autocorrelated residuals (v_{it}). It has been found that the correlation between the residuals of consecutive 1-year periods is 0.4–0.7 and the correlation between 5-year residuals is about half of it (Henttonen 1990; Miina 1993; Kangas 1997; Pasanen 1998). In this study, the autocorrelation coefficient of residuals (ρ) was assumed to be 0.300 for all species. The total variance of residual was 0.254 for pine, 0.283 for spruce and 0.228 for birch (Pukkala et al. 2013). The random numbers (e_{it}) generated for the trees in a particular 5-year growth period were assumed to be correlated (Pasanen 1998), resulting in both auto- and cross-correlated time series of growth residuals (Figure 1). In simulation, the stochastic residuals were added to the predicted diameter increment. As a result, the simulated differentiation of tree size was faster than it would be in deterministic simulation.

The diameter increments obtained from the diameter increment model were multiplied with a multiplier that describes the effect of climate change on tree growth (Pukkala and Kellomäki 2012). The climate-induced growth trend is based on a process-based model (Kellomäki and Väisänen 1997; Ge et al. 2010) and corresponds to climate change scenario A1B. The effect of changing climate on diameter increment depends on tree species and growing site. The trends are linear and growth will improve approximately 20% in 50 years. In this study it was assumed that the influence of climate change on diameter increment is not known with certainty. Therefore, the slope of the trend line was assumed to be stochastic, with standard deviation equal to 0.1 times the slope coefficient.

Ingrowth was defined as the number of new trees per hectare that reach the 1.3 m height during a 5-year period. Pukkala et al. (2013) modelled the residuals of the ingrowth model as follows

$$dev_{s,t} = \rho_s dev_{s,t-1} + se_s e_{s,t} \quad (3)$$

where $dev_{s,t}$ is the deviation from the deterministic logarithmic model for species s and 5-year period t, ρ_s is the autocorrelation coefficient of successive 5-year periods for species s, se_s is the standard deviation of the stochastic annual component and $e_{s,t}$ are multi-normally distributed correlated random numbers (N(0,1)) for pine, spruce, birch and hardwood other than birch. Correlated random numbers ($e_{s,t}$) were obtained by using the Cholesky decomposition of the covariance matrix of the residuals of the species-specific ingrowth models (Pukkala et al. 2013). Correlations between the residuals of different species are not high, but for instance the unexplained variation in the ingrowth of spruce is positively correlated with the residual for birch. On the contrary, the correlations between successive five year periods are rather high, 0.670 for pine, 0.577 for spruce, 0.657 for birch and 0.637 for hardwood other than birch. The main reason for the positive autocorrelation is most

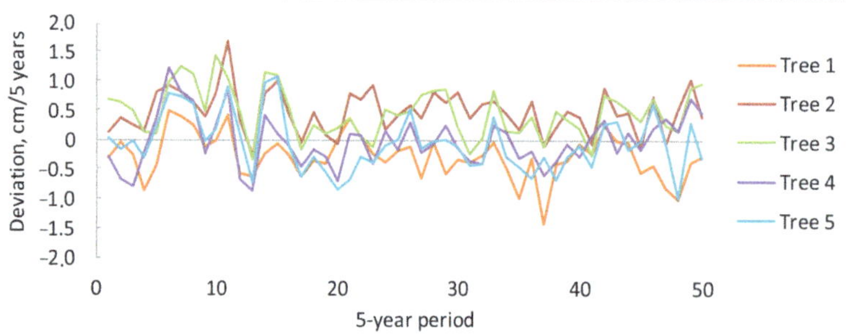

Figure 1 A diameter increment scenario. Sequences of stochastic deviations from deterministic model prediction for five trees and fifty 5-year periods. Each sequence consists of a tree factor and cross- and autocorrelated stochastic temporal components. Tree 2 is a fast-growing individual and Tree 1 is a slow-growing individual.

probably that a single good regeneration year (good seed crop with low seed predation and high germination rate) increases the ingrowth in several coming years. The standard deviation of the stochastic annual component (*se*) is 0.526 for pine, 0.990 for spruce, 1.027 for birch and 0.938 for hardwood other than birch. Stochastic ingrowth scenarios are produced by adding the simulated residuals to the deterministic logarithmic ingrowth model and converting the result to a non-logarithmic value. Figure 2 shows examples of ingrowth scenarios when the model prediction is 10 new trees per hectare. It can be seen that the resulting ingrowth scenarios are very erratic, reflecting to what happens in reality.

Leskinen and Kangas (1998) described the annual variation in timber prices with a set of models where the logarithmic price of a certain timber assortment depends on the price of the previous year plus a stochastic annual component

$$p_t - \bar{p} = \alpha(p_{t-1} - \bar{p}) + e_t \quad (4)$$

where p_t is the logarithmic price in year t, α is parameter ranging from 0.45 to 0.89 for different timber assortments and e is normally distributed random number. Correlated random numbers for different assortments were produced with the help of Cholesky decomposition. The model has been estimated from the historical timber price statistics of Finland. Figure 3 shows an example timber price scenario for six assortments. It can be seen that the prices of successive years are positively correlated and the prices of different assortments are also correlated.

Case study stands

Calculations were done for an uneven-aged spruce stand, mixed stands of pine, spruce and birch, and pure pine and spruce stands (Table 1). Each stand was assumed to grow on a typical growing site for the species. The stands represent typical and common stand structures in the managed forests of Finland. The stands were assumed to grow in Central Finland.

Each species and canopy layer was initially described by basal area, mean diameter, mean height and minimum and maximum of the diameter distribution. Stand basal area and the three diameters were used to predict the diameter distribution of each stratum (species or canopy layer) present in the stand. The predicted diameter distribution was divided into 10 classes of equal width, and 5 trees were taken to represent each class. The random tree factors of the residuals of the diameter increment model were generated at this point (a_i of Equation 1). As a result, each stratum of the stand was represented by 50 "representative trees" varying in size and inherent growth potential.

Growth, survival and ingrowth were simulated using 5-year time steps. If there was ingrowth, a new representative tree was generated for every 10 new conifers or 50 hardwoods (each new tree represented 10 or 50 trees per hectare). The random tree factors of the residuals of diameter growth models were drawn from normal distribution for each new representative tree. Mortality was

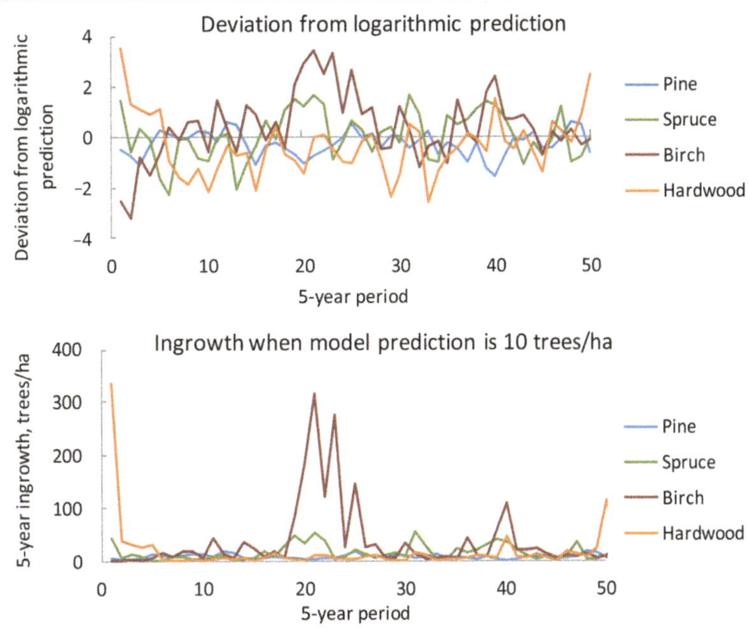

Figure 2 A stochastic ingrowth scenario. Cross- and autocorrelated logarithmic residuals are generated (top) and added to the logarithmic ingrowth prediction, which is then converted to non-logarithmic value (bottom).

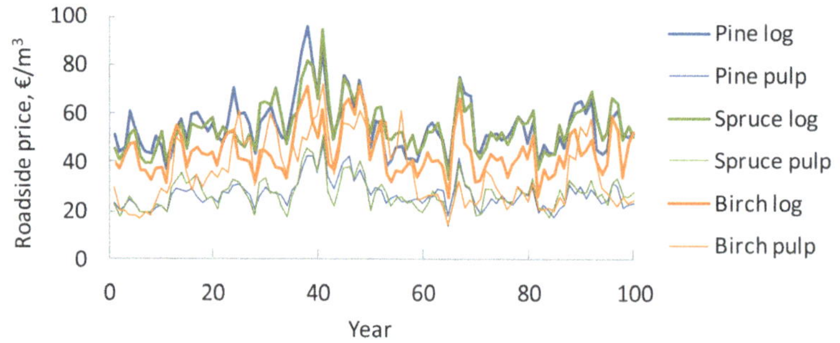

Figure 3 A stochastic timber price scenario for 100 years. The prices of different assortments are auto- and cross-correlated.

simulated by multiplying the frequency of the representative tree by its survival probability.

Optimization

The objective variable was the net present value of all future net incomes, calculated with 3% discount rate. The next three cuttings were optimized for all stands. The net present value of the remaining growing stock (after the 3rd cutting) was calculated with species-specific models using stand basal area, mean dbh, discount rate, site variables and timber prices as predictors (Pukkala 2005). These models explain 90%–95% of the variation of the NPV of the optimal management schedule, depending on tree species. Because of discounting, the value of the ending growing stock, i.e. the discounted value of predicted net present value of all cuttings conducted later than the third cutting, had only a small effect on the total NPV. Preliminary tests indicated that optimizing three first cuttings was enough to have a reliable estimate of the total NPV and to know how the stand should be managed in the near future (Figure 4). For example, when optimizing one to five next cuttings and using model prediction to calculate the NPV of the residual stand, the following total NPVs (calculated with 3% discount rate) were obtained for an uneven-aged spruce stand: 11426 € $\cdot$ ha^{-1} (1 cutting optimized), 11897 € $\cdot$ ha^{-1} (2 cuttings optimized), 11917 € $\cdot$ ha^{-1} (3 cuttings optimized), 11884 € $\cdot$ ha^{-1} (4 cuttings optimized), 11879 € $\cdot$ ha^{-1} (5 cuttings optimized).

In anticipatory optimization the decision variables for each cutting were as follows:

- Cutting year (exactly: number of years since the start or since previous cutting)
- Parameters of the thinning intensity curve, which was defined separately for each species present in the initial stand

Thinning intensity was first described with the following logistic function (Pukkala et al. 2014):

$$h(d) = \frac{1}{1 + a_3 \times \exp[a_1(a_2 - d)]^{1/a_3}} \quad (5)$$

where $h(d)$ is the proportion of harvested trees at dbh d and a_1, a_2 and a_3 are parameters to be optimized. This simple function has been found to result in almost as good solutions (in terms of NPV) as optimizing the

Table 1 Case study stands

Stand	Site	Strata	BA	Height	D_{min}	D_{mean}	D_{max}
Uneven spruce	MT	Spruce	18	21	17	22	28
		Spruce	7.6	6	1	8	16
Mature mixed	MT	Pine	6	21	13	22	28
		Spruce	6	14	3	16	22
		Birch	6	20	13	21	27
Young mixed	MT	Pine	4	17	13	18	22
		Spruce	3	14	3	16	22
		Birch	4	16	13	17	20
Young spruce	OMT	Spruce	15	11	5	12	18
Mature spruce	OMT	Spruce	28	21	15	23	29
Young pine	VT	Pine	15	11	5	12	18
Mature pine	VT	Pine	25	20	15	21	27

MT = mesic site; OMT = herb-rich site; VT = sub-xeric site; BA = stand basal area (m^2 $\cdot$ ha^{-1}); Height = mean tree height (m); D_{min} = minimum diameter (cm); D_{mean} = mean diameter (cm); D_{max} = maximum diameter (cm).

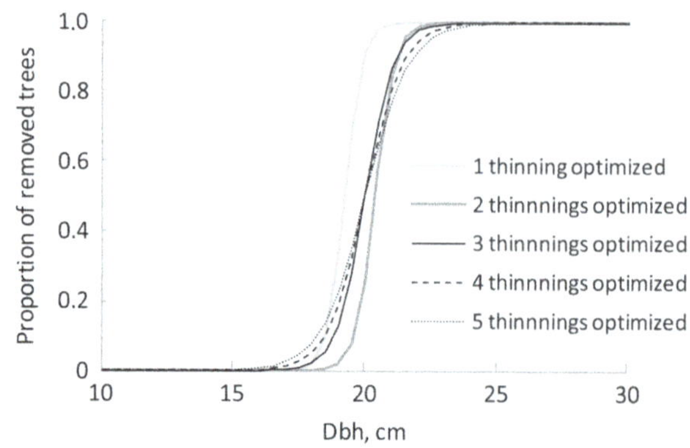

Figure 4 Effect of optimizing 1 to 5 next cuttings on the optimal thinning intensity curve of the first cutting in an uneven-aged spruce stand.

harvest intensities of different diameter classes separately (Pukkala et al. 2014). Moreover, preliminary analyses showed that parameter a_3 could be fixed to $a_3 = 1$ (i.e. a_3 could be removed) without any notable deterioration of the NPV of the optimal solution. Therefore, the following simplified thinning intensity model was used in this study:

$$h(d) = \frac{1}{1 + \exp[a_1(a_2 - d)]} \tag{6}$$

Parameter a_2 gives the diameter at which thinning intensity is 0.5, and a_1 defines the type of thinning. If a_1 is negative, small trees are thinned more than large ones, resulting in low thinning. When a_1 is positive, the thinning represents high thinning while $a_1$equal to 0 results in uniform thinning. As a result, the number of optimized variables was $3(1 + 2) = 9$ for one-species stand and $3(1 + 3 \times 2) = 21$ for a mixture of pine, spruce and birch.

In adaptive optimization, cutting years were replaced by a reservation price function. The following form was assumed, based on previous research (e.g. Pukkala and Kellomäki 2012), preliminary analyses and known relationships between stand basal area, mean tree diameter and financial maturity:

$$RP = \exp\left(b_1 + b_2\sqrt{D} + b_3\sqrt{G}\right) \tag{7}$$

where RP is the price of saw log (roadside price) that activates a cutting treatment and b_1, b_2 and b_3 are optimized parameters that define how the reservation price depends of stand basal area and mean tree diameter. The same reservation price was used in all cuttings. In a mixed stand the current timber price, which was compared to the reservation price, was computed as the weighted average of the saw log prices of all species present in the stand, using basal area as the weight variable.

The intensity and type of cutting were defined with the same logistic function that was used in anticipatory optimization. However, in adaptive optimization cutting may be postponed if timber price is not good enough. Using the same thinning intensity curve with varying cutting years may lead to situations in which the thinning is too heavy or too light, depending on how much and to which direction the cutting year is moved. To avoid this from happening, the problem formulation was changed so that parameter a_2 (location of thinning intensity curve) was calculated with a model, and only parameter a_1 (thinning type) was optimized. This resulted in problem formulations containing $3 + 3 \times 1 = 6$ decision variables in one-species stands, and $3 + 3 \times 3 \times 1 = 12$ decision variables in the mixture on pine, spruce and birch (the type of thinning was optimized separately for each species).

Several deterministic optimizations were conducted for different species on different growing sites to find the relationship between parameter a_2 (location) of the thinning intensity curve and the stand characteristics (Figure 5). On the basis of these optimizations, the following model was fitted to the diameter at which thinning intensity is 50%:

$$a_2 = 8.738 - 0.156G + 0.771D - 1.906CT \tag{8}$$

where D is basal-area-weighted mean diameter of the trees (cm), G is stand basal area ($m^2 \cdot ha^{-1}$) and CT is an indicator variable for xeric growing sites ($CT = 1$ for *Calluna* type and poorer sites, and 0 otherwise). The model explained 82% of the variation of a_2.

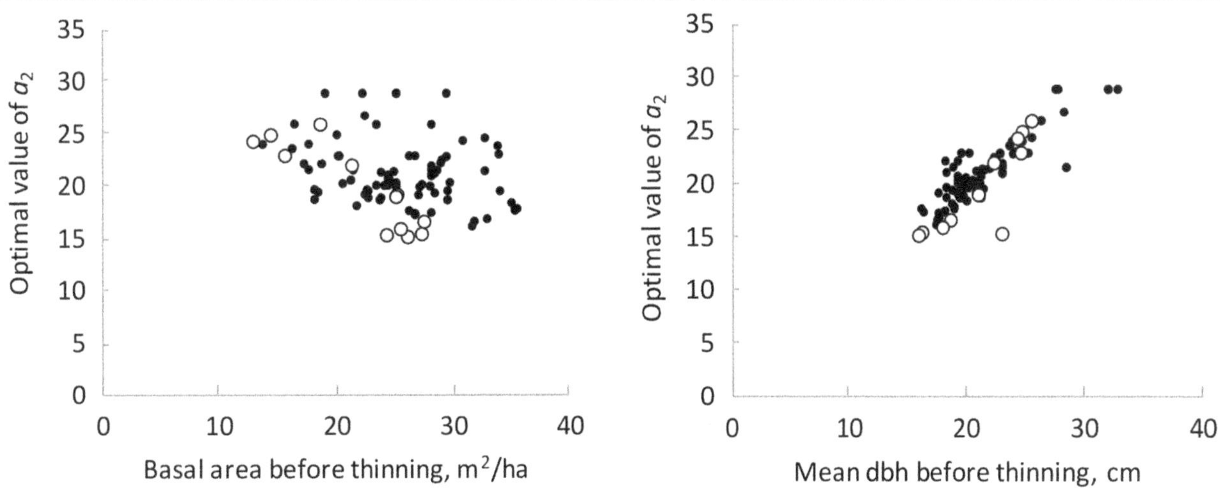

Figure 5 Dependence of parameter a_2 of the thinning intensity curve (Equation 5) on stand basal area and mean tree diameter. Open circles represent xeric (dry) growing sites.

The current forestry legislation of Finland does not allow the landowner to thin the stand below a certain minimum residual basal area (typically around 10 $m^2 \cdot ha^{-1}$). If the minimum basal area requirement is not met, the landowner is obliged to regenerate the stand within a certain time frame. In this study, any solution in which the minimum basal area was not met was penalized with the consequence that the selected schedules were better in line with the current forestry legislation.

Each management schedule evaluated during an optimisation run was simulated 600 times, and the mean NPV of the 600 stochastic outcomes was passed to the optimization algorithm. The results therefore represent the optimal management for risk neutral decision makers. When the effect of risk attitude was analysed the 10% accumulation point of the distribution of outcomes was used as the objective variable for a risk avoider, leading to the selection of such a management schedule for which the worst outcomes are as good as possible (Pukkala and Kangas 1996). The corresponding accumulation point for a risk seeker was 90%. The used optimization method was the direct search algorithm of Hooke and Jeeves (1961). Afterwards, all optimal solutions – also the deterministic ones – were simulated 1000 times with stochastic variation in tree growth, growth trend, in growth and timber price. The reported results on NPV, removals etc. are based on these simulations.

Results and discussion

Effect of risk factors

The effect of adding different stochastic components to simulation and anticipatory optimization was inspected in the uneven-aged spruce stand. Management was optimized without any stochasticity and with stochasticity in growth, ingrowth or timber price. The distributions of net present values produced by the optimal anticipatory solution are shown in Figure 6. It can be seen that when only ingrowth or only climate-induced growth trend is stochastic the distribution of outcomes is very narrow, indicating that these factors do not bring much uncertainty to decision-making. Stochastic variation in tree growth brought much more uncertainty in NPV than stochasticity in ingrowth or climate-induced growth trend. When timber price was stochastic the distribution of outcomes was much wider indicating that timber price is a more significant source of uncertainty than the biological growth process of trees.

Deterministic optimization and simulation resulted in the NPV of 11775 $€ \cdot ha^{-1}$. The average NPVs of the outcomes of stochastic anticipatory optima were almost the same. Also the optimal cutting years of the uneven-aged spruce stand were the same in all optimizations: the first cutting immediately, the second after 15 years and the third 10 years later. However, the way in which cuttings were conducted depended on the degree of stochasticity. The deterministic optimum advised the landowner to remove all trees larger than 20.2 cm in dbh. When stochastic factors were added to simulation and optimization, more and more trees larger than 20 cm were retained, and more and more trees less than 20 cm in dbh were removed, which means that stochastic optimization leads to higher dbh-variation in the post-cutting stand (Figure 7). The same trend was observed also in pure even-aged conifer stands, except mature pine stand (Figure 8). However, also in this stand the second and third thinnings showed similar differences between deterministic and stochastic optima as obtained for the other stands. Similar differences between deterministic and stochastic optima were obtained also for the mixed stands (results not shown). Rollin et al. (2005) found that counting for risk leads to clearly more diverse stand structures than suggested by deterministic solutions.

The total removal of the three cuttings was 6%–23% lower in the stochastic anticipatory optima than in the deterministic optima. The interval between the 1st and the 3rd cutting was 5–20 years shorter in the stochastic optima. These are indications of risk sharing behavior: in a risky situation it is optimal to cut more often but remove a smaller volume at a time.

Figure 7 shows that the cutting intensity curve is located at larger diameters in mature stands. Because the mature stands are to be cut immediately, the result suggests that the cutting may already be late. Another partial explanation for the difference between young and mature stands is that the basal area of the young stands would increase too much without cutting, decreasing the relative value increment of the stand (see Figure 5). Table 2 shows that the stand basal area at cutting is larger for the young initial stands but the mean tree diameter is smaller, suggesting that high stand densities call for earlier cuttings, which is a logical result. In general, the higher was the mean tree size the lower was the pre- and post-cutting stand basal area.

Effect of risk attitude

The effect of risk attitude on optimal management was analyzed in the mixed stands with the hypothesis that a risk avoider maintains a more diverse stand structure than a risk seeker. However, the thinning intensity curves were very similar for both risk attitudes suggesting that the post-cutting diameter distributions were also similar for both attitudes. The same difference as in pure stands was observed between deterministic and stochastic anticipatory optima: the deterministic optima proposed diameter-limit cutting with a narrower post-cutting diameter distribution than obtained in stochastic anticipatory optimization.

The proportions of different species after the first cutting were more uniform for risk avoider than for risk seeker (Figure 9), i.e. risk aversion led to more

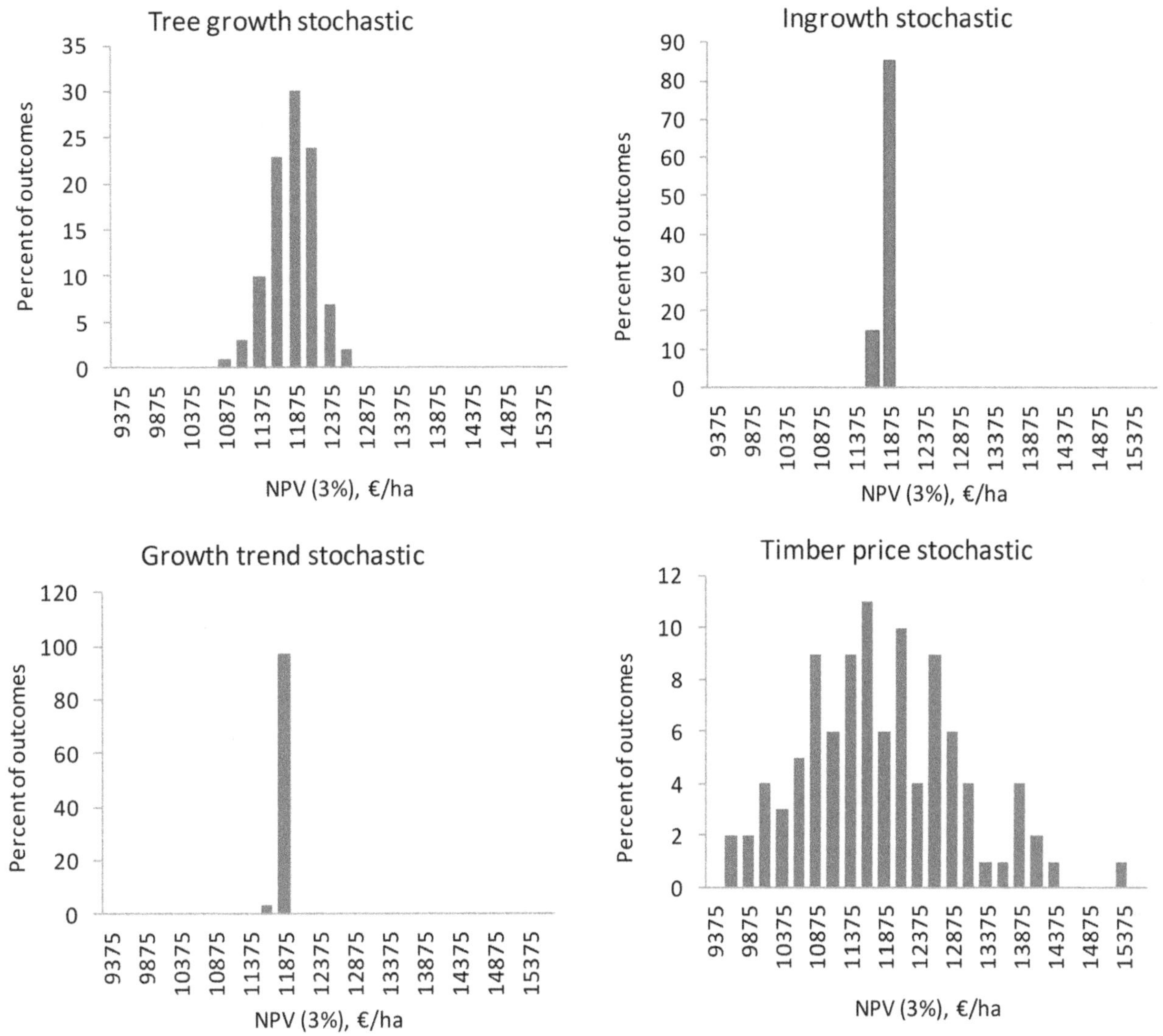

Figure 6 Distributions of net present value in the optimal solution when one factor at a time is stochastic (growth, ingrowth, growth trend, or timber price).

mixed stand. In the young mixed stand the risk seeker removed all pines in the first cutting whereas the risk avoider left all species in the residual stand. In the mature stand the post-cutting stand was more spruce dominated for the risk seeker, the risk avoider maintaining more birch and slightly more pines than the risk seeker. The differences were in line with the hypothesis, but they were small. The reason for the small differences may be that the legal limits force the landowner to keep more than species in the first cutting since otherwise the stand balsa area would be too low. Another reason is the fact that since pines and birches were clearly larger than spruces, it was optimal to gradually remove them irrespective of risk attitude. In addition, since the prices of different tree species correlate (Figure 3), increasing species diversity does not decrease the financial risk very much.

In the mature mixed stand, the cuttings were the earlier the more risk-averse the decision-maker was (Table 3). This is in line with Gong (1998) who concluded that risk avoiders should have the final felling earlier than risk-neutral forest landowners. The removed volume increased towards increasing risk tolerance (Gong 1998; Lu and Gong 2003). In the young mixed stand the removal was larger for the risk-neutral decision-maker than for risk avoider, but the seeker cuts less, most probably because the third cutting was 10 years earlier for the risk seeker than for other risk attitudes.

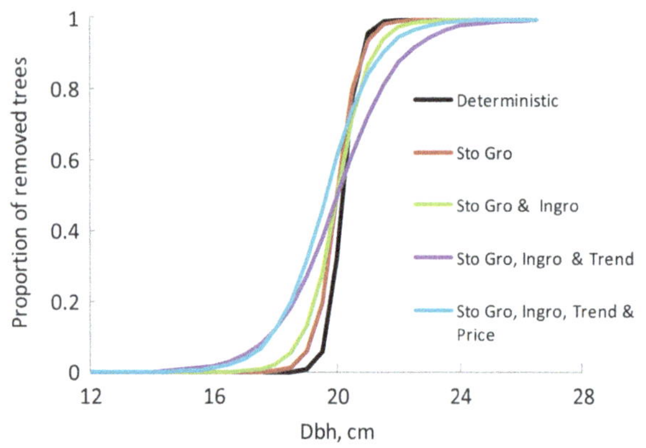

Figure 7 Optimal first cutting in deterministic solution and in stochastic anticiparoty optima with different sources of stochasticity (Gro = growth, Ingro = ingrowth, Trend = climate-induced growth trend, Price = timber price).

Adaptive optima

In adaptive optimization, cutting years were replaced by the reservation price function, resulting in cutting years that may be different in repeated stochastic simulations, depending on the realized stand development and timber price. To make the thinning intensity curve sensitive to changes in cutting year, the "location" parameter of the curve (a_2, dbh at which thinning intensity is 0.5) was calculated with a model (Equation 8) and only the type of thinning (low, uniform or high depending on parameter a_1 of Equation 6) was optimized.

The optimal reservation price functions were very similar for all initial stands (examples shown in Figure 10). As expected, the mean net present values of several repeated stochastic simulations with the optimal parameters were clearly better for the adaptive optima (Figure 11), the advantage of adaptive optimization and management being 6%–14%. There were no systematic differences in the average cutting years or removals between anticipatory and adaptive optima (Table 2). In mature stands, the average cutting year suggested by the adaptive optima was about 5 years later than in the anticipatory optima. However, the reason for this difference is most probably technical: it was possible to only postpone the first cutting from year zero, not have it earlier.

The solutions of the adaptive optimization problems were also simulated so that the optimized value of parameter a_1 (thinning type) of the thinning intensity curve (Equation 6) was replaced by 1, corresponding to high thinning. The average NPVs of 1000 simulations were nearly the same as obtained with the optimized values of parameter a_2, except for the mature mixed stand. The result indicates that nearly optimal adaptive management can be found when optimizing only the reservation price function and calculating the thinning intensity curve with model, fixing parameter a_1 to 1. The whole management schedule can be defined and optimized only by three decision variables, namely the parameters of the reservation price function. In the anticipatory optima for mixed stands there are 21 decision variables and yet the expected NPV is clearly better for the adaptive solution defined by only 3 decision variables.

The average roadside price obtained from saw log was about 20% higher in adaptive optima than in deterministic or stochastic anticipatory optima (Table 2). The difference was smaller in the first cutting of the mature stands, due to the high opportunity cost of the growing stock (high financial maturity of the initial stand). The results are in agreement with the assumptions made about the shape of the reservation price function.

Conclusions

All the hypotheses of the study were supported by the results. However, the effect of risk attitude on optimal management was very small, which may be related to the current forestry legislation which ruled out a part on the management options. Another reason may be the size differences of the species of mixed stands, which

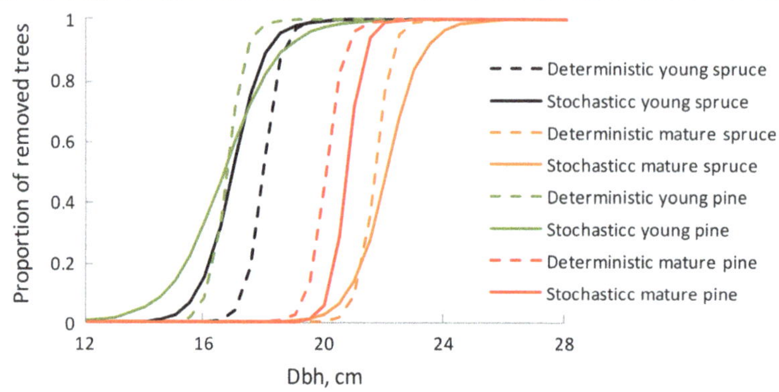

Figure 8 Thinning intensity curves in the first cutting of pure conifer stands in deterministic (dashed lines) and stochastic anticipatory optimization (solid line).

Table 2 Results calculated from 1000 stochastic simulations with the optimal values of decision variables for a risk neutral decision maker in different problem formulations when tree growth, ingrowth and timber price are stochastic (Det = deterministic optimization, Anti = stochastic anticipatory optimization, Ada = stochastic adaptive optimization)

Number of the cutting	Young spruce			Mature spruce			Young pine			Mature pine		
	Det	Anti	Ada	Det	Anti	Ada	Det	Anti	Ada	Det	Anti	Ada
	Cutting year											
1st	20	20	21.9	0	0	4.6	20	20	25.3	0	0	5.6
2nd	35	35	42.1	20	15	19.6	35	35	47.4	15	15	21.2
3rd	55	50	62.9	55	45	30.8	55	50	70.3	45	25	39.2
	Diameter before cutting (cm)											
1st	19.1	19.1	19.6	23.0	23.0	24.4	17.5	17.6	18.6	21.0	21.0	22.2
2nd	19.9	19.7	21.9	28.3	26.6	27.7	17.8	18.9	20.0	23.5	23.5	23.8
3rd	23.3	22.3	25.3	31.5	30.5	29.9	19.9	19.1	21.6	24.8	24.3	23.7
	Basal area before cutting ($m^2 \cdot ha^{-1}$)											
1st	35.7	35.7	37.1	28.1	28.1	31.3	32.1	32.2	35.8	25.1	25.1	28.1
2nd	29.5	27.1	33.6	21.6	19.8	19.7	26.7	26.6	33.2	16.4	19.4	19.5
3rd	27.6	25.7	29.0	26.7	21.6	14.9	27.6	26.0	32.0	23.9	14.1	19.6
	Basal area after cutting ($m^2 \cdot ha^{-1}$)											
1st	14.6	12.4	14.5	9.8	10.8	11.1	13.5	13.4	14.4	9.3	11.8	11.5
2nd	10.6	12.7	12.5	6.4	5.5	9.2	10.5	12.8	12.5	7.1	9.4	8.7
3rd	10.5	12.6	10.1	11.6	7.2	5.1	11.3	12.1	11.2	9.2	6.2	9.3
	Removed volume ($m^3 \cdot ha^{-1}$)											
1st	176	192	192	197	187	219	140	141	167	150	126	162
2nd	163	126	190	164	153	113	129	111	175	93	99	109
3rd	160	122	182	167	157	109	141	118	181	153	78	93
Total	499	440	564	528	497	441	410	370	523	396	303	364
	Average roadside saw log price obtained ($€ \cdot m^{-3}$)											
1st	56.8	55.8	64.6	56.1	56.3	61.9	56.7	56.4	66.8	55.7	56.3	62.9
2nd	56.2	56.5	65.0	56.5	56.4	66.2	56.1	55.8	67.0	56.2	56.5	67.5
3rd	56.5	55.6	65.1	54.5	55.1	67.4	55.6	55.8	65.2	54.5	55.7	68.1

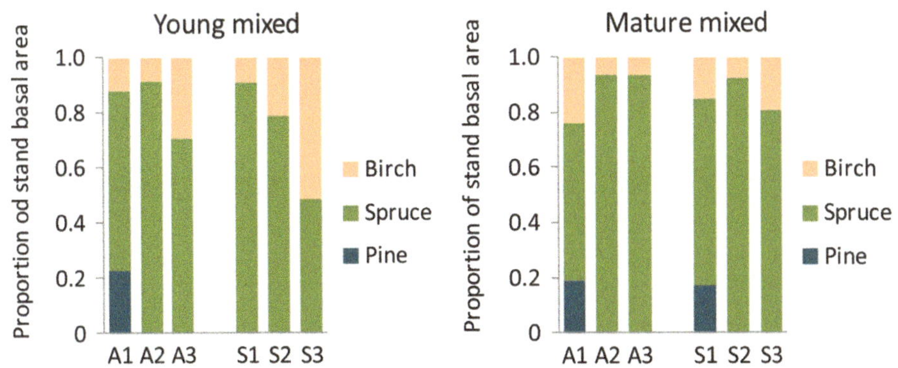

Figure 9 Proportions of different tree species in the post-cutting stands according to stochastic anticipatory optima for risk avoider (A) and risk seeker (S). The number after A or S is the number of the cutting.

Table 3 Optimal cutting years in anticipatory stochastic optima for different risk attitudes

No. of the cutting	Young mixed stand			Mature mixed stand		
	Avoider	Neutral	Seeker	Avoider	Neutral	Seeker
	Cutting year					
1st	20	20	20	0	5	10
2nd	35	35	35	15	20	25
3rd	65	65	55	45	50	55
	Removed volume ($m^3 \cdot ha^{-1}$)					
1st	142	145	162	77	107	139
2nd	110	107	83	102	130	115
3rd	193	215	127	229	210	243
Total	446	467	372	408	448	497

had a greater impact on the results than risk attitude. Positive correlation between timber prices of different tree species (Figure 3) also decreases the possibilities to reduce financial risk by increased species diversity. Roessiger et al. (2011) concluded that the optimal management for a cautious risk-avoiding forest landowner uses tree species diversification, avoiding clear-cutting and mono-species forest composition.

All thinnings of all solutions were high thinnings. The very high stochastic variation of ingrowth did not affect the expected NPV of the management schedule and it did not bring much uncertainty in decision-making. This is because the removals and incomes of the first three cuttings were obtained from trees that already existed in the initial stands. Ingrowth affects the incomes of distant cuttings whose effect on NPV is very small when the discount rate is 3% or higher. In addition, infrequent regeneration and ingrowth, combined with uneven growth rate of the ingrowth trees may provide a continuous enough supply of trees to larger diameter classes.

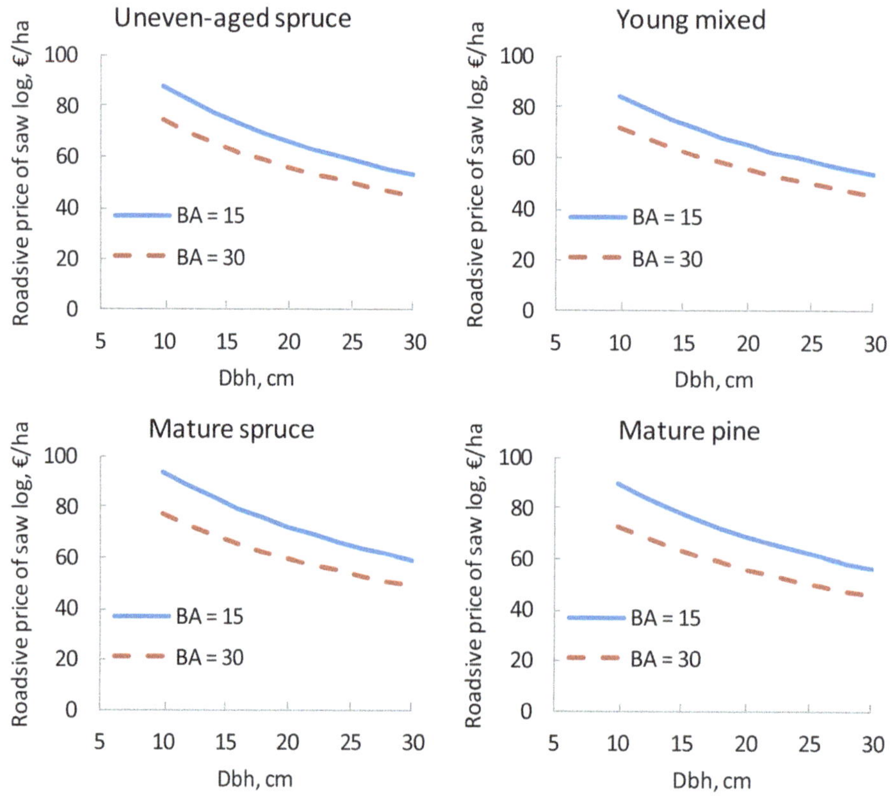

Figure 10 Examples of optimal reservation price functions obtained in adaptive optimization. Reservation price is the minimum price (in this case the minimum roadside price of saw log) which must be obtained to sell timber.

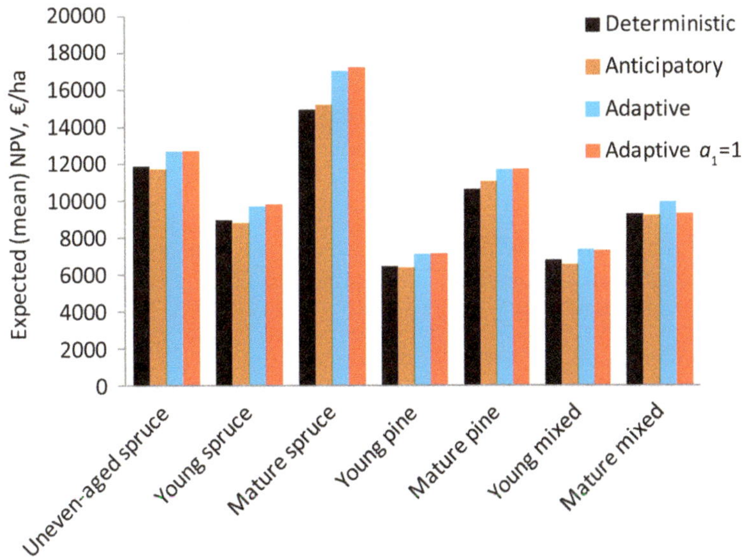

Figure 11 Average net present values of 1000 stochastic simulations with the optimal values of decision variables obtained in different problem formulations. "Adaptive a1 = 1" is a simulation in which the optimized value of parameter a_1 of the tinning intensity curve was replaced by a constant value (a_1 = 1).

Timber price was by far the most significant source of risk and uncertainty.

By looking at the average NPVs of 1000 stochastic simulations conducted with different optimal solutions (Figure 11) it can be concluded that there is also some uncertainty related to the optimality of the found solutions. Theoretically, stochastic anticipatory optima should produce better results than 1000 stochastic simulations with the deterministic optima, but this was not always the case. Correspondingly, fixing parameter a_1 to 1 should decrease the simulated NPVs, compared to adaptive solutions where a_1 was optimized, but this did not happen always. The results suggest that stochastic problems are more difficult to solve than the deterministic ones. Simulating each schedule clearly more than 600 times in optimization (600 realizations were used in optimization runs) would most probably partially solve the problem, but with a high computational cost. An alternative approach, namely the Markovian decision process model, would be better from the computational point of view (Kaya and Buongiorno 1987).

Corresponding to the hypotheses and previous studies (Gong 1998; Lu and Gong 2003; Pukkala and Kellomäki 2012), adaptive optimization led to higher NPVs than anticipatory optima. However, the differences were smaller than what could be expected on the basis of some earlier studies (Gong and Yin 2004; Pukkala and Kellomäki 2012). This was partly because the growth interval was always 5 years although the truly optimal cutting year might be one of the years within the 5-year time step used in simulation. This most probably decreased the NPVs more in adaptive optimization since it was not possible to pick the year of the 5-year period that had the highest timber price. Therefore, the results of this study can be interpreted so that the benefit of adaptive optimization is at least 6%–14% but it can be also higher. Zhou et al. (2008) found a 17% higher NPV for adaptive strategy compared to fixed strategy, with little difference in length of cutting cycle.

The adaptive approach facilitates very simple management rules. The optimal future management can be described with only 3 parameters, namely the coefficients of the reservation price function. A thinning treatment should be conducted when the actual price is higher than the reservation price. The thinning intensity of different diameter classes is calculated with Equation 6. Parameter a_1 of the equation can be taken as 1, and parameter a_2 is calculated with Equation 8. If the use of equations is difficult to the forest manager, the equations can be converted to diagrams that show the optimal management in a changing environment.

Competing interests
The author declares that he has no competing interests.

Authors' contributions
TP conducted the analyses and wrote the manuscript. The author read and approved the final manuscript.

References

Brazee R, Mendelsohn R (1988) Timber harvesting with fluctuating prices. For Sci 34(2):359–372

Ge Z-M, Zhou X, Kellomäki S, Wang K-Y, Peltola H, Väisänen H, Strandman H (2010) Effects of changing climate on water and nitrogen availability with implications on the productivity of Norway spruce stands in southern Finland. Ecol Modell 221(13–14):1731–1743

Gong P (1998) Risk preferences and adaptive harvest policies for even-aged stands management. For Sci 44(4):496–506
Gong P, Yin R (2004) Optimal harvest strategy for slash pine plantations: the impact of autocorrelated prices for multiple products. For Sci 50(1):10–19
Henttonen H (1990) Kuusen rinnankorkeusläpimitan kasvun vaihtelu Etelä-Suomessa. Summary: Variation in the diameter growth of Norway spruce in Southern Finland. University of Helsinki, Department of Forest Mensuration and Management. Res Notes 25:1–88
Hooke R, Jeeves TA (1961) "Direct search" solution of numerical and statistical problems. J Assoc Comput Mach 8:212–229
Kangas A (1997) On the prediction bias and variance in long-term growth projections. For Ecol Manage 96:207–216
Kaya I, Buongiorno J (1987) Economic harvesting of uneven-aged northern hardwood stands under risk: a Markovian decision model. For Sci 33(4):889–907
Kellomäki S, Väisänen H (1997) Modelling the dynamics of the forest ecosystem for climate change studies in the boreal conditions. Ecol Model 97:121–140
Laasasenaho J (1982) Taper curve and volume equations for pine spruce and birch. Communicationes Instuti Forestalis Fenniae 108:1–74
Lembersky MR, Johnson KN (1975) Optimal policies for managed stands: an infinite horizon Markov decision process approach. For Sci 21(2):109–122
Leskinen P, Kangas J (1998) Modelling and simulation of timber prices for forest planning calculations. Scand J For Res 13:469–476
Lohmander P (1995) Reservation price models in forest management: errors in the estimation of probability density function parameters and optimal adjustment of the bias free point estimates. Management systems for a global forest economy with global resource concerns. Society of American Foresters, USA, pp 439–456
Lohmander P (2007) Adaptive optimization of forest management in a stochastic world. In: Weintraib A, Romero C, Bjørndal T, Epstein R, Miranda J (eds) Handbook of operations research in natural resources. International Series in Operations Research and Management Science, Vol. 99. Springer Science + Business Media B.V., pp 525–543.
Lu F, Gong P (2003) Optimal stocking level and final harvest age with stochastic prices. J For Econ 9:119–136
Miina J (1993) Residual variation in diameter growth in a stand of Scots pine and Norway spruce. For Ecol Manage 58:111–128
Pasanen K (1998) Integrating variation in tree growth into forest planning. Silva Fennica 32(1):11–25
Pukkala T (2005) Metsikön tuottoarvon ennustemallit kivennäismaan männiköille, kuusikoille ja rauduskoivikoillle. Metsätieteen aikakauskirja 3(2005):311–322
Pukkala T, Kangas J (1996) A method for integrating risk and attitude toward risk into forest planning. For Sci 42(2):198–205
Pukkala T, Kellomäki S (2012) Anticipatory vs. adaptive optimization of stand management when tree growth and timber prices are stochastic. Forestry 85 (4):463–472
Pukkala T, Lähde E, Laiho O (2009) Growth and yield models for uneven-sized forest stands in Finland. For Ecol Manage 258:207–216
Pukkala T, Lähde E, Laiho O (2013) Species interactions in the dynamics of even- and uneven-aged boreal forests. J Sustainable For 32:1–33
Pukkala T, Lähde E, Laiho O (2014) Optimizing any-aged management of mixed boreal forest under residual basal area constraint. J For Res 25(3):627–636
Roessiger J, Griess VC, Knoke T (2011) May risk aversion lead to near-natural forestry? A simulation study. Forestry 84(5):527–537
Rollin F, Buongiorno J, Zhou M, Peyron J-L (2005) Management of mixed-species, uneven-aged forests in the French Jura: from stochastic growth and price models to decision tables. For Sci 51(1):64–75
Valsta L (1992) A scenario approach to stochastic anticipatory optimization in stand management. For Sci 38:430–447
Vanclay JK (1994) Modelling forest growth and yield: applications to mixed tropical forests. CAB International, United Kingdom
Zhou M, Buongiorno J (2006) Forest landscape management in a stochastic environment, with an application to mixed loblolly pine–hardwood forests. For Ecol Manage 223:170–182
Zhou M, Liang J, Buongiorno J (2008) Adaptive versus fixed policies for economic or ecological objectives in forest management. For Ecol Manage 254:178–187

3

Estimating forest aboveground biomass by low density lidar data in mixed broad-leaved forests in the Italian Pre-Alps

Antonio Montagnoli[1*], Sara Fusco[1], Mattia Terzaghi[1], Alan Kirschbaum[3], Dirk Pflugmacher[2], Warren B Cohen[4], Gabriella S Scippa[5] and Donato Chiatante[1]

Abstract

Background: Estimation of forest biomass on the regional and global scale is of great importance. Many studies have demonstrated that lidar is an accurate tool for estimating forest aboveground biomass. However, results vary with forest types, terrain conditions and the quality of the lidar data.

Methods: In this study, we investigated the utility of low density lidar data (<2 points·m^{-2}) for estimating forest aboveground biomass in the mountainous forests of northern Italy. As a study site we selected a 4 km^2 area in the Valsassina mountains in Lombardy Region. The site is characterized by mixed and broad-leaved forests with variable stand densities and tree species compositions, being representative for the entire Pre-Alps region in terms of type of forest and geomorphology. We measured and determined tree height, DBH and tree species for 27 randomly located circular plots (radius =10 m) in May 2008. We used allometric equations to calculate total aboveground tree biomass and subsequently plot-level aboveground biomass (mg·ha^{-1}). Lidar data were collected in June 2004.

Results: Our results indicate that low density lidar data can be used to estimate forest aboveground biomass with acceptable accuracies. The best height results show a R^2 = 0.87 from final model and the root mean square error (RMSE) 1.02 m (8.3% of the mean). The best biomass model explained 59% of the variance in the field biomass. Leave-one-out cross validation yielded an RMSE of 30.6 mg·ha^{-1} (20.9% of the mean).

Conclusions: Low-density lidar data can be used to develop a forest aboveground biomass model from plot-level lidar height measurements with acceptable accuracies. In order to monitoring the National Forest Inventory, and respond to Kyoto protocol requirements, this analysis might be applied to a larger area.

Keywords: LiDAR; Allometric equations; Plant height; Mixed forest

Background

Forest biomass is a key biophysical property that describes the carbon content of vegetation. Quantification at various scales, from root system (Montagnoli et al. 2012a, 2014) to above-ground organs, is critical for understanding the stocks and fluxes associated with forest clearance, degradation, and regeneration, particularly given current concerns regarding global climate change (Barrett et al. 2001; Palombo et al. 2014). Knowledge of carbon dynamics (Montagnoli et al. 2012b) is crucial when addressing issues relating to carbon accounting, including quantifying carbon for credit schemes (Patenaude et al. 2004; Kim et al. 2009). National reporting of carbon sources and sinks is also required to fulfill obligations to international agreements such as the United Nations Framework Convention on Climate Change (Rosenqvist et al. 2003). Despite these requirements, there is still much uncertainty in biomass estimation at a range of scales and in particular on how much carbon is cycled through the Earth's forests. Scenario development to assess whether this cycling might change as a result of forest alteration (e.g. degradation induced by climate change) is also needed and is becoming increasingly important as a research field (Brack et al. 2006; Lucas et al. 2008). A better knowledge of forest

* Correspondence: antonio.montagnoli@uninsubria.it
[1]Department of Biotechnology and Life Science, University of Insubria, Via Dunant, 3, 21100 Varese, Italy
Full list of author information is available at the end of the article

ecosystems is critical to greenhouse gas control and biodiversity conservation. In fact, the Kyoto Protocol accounts for sinks of carbon emissions associated to vegetation growth and expansion (UNFCCC 1997; Rosenqvist et al. 2003). This is the reason why sustainable forest management is assuming an increasing importance and represents the second line of action of every governmental institution to be added to their commitment to reduce CO_2 emission. In this context biomass is important to monitor, whether associated with land use change, afforestation, reforestation or deforestation (Schulze et al. 2002). It is not surprising that assessment of vegetated land characteristics by remote sensing is highlighted as a recommended tool in a number of political charters and treaties of different nations (Almeida et al. 2014). A detailed knowledge covering the large areas of variability, which characterize forest biomass and biophysical structure, the properties and the state of evolution of vegetation, is frequently critical given that it requires significant campaign operations in terms of time/operator and cost (Chen et al. 2007; Popescu 2007; Wallerman and Holmgren 2007). Remote sensing in general is an excellent tool for monitoring the environmental state of a vegetation canopy over space and time. It provides spatially continuous and temporally frequent information products over extended areas (Coops et al. 2007; Ota et al. 2014). To date many methods have been used to study the vegetation with remote sensing and different spectral indices have been proposed (e.g. NDVI and EVI) (Glenn et al. 2008; Kouadio et al. 2014; White et al. 2014). Despite recent steady advancements in remote sensing techniques, there still remains an inability to reliably quantify plant diversity and totally eliminate environmental interferences (Wang et al. 2010; Pettorelli et al. 2014). Remote sensing is basically the measurement and interpretation of spatially distributed radiation fluxes reflected or emitted from the Earth surface. The measured radiation fluxes are driven by radiative transfer processes, such as scattering, absorption and emission, intrinsically related to the properties of the observed surface (Campbell 1996;

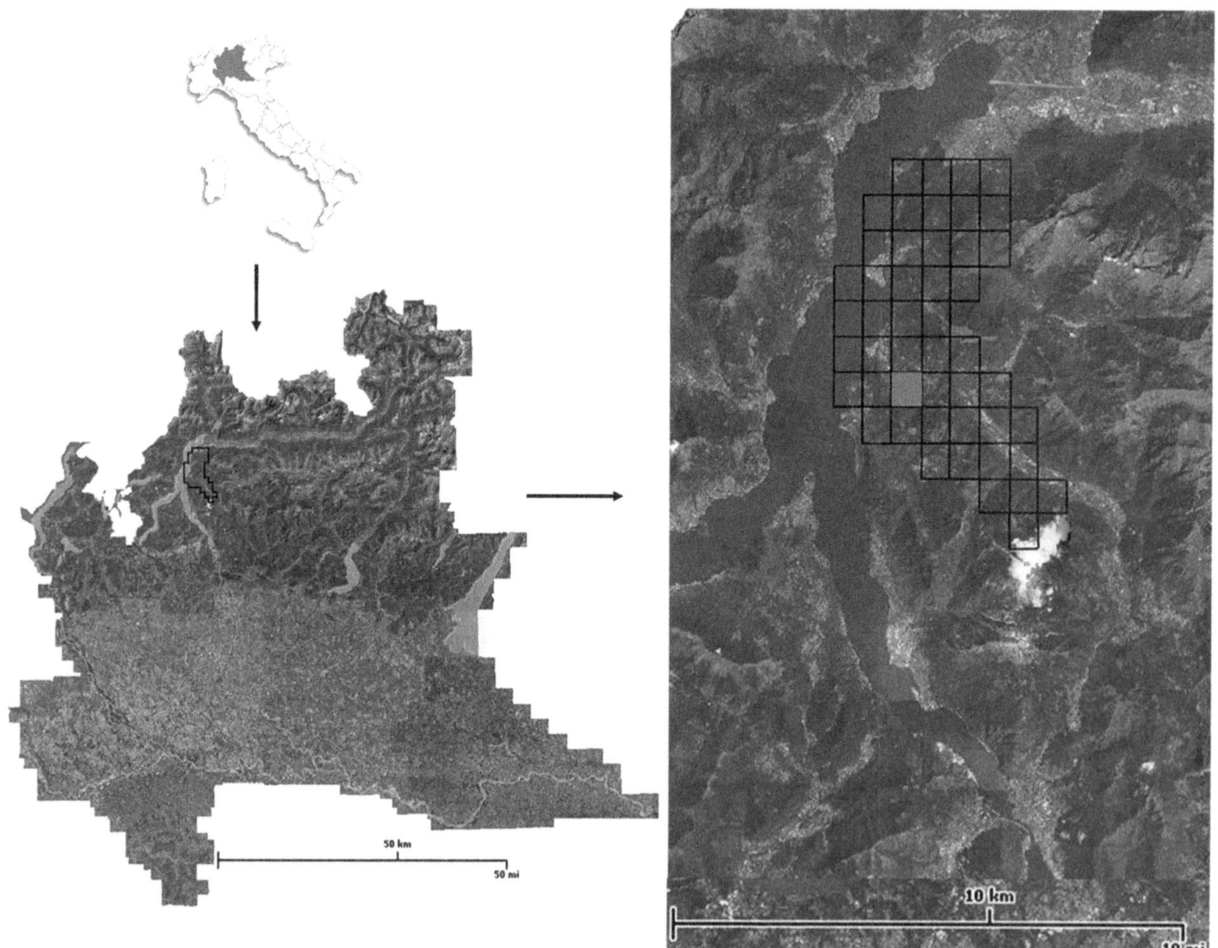

Figure 1 Map of the Lombardia region (larger area, left) and study area location (small black polygon, left). Grid area (right) indicates the LiDAR data survey. Aerial photos from free access Geoportale Lombardia. Gray cell (right) indicate location of field plots.

Jensen 2006). However, the variables controlling the radiative transfer and thus also remotely sensed data are not necessarily directly related to the surface properties of ultimate interest, like vegetation cover or bare soil (Verstraete et al. 1996). Due to the indirect and in the case of vegetation, mostly underdetermined character of this relationship, interpretation of remote sensing data should thus rely on as many independent observations as possible. In addition, the knowledge of the physical and biological processes involved needs to be considered in the interpretation of remote sensing data. Considering the limited amount of measurements generally provided by remote sensing and the high number of open variables, the problem of estimating vegetation properties based on remote sensing data is underdetermined (Wang and Sassen 2001; Combal et al. 2003; Kimes et al. 2006). A reliable retrieval is thus only possible if additional assumptions, constraints or further independent observations (e.g. field data, other sensors) are introduced (Verstraete et al. 1996). Assumptions and constraints are often used to simplify the problem, but are also limiting the retrieval in its transferability since they are generally only applicable for a specific problem (Duggin and Robinove 1990; Kötz et al. 2004). In this context, the aerial territorial survey technique, using laser scanning instruments (lidar) is particularly promising and, in some aspects, represents an alternative to satellite or aerial remote sensing because it makes it possible to directly survey the three-dimensional structure of the vegetation (Lindberg et al. 2012; Korpela and Hovi 2013). Lidar systems are an active remote sensing device that measure the time of travel needed for a pulse of laser energy sent from the airborne system to reach the ground and reflect back to the sensor. The time measured is converted into a distance measurement that is used to derive a precise three-dimensional characterization of the reflecting ground surface (Lim et al. 2003). In fact, the active optical remote sensing system, light detection and ranging (lidar), provides direct measurements on the vertical distribution of canopy elements within a vegetation canopy (Næsset and Bjerknes 2001; Lefsky et al. 2002). The measurement principle of lidar relies on laser pulses propagating vertically through the canopy, while scattering events are recorded as function of time. The remote sensing technique lidar is thus particularly suited to derive vegetation properties such as tree elevation, the vertical profile of foliage and terrain height (Harding et al. 2001). Usually the high resolution of small footprint lidar even allows for the three dimensional geometric reconstruction of single trees within a forest (Hyyppä et al. 2001; Morsdorf et al. 2004; Wang et al. 2008). Furthermore, lidar has also been shown to be an innovative tool for the study of vegetation at large scale, in particular for canopy structure, plant height and biomass (Andersen et al. 2005; Lefsky et al. 2002, 2005, 2010; Hansen et al. 2014). In this study, we investigated the utility of low-density lidar data (<2 points·m^{-2}) whether this type of data are useful for measuring forest attributes, such as height and above ground biomass (AGB) in the mountainous forests of northern Italy. The literature (Bortolot and Wynne 2005; Pilli et al. 2006) shows high correlations between forest AGB and tree height. The objectives of this work were: (i) to develop models of forest AGB from plot-level lidar height metrics and (ii) to understand if low density lidar is accurate enough in such conditions to produce a map of forest AGB for the region.

Methods

Study area

The low density lidar data cover a total area of approximately 80 km^2 located in northern Italy in the Lombardy Pre-Alps (Valsassina; small black box in the larger are,

Table 1 Forest tree density and characteristics for each forest plot of the study area

Plot number	Number of stem	Height (m)	DBH* (cm)
1	40.5	9.3 (0.5)	14.3 (1.4)
2	20	19.4 (0.7)	32.5 (2.4)
3	32	13.5 (0.9)	20.4 (1.8)
4	27	13.8 (1.1)	27.0 (2.7)
5	53	11.6 (0.7)	13.3 (0.7)
6	64	11.9 (0.4)	11.0 (0.5)
7	22	18.4 (1.3)	28.7 (3.1)
8	40	11.0 (0.5)	16.6 (1.1)
9	59	9.4 (0.5)	11.6 (0.9)
10	46	13.3 (0.6)	14.6 (0.9)
11	31	13.8 (0.3)	20.0 (1)
12	80	10.5 (0.4)	12.1 (0.8)
13	56	9.6 (0.5)	13.0 (0.8)
14	36	15.5 (0.9)	19.1 (1.4)
15	47	12.1 (0.6)	14.3 (1.3)
16	32	14.8 (0.8)	19.0 (1.3)
17	38	15.9 (0.6)	25.7 (2)
18	42	11.7 (0.8)	15.8 (1.2)
19	59	12.7 (0.4)	12.0 (0.5)
20	25	11.2 (0.6)	18.8 (1.4)
21	55	15.6 (0.6)	15.3 (0.9)
22	92	9.4 (0.2)	10.9 (0.5)
23	93	10.6 (0.3)	11.0 (0.6)
24	90	9.0 (0.2)	10.4 (0.5)
25	119	8.9 (0.2)	10.4 (0.3)
26	41	9.5 (0.4)	14.4 (1.2)
27	79	10.0 (0.2)	10.2 (0.3)

*DBH (diameter at breast height), height and DBH values are mean of the number of stem (±SE).

Figure 1, left) (46°00′N, 9°23′E centre area coordinates). In the present study, to test if these lidar data are reliable for forest AGB estimation, a sample area of 8 km^2 was analysed (gray cell, Figure 1, right). The study area topography is characterised by a mean elevation of 890 m with minimum elevation of 590 m and maximum of 1293 m. The study area includes mixed and broadleaved forest with variable stand densities and tree species compositions. The site is representative for the entire Pre-Alps region in terms of type of forest and geomorphology. The main forest types are coppice management with plantations of chestnut (*Castanea sativa*) together with beech (*Fagus sylvatica*), birch (*Betula pendula*), linden (*Tilia cordata*), ash (*Fraxinus excelsior*), poplar (*Populus tremula*), field maple (*Acer campestre*), hazel (*Corylus avellana*), European Hop Hornbeam (*Ostrya carpinifolia*), wild cherry (*Prunus avium*) and natural stands of oak (*Quercus* spp.).

Lidar data

Large-footprint discrete first and last returns lidar data were acquired in October 2003 when the canopy first started to change colour in the fall. Acquisition was made by Compagnia Generale Riprese aree SpA, Parma – CGR and property of Regione Lombardia, using an Optech ALTM 3033 scanner Airborne, at flying height of 2000 m with a swath width of 1450 m. Scan angle was 20° with an approximate footprint of 50 cm, and an average pulse spacing of 1.75 m and pulse rate of 33 MHz.

Field measurements

In a selected 2 km × 2 km area (Figure 1, right) 27 circular plots (radius = 10 m) were randomly located. During May 2008 tree number, height and diameter at breast height (DBH) were measured for all tree species within each plot (Table 1). Possible lidar data underestimation of tree height due the 5 years growth discrepancy between lidar data collection (2003) and field measurements was considered minimal. Analysed forests were at the mature stage (20–30 years old) therefore minimal growth was assumed (Brassard et al. 2009; Franceschini and Schneider 2014). Trees with diameters smaller than 5 cm were excluded. A total of 1417 trees were measured on these plots. On each plot the tree heights were measured using a Vertex Laser VL-400 - telemeter/hypsometer (Haglöf), and DBH with a Forestry Suppliers Metric Fabric Diameter Tape. Although it can be difficult to clearly distinguish the treetop in dense forests, the measured forests were not dense enough to completely block the treetops. Moreover, effort was taken in the field to move around in the forest until a spot was found that did not obscure the treetop. Geographic coordinates were recorded at the centre of each plot with a Trimble® GeoXM™ GPS with 1–3 meter accuracy. Specific allometric equations (Leonardi et al. 1996; Hamburg et al. 1997; Gasparini et al. 1998; Zianis et al. 2005; Alberti et al. 2006; Tabacchi et al., 2011) were used to calculate total aboveground tree biomass and subsequently plot-level biomass (mg·ha^{-1}) (see Table 2).

Data analysis

The result of a laser scan is a cloud composed by geographically located points (raw data) corresponding to all the elements composing the scanned surface. The first step in data processing was to identify and exclude (filtering) all outliers due to their distance from the mean surface. The TerraScan™ of Terrasolid software was used for this automated process (Axelsson 1999). Afterwards analysis parameters were manually corrected according to the different geo-morphology characteristics and an additional

Table 2 Biomass equations by tree species

Species name	Equation	Parameters			Reference
		a	*b*	*c*	
Acer campestre (L.)	$a \cdot (DBH)^b$	0.05	2.67		Alberti et al. 2006
Betula pendula (Roth)	$a \cdot [(DBH)^2 \cdot H]^b$	0.5443	0.65270		Hamburg et al. 1997
Castanea sativa (Miller)	$a \cdot (DBH)^b$	0.137	2.247		Leonardi et al. 1996
Corylus avellana (L.)	$a \cdot (H)^b$	0.0768	1.8329		Hamburg et al. 1997
Fagus sylvatica (L.)	$a + b \cdot (DBH)^2 \cdot H$	1.6409	0.030775		Tabacchi et al. 2011
Fraxinus excelsior (L.)	$a \cdot (DBH)^b$	0.11	2.49		Alberti et al. 2006
Populus tremula (L.)	$a \cdot (DBH)^b$	0.0519	2.545		Zianis et al. 2005
Prunus avium (L.)	$a \cdot (DBH)^b$	0.12	2.33		Alberti et al. 2006
Quercus petraea (Matt.) Liebl.	$a \cdot (DBH)^b$	0.2176	2.0513		Alberti et al. 2006
Tilia cordata (Miller)	$\ln(ABW) = a + b \cdot \ln(DBH)$	−2.6788	2.4542		Zianis et al. 2005
Ostrya carpinifolia (Scop.)	$a + b \cdot DBH \cdot H + c \cdot DBH^2$	−4.5877	$5.2638 \cdot 10^{-3}$	$4.09 \cdot 10^{-1}$	Gasparini et al. 1998

DBH = Diameter at breast height; *H* = Plant height; ln = natural logarithm; ABW = Total aboveground woody biomass.
The format of the biomass equation is given in the column labelled Equation, and *a*, *b*, and *c* are parameter values. References to the original papers according to authors.

filtering was applied (Barilotti et al. 2007). Interpolating lidar point elevation to a regular grid with a 5 m resolution created a Digital Surface Model (DSM) and Digital Terrain Model (DTM; 6–9 pulses per 5 m^2). The characterization of the top canopy surface (DSM) included only the highest laser reflection points, while lowest laser reflection points were used to estimate the ground level (DTM) (Lefsky et al. 1999; Holmgren and Persson 2004; Popescu et al. 2004; Popescu 2007).

The low density point clouds may miss the tree tops, because naturally tree tops have fewer hits compared to the tree crowns. The same problem exists in identifying the ground (Suárez et al. 2005), especially in a steep terrain (Estornell et al. 2011) like the Alps. Therefore, the elevation recorded with GPS units was compared with those identified in the lidar point cloud. Lidar returns were extracted for each plot using the geographic coordinates taken in the field. Returns above a threshold of 2 m were considered vegetation returns, and returns below that threshold were considered ground returns. Height threshold was chosen according to plot analysis where all trees below the 2 m cut off were smaller than 3 cm diameter and not considered for measurement. Then data were normalized to height-above-ground using the DTM (spatial resolution 5 meter; Figure 2b) supplied by the lidar vendor. Furthermore, points below two meters were eliminated in the plot to omit returns from understory or falsely classified ground returns (Figure 2c). The software used for these analyses was the free FUSION/LDV developed by Robert J. McGaughey (U.S. Forest Service Pacific Northwest Research Station, Oregon).

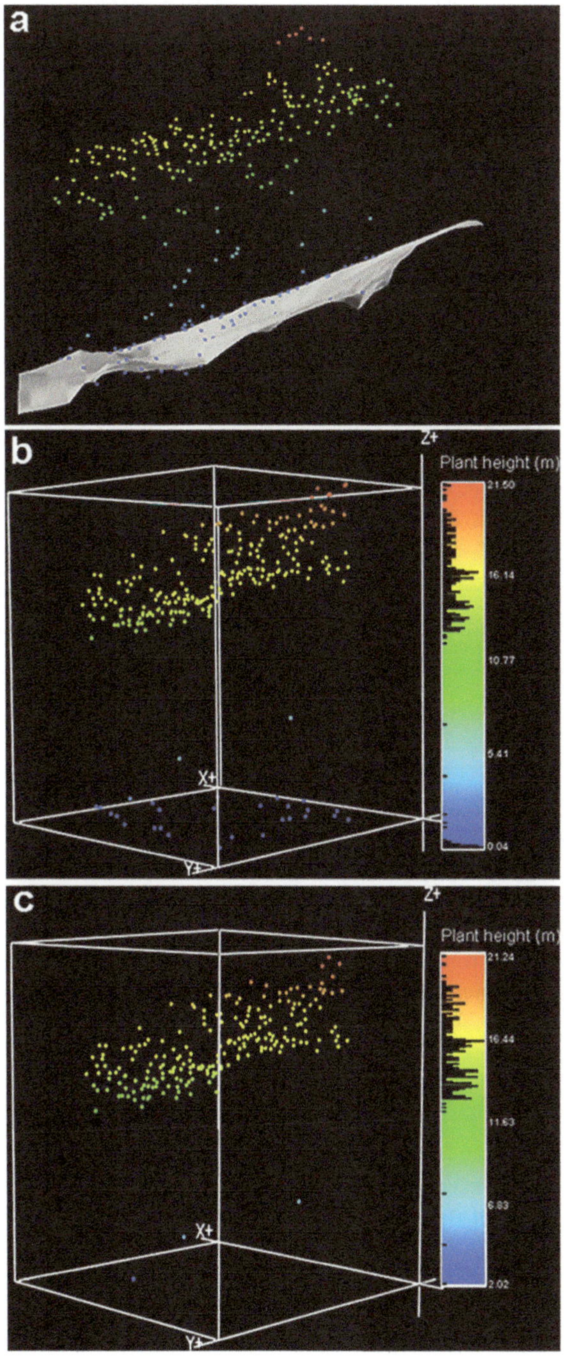

Figure 2 Examples of vertical plot LiDAR return distributions: **(a)** Raw LiDAR data plot; **(b)** data normalized by digital terrain model; **(c)** data points above 2 m from soil surface.

Regression analysis

To investigate the relationship of low density returns and AGB, regression models were used to develop equations relating lidar-derived tree height with field inventory tree height and field-based estimates of aboveground biomass for individual plots. In particular, the first relationship analysed was between lidar height metrics and field-measured height. Secondly the relationships between lidar height metrics and AGB from field data was considered at the plot scale (Pflugmacher et al. 2012). Analysis of variances for linear regression was carried out on results of each relationship to test for significance of the slopes at 95% significance level.

Results and discussion

Plant height

We report the best regression model developed to explain the relationship between lidar height metrics and field-measured height at the plot level. Linear regression indicated a multiple R^2 of 0.87 (Figure 3a; slope test $p < 0.001$) and a root mean square error (RMSE) of 1.02 m (BIAS of 0; Figure 3b), which is approximately 8.3% of the average plant height of all measured trees. Cross validation showed a RMSE of 2.02 m (16.4% of the mean) and a BIAS of 0.02. The final model derived from height distribution data was based on the following multiple percentile of tree height distribution: Height 25%; Height 20%; Height 30%; Height 10%; Height 80%. Our results are in

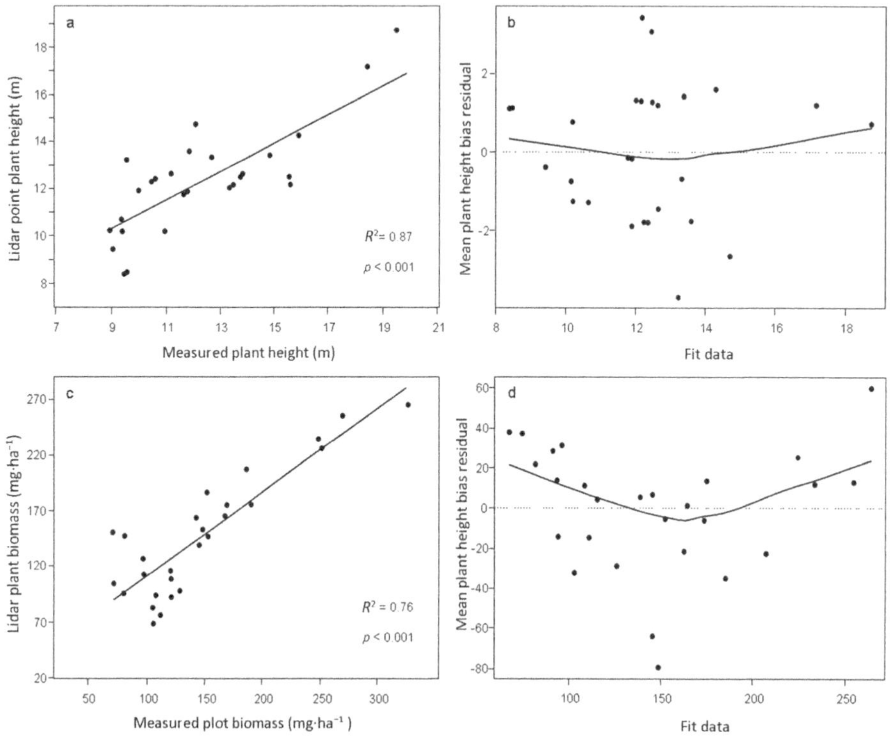

Figure 3 The best regression model developed to explain: **(a)** Relationship between LiDAR data and mean plant height measured in the field; **(b)** plant height BIAS against fitted height data LiDAR derived; **(c)** relationship between LiDAR data and mean biomass measured in the field; **(d)** plant biomass BIAS against fitted height data LiDAR derived.

line with previous studies (Coops et al. 2007; Popescu 2007; Stepper et al. 2014) in which a good correlation between tree height lidar-estimated and field-measured for both needle and broad leaved trees has been reported.

Plant biomass

The best model for the relationship between field AGB and lidar height explained 76% (multiple R^2) of the variance in the field biomass (Figure 3c; slope test $p < 0.001$) with a RMSE of 30.56 mg·ha^{-1} (20.9% of the mean) and a BIAS of 0 (Figure 3d). Leave-one-out cross validation yielded an RMSE of 53.7 mg·ha^{-1} (36.8% of the mean) and a BIAS of 3.7. The final model selected was based on the following multiple percentile of tree height distribution: Height 0; Height 5%; Height 25%; Height 55%; Height 70%; Height 80%; Height 95%; Height 100%.

Potential source of errors can include statistical error associated with estimating coefficients and form of selected equation. Moreover, errors may occur from both field measurement and data processing as well as errors associated with developing wide scale equation by compiling species- and site-specific equation that may be biased in favour of species for which published equations exist (Sileshi 2014). With the present approach, part of the unexplained variance when estimating aboveground biomass is associated with error of estimating biomass with field measurements of DBH and height, error associated with lidar-measured height and GPS misregistration errors. Selected variables for lidar height with measured field AGB (Figure 4a) were applied to the entire sample area of 8 km^2 in order to obtain a biomass map (Figure 4b). Previous studies have successfully estimated AGB or tree volume from lidar-derived vegetation-height statistical metrics in different boreal and temperate forests (Næsset 1997; Magnussen and Boudewyn 1998; Means et al. 2000; Popescu et al. 2004; Hall et al. 2005; Popescu 2007; Alberti et al. 2012). The average AGB and plant height estimated by lidar data of our forests were 146 mg·ha^{-1} and 12.28 m respectively. As demonstrated in previous works (Raber et al. 2002; Clark et al. 2004; Estornell et al. 2011) for relatively dense and structural deciduous forest on steep slope conditions, data characteristics such as scan angle most often causes DTM inaccuracy that effect derived trees canopy height (Raber et al. 2002; Clark et al. 2004). In the present study, plant height normalization performed with a low spatial resolution DTM, might have been another cause of error in lidar metrics calculation. Although these possible effects were not analysed, our results were similar to values of above ground biomass (123.7 mg·ha^{-1}) found in a Pre-alpine beech forest (Montagnoli et al. 2012b). Even though these forests have different species composition, these values

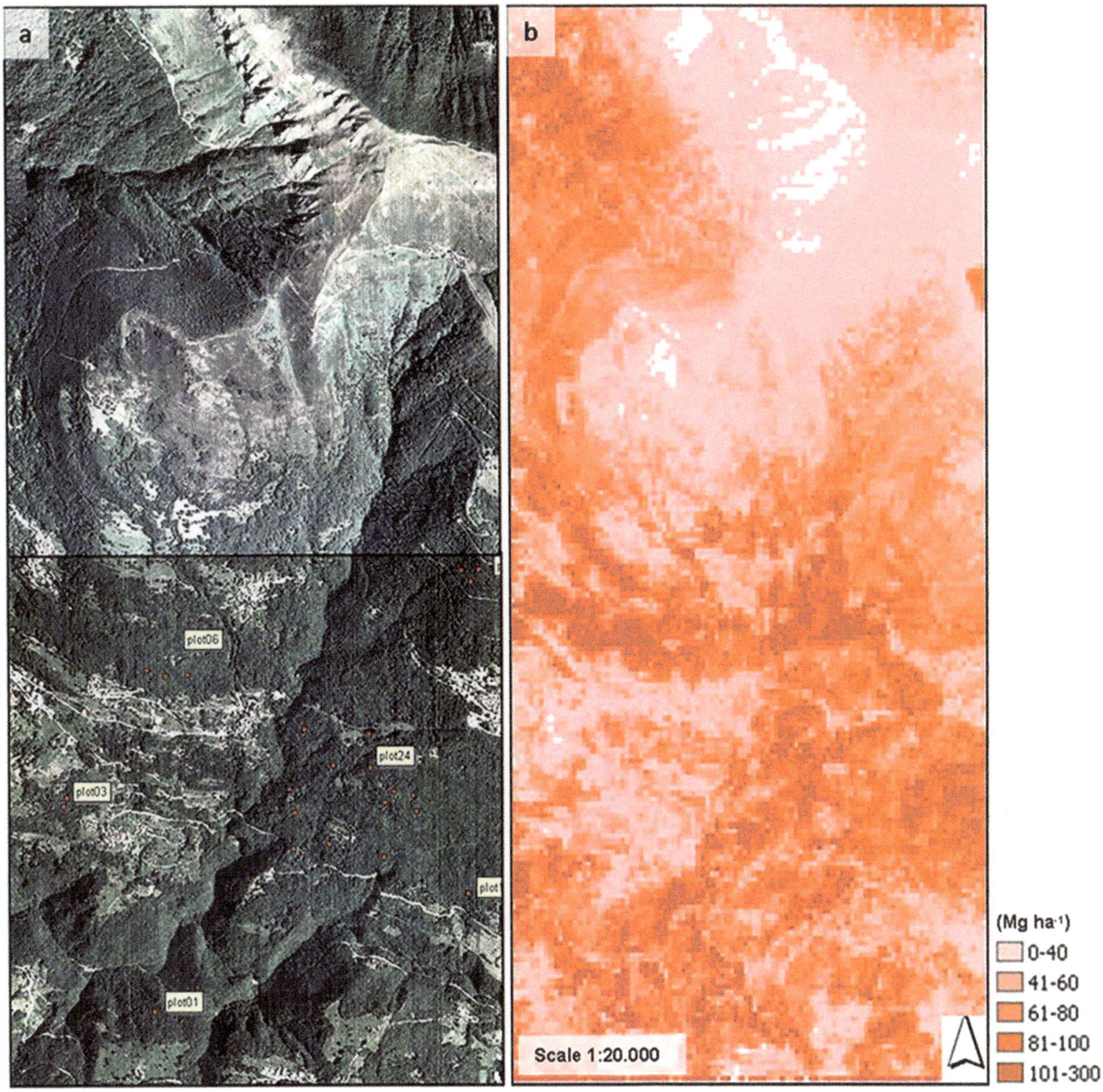

Figure 4 Example of application for plant biomass model by plant height measured by LiDAR data. **(a)** Digital orthophoto, in the lower part showed plots for field plant measurements. **(b)** Example of biomass map of the same area. Plant biomass values, represented by a multiple colours legend, increase from pink (lower value) to red (higher value). Pixel (1 m^2).

are comparable, since our study sites are in the transition zone of lowland and montane forests. Forest estimates (of height, volume and biomass), using laser data, are often based on linear regression models of forest canopy height (Nilsson 1994, 1996) and statistical measures derived from the distribution of laser point data (Lefsky et al. 1999; Næsset and Gobakken 2005). Several studies have noted that measures of canopy characteristics obtained from the laser height distribution, together with selected laser height percentiles, have proven useful for estimating timber volume (Means et al. 1999; Næsset & Økland 2002). Our results also show that mean height was the most reliable estimator of AGB (RMSE = 53.7 mg·ha^{-1}, corresponding to 36.8% of the mean) in single regression analysis. Finally, in our case we also demonstrated a good fitting model even with a difference in time between lidar data collection and field measurements.

Conclusion

Our results, in mixed broad leaved forests growing on patchy slope conditions, indicate that low-density lidar data can be used to develop a forest AGB model from plot-level lidar height measurements in the study area with acceptable accuracies. Moreover, these results highlight the opportunity to apply this analysis to a larger area, with the aim of monitoring the National Forest Inventory, and create a database of the forest carbon content in order to respond to requirements of the Kyoto protocol. The biomass map derived from the selected regression model and the potential for integrating lidar with co-registered multi and hyperspectral digital imagery, make lidar a realistic alternative to traditional forest measurements.

Competing interests
The authors declare that they have no competing interests.

Authors' contributions
Montagnoli make substantial contributions to the study concept and design, to field data collection and relative interpretation. Montagnoli co-writes the draft paper and dealt with manuscript process, improvements and revisions. Fusco leads the research between the Italian and American Labs. She participate to all works aspects such as concept and design, fieldwork, both field and lidar data collection, processing and interpretation. Fusco and Montagnoli were co-writing the draft paper. Terzaghi mainly contributed to the field work. He also partially contributed to the other aspects of the work carried in the Italian lab. Pflugmacher make substantial contribution to the lidar data analysis. Kirschbaum make substantial contribution to the field work concept, design and data collection. Slightly revised the manuscript. Cohen supervised the research and make a substantial contribution to all works aspects. Participate in drafting the article and revising it critically for important intellectual content. Scippa make substantial contribute to experiment and process supervision. Slightly revised the manuscript. Chiatante conceived and supervised the research in all aspects. Contributed to the paper processing. All authors read and approved the final manuscript.

Acknowledgements
This work was supported in part by a grant awarded by MIUR (PRIN 2008 n. 223), grant by University of Insubria (FAR) and EC FP7 Project ZEPHYR-308313. We are grateful to ERSAF and Regione Lombardia for providing lidar data. The authors are also in debt with Dr. Antonino Di Iorio and Dr. Barbara Baesso for helping with biomass equations, Carlo Fraquelli for helping in field measurements.

Author details
[1]Department of Biotechnology and Life Science, University of Insubria, Via Dunant, 3, 21100 Varese, Italy. [2]Department of Forest Ecosystems and Society, Oregon State University, 321 Richardson Hall, Corvallis, OR 97331, USA. [3]National Park Service, Great Lakes Inventory and Monitoring Network, 2800 Lake Shore Drive East, Ashland, WI 54806, USA. [4]USDA Forest Service, Pacific Northwest Research Station, Forestry Sciences Laboratory, 3200 SW Jefferson Way, Corvallis, OR 97331, USA. [5]Department of Biosciences and Territory, University of Molise, 86090 Pesche, Italy.

References
Alberti G, Marelli A, Piovesana D, Peressotti A, Zerbi G, Gottardo E, Bidese F (2006) Carbon stocks and productivity in forest plantations (Kyoto forests) in Friuli Venezia Giulia (Italy). Forest@ 3:488–495

Alberti G, Boscutti F, Pirotti F, Bertacco C, De Simon G, Sigura M, Cazorzi F, Bonfanti P (2012) A LiDAR-based approach for a multi-purpose characterization of Alpine forests: an Italian case study." iForest - Biogeosciences and Forestry 6: 156–168. http://www.sisef.it/iforest/contents/?id=ifor0876-006 doi:10.3832/ifor0876-006

Almeida P, Altobelli A, D'Aietti L, Feoli E, Ganis P, Giordano F, Napolitano R, Simonetti C (2014) The role of vegetation analysis by remote sensing and GIS technology for planning sustainable development: A case study for the Santos estuary drainage basin (Brazil). Plant Biosyst 148:540–546

Andersen H-E, McGaughey RJ, Reutebuch SE (2005) Estimating forest canopy fuel parameters using LIDAR data. Remote Sens Environ 94:441–449

Axelsson P (1999) Processing of laser scanner data – algorithms and applications. ISPRS J Photogramm Remote Sens 54:138–147

Barilotti A, Sepic F, Abramo E, Crosilla F (2007) Improving the morphological analysis for tree extraction: a dynamic approach to lidar data. In: Proceedings of the ISPRS Workshop on 'Laser Scanning 2007 and SilviLaser 2007' Espoo, Finland, 12–14 September 2007. Volume XXXVI, part 3/W52. Published by ISPRS Working Groups, ASPRS Lidar Committee, Finnish Geodetic Institute Institute of Photogrammetry and Remote Sensing, Helsinki University of Technology (TKK).

Barrett DJ, Galbally IE, Graetz RD (2001) Quantifying uncertainty in estimates of C emissions from above-ground biomass due to historic land-use change to cropping in Australia. Glob Change Biol 7:883–902

Bortolot ZJ, Wynne RH (2005) Estimating forest biomass using footprint LiDAR data: An individual tree-based approach that incorporates training data. J Phot & Remote Sens 59:342–360

Brack CL, Richards G, Waterworth R (2006) Integrated and comprehensive estimation of greenhouse gas emissions from land systems. Sustain Sci 1:91–106

Brassard BW, Chen HYH, Bergeron Y (2009) Influence of environmental variability on root dynamics in northern forests. Cr Rev Plant Sci 28:179–197

Campbell JB (1996) Introduction to Remote Sensing (2nd Ed). Taylor and Francis, London

Chen Q, Gong P, Baldocchi D, Tian YQ (2007) Estimating basal area and stem volume for individual trees from lidar data. Photogramm Eng Remote Sensing 73:1355–1365

Clark ML, Clark DB, Roberts DA (2004) Small-footprint lidar estimation of sub-canopy elevation and tree height in a tropical rain forest landscape. Remote Sens Environment 91:68–89

Combal B, Baret F, Weiss M, Trubuil A, Mace D, Pragnere A, Myneni R, Knyazikhin Y, Wang L (2003) Retrieval of canopy biophysical variables from bidirectional reflectance — Using prior information to solve the ill-posed inverse problem. Remote Sens Environ 84:1–15

Coops NC, Thomas H, Wulder MA, St-Onge B, Newnham G, Siggins A, Trofymow JAT (2007) Estimating canopy structure of Douglas-fir forest stands from discrete-return LiDAR. Trees 21:295–310

Duggin MJ, Robinove CJ (1990) Assumptions implicit in remote-sensing data acquisition and analysis. Int J Remote Sens 11:1669–1694

Estornell J, Ruiz LA, Velázquez-Martí HT (2011) Analysis of the factors affecting LiDAR DTM accuracy in a steep shrub area. Int J Digital Earth 4:521–538

Franceschini T, Schneider R (2014) Influence of shade tolerance and development stage on the allometry of ten temperate tree species. Oecologia 176:739–749

Gasparini P, Nocetti M, Tabacchi G, Tosi V (1998) Biomass equations and data for forest stands and shrublands of the Eastern Alps. Forest and Range Management Research Institute. I.S.A.F.A. - C.R.A.-, Villazzano, Trento, Italy

Glenn EP, Huete AR, Nagler PL, Nelson SG (2008) Relationship between remotely-sensed vegetation indices, canopy attributes and plant physiological processes: what vegetation indices can and cannot tell us about the landscape. Sensors 8:2136–2160

Hall SA, Burke IC, Box DO, Kaufmann MR, Stoker JM (2005) Estimating stand structure using discrete-return lidar: an example from low density, fire prone ponderosa pine forests. For Ecol Manage 208:189–209

Hamburg SP, Zamolodchikov DG, Korovin GN, Nefedjev V, Utkin AI, Gulbe T (1997) Estimating the carbon content of Russian forests; a comparison of phytomass/volume and allometric projections. Mitig adapt strategies glob chang 2:247–265

Hansen AJ, Phillips LB, Dubayah R, Goetz S, Hofton M (2014) Regional-scale application of lidar: Variation in forest canopy structure across the southeastern US. Forest Ecol Manag 329:214–226

Harding DJ, Lefsky MA, Parker GG, Blair JB (2001) Laser altimeter canopy height profiles — Methods and validation for closed-canopy, broadleaf forests. Remote Sens Environ 76:283–297

Holmgren J, Persson Å (2004) Identifying species of individual trees using airborne laser scanner. Remote Sens Environ 90:415–423

Hyyppä J, Kelle O, Lehikoinen M, Inkinen M (2001) A segmentation-based method to retrieve stem volume estimates from 3-D tree height models produced by laser scanners. IEEE T Geosci Remote 39:969–975

Jensen JR (2006) Remote sensing of the environment: an earth resource perspective, 2nd edn. Prentice Hall, New Jersey

Kim Y, Yang Z, Cohen WB, Pflugmacher D, Lauver CL, Vankat JL (2009) Distinguishing between live and dead standing tree biomass on the North Rim of Grand Canyon National Park, USA using small-footprint lidar data. Remote Sens Environ 113:2499–2510

Kimes DS, Ranson KJ, Sun G, Blair JB (2006) Predicting lidar measured forest vertical structure from multi-angle spectral data. Remote Sens Environ 100:503–511
Korpela I, Hovi A (2013) Korhonen L (2013) Backscattering of individual LiDAR pulses from forest canopies explained by photogrammetrically derived vegetation structure. Isprs J Photogramm 83:81–93
Kötz B, Schaepman M, Morsdorf F, Bowyer P, Itten K, Allgöwerd B (2004) Radiative transfer modeling within a heterogeneous canopy for estimation of forest fire fuel properties. Remote Sens Environ 92:332–344
Kouadio L, Newlands NK, Davidson A, Zhang Y, Chipanshi A (2014) Assessing the performance of MODIS NDVI and EVI for seasonal crop yield forecasting at the ecodistrict scale. Remote Sens 6:10193–10214
Lefsky MA (2010) A global forest canopy height map from the moderate resolution imaging spectroradiometer and the geoscience laser altimeter system. Geophys Res Lett 37:L15401, http://dx.doi.org/10.1029/2010GL043622
Lefsky MA, Harding D, Cohen WB, Parker G, Shugart HH (1999) Surface lidar remote sensing of basal area and biomass in deciduous forests of eastern maryland, USA. Remote Sens Environ 67:83–98
Lefsky MA, Cohen WB, PARKER GG, Harding DJ (2002) Lidar Remote Sensing for Ecosystem Studies. Bioscience 52:19–30
Lefsky M, Harding D, Keller M, Cohen W, Carabajal C, Espirito-Santo F, Hunter M, de Oliveira R (2005). Estimates of forest canopy height and aboveground biomass using ICESat. Geophys Res Lett 32:L22S02. http://dx.doi.org/10.1029/2005GL023971
Leonardi S, Santa Regina I, Rapp M, Gallego HA, Rico M (1996) Biomass, litterfall and nutrient content in *Castanea sativa* coppice stands of southern Europe. Ann For Sci 53:1071–1081
Lim K, Treitz P, Wulder M, St-Onge B, Flood M (2003) Lidar remote sensing of forest structure. Prog Phys Geog 27:88–106
Lindberg E, Olofsson K, Holmgren J, Olsson H (2012) Estimation of 3D vegetation structure from waveform and discrete return airborne laser scanning data. Remote Sens Environ 118:151–161
Lucas R, Lee A, Armston J, Breyer J, Bunting P, Carreiras J (2008) Advances in forest characterisation, mapping and monitoring through integration of LiDAR and other remote sensing datasets. SilviLaser, Edinburgh, UK
Magnussen S, Boudewyn P (1998) Derivations of stand heights from airborne laser scanner data with canopy-based quantile estimators. Can J For Res 28:1016–1031
Means JE, Acker SA, Harding DJ, Blair JB, Lefsky MA, Cohen WB, Harmon ME, McKee WA (1999) Use of large-footprint scanning airborne lidar to estimate forest stand characteristics in the western cascades of Oregon. Remote Sens Environ 67:298–308
Means JE, Acker SA, Brandon JF, Renslow M, Emerson L, Hendrix CJ (2000) Predicting forest stand characteristics with airborne scanning lidar. Photogramm Eng Remote Sens 66:1367–1371
Montagnoli A, Terzaghi M, Di Iorio A, Scippa GS, Chiatante D (2012a) Fine-root morphological and growth traits in a Turkey-oak stand in relation to seasonal changes in soil moisture in the southern Apennines, Italy. Ecol Res 27:725–733
Montagnoli A, Terzaghi M, Di Iorio A, Scippa GS, Chiatante D (2012b) Fine-root seasonal pattern, production and turnover rate of European beech (*Fagus sylvatica* L.) stands in Italy Prealps: Possible implications of coppice conversion to high forest. Plant Biosyst 146:1012–1022
Montagnoli A, Di Iorio A, Terzaghi M, Trupiano D, Scippa GS, Chiatante D (2014) Influence of soil temperature and water content on fine-root seasonal growth of European beech natural forest in Southern Alps, Italy. Eur J Forest Res, doi:10.1007/s10342-014-0814-6
Morsdorf F, Meier E, Kötz B, Itten KI, Dobbertin M, Allgöwer B (2004) LIDAR-based geometric reconstruction of boreal type forest stands at single tree level for forest and wild land fire management. Remote Sens Environ 92:353–362
Næsset E (1997) Estimating timber volume of forest stands using airborne laser scanner data. Remote Sens Environ 61:246–253
Næsset E, Bjerknes KO (2001) Estimating tree heights and number of stems in young forest stands using airborne laser scanner. Remote Sens Environ 78:328–340
Næsset E, Gobakken T (2005) Estimating forest growth using canopy metrics derived from airborne laser scanner data. Remote Sens Environ 96:453–465
Næsset E, Økland T (2002) Estimating tree height and tree crown properties using airborne scanning laser in a boreal nature reserve. Remote Sens Environ 79:105–115
Nilsson M (1994) Estimation of tree heights and stand volume using airborne lidar system. In: Report 57, Dept of Forest Survey. Swedish Univ of Agric Sciences, Umeå, p 59
Nilsson M (1996) Estimation of tree heights and stand volume using an airborne lidar system. Remote Sens Environ 56:1–7
Ota T, Ahmed OS, Franklin SE, Wulder MA, Kajisa T, Mizoue N, Yoshida S, Takao G, Hirata Y, Furuya N, Sano T, Heng S, Vuthy M (2014) Estimation of airborne lidar-derived tropical forest canopy height using landsat time series in Cambodia. Remote Sens 6:10751–10772
Palombo C, Marchetti M, Tognetti R (2014) Mountain vegetation at risk: Current perspectives and research reeds. Plant Biosys 148:35–41
Patenaude G, Hill R, Milne R, Gaveau D, Briggs B, Dawson T (2004) Quantifying forest above ground carbon content using lidar remote sensing. Remote Sens Environ 93:368–380
Pettorelli N, Safi K, Turner W (2014) Satellite remote sensing, biodiversity research and conservation of the future. Philos Trans R Soc Lond B Biol Sci 369:20130190, doi:10.1098/rstb.2013.0190
Pflugmacher D, Cohen WB, Kennedy RE, Yang Z (2012) Using Landsat-derived disturbance history (1972–2010) to predict current forest structure. Remote Sens Environ 122:146–165
Pilli R, Anfodillo R, Carrer M (2006) Toward a functional and simplified allometry for estimating forest biomass. Forest Ecol Manag 237:583–593
Popescu SC (2007) Estimating biomass of individual pine trees using airborne LiDAR. Biomass Bioenerg 31:646–655
Popescu SC, Wynne RH, Scrivani JA (2004) Fusion of small foot print LiDAR and multispectral data to estimate plot-level volume and biomass in deciduous and pine forests in Virginia, USA. Forest Sci 50:551–565
Raber GT, Jensen JR, Schill SR, Schuckman K (2002) Creation of digital terrain models using an adaptive lidar vegetation point removal process. Photogramm Eng Remote Sens 68:1307–1315
Rosenqvist Å, Milne A, Lucas R, Imhoff M, Dobson C (2003) A review of remote sensing technology in support of the Kyoto Protocol. Environ Sci Policy 6:441–455
Schulze ED, Valentini R, Sanz MJ (2002) The long way from Kyoto to Marrakesh: Implications of the Kyoto Protocol negotiations for global ecology. Glob Change Biol 8:505–518
Sileshi GW (2014) A critical review of forest biomass estimation models, common mistakes and corrective measures. For Ecol Manag 329:237–254
Stepper C, Straub C, Pretzsch H (2014) Assessing height changes in a highly structured forest using regularly acquired aerial image data. Forestry 0: 1–13, doi:10.1093/forestry/cpu050
Suárez JC, Ontiveros C, Smith S, Snape S (2005) Use of airborne LiDAR and aerial photography in the estimation of individual tree heights in forestry. Computers & Geosciences 31:253–262
Tabacchi G, Di Cosmo L, Gasparini P (2011) Aboveground tree volume and phytomass prediction equations for forest species in Italy. Eur J Forest Res 130:911–934
UNFCCC (1997) Kyoto Protocol to the United Nations Framework Convention on Climate Change adopted at COP3 in Kyoto, Japan, on 11 December 1997
Verstraete MM, Pinty B, Myneni R (1996) Potential and limitations of information extraction on the terrestrial biosphere from satellite remote sensing. Remote Sens Environ 58:201–214
Wallerman J, Holmgren J (2007) Estimating field-plot data forest stands using airborne laser scanning and SPOT HRG data. Remote Sens Environ 110:501–508
Wang Z, Sassen K (2001) Cloud type and property retrieval using multiple remote sensors. J Appl Meteorol 40:1665–1682
Wang Y, Weinacker H, Koch B (2008) A lidar point cloud based procedure for vertical canopy structure analysis and 3D single tree modelling in forest. Sensors 8:3938–3951
Wang K, Franklin SE, Guo X, Cattet M (2010) Remote sensing of ecology, biodiversity and conservation: a review from the perspective of remote sensing specialists. Sensors 10:9647–9667, doi:10.3390/s101109647
White K, Pontius J, Schaberg P (2014) Remote sensing of spring phenology in northeastern forests: a comparison of methods, field metrics and sources of uncertainty. Remote Sens Environ 148:97–107
Zianis D, Muukkonen P, Mäkipääand R, Mencuccini M (2005) Biomass and stem volume equations for tree species in Europe. Silva Fenn Monogr 4(1–2):5–63

4

Forecasting the development of boreal paludified forests in response to climate change: a case study using Ontario ecosite classification

Benoit Lafleur[1,2*], Nicole J Fenton[1] and Yves Bergeron[1]

Abstract

Background: Successional paludification, a dynamic process that leads to the formation of peatlands, is influenced by climatic factors and site features such as surficial deposits and soil texture. In boreal regions, projected climate change and corresponding modifications in natural fire regimes are expected to influence the paludification process and forest development. The objective of this study was to forecast the development of boreal paludified forests in northeastern North America in relation to climate change and modifications in the natural fire regime for the period 2011–2100.

Methods: A paludification index was built using static (e.g. surficial deposits and soil texture) and dynamic (e.g. moisture regime and soil organic layer thickness) stand scale factors available from forest maps. The index considered the effects of three temperature increase scenarios (i.e. +1°C, +3°C and +6°C) and progressively decreasing fire cycle (from 300 years for 2011–2041, to 200 years for 2071–2100) on peat accumulation rate and soil organic layer (SOL) thickness at the stand level, and paludification at the landscape level.

Results: Our index show that in the context where in the absence of fire the landscape continues to paludify, the negative effect of climate change on peat accumulation resulted in little modification to SOL thickness at the stand level, and no change in the paludification level of the study area between 2011 and 2100. However, including decreasing fire cycle to the index resulted in declines in paludified area. Overall, the index predicts a slight to moderate decrease in the area covered by paludified forests in 2100, with slower rates of paludification.

Conclusions: Slower paludification rates imply greater forest productivity and a greater potential for forest harvest, but also a gradual loss of open paludified stands, which could impact the carbon balance in paludified landscapes. Nonetheless, as the thick *Sphagnum* layer typical of paludified forests may protect soil organic layer from drought and deep burns, a significant proportion of the territory has high potential to remain a carbon sink.

Keywords: Boreal forest; Canada; Carbon sequestration; Paludification; Peat forests; Forest harvest

Background

In boreal forest ecosystems, successional paludification is described as a dynamic process driven by forest succession between fire events that leads to peat accumulation, and a concomitant thickening of the soil organic layer (SOL), and the formation of waterlogged conditions on a formerly dry mineral soil (Simard et al. 2007). Paludification is influenced by climatic factors and permanent site features, such as surficial deposits and soil texture, as well as by natural fire regimes (Lecomte et al. 2006; Simard et al. 2009; Payette et al. 2013). In boreal regions, in the extended absence of fire, paludification leads to the formation of paludified forests and can reduce forest productivity by up to 50%–80% (Simard et al. 2007).

According to the most recent report of the Intergovernmental Panel on Climate Change, warming of the climate system is unequivocal (IPCC 2013). This changing climate is expected to increase drought severity in boreal regions (Girardin and Mudelsee 2008), and therefore to

* Correspondence: benoit.lafleur@uqat.ca
[1]Institut de recherche sur les forêts, Université du Québec en Abitibi-Témiscamingue, 445 boul. de l'Université, Rouyn-Noranda, QC J9X 5E4, Canada
[2]Centre d'étude de la forêt, Université du Québec à Montréal, 141 Avenue du Président-Kennedy, Montréal, QC H2X 1Y4, Canada

influence the natural fire regime, resulting in an increase of fire severity and burn rate (Flannigan et al. 2005; de Groot et al. 2009; Bergeron et al. 2010; van Bellen et al. 2010). Fire plays an important role in landscape level paludification processes, as fire can "depaludify" forest stands if most of the SOL is burnt (Dyrness and Norum 1983; Greene et al. 2005). However, if the fire is not severe and a relatively thick SOL remains after fire, the regenerating forest stands may remain paludified (Lecomte et al. 2005). Because boreal peatlands represent important carbon reservoirs (it is estimated that boreal peatlands, including paludified forests, store 455 Pg of carbon, i.e. approximately 15% of the Earth's terrestrial carbon (Gorham 1991; Lavoie et al. 2005)) any modification to the fire cycle may have important consequences on the carbon cycle and the global climate.

Throughout the boreal region, paludified forests support important forest industries. It is expected that any modifications to the climate and natural fire regimes will, in all likelihood, require industries to adapt to new ecosystem conditions and, presumably, to modify their practices. Depending on the type of harvest practices, forest harvest in paludified forests can both promote or reduce paludification (Lafleur et al. 2010a, 2010b), and, as is the case for fire, have important effects on the C budget at the landscape level. This context provides strong incentives for the development of simple tools that can be used to rapidly and easily forecast the combined effects of climate change and modified fire regime on paludification and forest development.

In North America, forest mapping is commonly used to describe the forest mosaic at the regional scale. Forest maps provide information on stand scale environmental factors (e.g. surficial deposits, soil texture, moisture regime, slope), as well as on stand species composition, height and density. This information presents a great potential for research in forest ecology and management. For instance, it can be used by forest managers to forecast the effects of silvicultural practices or wildfire on stand regeneration, composition and productivity. Because some stand scale environmental factors provided in forest maps are intrinsically dynamic (e.g. soil moisture regime and SOL thickness) and potentially influenced by climate variables, forest maps could also be used to forecast the effects of climate change on forest development.

In this context, the main objective of this study was to use information commonly available on forest maps in Canada in order to evaluate the potential for paludification of boreal forest stands in relation to climate change and modifications to the natural fire regime. To achieve this objective we developed a dynamic paludification index. First, we developed a base index projecting the development of paludified forests over time without considering climate change. Then, we added the effects of climate change on the thickening on SOL to this base index in order to forecast the development of paludified forests in response to projected climate change. Finally, forecasted climate change and modifications to the natural fire regime were added to the base index to further explore the development of paludified forests in the context of climate change. Based on information commonly available on forest maps, the paludification index can therefore estimate the effects of both climate change and natural fire regimes on the development of paludified forests at both the local and regional scales.

Methods

Study area

A territory in eastern Canada was used to model the development of paludified forests in relation to climate change and changes in the natural fire regime. The Gordon Cosens Forest is a 20 000 km^2 (17 360 km^2 of which is covered in forest, and the remaining area is water bodies) forest management unit located in the Ontario ecodistrict 3E-1, a region also known as the Clay Belt (Figure 1). Cold climate, flat topography, and surficial deposits that are resistant to water penetration all make this region favorable for the development of paludified forests (Jeglum 1991; Riley 1994). The southern part of the Clay Belt is covered by thick (>10 m) glaciolacustrine clay and silt deposited by glacial Lake Ojibway, while the northern part, known as the Cochrane till, is covered by a compact till made up of a mixture of clay and gravel, created by a southward ice flow approximately 8000 years BP (Veillette 1994). Soils of the study area are mostly classified as Gleysols and Luvisols (Soil Classification Working Group 1998). Nonetheless, organic deposits (i.e. a surficial deposit consisting of a SOL > 40 cm thick) are found in many locations in both the southern and northern parts of the study area. Black spruce (*Picea mariana* [Mill.] BSP.) is the dominant tree species of the study area.

According to the local weather station (Kapuskasing, Ontario), from 1981 to 2010, the average annual temperature was 1.3°C and the average annual precipitation was 830 mm, with 30% falling during the growing season (Environment Canada 2014). The average number of degree-days (>5°C) is 1430, and the frost-free season lasts about 100 days; frost can occasionally occur during the growing season. In this region, according to different scenarios (A1 and B2) and simulation models (Canadian Center for Climate Modelling and Analysis [CGCM 3.1], Australian-based Commonwealth Scientific and Industrial Research Organization [CSIROMk3.5], National Institute for Environmental Studies [MIROC 3.2 medres], and National Center for Atmospheric Research [NCAR-CC SM3]), climate warming is projected to be between 3–6°C

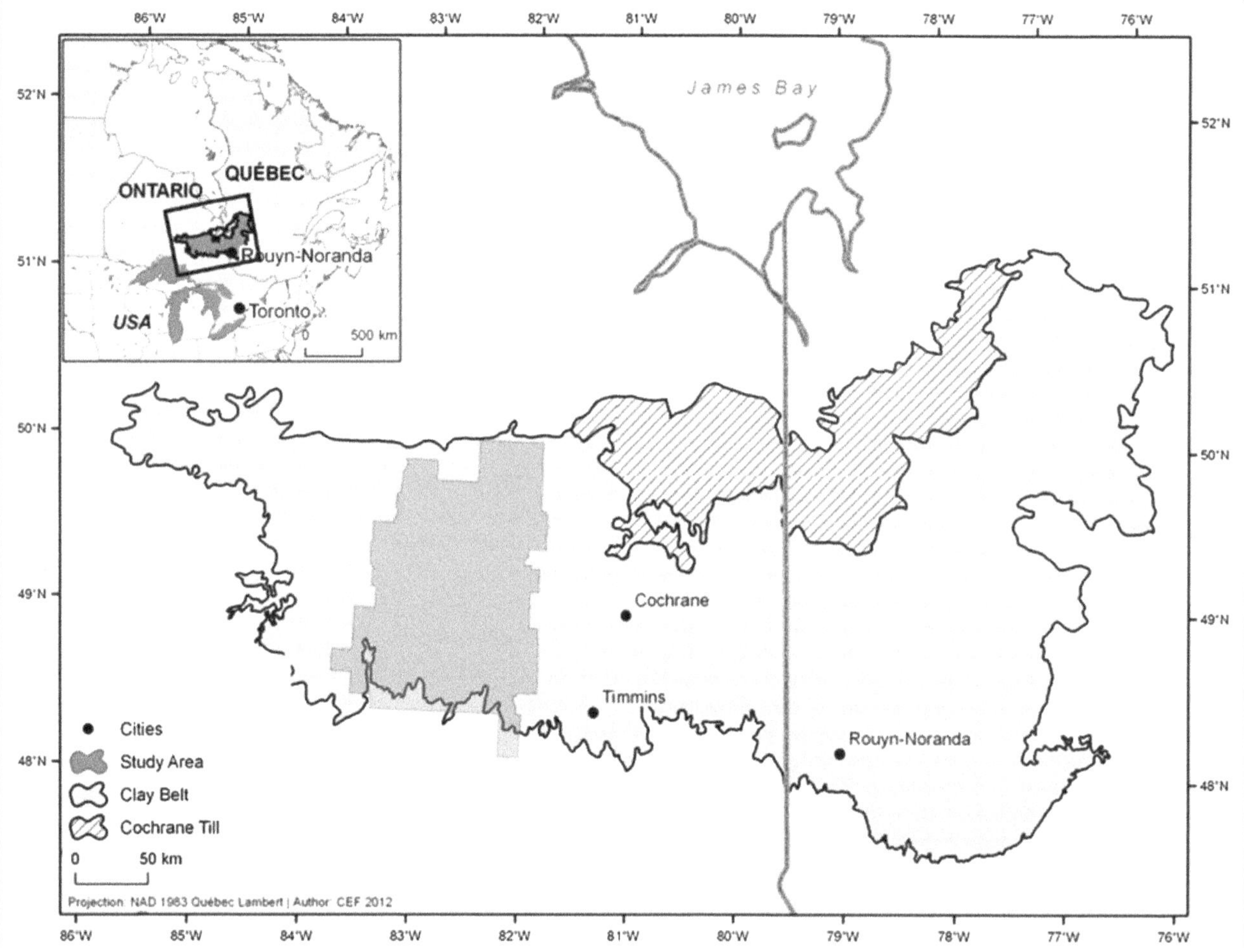

Figure 1 Location of the Clay Belt and the Gordon Cosens Forest in northeastern Ontario (inset).

and precipitation is projected to increase by 10%–20% by the end of the 21st century (McKenney et al. 2010).

In the study area, fire frequency has diminished from a 100-year cycle to an approximately 400-year cycle since the little Ice Age (ca. 1850; Bergeron et al. 2004). As a result of anthropogenic climate change, Bergeron et al. (2010) predict a doubling of fire frequency in this region by the end of the 21st century.

Map data

Data for individual forest stands polygons, used to forecast the development of boreal forested peatlands, were retrieved from the Ontario ecosite classification system (Taylor et al. 2000). This classification seeks to classify the province's ecosystems, such as non forested uplands (prairie, cliff top, dunes), forested ecosystems (both upland and lowland), and non forested wetlands (marshes, swamps, fen, bogs). For forest polygons, ecosites are defined as homogeneous landscape areas (i.e. polygons typically 10–100 hectares) of common surficial deposits, soil moisture regime, soil texture, SOL thickness, humus form, and tree cover. Figure 2 illustrates the forest stand mosaic of the Gordon Cosens Forest, whereas Table 1 shows the size distribution of forest stand polygons. Ecosite classification is meant to be a practical tool for resource managers, and can be used for a variety of forest and site level applications including timber supply analysis, harvest planning, wildlife habitat studies and assessments, and successional studies.

The forest stand polygon data used in this study were retrieved from the 2004 forest inventory reassessment, the most recent reassessment available for the study area. The analysis of this inventory reveals that in 2004, organic deposits (i.e. surficial deposit consisting of a SOL > 40 cm thick) covered 42% of the study area, and nearly 70% of the area had moist to wet soils. While black spruce stands covered 55% of the study area, trembling aspen (*Populus tremuloides* Michx.) and mixed black spruce-trembling aspen stands covered 5% and 25% of the area, respectively. Jack pine (*Pinus banksiana*

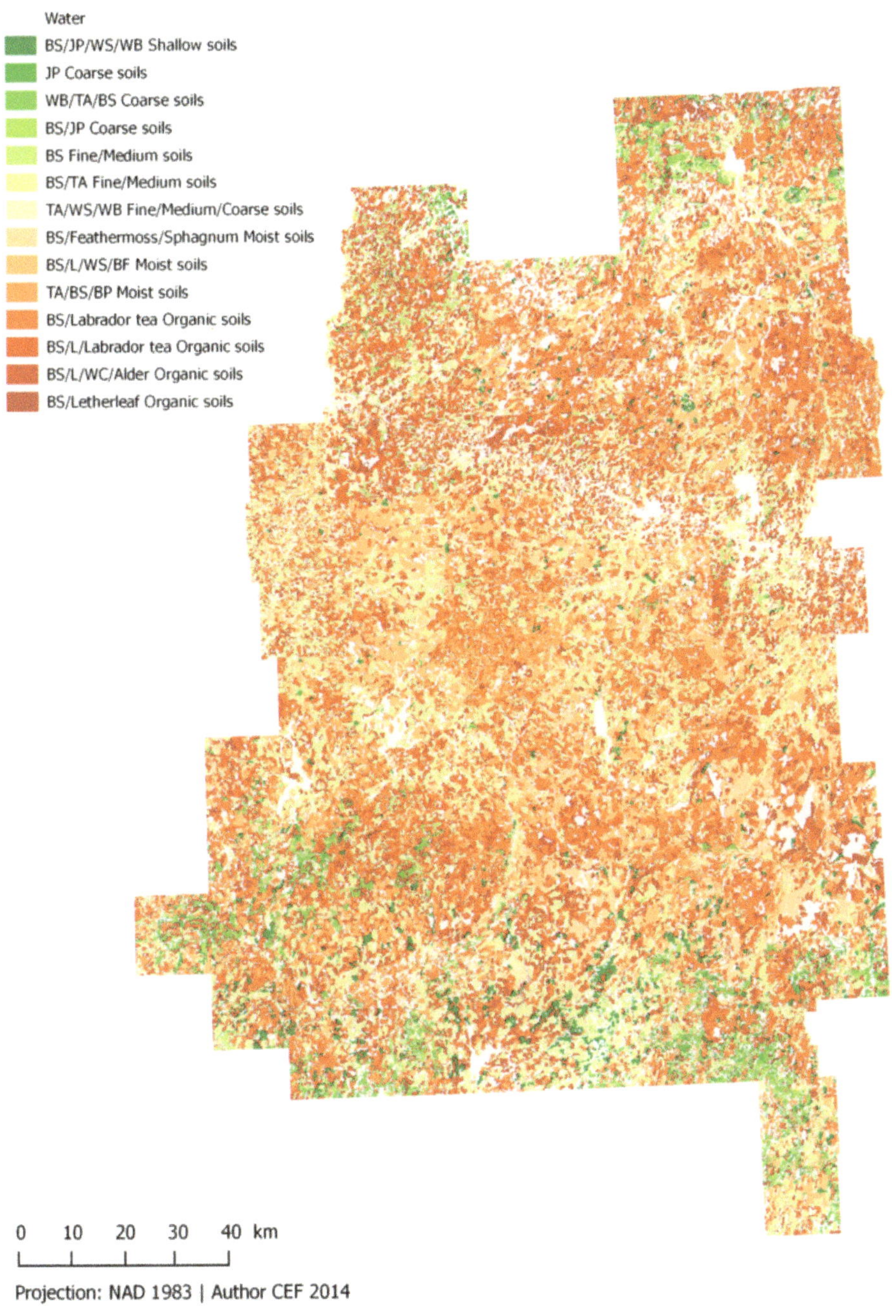

Figure 2 Forest mosaic of the Gordon Cosens Forest. BF = balsam fir, BP = balsam poplar, BS = black spruce, JP = jack pine, L = larch, TA = trembling aspen, WB = white birch, WC = white cedar, WS = white spruce.

Table 1 Size and area distribution of polygons of the Gordon Cosens Forest map

Polygon area (ha)	Number of polygons	Total area (km^2)
<1	9402 (11%)	29 (0.2%)
1–10	36030 (41%)	1702 (9.8%)
10–100	38694 (45%)	11631 (67.0%)
>100	2489 (3%)	1702 (23.0%)

Lamb.), white spruce (*Picea glauca* Moench), and white birch (*Betula papyrifera* Marsh.) were the other dominant species of the regional landscape.

Development of the paludification index

Using static (e.g. surficial deposit and soil texture) and dynamic (e.g. moisture regime and SOL thickness) stand scale factors available from forest maps (Table 2; Taylor et al. 2000), we developed a dynamic paludification index

Table 2 Paludification scores for static and dynamic stand scale factors

Surficial deposit (D)		Soil texture (T)		Humus form (H)		Soil organic layer thickness (SOL)		Moisture regime (M)		Overstorey (O)	
Class	Score	Class[1]	Score	Class[1]	Score	Class[2]	Score	Class[3]	Score	Class	Score
Rock	0	Rock	0	Mull	0	0–9	0	Dry	0	Other spp.	0
Aeolian	1	Sandy	0	Moder	0	10–19	1	Moderately fresh	0	Black spruce	1
Fluviatil	1	Coarse loam	0	Humic mor	1	20–29	1	Fresh	0		
Fluvial till	1	Medium loam	1	Fibric mor	1	30–39	2	Very fresh	0		
Clay till	2	Silty	1	Humic	2	40–120	2	Moderately moist	1		
Lacustrine	2	Fine loam	1	Mesic	2	>120	2	Moist	2		
Organic	2	Clay	2	Fibric	2			Very moist	3		
								Wet	4		
								Very wet	5		

Each static and dynamic factor was divided into classes, each of which was attributed a score related to its paludification "power"; 0 = null "power", 1 = low "power", 5 = high "power". Adding up the score of each factor, gives a Paludification Index that estimates the liability of each stand to paludification.
[1]According to Sims and Baldwin (1996).
[2]In cm. In our paludification index, we used the median value of each class.
[3]According to Sims et al. (1989).

for each forest stand polygon. Each static and dynamic factor was divided into classes (2 to 9 according to site feature), each of which was attributed a score related to its paludification "power" (between 0 and 5; 0 = null paludification "power" and 5 = high paludification "power"). Table 2 lists the factor classes with their related paludification "power" score. Adding up the score of each factor, gives the paludification index, which describes the level of paludification of each forest stand polygon (hereafter referred to as stand). The maximum score an ecosite could achieve is 14. The paludification index was calculated as follows:

$$\mathrm{PI} = \text{Static factors} + \text{Dynamic factors} \tag{1}$$

or

$$\mathrm{PI} = (D + T + H) + (SOL + M + O), \tag{2}$$

where PI is the paludification index, D is surficial deposit, T is soil texture, H is humus form, SOL is soil organic layer thickness, M is moisture regime, and O is overstorey composition. Although we recognize the dynamic nature of surficial deposits and soil texture, we treated these factors as static because the transformation rates of these variables are slow relative to the projection time (i.e. 100 years) used in this study. Similarly, humus form was considered to be static because of its slow rate of transformation (Yu et al. 2001). Stands with a $PI \geq 13$ were classified as *paludified,* whereas stands with a PI between ≥ 7 and < 13 were classified as *nearly paludified.* Stands with a $PI \leq 6$ were classified as *not paludified.* Although we recognize the possibility of interactions among variables and nonlinearity in the effects, the use of a simple index with linear effects was justified because of the lack of information about possible interactions among some variables.

Dynamic factors were allowed to vary over time according to certain rules. In the base index (i.e. the index not considering climate change or fire), SOL thickness increased with time, following peat accumulation rates determined by Lecomte et al. (2006) for nearby sites located in the Clay Belt of Quebec. According to Lecomte et al. (2006), for the past ca. 200 years the peat accumulation rate varied between 10 and 20 cm per century. Lecomte et al. (2006) also showed that peat tends to accumulate at a faster rate where SOL is > 20 cm deep. As a result, the base index was adjusted to allow stands with a median SOL depth > 20 cm to accumulate peat at a rate of 20 cm per century, whereas stands with SOL median depth < 20 cm we allowed to accumulate peat at a rate of 10 cm per century. In the same vein, stands with an initial SOL depth < 20 cm had their peat accumulation rate adjusted to 20 cm per century when their SOL depth reached 20 cm. Simard et al. (2009) and Drobyshev et al. (2010) observed an important decline in tree growth when SOL depth is 20 cm or greater, further confirming the pertinence of the 20 cm cut off point. Furthermore, in the base index, moisture regime was allowed to vary in stands where soil texture was finer than medium loam, only if SOL depth was ≥ 20 cm. Hence, in all cases when SOL depth reached ≥ 20 cm, moisture regime stepped one class ahead. For example, stands with a fine loam soil texture and a moisture regime classified as *Moist* had their moisture regime changed to *Very moist* when their SOL depth reached 20 cm.

In the index considering climate change, peat accumulation rate (hence SOL thickness) was allowed to vary according to an adaptation of the Peat Accumulation Model (PAM; Hilbert et al. 2000) made by Wu (2012).

Wu (2012) modified PAM to study the response of peatland development and carbon cycling to climate change and to answer several research questions, among which was the following: How does peat accumulation respond to changes in precipitation and temperature? In its most basic form, PAM considers peat accumulation (i.e. SOL thickness) as a simple equation: "peat production minus the sum of oxic decomposition and anoxic decomposition equals peat accumulation" (i.e. change in SOL thickness = peat production − peat decomposition). In Wu's (2012) adaptation of PAM, the peat accumulation rate in ombrotrophic bogs could decrease by up to 70% within the first 100 years following the initiation of climate change. In fact, Wu's (2012) estimation suggested that for the 30 year period between 2011 and 2040, which corresponds to a 1°C increase in temperature, peat accumulation rate would drop by 15%. For the 30 year period between 2041 and 2070 (2°C increase in temperature) peat accumulation rate would drop by an additional 25%. Finally, for the 30 year period between 2071 and 2100 (3°C increase in temperature) peat accumulation rate would drop by an additional 30%. Hence, according to our estimations based on Wu's (2012) relationship between temperature increase and decrease in peat accumulation rate, this relationship fits the following equation:

$$\mathrm{Peat}_{\mathrm{AR}} = (16.0414 \times T) + \left(-5.7383 \times T^2\right) + \left(2.7230 \times T^3\right) \quad (3)$$

where $\mathrm{Peat}_{\mathrm{AR}}$ is the accumulation rate of peat in $\mathrm{cm} \cdot \mathrm{yr}^{-1}$ and T the temperature (°C) increase. As a result, the peat accumulation rate decreases rapidly to the point where the accumulation rate was negative (i.e. peat decomposition rate was greater than production rate). Consequently, a temperature increase between 3°C and 4°C results in a reduction of SOL thickness of approximately 0.5 $\mathrm{mm} \cdot \mathrm{yr}^{-1}$, between 4°C and 5°C a reduction of 1.8 $\mathrm{mm} \cdot \mathrm{yr}^{-1}$, and between 5°C and 6°C a reduction of 3.8 $\mathrm{mm} \cdot \mathrm{yr}^{-1}$. Hence, when a 4°C increase is reached, peat decomposition rate is greater than its production rate, leading to a reduction in SOL thickness. Although we recognize that this extension is simplistic and should be interpreted with greater caution, we believe this extension should be representative of the effect of temperatures increases beyond 3°C on peat accumulation rate and its effect on SOL thickness. Furthermore, although we acknowledge that ombrotrophic bogs and paludified forest support a different aboveground vegetation structure (i.e. shrubs- vs. tree-dominated aboveground vegetation for ombrotrophic bogs and paludified forests, respectively), we assumed that peat decomposition in ombrotrophic bogs and paludified forests would show similar responses to temperature increase.

From this information, we projected peat accumulation and SOL thickness according to three temperature increase scenarios (i.e. +1°C, +3°C and +6°C, which respectively correspond to 5%, 40% and 70% reduction in peat accumulation rate) and for four initial SOL thickness (5 cm, 10 cm, 20 cm, and 40 cm). For each scenario, average temperature was progressively increased during the simulations so that by 2100 temperature increases amount to +1°C, +3°C and +6°C relative to the average temperature as of 2004. For these four initial SOL thicknesses, peat accumulation rate varied according to Lecomte et al. (2006), i.e. where SOL is < 20 cm thick, peat accumulated at a rate of 10 cm per century, whereas where SOL is > 20 cm thick, peat accumulated at a rate of 20 cm per century. At each time step, SOL thickness was estimated as follow:

$$\mathrm{SOL}_{\mathrm{t}} = \mathrm{SOL}_{\mathrm{i}} + (30 \times \mathrm{Peat}_{\mathrm{ARt}}) \quad (4)$$

where $\mathrm{SOL}_{\mathrm{i}}$ is the thickness of the soil organic layer at the beginning of the reference period and $\mathrm{Peat}_{\mathrm{ARt}}$ is the accumulation rate of peat for the reference period. $\mathrm{Peat}_{\mathrm{ARt}}$ was calculated as follow:

$$\mathrm{Peat}_{\mathrm{ARt}} = \mathrm{Peat}_{\mathrm{ARi}} - (\mathrm{Peat}_{\mathrm{ARi}} \times (R_{\mathrm{t}}/100) \quad (5)$$

where $\mathrm{Peat}_{\mathrm{ARi}}$ is the initial accumulation rate of peat and R_{t} is the reduction of peat accumulation rate for the period of reference, i.e. accumulation rate for each 30 year period. For each SOL thickness class (i.e. 5 cm, 10 cm, 20, and 40 cm), we calculated peat accumulation rate and the SOL thickness for each time step under the three climatic scenarios (i.e. +1°C, +3°C and +6°C).

Finally, in the index considering both climate change and natural fire regime, we first projected forest paludification considering the current fire cycle of 400 years as determined by Bergeron et al. (2004). Then we projected forest paludification allowing the fire cycle to decrease according to projections made by Bergeron et al. (2010). In this sub-index, the fire cycle decreased to 300 years for 2011–2041, to 250 years for 2041–2071, and to 200 years for 2071–2100. Furthermore, based on our own observations, we considered that currently 50% of fire ignitions resulted in high-severity fires, i.e. fires that left < 5 cm of residual SOL over the mineral soil (Simard et al. 2007). In light of greater uncertainty surrounding the impact of the climate change on fire severity, we ran our projections considering three proportions of high-severity fires, i.e. 25%, 50% and 75%. Furthermore, we considered that stands with SOL > 120 cm could not be submitted to high-severity fires because it is highly unlikely that a fire would leave < 5 cm of residual SOL over the mineral soil in such stands. At each time step, fire events and severity were randomly attributed to the forest stands of the study area.

The three indices were run in the following sequence: (1) base index, (2) base index + climate change, and (3) base index + climate change + natural fire regime modifications. This sequence allowed us to first explore the effect of climate change alone and then the combined effects of climate change and modifications of the fire regime on the potential for paludification of the forest stands of the study area.

Results

Base paludification index

According to the Ontario forest map data, 42.2% of the area of the Gordon Cosens Forest was already *paludified* in 2004, whereas 57.8% of the area was *not paludified* (Table 3).

As paludification is a relatively slow process, the index did not forecast any change in the level of paludification of the territory for the 2041 time step compared to the current state of the Gordon Cosens Forest (Table 3). As an illustration, Figure 3 shows the slow but constant thickening of the SOL for four different initial SOL thickness scenarios. When initial SOL thickness was set at 5 cm or 10 cm, SOL thickness projected for 2100 remained below 20 cm (Figure 3). However, when initial SOL thickness was set at 20 cm, SOL thickness reached 30 cm around 2060 and nearly 40 cm in 2100 (Figure 2). Similarly, when initial SOL thickness was set at 40 cm, SOL thickness reached 50 cm around 2060 and nearly 60 cm in 2100 (Figure 3). As a result, for the 2071 time step, the proportion of the study area occupied by *nearly paludified* stands increased from 0 to 11.7%, whereas as that occupied by *not paludified* stands decreased to 46.1% (Table 3); the proportion of the study area classified as *paludified* did not change. For the 2100 time step, the index forecasted that the proportion of the study area classified as *nearly paludified* increased to 26.4% (Table 3), while the proportion of area classified as *not paludified* decreased to 31.4%. At this time step, nearly 70% of the forest stands of the Gordon Cosens Forest could be classified as *paludified* or *nearly paludified*, which corresponds to a 166% increase in the cover of *paludified* or *nearly paludified* areas.

SOL thickness and climate change

For every initial SOL thickness (i.e. 5 cm, 10 cm, 20 cm and 40 cm), SOL thickness tended to increase with time for the +1°C and +3°C scenarios (Figure 4), the increase being greater for the +1°C scenario. For the +6°C scenario, an increase was observed until ca. 2060 when SOL thickness started to decrease. This tipping point corresponds to the moment where the temperature increase was ca. 4°C. For the stands where initial SOL thickness was 5 cm or 10 cm, the +6°C scenario led to an almost complete disappearance of the SOL by 2100. For the stands where initial SOL thickness was 20 cm or 40 cm, the same scenario ended with SOL thickness similar to what was observed initially.

These negatives effect of climate change on peat accumulation resulted in no change in the proportion of the study area classified as *paludified, nearly paludified* and *not paludified* between 2011 and 2100 (Table 3).

Combining climate change with natural fire regime

The inclusion of the current fire cycle (400 years) in our index resulted in a slight decrease (<6%) in the paludified area, regardless of the proportion of burnt stands submitted to high-severity fires (Figure 5a). Nonetheless, the decrease in paludified area was lower (ca. 2%) when 25% of the burned stands were submitted to high-severity fires and steeper (ca. 6%) when 75% were submitted to high-severity fires.

When the fire cycle was allowed to decrease according to the projections made by Bergeron et al. (2010), declines in paludified area was steeper for the three proportions of high-severity fires (Figure 4b). Declines in paludified area were ca. 9%, 25% and 40% for the 25%, 50% and 75% high-severity fires, respectively (Figure 5b).

Discussion

The boreal forest represents one of the Earth's largest biome, encompassing an area of approximately 14×10^6 km^2 (Wieder et al. 2006). According to Bhatti et al. (2003), it is also one of the Earth's biomes most affected by global warming. About 25% of the boreal forest region is occupied by peatlands (Wieder et al. 2006), a large proportion of which are forested peatlands. Overall, peatlands store approximately 15% of the Earth's terrestrial carbon (Gorham 1991; Lavoie et al. 2005). Consequently, at the global scale, any modifications to the natural fire regime in response to climate change may have a significant

Table 3 Paludification level (% of the territory) of the Gordon Cosens forest according to the base index and the combination of the base index with climate change

Paludification level		Base index			Base index + climate change		
	Current[1]	2041	2071	2100	2041	2071	2100
Not paludified	57.8	57.8	46.1	31.4	57.8	57.8	57.8
Nearly paludified	0.0	0.0	11.7	26.4	0.0	0.0	0.0
Paludified	42.2	42.2	42.2	42.2	42.2	42.2	42.2

[1]Based on the Ontario's 2004 Forest Resource Inventory.

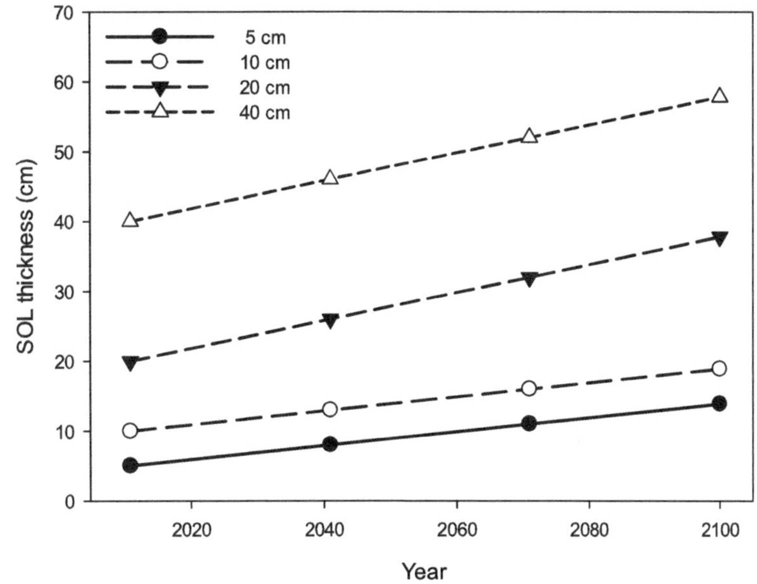

Figure 3 Soil organic layer (SOL) thickening with time without considering the effects of climate change. Initial SOL thickness (cm) is illustrated for four different scenarios: 5 cm, 10 cm, 20 cm, and 40 cm.

impact on the carbon sequestered in paludified forests. Our results suggest that in this context paludified forests may turn from C sink to atmospheric C source (in turn increasing atmospheric CO_2 concentration and providing a positive feedback on climate warming) if the increase in forest productivity does not compensate for carbon losses. However, it is important to note that the thick *Sphagnum* layer typical of paludified forests may protect SOL from drought and potential increasing depth of burn (and hence CO_2 emissions), and therefore that a significant proportion of the territory has high potential to remain paludified (Magnan et al. 2012; Terrier et al. 2014a).

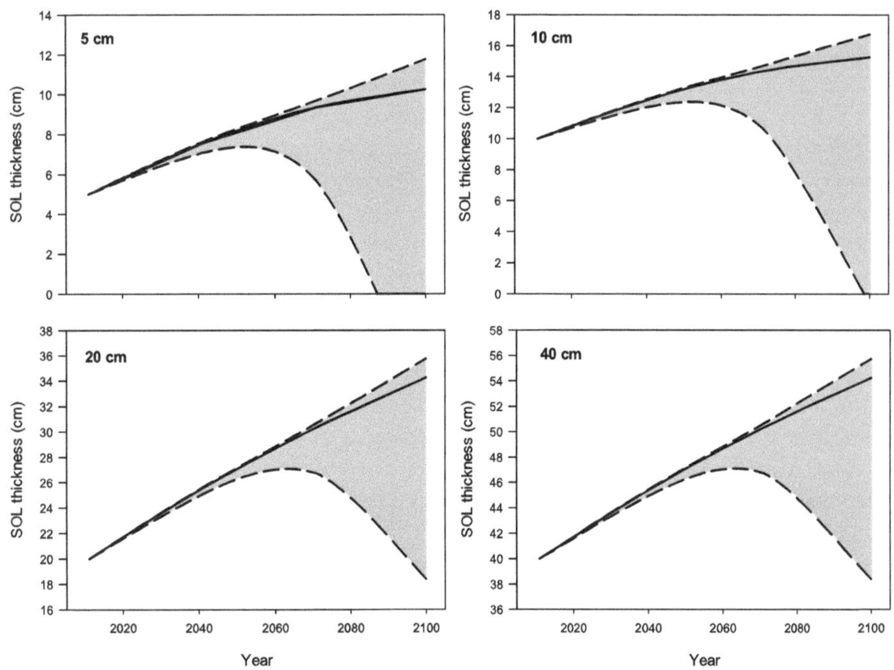

Figure 4 Soil organic layer (SOL) thickness (cm) variation with time in response to climate change. Each panel represents a different initial SOL thickness (i.e. 5 cm, 10 cm, 20 cm, and 40 cm) at the start of the projection. For each panel, upper dashed line = +1°C scenario; middle solid line = + 3°C scenario; lower dashed line = + 6°C scenario.

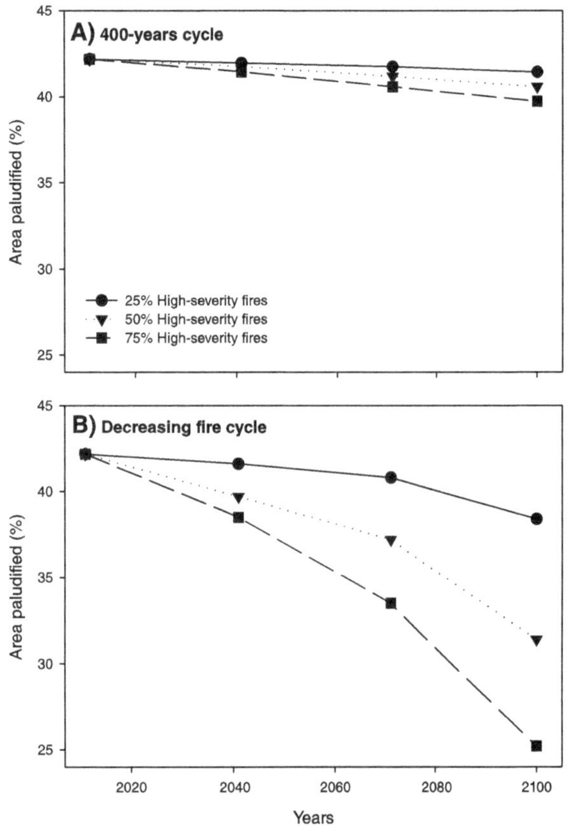

Figure 5 Projected effects of climate change and fire severity on paludified area for the period 2011–2100. A) Fire cycle was maintained at 400 years throughout the projection. **B)** Fire cycle was adjusted according to projections by Bergeron et al. (2010) for a region covering northeastern Ontario and northwestern Quebec; in 2041, fire cycle = 300 years; in 2071, fire cycle = 250 years; in 2100, fire cycle = 200 years.

The base paludification index produced slow but noticeable increases in proportion of the study area classified as *paludified* or *nearly paludified*. In the absence of fire in 2100 nearly 70% of the area of the Gordon Cosens Forest could be classified as nearly paludified (26%) or paludified (42%). This change is attributable to increased SOL thickness and concomitant change in moisture regime. Stands most susceptible to paludification were generally located on lacustrine surficial deposits or clay till, had fine textured soil (clay to coarse loam), a moisture regime classified as *Moist* to *Very moist*, SOL median depth of 20 cm, and an overstorey comprised of black spruce (sometimes accompanied by trembling aspen) (data not shown). Over the projected time frame considered with this base index, most of the stands of the study area were classified as *nearly paludified* or *paludified*.

Furthermore, as expected, temperature increase and the concomitant reduction in peat accumulation rate had an effect on SOL thickness. For the +1°C and +3°C scenarios, SOL thickness increased with time, although at a slower rate than in the model not considering climate change. For the +6°C scenario, however, peat accumulation ceased and SOL thickness started to decrease around the year 2060, indicating that the peat production rate was subsequently lower than its decomposition rate. In a long-term simulation, Ise et al. (2008) observed that a 4°C air temperature rise caused a 15% loss of soil organic carbon (measured in kg $C \cdot m^{-2}$) for a reference period of 100 years, which corresponds to a 10 cm reduction in SOL thickness. This is similar to the difference we observed (ca. 15 cm) between the +1°C and the +6°C scenarios. These results suggest that paludified stands (i.e. stands where SOL thickness > 40 cm) will remain paludified in 2100.

In consequence, the introduction of the effects of temperature increase in the paludification index did not produce any modifications in the proportion of the study area whether classified as *paludified, nearly paludified* or *not paludified* compared to the base index. Hence, despite increasing temperature between 2011 and 2100, the level of paludification of the study area did not change over time. Although our model did not consider the effects of changes in precipitation on paludification, a recent modeling study suggested that in the boreal forest of eastern Canada precipitation could increase by 10%–20% by the end of the 21st century (McKenney et al. 2010). An increase in precipitation would logically lead to an increase in peat accumulation rate via the influence of higher water tables and a subsequent decrease in peat decomposition rate (Silvola et al. 1996; Ise et al. 2008). However, increased temperature is expected to lead to an increase in evapotranspiration, therefore leaving less moisture in the system (Soja et al. 2007; Wu 2012) and increasing risks of drought. A decrease in moisture availability in the system would, in turn, induce a lowering of the water table, a thickening of the oxic layer, and an increase in substrate temperature. Together these modifications would depress the peat accumulation rate, but more importantly, increase the peat decomposition rate, resulting in a steady SOL thickness over time. Therefore as two factors are acting in opposite directions (i.e. peat accumulation with time vs. increase in peat decomposition rate in response to climate change), the expected effects of climate change (not considering modifications to the current fire cycle) in our study area could result in a territory similar to the present in terms of paludified area.

Introducing wildfire in the index produced a quite different picture of the study area. Keeping the fire cycle constant and at the current level (i.e. 400 years) produced only a slight decrease in paludified area. However, allowing the fire cycle to decrease with time (as projected by Bergeron et al. 2010) reduced the paludified area by between 10% and 40% under the high-fire severity scenario. Decreased fire cycle and reduced paludified area could be attributed to the fact that the projected increase in precipitation (McKenney et al. 2010) may not fully compensate for increase in temperature, thus creating conditions that are more prone to fire occurrence (Terrier et al. 2013). Yet, we feel that an increase in the proportion of high-severity fires in areas already paludified is unlikely as a recent study conducted in the forested peatlands of the Clay Belt during an extreme drought year failed to detect any effect of drought on soil moisture (measured as gravimetric water content [%]; Terrier et al. 2014a) where thick SOL (>40 cm) occurs. In the same study, the authors projected for the period 2071–2100 the effects of extreme drought on potential SOL depth of burn, and concluded that increase in drought conditions should not be sufficient to greatly modify SOL depth of burn in areas where thick SOL prevail with potential depth of burn up to 0.7 cm (Terrier et al. 2014a) for any individual fire event occurring between early spring and late fall. This resistance of SOL to drought and burn is related to the presence of *Sphagnum* species which hyaline cells stock large amounts of water (Silvola 1991) and limit its evaporation (Busby and Whitfield 1978), and therefore reduce potential depth of burn. However, Terrier's study (Terrier et al. 2014a) also suggested that in areas where SOL is < 40 cm and where *Pleurozium schreberi* is the dominant ground-covering moss, drought was able to depress soil moisture to the extent where they projected a potential depth of burn up to 3.2 cm for the period 2071–2100. These results suggest that areas that are in the process of being paludified exhibit a relatively high potential to remain unpaludified, and that those areas that are already paludified areas show a low potential for depaludification. These results are also in accordance with Magnan et al. (2012) who found no major changes in boreal peatlands despite evidence of slowed peat accumulation rates due to fire, and with Magnan et al. (2014) who showed that *Sphagnum*-dominated bogs located in a maritime environment have persisted over millennia and that fires had few impacts on their vegetation dynamics.

Thus, at the landscape level, when fire cycle is kept constant at the current level, the slow rate of paludification results in a slight decrease in paludified area regardless of the proportion of high severity fires, suggesting either that the rate of peat accumulation in the index is too conservative, or that the current highly paludified landscape is not completely in balance with the current climate and fire regime (Payette 2001). Similarly, a decreasing fire cycle is likely to limit the development of paludified stands.

Conclusions

Overall, our models predict a slight (no modification to the current fire cycle) to moderate (decreasing fire cycle over time) decrease in area covered by paludified forests within the Gordon Cosens Forest in 2100. This might have several impacts for the forest industries of the Clay Belt as slower paludification rates imply greater forest productivity and potentially a greater potential for forest harvest.

Furthermore, at the global scale, increasing fire frequency in boreal paludified forests may have important consequences on carbon storage and climate if increases in forest productivity do not compensate for carbon losses. However, as the thick *Sphagnum* layer typical of paludified forests may protect soil organic layer from drought and deep burns, a significant proportion of the territory has high potential to remain a carbon sink.

In this context, our findings are supported by those of Terrier et al. (2014b) who in an area adjacent to our study area modeled the impacts of climate change on fire regime, vegetation dynamics and SOL depth of burn. They concluded that although climate change is likely to increase burn rates, the moist and cool conditions in these forests would prevent high depth of burn and the landscape would remain paludified.

This simple tool could be used by forest managers to forecast the development of paludified forests and to plan forest operations and conservation areas, and by policy makers to plan carbon management at the regional scale. Nonetheless, we recognize that in order to strengthen the predictions made by our index, the next step would be to construct a mechanistic model that includes sensitivity analysis as well as fire sub-indices that take into account drought and wildfire risk.

Competing interests

The authors declare that they have no competing interests.

Authors' contributions

BL developed the models, carried out the analysis, and wrote the manuscript, NJF and YB were instrumental in conceptualizing and refining the models, and in revising the manuscript. All authors read and approved the final manuscript.

Acknowledgements

This study was made possible by funding from the Ontario Ministry of Natural Resource. We thank Rachelle Lalonde for providing forest maps, Mélanie Desrochers for map production, and Aurélie Terrier for helpful discussion on fire modeling.

References

Bergeron Y, Gauthier S, Flannigan M, Kafka V (2004) Fire regimes at the transition between mixedwood and coniferous boreal forest in northwestern Quebec. Ecology 85:1916–1932

Bergeron Y, Cyr D, Girardin MP, Carcaillet C (2010) Will climate change drive 21st century burn rates in Canadian boreal forest outside of its natural variability: collating global climate model experiments with sedimentary charcoal data. Int J Wildl Fire 19:1127–1139

Bhatti JS, Van Kooten GC, Apps MJ, Laird LD, Campbell ID, Campbell C, Turetsky MR, Yu Z, Banfield E (2003) Forest management planning based on natural disturbance and forest dynamics. In: Burton PJ, Messier C, Smith DW, Adamowicz WL (eds) Towards sustainable management of the boreal forest. NRC Research Press, Ottawa, pp 799–855

Busby JR, Whitfield DWA (1978) Water potential, water content, and net assimilation of some boreal forest mosses. Can J Bot 56:1551–1558 de Groot WJ, Pritchard JM, Lynham TJ (2009) Forest floor fuel consumption and carbon emissions in Canadian boreal forest fires. Can J For Res 39:367–382

Drobyshev I, Simard M, Bergeron Y, Hofgaard A (2010) Does soil organic layer thickness affect climate-growth relationships in the black spruce boreal ecosystem? Ecosystems 13:556–574

Dyrness CT, Norum RA (1983) The effects of experimental fires on black spruce forest floors in interior Alaska. Can J For Res 13:879–893

Environment Canada (2014) Canadian climate normals 1981–2010. http://www.climate.weather.gc.ca/climate_normals/index_e.html. Accessed 20 Oct 2014

Flannigan M, Logan KA, Amiro B, Skinner W, Stocks BJ (2005) Future area burned in Canada. Clim Change 72:1–16

Girardin MP, Mudelsee M (2008) Past and future changes in Canadian boreal wildfire activity. Ecol Appl 18:391–406

Gorham E (1991) Northern peatlands: role in the carbon cycle and probable responses to climatic warming. Ecol Appl 1:182–195

Greene DF, Macdonald SE, Cumming S, Swift L (2005) Seedbed variation from the interior through the edge of a large wildfire in Alberta. Can J For Res 35:1640–1647

Hilbert DW, Roulet N, Moore T (2000) Modelling and analysis of peatlands as dynamical systems. J Ecol 88:230–242

IPCC (2013) Climate change 2013: The physical science basis. In: Stocker TF, Qin D, Plattner G-K, Tignor M, Allen SK, Bashung J, Nauels A, Xia Y, Bex V, Midgley PM (eds) Contribution of Working Group I to the Fifth Assessment Report of the Intergovernmental Panel on Climate Change. Cambridge University Press, Cambridge, UK

Ise T, Dunn AL, Wofsy SC, Moorcroft PR (2008) High sensitivity of peat decomposition to climate change through water-table feedback. Nat Geosci 1:763–766

Jeglum JK (1991) Definition of trophic classes in wooded peatlands by means of vegetation types and plant indicators. Ann Bot Fenn 28:175–192

Lafleur B, Fenton NJ, Paré D, Simard M, Bergeron Y (2010a) Contrasting effects of season and method of harvest on soil properties and the growth of black spruce regeneration in the boreal forested peatlands of eastern Canada. Silva Fenn 45:799–813

Lafleur B, Paré D, Fenton NJ, Bergeron Y (2010b) Do harvest and soil type impact the regeneration and growth of black spruce stands in northwestern Quebec? Can J For Res 40:1843–1851

Lavoie M, Paré D, Bergeron Y (2005) Impact of global change and forest management on carbon sequestration on northern forested peatlands. Environ Rev 13:199–240

Lecomte N, Simard M, Bergeron Y, Larouche A, Asnong H, Richard PJH (2005) Effects of fire severity and initial tree composition on understorey vegetation dynamics in a boreal landscape inferred from chronosequence and paleoecological data. J Veg Sci 16:665–674

Lecomte N, Simard M, Fenton N, Bergeron Y (2006) Fire severity and long-term ecosystem biomass dynamics in coniferous boral forests of eastern Canada. Ecosystems 9:1215–1230

Magnan G, Lavoie M, Payette S (2012) Impact of long-term vegetation dynamics of ombrotrophic peatlands in northwestern Québec, Canada. Quat Res 77:110–121

Magnan G, Garneau M, Payette S (2014) Holocene development of maritime ombrotrophic peatlands of the St. Lawrence North Shore in eastern Canada. Quat Res 82:96–106

McKenney DW, Pedlar JH, Lawrence K, Gray PA, Colombo SJ, Crins WJ (2010) Current and projected future climatic conditions for ecoregions and selected natural heritage areas in Ontario, Climate Change Res. Rep. CCRR-016. Ontario Forest Research Institute, Ministry of Natural Resources, Sault Ste. Marie, ON, Canada

Payette S (2001) Les principaux types de tourbières. In: Payette S, Rochefort L (eds) Écologie des tourbières du Québec-Labrador. Les Presses de l'Université Laval, Québec, Québec, pp 39–89

Payette S, Garneau M, Delwaide A, Schaffhauser A (2013) Forest soil paludification and mid-Holocene retreat of jack pine in easternmost North America: Evidence for a climatic shift from fire-prone to peat-prone conditions. The Holocene 23:494–503

Riley JL (1994) Peat and peatland resources of northeastern Ontario. Ministry of Northern Development and Mines, Ontario Geological Survey. Misc. Paper No, 153

Silvola J (1991) Moisture dependence of CO_2 exchange and its recovery after drying in certain boreal forest and peat mosses. Lindergia 17:5–10

Silvola J, Alm J, Ahlholm U, Nykanen H, Martikainen PJ (1996) CO_2-fluxes from peat in boreal mires under varying temperature and moisture conditions. J Ecol 84:219–228

Simard M, Lecomte N, Bergeron Y, Bernier PY, Paré D (2007) Forest productivity decline caused by successional paludification of boreal soils. Ecol Appl 17:1619–1637

Simard M, Bernier PY, Bergeron Y, Paré D, Guérine L (2009) Paludification dynamics in the boreal forest of the James Bay Lowlands: effect of time since fire and topography. Can J For Res 39:546–552

Sims RA, Baldwin KA (1996) Forest humus forms in northwestern Ontario. Natural Resources Canada, Canadian Forest Service, Great Lakes Forestry Centre, Technical Report TR-28

Sims RA, Towill WD, Baldwin KA, Wickware GM (1989) Field guide to the forest ecosystem classification for northwestern Ontario, Forest Canada and Ontario Ministry of Natural Resources., p 191

Soil Classification Working Group (1998) The canadian system of soil classification, 3rd edn. Agriculture and Agri-Food Canada Publication 1646, Canada

Soja AJ, Tchebakova NM, French NHF, Flannigan MD, Shugart HH, Stocks BJ, Sukhinin AI, Parfenova EI, Chapin FS III, Stackhouse PW (2007) Climate-induced boreal forest change: Predictions versus current observations. Global Planet Change 56:274–296

Taylor KC, Arnup RW, Merchant BG, Parton WJ, Nieppola J (2000) A field guide to forest ecosystems of Northeastern Ontario, 2nd edn. Queen's Printer for Ontario, Canada

Terrier A, Girardin MP, Périé C, Legendre P, Bergeron Y (2013) Potential changes in forest composition could reduce impacts of climate change on boreal wildfires. Ecol Appl 23:21–35

Terrier A, de Groot WJ, Girardin MP, Bergeron Y (2014a) Dynamics of moisture content in spruce-feather moss and spruce-*Sphagnum* organic layers during an extreme fire season and implications for future depths of burn in Clay Belt black spruce forests. Int J Wildl Fire 23:490–502

Terrier A, Girardin MP, Cantin A, de Groot WJ, Anyomi KA, Gauthier S, Bergeron Y (2014b) Disturbance legacies and paludification mediate the ecological impact of an intensifying wildfire regime in the Clay Belt boreal forest of eastern North America. J Veg Sci doi:10.1111/jvs.12250.

van Bellen S, Garneau M, Bergeron Y (2010) Impact of climate change on forest fire severity and and consequences for carbon stocks in boreal forest stands of Quebec, Canada: A synthesis. Fire Ecol 6:16–44

Veillette JJ (1994) Evolution and paleohydrology of glacial lakes Barlow and Ojibway. Quat Sci Rev 13:945–971

Wieder RK, Vitt DH, Benscoter BW (2006) Peatlands and the boreal forest. In: Wieder RK, Vitt DH (eds) Boreal peatland ecosystems. Springer, New York, NY, pp 1–8

Wu J (2012) Response of peatland development and carbon cycling to climate change: a dynamic system modeling approach. Environ Earth Sci 65:141–151

Yu Z, Campbell ID, Vitt DH, Apps MJ (2001) Modelling long-term peatland dynamics. I. Concepts, review, and proposed design. Ecol Model 145:197–210

5

Herbivory of sympatric elk and cattle on Lincoln National Forest, south-central New Mexico

Heather H. Halbritter and Louis C. Bender*

Abstract

Background: Wildlife and livestock grazing are important products of forest ecosystems, but can be controversial. Herbivory by North American elk and domestic cattle is a contentious management issue throughout western North America, often driving management proposals to decrease cattle and elk numbers based on perceived overutilization of forages. Such observations are often site level rather than landscape, and may confuse ecological sustainability with desired conditions.

Methods: We used line transects to document vegetation composition, structure, and grazing and browsing utilization for 4 key habitat types: mountain meadows, aspen, thinned conifer, and burned conifer on Lincoln National Forest, New Mexico, USA. We documented relative habitat use of these types by elk, mule deer, and cattle and modeled relative use on residual grass biomass of mountain meadows and browse utilization of forested types. We determined diets and diet quality of elk and cattle to assess degree of competition.

Results: Use of grasses in meadows was below management thresholds, and combined elk, cattle, and deer relative habitat use accounted for < 14 % of the variance in residual stubble height of *Poa pratensis*, the most abundant grass. Palatable browse was limited in habitat types (< 107 stems·ha^{-1}), use was generally high, and elk presence was correlated with the majority of browsing. Elk and cattle diets did not significantly overlap (Schoener's index 0.54–0.57); elk fed primarily on deciduous shrubs (34 %–55 % of annual diets) and cattle on grass (72 %–77 %). Digestibility and crude protein levels of cattle diets and body condition of elk indicated high quality diets for cattle and marginal–good quality diets for elk.

Conclusions: At observed stocking levels and densities, cattle and elk were not competing for forage based on diet similarity, nor were key habitat types being used beyond sustainable levels. Low browse availability indicates that opportunity exists to increase forage availability on Lincoln National Forest, and thus maintain or increase populations without increasing impacts to plant communities. Management actions that address increased stand densities and forest encroachment of meadows, a result of altered historical disturbance regimes, could increase distribution, quantity, and quality of forage.

Keywords: Browsing; Cattle; Carrying capacity; Elk; Grazing; Herbivory; Nutrition

Background

Wildlife and livestock grazing are important products of forest ecosystems, but conflicts can arise when co-use is perceived to negatively affect users or the ecosystem. Balancing elk (*Cervus elaphus*), cattle (*Bos taurus*), and habitat conditions is a challenging management issue throughout the western United States (Kay 1997; National Research Council 2002; Vavra 2005), where management for the full continuum of forest resources, including wildlife and livestock forage, is mandated on National Forests by the Multiple-Use Sustained-Yield Act of 1960 (Public Law 86–517), the Forest and Rangeland Renewable Resources Planning Act of 1974 (Public Law 93–378), and the National Forest Management Act of 1976 (Public Law 94–588). Through foraging activities, elk and livestock can significantly impact plant community structure, function, health, and composition (Chadde and Kay 1991; Fleischner 1994; Ripple and Larsen 2000; Ripple et al. 2001; National Research Council 2002; Peinetti et al. 2002; Vavra 2005). These changes can be positive (McNaughton 1976; Vavra 2005; Stewart et al. 2006) or negative (Severson

* Correspondence: lbender@nmsu.edu
Extension Animal Sciences and Natural Resources, New Mexico State University, Las Cruces, NM, USA

and Medina 1983; Riggs et al. 2005). This distinction depends in part on the intensity of use and characteristics of the plant community (Georgiadis et al. 1989; Hamilton et al. 1998; De Mazancourt et al. 1998; Alstad et al. 1999; Stewart et al. 2006).

Much controversy regarding herbivory has arisen because of negative impacts on vegetation communities. Recruitment of preferred browse species such as *Salix* spp. and *Populus tremuloides* can be eliminated by large populations of elk (Chadde and Kay 1991; Relva and Veblen 1998; Kay and Bartos 2000). Similarly, cattle herbivory can affect sustainability of herbaceous and woody forages (Fleischner 1994; Vavra 2005). Foraging by large herbivores can be especially contentious in areas where forage is managed for multiple uses, including livestock forage and wildlife habitat (Kay 1997; National Research Council 2002; Vavra 2005).

Such a situation exists on Lincoln National Forest (LNF) in the Sacramento Mountains of south-central New Mexico, where perceived impacts of herbivory have resulted in management recommendations to lower cattle and elk numbers in conflict areas (Hurd 2002; Hurd et al. 2004; Halbritter and Bender 2011b). Conflicts involve perceived inadequate residual biomass of grasses in mountain meadow communities, which is attributed to elk grazing, cattle grazing, or both, dependent upon user group. Further, the issue is complicated by hypothesized negative impacts of inadequate residual herbaceous biomass on habitat quality of prey of the endangered Mexican spotted owl (*Strix occidentalis*) (Hurd 2002; Hurd et al. 2004). Thus, the conflict includes both ecological carrying capacity (the maximum number of herbivores supportable by the standing plant biomass; in this case, are elk and cattle overabundant with regard to sustainability of plant communities and are they limiting performance of each other?) and sociological carrying capacity (the maximum number of animals tolerable for some desired ecological condition; in this case, is combined large herbivore grazing – regardless of level or sustainability – negatively impacting Mexican spotted owls?) (definitions following Caughley 1979 and Carpenter et al. 2000). Despite this, arguments are invariably couched in the context of unsustainable utilization of meadows by large herbivores, with consequent claims that one or both species are "over carrying capacity" in the ecological context.

Much of the data driving these conflicts and resultant management actions on LNF were based upon only a few indicator sites and included no direct knowledge of individual herbivore contributions to perceived impacts or preferences for key foraging habitats. Moreover, while concurrent use by elk and cattle is anecdotally believed to decrease forage quantity and/or quality available for the other herbivore, no data is available on comparative habitat use or diets attained by elk or cattle in LNF. Thus, to address the issue of ecological carrying capacity only, our goal was to identify the degree of use and impacts of elk and cattle foraging across key habitat cover types in LNF. Our objectives included: 1) determine plant community composition and relative use by large herbivores, 2) determine levels of browsing of key woody plants, 3) quantify degree of grazing on herbaceous communities, and 4) determine diets and relative diet quality of elk and cattle.

Methods

Study area

Our study area encompassed much of the Sacramento Ranger District of Lincoln National Forest (LNF), located in the Sacramento Mountains of south-central New Mexico (approximately 32°49′59′′N, 105°43′01′′ W; Fig. 1). The Sacramento Ranger District manages 182,278 ha of the LNF. Average high temperature in June is 23 °C and average low January temperature is −7 °C. Annual precipitation averages 67 cm with > 50 % falling June–September. Annual snowfall averages 183 cm. Annual precipitation was 0.94, 0.82, and 1.36 of normal, 2004–2006, respectively. Precipitation during the May–October growing and cattle grazing season was 0.69, 0.77, and 1.82 of normal, 2004–2006, respectively.

Vegetation communities in the Sacramento Mountains are very diverse due to differences in elevation and precipitation. Desert scrub-grassland habitat covers the lower foothills from 1380 to 1830 m (Kaufmann et al. 1998). *Pinus edulis-Juniperus monosperma* woodland occurs along with areas of dense *Quercus gambelii* above 1700–1800 m, usually along canyons and ridges. *Pinus ponderosa* is found above 2100 m typically mixed with *Pinus edulis-Juniperus monosperma* or *Pseudotsuga menziesii*. Above 2200–2500 m, the mountains are dominated by a mixed conifer forest, consisting of *Abies concolor, Pseudotsuga menziesii, Pinus ponderosa, Pinus strobiformis*, and occasionally *Picea pungens* or *Picea engelmannii*. Pockets of *Populus tremuloides* are found mixed within the conifers, rarely in pure stands. Mountain meadows are numerous at high elevations and comprise < 5 % of the landscape. Mountain meadows and other open habitats are much reduced in LNF as compared to historical coverage primarily because of altered fire regimes and reduced timber harvest (Kaufmann et al. 1998; Frost et al. 2007). Similarly, stem densities in forested stands are generally much higher than historically again because of altered natural disturbance regimes and reduced timber harvest (Kaufmann et al. 1998; Frost et al. 2007).

Annual population estimates for elk in the surrounding game management unit have varied greatly, 1998–2006. Corroboration of estimators from 2003 to 2007 suggested an elk density of 0.7–0.9+ elk$\cdot$ km^{-2} (Bender 2007). During our study the Sacramento Ranger District managed 38

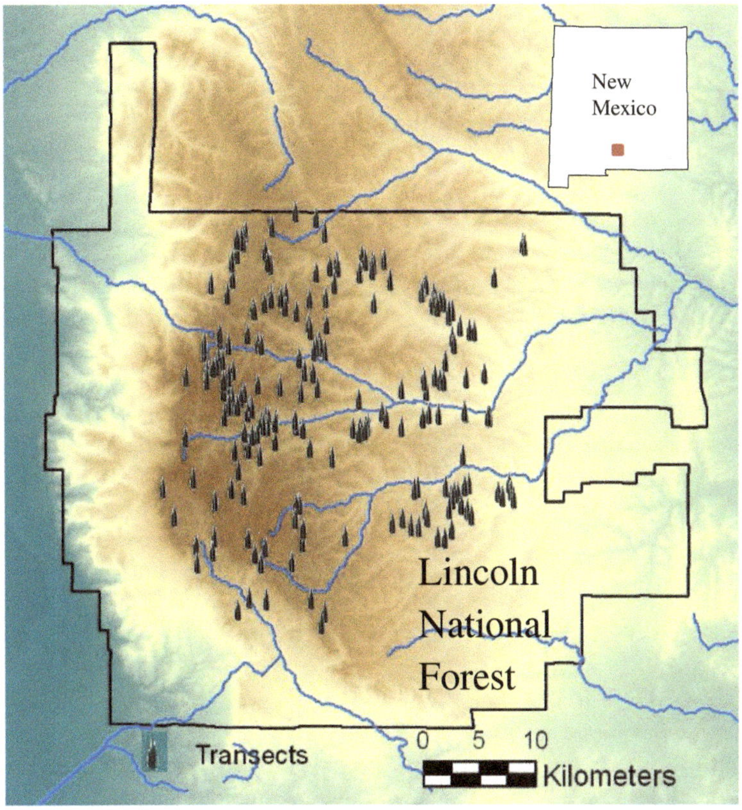

Fig. 1 Digital elevation map of the Sacramento Ranger District (solid line) showing locations of sampling transects, Lincoln National Forest, south-central New Mexico, USA. Dashed inset shows location of figure (red polygon) within the state of New Mexico

grazing allotments with 34 permittees. The majority of our study area was in the spring–autumn range of the Sacramento allotment, which encompassed approximately 44,744 ha. In 2004–2006, approximately 330 permitted cattle grazed the allotment for a density of 0.7 cows·km^{-2}. Grazing followed a deferred rotation system from mid-May through October, with 2 pastures alternating annually between early summer (mid-May through early August) and late summer (early August through October) grazing. Each pasture was grazed by cattle during 1 of the periods each year, but not the same period in consecutive years.

Vegetation surveys

We surveyed vegetation composition and structure during mid-late September, 2005 and 2006, which was approximately the peak of herbaceous species growth. We defined vegetation types from a GIS database, then randomly selected 25–30 stands of each of 4 key vegetation types: mountain meadow, aspen (*Populus tremuloides*), thinned conifer, and burned conifer. We selected these types based on preferences for elk and cattle foraging use in the montane habitats of LNF (Wright 2000; Hurd 2002).

We established a single random 100-m transect per stand replicate, for a total of 25–30 replicates per habitat type. We tallied ground cover (i.e., soil, rock, plant, litter) every 1 m along transects. If cover was a plant, we identified it to species; if a grass, we measured its average leaf height ("stubble height") by pulling the blades of grass up and measuring their height (Allison et al. 2007). We determined mean residual stubble heights (RSH) by species and compared these to published standards to determine grazing intensity (Allison et al. 2007). If the cover was not a grass, we located the nearest grass plant to the point and recorded its species and stubble height. We calculated 90 % CIs around mean RSH to see if mean RSH differed from standards; if the standard was included in 90 % CIs, there was no difference. We used a linear mixed model (West et al. 2014) with year as a random effect to test for differences among ground cover type, habitat type, and year, specifically testing the ground cover type × habitat type × year interaction.

We determined densities of woody shrubs (<5.1 cm dbh) and trees (≥5.1 cm dbh) using the point-centered quarter method (Cottam and Curtis 1956), with points occurring every 25 m along transects. We determined percent canopy cover using a spherical densitometer (Lemmon 1956) every 25 m along the 100-m transect. We took 4 estimates at each point, 1 for each cardinal direction, and estimated percent canopy cover by averaging the estimates. We modeled total plant cover and woody stem densities as a function of percent canopy cover using linear regressions (Zar 1996). We used a linear mixed model (West et al. 2014) with year as a random effect to test for differences in tree and shrub total density among habitat type and year, specifically testing the habitat type × year interaction.

We used stepwise multiple regression (Zar 1996) to model the proportional effects of relative habitat use by elk, cattle, and deer (i.e., number of pellet groups per stand; see below) in mountain meadows on RSH of *Poa pratensis*, which dominated the majority of meadows in LNF. We used stepwise regression because it is more conservative than information-theoretic model selection procedures (Murtaugh 2009). We included year as a dummy variable in models and used partial R^2 values to interpret how much each variable contributed to the overall variance of the model.

Browse surveys

We conducted browse surveys in April 2005 and 2006 prior to leaf-out of deciduous woody species. We randomly selected 25 replicate stands in mountain meadow, aspen, and thinned conifer, and 15 each in 2 large burns. As above, we selected these types based on preferences for elk and cattle foraging use in the montane habitats of LNF (Wright 2000; Hurd 2002).

We placed a single random transect within each stand replicate, parallel to the longest dimension of the stand. We used the Wyoming browse survey method (Wyoming Game and Fish Department 1982) to assess browse levels on key species. This involved sampling successive shrubs from the first shrub (closest to the random point), with each successive shrub being the closest shrub to the previous in the forward-facing 180 degrees (Wyoming Game and Fish Department 1982). We sampled successive shrubs until 50 live shrubs ≥ 0.25 m in height but < 2.54 cm dbh were examined per transect or until the stand had been completely traversed.

We identified all shrubs to species. We assessed percent use by examining all current annual growth (previous years) on each branch of the shrub between 0.25 and 2 m high, as that height range is within ungulate reach but usually too high for non-ungulate herbivores. We tallied the number of browsed shoots out of the total number of shoots examined for each shrub to determine frequency of use of browse. We calculated mean browse levels for all species encountered and used these to identify key browse species. We used a linear mixed model (West et al. 2014) with year as a random effect to test for differences among browse species, habitat types, and year, specifically testing the species × habitat type × year interaction. We used stepwise multiple regression (Zar 1996) as above to determine proportional use of browse species by elk, cattle, and mule deer with numbers of pellet groups per stand of each species used as predictor variables (see below) and browse levels as the outcome variable.

Relative habitat use

We assessed annual relative habitat use of elk, deer, and cattle by counting elk and deer pellet groups and cattle pats (hereafter, pellet groups) encountered within 2 m of a straight line between shrubs used in browse surveys until we covered 100 m (i.e., a 2 × 100-m belt transect). Pellet-group transects are adequate to assess relative use of habitats by wild and domestic ungulates (e.g., Cook 1966; Neff 1968; Collins and Urness 1984; Loft and Kie 1988; Halbritter and Bender 2011b) as shown by comparisons with radio telemetry-based approaches (Loft and Kie 1988). We also ran 2 × 100-m pellet-group belt transects concurrently with vegetation surveys in autumn, and used these surveys to determine the degree of variation in use within key habitat types over the spring–autumn grazing period only. We used a linear mixed model (West et al. 2014) with year as a random effect to test for differences among species, habitat type, and year, specifically testing the species × habitat type × year interaction. We also used these data as inputs in regression models of browsing and grazing impacts (see above).

Diets and diet quality

We collected ≥ 10 fecal samples per species per month, 1 May–31 October, 2004–2006, for analysis of diets and diet quality of elk and cattle. This period coincided with cattle use of LNF allottments. We collected only fresh fecal samples, usually those deposited within 24 h. Diet composition samples were sent to the Wildlife Habitat Nutrition Lab (Washington State University, Pullman, WA, USA) for microhistological analysis (Sparks and Malechek 1968). We compared elk and cattle diets by major class (shrubs, conifer, forb, grass, sedge/rush) for the May–October period using Mann–Whitney tests (Zar 1996) and calculated Schoener's similarity index (Schoener 1970) for forage classes and for species comprising ≥ 1 % of either elk or cattle spring–autumn diets.

We sent diet quality samples for cattle to the Grazing Animal Nutrition Lab (Texas A&M University, College Station, TX, USA) for analysis using fecal near infrared reflectance spectroscopy (NIRS) to determine dietary digestibility (DD) and crude protein (CP) of cattle diets (Lyons and Stuth 1992). We analyzed diet quality by calculating monthly means for each year. We tested for differences among years using Kruskal-Wallis tests (Zar 1996).

We used percent ingesta-free body fat (BF) accrued by lactating female elk to estimate diet quality of elk during the spring–autumn season (Halbritter and Bender 2011a).

Percent BF that lactating elk are able to accrue is strongly related to digestible energy content of diets and dietary thresholds for levels of BF have been established (Cook et al. 2004). We estimated BF for 16–23 radio-collared elk captured in late autumn at the approximate peak of accrual of BF as detailed in Halbritter and Bender (2011a, c). We estimated BF levels by measuring maximum subcutaneous fat thickness at the rump using a SonoVet 2000 ultrasound (Medison, Seoul, South Korea) and by determining body condition scores at the rump (Cook et al. 2001; Halbritter and Bender 2011a, c). We then extrapolated approximate digestible energy content of spring–autumn diets of elk based on mean level of BF from Cook et al. (2004).

To test for the influence of precipitation on diet quality we obtained monthly precipitation data for a weather station located in the northern quarter of our study area (Cloudcroft, Otero County, New Mexico, USA) from the Western Regional Climate Center (2007). We used linear regression (Zar 1996) to assess relationships between precipitation and cattle diet quality by modeling mean monthly CP and DD as a function of monthly precipitation. We also related the prior monthly precipitation to mean CP and DD for each month to determine if forage quality was affected by the amount of precipitation received the previous month.

Results

Relative habitat use

Relative habitat use of large herbivores differed ($F_{22,585}$ = 24.1; $P < 0.001$), with elk ($P < 0.001$), cattle ($P < 0.001$), and deer ($P \leq 0.002$) each showing differential use of habitat types (Table 1). In general, elk primarily used meadows and thinned conifer, cattle used meadows, and deer used burns and thinned conifer (Table 1). Cattle and deer use of burns was generally similar to use of other sites whereas elk use of burns was generally lower than meadows or thinned conifer (Table 1).

Spring–autumn pellet group surveys were conducted to determine the degree of variation in use within key habitat types over the spring–autumn grazing period only. Although we present data on relative use of habitat types in this period by species, the annual use data above should be used for stronger inferences of relative use by species. Similar to spring surveys, relative habitat use of large herbivores differed ($F_{25,646}$ = 6.3; $P < 0.001$). In 2005, both elk and cattle use was significantly ($P \leq 0.008$) greater in mountain meadows than in thinned conifer (Table 1). In 2005, elk use did not differ ($P > 0.621$) among types, whereas cattle ($P = 0.005$) primarily used meadows and deer ($P < 0.001$) primarily used burns and thinned conifer (Table 1). In 2006, elk ($P < 0.001$) and cattle ($P < 0.001$) primarily used meadows and thinned conifer, while deer ($P = 0.010$) used thinned conifer and burns (Table 1).

Substantial variation existed within each habitat type in degree of relative habitat use. For all years, relative elk use ranged from 0 to 86 groups·stand^{-1} in mountain meadows, 0–22 groups·stand^{-1} in thinned conifer, 0–32 groups·stand^{-1} in burns, and 0–35 groups·stand^{-1} in aspen. Comparable groups·stand^{-1} for cattle were 0–34, 0–13, 0–25, and 0–8, respectively. For deer, ranges were 0–2, 0–12, 0–11, and 0–4, respectively.

Vegetation composition and cover

Ground cover differed ($F_{38,910}$ = 162.7; $P < 0.001$) in LNF; herbaceous plants ($P < 0.001$), bare soil ($P < 0.065$), litter ($P < 0.001$), rock ($P < 0.068$), and woody debris ($P < 0.001$) all varied among habitat types, except for woody debris in 2006 ($P > 0.151$). Meadows had the greatest plant cover each year while burns had the greatest plant cover among forested types (Table 2). Cover of herbaceous plants, soil, litter, rock, and woody debris also varied within habitat types in both 2005 ($P \leq 0.001$) and 2006 ($P \leq 0.001$). Herbaceous plant cover was greater in 2006 than 2005 in thinned conifer ($P = 0.019$), burned ($P = 0.081$), and meadow ($P = 0.001$) habitat types (Table 2). Plant cover

Table 1 Mean number ± SE of elk, cattle, and mule deer pellet groups by habitat types and season, Lincoln National Forest, 2004–2006

Season	Cover type	2004		2005			2006		
		Elk	Cattle	Elk	Cattle	Deer	Elk	Cattle	Deer
Annual	Meadow	–	–	12.8 ± 1.8X	3.4 ± 1.0X	0.0 ± 0.0Y	29.5 ± 4.6X	10.6 ± 1.9X	0.2 ± 0.1X
	Aspen	–	–	4.9 ± 1.1Y	0.1 ± 0.1Y	0.6 ± 0.2Y	7.6 ± 1.4Y	0.6 ± 0.4Y	0.6 ± 0.2X
	Thinned conifer	–	–	11.9 ± 2.1X	1.0 ± 0.4Y	1.4 ± 0.5X	12.4 ± 1.9Y	0.3 ± 0.2Y	1.5 ± 0.4Y
	Burn	–	–	7.6 ± 1.3XY	3.3 ± 1.1X	1.8 ± 0.4X	7.4 ± 1.4Y	0.9 ± 0.4Y	2.4 ± 0.7Y
Spring-autumn	Meadow	12.3 ± 3.4A	8.8 ± 1.5A	10.4 ± 2.9	7.0 ± 1.9A	0.2 ± 0.1A	7.1 ± 1.5A	5.8 ± 1.3A	<0.1 ± 0.1A
	Aspen	–	–	4.8 ± 0.9	0.5 ± 0.2B	0.4 ± 0.2A	3.6 ± 0.7AB	0.1 ± 0.6B	0.4 ± 0.2AB
	Thinned conifer	3.3 ± 0.9B	2.4 ± 0.8B	4.7 ± 1.0	1.0 ± 0.4B	1.7 ± 0.3B	5.2 ± 0.5A	0.6 ± 0.5B	1.0 ± 0.2C
	Burns	–	–	4.3 ± 0.7	1.6 ± 0.6B	2.7 ± 0.6B	1.4 ± 0.4B	0.4 ± 0.2B	0.8 ± 0.4BC

Means with different letters (Annual: XYZ; Spring–autumn: ABC) within a COLUMN and SEASON differ ($P < 0.10$)

Table 2 Mean ± SE percent ground cover by habitat type, Lincoln National Forest, autumn 2005–2006

Habitat type	2005					2006				
	Plant	Litter	Wood	Soil	Rock	Plant	Litter	Wood	Soil	Rock
Aspen	27.4 ± 4.0DX	60.5 ± 4.2CX	7.6 ± 1.0BX	3.5 ± 1.2ABX	0.9 ± 0.4AX	28.7 ± 3.9cx	59.2 ± 3.9bx	11.0 ± 1.3d	1.3 ± 0.1ax	2.0 ± 0.1ax
Thinned	23.9 ± 2.6DX	55.4 ± 2.6AX	14.0 ± 2.1EY	4.2 ± 0.7AX	2.5 ± 0.5BX	34.5 ± 3.5cx	48.2 ± 3.3by	11.0 ± 1.2d	6.3 ± 1.0ay	4.8 ± 0.6ay
Burned	58.3 ± 2.7DY	11.2 ± 1.3ACY	5.6 ± 0.9BZ	15.8 ± 1.9AY	9.3 ± 1.4ABY	70.1 ± 2.7by	6.2 ± 0.8az	8.1 ± 1.1a	7.6 ± 1.1az	11.0 ± 2.0az
Meadow	84.2 ± 1.7BZ	2.1 ± 0.8AZ	0.5 ± 0.2AW	12.1 ± 1.3BZ	1.0 ± 0.7AX	91.7 ± 1.2cz	2.6 ± 0.4bz	1.7 ± 0.2b	7.1 ± 0.8az	5.0 ± 1.0by

ABC and abc Means within a ROW and YEAR sharing a letter do not differ ($P > 0.10$); XYZ and xyz Means within a column sharing a letter do not differ ($P > 0.10$)

was positively related to the amount of annual precipitation for thinned ($F_{1,48} = 6.0$, $P = 0.018$, $R^2 = 0.11$; $\beta = 0.465$ [SE = 0.192]), burned ($F_{1,38} = 8.7$, $P = 0.005$, $R^2 = 0.19$; $\beta = 0.553$ [SE = 0.188]), and meadow habitat types ($F_{1,48} = 12.2$, $P = 0.001$, $R^2 = 0.20$; $\beta = 0.329$ [SE = 0.094]).

Poa pratensis was the most common plant species in all but the burned habitat type and comprised > 47 % of the vegetation cover on mountain meadows (Table 3). Understory vegetation in burns was comprised primarily of *Quercus* spp. and various grasses, primarily *Dactylis glomerata* (2005) and *Bromus inermis* (2006) (Table 3).

Mean RSH of grasses were at or above recommended minimum stubble heights (Tables 4 and 5) except for *Poa pratensis* in mountain meadows in 2005 (6.2 cm (90 % CI = 4.8–7.6)) (Table 4), which was below the standard of 6.4 cm, and *Bromus* spp. in mountain meadows in 2006 (9.9 cm (90 % CI = 7.8–12.0) compared to a recommendation of 10.2 cm) (Table 5). Neither of these deviations was statistically significant based on inclusion of recommended RSHs within 90 % CIs of observed RSHs (Tables 4 and 5).

Habitat types did not vary ($F_{4,131} = 1.3$; $P = 0.264$) in total shrub density due to extremely high variation among stand replicates. Burned conifer had the highest mean number of shrubs in 2005, but was dominated by less palatable shrubs including *Quercus* spp. and *Robinia neomexicana*. In 2006, greatest total shrub densities were in aspen stands (Table 6).

Habitat types differed in total tree density ($F_{4,102} = 22.3$; $P < 0.001$) (Table 7). Aspen stands had the greatest mean tree density in both years ($P \leq 0.004$) when compared to other managed or disturbed forests, which did not differ ($P \geq 0.383$) (Table 7). Burned stands had the lowest mean tree density, as the majority of each burn was characterized by near-complete stand replacement in which only snags remained.

Total herbaceous cover was negatively related ($F_{1,98} = 111.8$, $P < 0.001$, $R^2 = 0.53$; $\beta = -1.122$ [SE = 0.106]) to percent canopy cover. Overall, conifer stands ($n = 60$) averaged 71.7 % canopy cover (range = 28.5–97.7, SE = 2.0).

Browsing impacts

We found a total of 5 browse species of moderate or higher palatability during browse transects: *Holodiscus dumosis*, *Populus tremuloides*, *Acer glabrum*, *Symphoricarpos rotundifolius*, and *Prunus virginiana* (Table 8). Palatable shrubs were rare in sampled stands, with only *Holodiscus dumosis* being found on ≥ 9 transects in any cover type. Because of this, few species had the desired 50 plants per transect and none were present on all 25 replicate stand transects. Consequently power to detect differences in use was low for all species. Browse use varied ($F_{16,122} = 3.7$; $P < 0.001$) among comparisons (Table 8), although only use of *Populus tremuloides*

Table 3 Percent cover ± SE of the most common plant species by habitat type, Lincoln National Forest, autumn 2005–2006

Habitat type	Species	% Cover	
		2005	2006
Aspen	*Abies concolor*	–	13.6 ± 4.0
	Aster spp.	13.5 ± 3.4	6.5 ± 2.0
	Fragaria vesca	11.2 ± 2.4	10.4 ± 1.1
	Moss	7.8 ± 2.5	8.3 ± 2.8
	Poa pratensis	15.0 ± 3.5	13.6 ± 4.0
	Populus tremuloides	8.9 ± 4.3	–
Burn	*Artemisia ludoviciana*	–	7.4 ± 3.2
	Bromus inermis	7.2 ± 2.7	12.8 ± 4.7
	Dactylis glomerata	15.8 ± 4.0	5.2 ± 1.2
	Piptochaetium fimbriatum	6.7 ± 2.4	–
	Poa pratensis	6.1 ± 2.6	4.4 ± 2.0
	Quercus undulata	13.3 ± 3.9	16.3 ± 5.5
Meadow	*Achillea millefolium*	7.3 ± 1.2	3.7 ± 0.1
	Bromus spp.	–	4.6 ± 3.5
	Bromus inermis	3.9 ± 2.1	–
	Poa pratensis	52.8 ± 3.9	47.4 ± 5.4
	Erodium cicutarium	5.0 ± 1.8	7.2 ± 2.4
	Trifolium spp.	3.4 ± 1.4	–
Thinned	*Aster* spp.	12.0 ± 2.9	13.8 ± 2.8
	Bromus spp.	7.3 ± 1.7	8.0 ± 1.8
	Fragaria vesca	8.9 ± 2.2	–
	Moss	5.4 ± 1.2	7.0 ± 2.9
	Poa pratensis	27.5 ± 4.6	17.3 ± 4.0
	Quercus gambelii	6.5 ± 3.0	8.0 ± 4.3

Table 4 Mean percent grass cover ± SE, residual stubble heights (RSH) of grasses ± 90 % CI, and recommended minimum RSH, Lincoln National Forest, autumn 2005, by habitat type

Habitat type	Species	% Cover	RSH (cm)	Minimum
Aspen	*Bromus* spp.	50.4 ± 6.9	18.0 ± 1.4	10.2
	Poa pratensis	35.0 ± 5.6	14.4 ± 1.7	6.4
	Bromus inermis	4.7 ± 3.6	29.7 ± 11.1	10.2
	Bromus ciliates	3.6 ± 3.6	20.3 ± 0.0	10.2
Burn	*Dactylis glomerata*	29.9 ± 6.6	18.6 ± 3.6	10.2
	Piptochaetium fimbriatum	14.2 ± 5.0	27.1 ± 2.8	–
	Poa pratensis	8.6 ± 3.7	10.8 ± 1.9	6.4
	Bromus inermis	8.3 ± 2.8	17.5 ± 3.9	10.2
	Piptochaetium pringlei	7.1 ± 3.9	20.1 ± 3.2	–
Meadow	*Poa pratensis*	78.8 ± 5.4	6.2 ± 1.4	6.4
	Bromus inermis	3.9 ± 2.2	19.5 ± 3.5	10.2
	Agropyron trachycaulum	3.8 ± 2.4	15.1 ± 5.3	10.2
Thinned	*Poa pratensis*	40.7 ± 5.2	11.6 ± 1.3	6.4
	Bromus spp.	38.3 ± 5.1	15.5 ± 1.6	10.2
	Achnatherum robustum	6.0 ± 3.9	17.3 ± 8.5	10.2

RSH standards from Allison et al. (2007)

showed a difference, with greater use in aspen stands than in thinned conifer in 2006 ($P = 0.020$). Too few browse plants were present in mountain meadows for analysis.

Year was not included in any model of relative use of browse in any habitat type ($P \geq 0.672$). Relative use by elk, cattle and deer was positively related ($F_{3,17} = 9.0$; $P =$ 0.001) to browse levels in burns. Collectively, large herbivore use accounted for 61 % of the variance in browse use with elk use accounting for 45 %, deer use 12 %, and cattle use 4 % of the total variance. In thinned conifer, elk, cattle, and deer use ($F_{3,31} = 1.21$; $P = 0.322$) was not related to use of primary browse species. Combined herbivory accounted for only 10 % of the total variation in browse

Table 5 Mean percent grass cover ± SE, residual stubble heights (RSH) of grasses ± 90 % CI, and recommended minimum RSH, Lincoln National Forest, autumn 2006, by habitat type

Habitat type	Species	% Cover	RSH (cm)	Minimum
Aspen	*Bromus* spp.	62.7 ± 8.9	19.8 ± 1.4	10.2
	Poa pratensis	36.0 ± 9.2	14.6 ± 1.6	6.4
	Agropyron trachycaulum	1.2 ± 1.2	15.8 ± 0.0	10.2
Burn	*Bromus inermis*	20.0 ± 6.1	17.3 ± 2.8	10.2
	Dactylis glomerata	16.6 ± 3.8	19.8 ± 2.3	10.2
	Poa pratensis	9.6 ± 4.3	12.7 ± 7.3	6.4
	Piptochaetium fimbriatum	9.4 ± 3.0	24.3 ± 2.6	–
	Piptochaetium pringlei	8.6 ± 3.6	32.8 ± 6.0	–
Meadow	*Poa pratensis*	70.8 ± 6.6	6.4 ± 1.4	6.4
	Bromus spp.	12.2 ± 3.5	9.9 ± 2.1	10.2
	Bromus inermis	5.0 ± 3.8	10.8 ± 27.4	10.2
	Muhlenbergia fragilis	4.4 ± 3.8	12.2 ± 2.7	-
	Achnatherum robustum	3.9 ± 2.4	46.0 ± 8.7	10.2
Thinned	*Bromus* spp.	39.5 ± 6.5	17.7 ± 1.0	10.2
	Poa pratensis	27.5 ± 6.0	12.7 ± 1.2	6.4
	Piptochaetium pringlei	5.7 ± 2.9	30.6 ± 1.2	–
	Piptochaetium fimbriatum	4.1 ± 3.6	28.7 ± 37.5	–

RSH standards from Allison et al. (2007)

Table 6 Mean total woody stems ≤ 2.54 cm dbh ± SE per hectare and mean total woody stems per hectare of the 5 most common species in aspen, burn, and thinned conifer habitat types, Lincoln National Forest, 2005–2006

Habitat type	Species	Stems · ha^{-1}	
		2005	2006
Aspen	*Abies concolor*	474	652
	Holodiscus dumosis	177	307
	Pseudotsuga menziesii	177	111
	Robinia neomexicana	111	–
	Ribes spp.	99	–
	Populus tremuloides	–	107
	Acer glabrum	–	93
	TOTAL	1332 ± 30	1640 ± 21
Burn	*Quercus undulata*	888	487
	Quercus gambelii	726	305
	Robinia neomexicana	300	250
	Juniperus deppeana	156	174
	Holodiscus dumosis	60	85
	TOTAL	2394 ± 26	1406 ± 46
Thinned	*Holodiscus dumosis*	203	361
	Quercus gambelii	180	241
	Abies concolor	138	138
	Pseudotsuga menziesii	135	118
	Robinia neomexicana	39	–
	Symphoricarpos rotundifolius	–	82
	TOTAL	822 ± 39	1262 ± 26

Table 7 Mean total number of trees per hectare ± SE and number of trees per hectare for the 5 most common species in aspen, burn, and thinned conifer habitat types, Lincoln National Forest, 2005– 2006

Habitat type	Species	Trees · ha^{-1}	
		2005	2006
Aspen	*Populus tremuloides*	559	512
	Abies concolor	222	396
	Pseudotsuga menziesii	158	196
	Quercus gambelii	67	73
	Robinia neomexicana	48	–
	Acer glabrum	–	30
	TOTAL	1149 ± 16	1254 ± 13
Burn	*Pinus ponderosa*	44	39
	Pseudotsuga menziesii	31	20
	Juniperus deppeana	21	39
	Quercus gambelii	15	–
	Abies concolor	11	7
	Pinus strobiformis	–	20
	TOTAL	137 ± 81	131 ± 49
Thinned	*Pseudotsuga menziesii*	158	174
	Abies concolor	89	136
	Populus tremuloides	28	–
	Quercus gambelii	13	25
	Pinus ponderosa	8	–
	Pinus strobiformis	–	39
	Picea engelmannii	–	28
	TOTAL	329 ± 30	454 ± 22

levels. Of this total, cattle use accounted for 6 %, elk 4 %, and deer < 1 %. In aspen stands, combined herbivore relative use was positively related ($F_{3,45} = 3.1$; $P = 0.038$) to browse use and accounted for a combined 17 % of the variance in total browse levels. Individually, elk, cattle, and deer use accounted for 12 %, 4 %, and 1 % of the total variance in browse use in aspen stands, respectively.

Grazing impacts

Year was not included in models of RSH of *Poa pratensis* ($F_{1,47} < 0.1$; $P = 0.894$). Residual stubble height of *Poa pratensis* was negatively related to relative use of all herbivores ($F_{3,45} = 2.4$; $P = 0.085$), which accounted for 14 % of the total variance in RSH of *Poa pratensis*. Cattle use accounted for 10 %, elk use 4 %, and deer use < 1 % of the variance in RSH of *Poa pratensis*.

Diets and diet quality

Across years, spring–autumn diets of elk were comprised of significantly more deciduous browse ($U = 1.7$; $P = 0.080$) and forbs ($U = 2.2$; $P = 0.032$) than were cattle diets, but less grass ($U = 1.7$; $P = 0.080$) and sedge ($U = 1.7$; $P = 0.080$). Use of conifers did not differ ($U = 1.1$; $P = 0.282$) (Table 9). Elk and cattle diets were not significantly similar (Schoener's index < 0.60) for either major forage classes (0.57) or individual forage species (0.54).

Digestibility of cattle diets was 69.2 % (SE = 0.5), 68.0 % (SE = 0.4), and 67.4 % (SE = 0.6) for May–October 2004, 2005, and 2006, respectively, and did not differ ($H_2 = 4.0$; $P = 0.132$) among years. CP levels varied annually ($H_2 = 12.7$; $P = 0.002$) in diets and were 14.1 % (SE = 0.5), 13.2 % (SE = 0.3), and 14.7 % (SE = 0.4) for 2004, 2005, and 2006, respectively. CP was higher ($P = 0.004$) in cattle diets in 2006 as compared to 2005; no other year comparisons differed ($P \geq 0.101$).

Monthly precipitation was positively related to CP in cattle diets ($F_{1,13} = 6.0$, $P = 0.029$, $R^2 = 0.32$; $\beta = 1.186$ [SE = 0.484]), but not DD of cattle diets ($F_{1,13} = 0.5$; $P = 0.506$). Precipitation lagged 1 month resulted in similar effects, showing a positive relationship with CP ($F_{1,13} = 6.2$, $P = 0.028$, $R^2 = 0.32$; $\beta = 1.220$ [SE = 0.492]) and no effect on DD ($F_{1,13} = 0.3$; $P = 0.612$).

Table 8 Percent utilization ± SE of key browse species on Lincoln National Forest by cover-type. Only use of *Populus tremuloides* in 2006 differed between aspen and thinned conifer habitat types

Year	Habitat type	Browse species				
		H. dumosis	*P. tremuloides*	*A. glabrum*	*P. virginiana*	*S. rotundifolius*
2005	Aspen	59.5 ± 4.1	87.5 ± 12.5	67.5 ± 12.0	73.1 ± 1.8	44.9 ± 3.6
	Thinned	64.0 ± 5.2	–	–	72.1 ± 27.9	18.9 ± 5.9
2006	Aspen	77.1 ± 2.1	75.4 ± 8.0	67.5 ± 12.0	57.9 ± 0.7	56.5 ± 7.2
	Thinned	73.4 ± 3.2	56.5 ± 7.2	69.0 ± 7.7	44.0 ± 23.8	43.4 ± 8.7

Lactating cow elk were able to accrue 10.5 % (SE = 0.6), 11.4 % (SE = 0.7), and 8.9 % (SE = 0.7) BF by autumn 2003, 2004, and 2005, respectively (Halbritter and Bender 2011a, c). Levels of BF that lactators were able to achieve differed ($H_2 = 5.9$, $P = 0.053$) among years; 2005 was lower than 2003 and 2004, which were similar. This level of condition corresponded to diet qualities in the marginal to good range (digestibility = 56–63 %) based on criteria from penned elk (Cook et al. 2004).

Discussion

Herbivory by large herbivores had varying impacts on LNF. Whereas high levels of browsing suggested that herbivores, primarily elk, could potentially deleteriously

Table 9 Mean ± SE percent occurrence in spring–autumn diets of elk and cattle by forage class, 2004–2006

Class	Species	Elk				Cattle			
		2004	2005	2006	Mean	2004	2005	2006	Mean
Shrubs	*Acer glabrum* leaf	3.2	1.5	1.0	1.9 ± 0.7	–	–	–	0
	Holodiscus dumosis, leaf	2.3	0.5	0.3	1.1 ± 0.6	–	–	–	0
	Quercus spp. leaf	24.8	30.6	20.2	25.2 ± 3.0	6.5	2.9	9.3	6.2 ± 1.9
	Salix spp. leaf	17.4	13.9	7.1	12.8 ± 3.0	–	–	–	0
	Salix spp. stem	1.9	0.9	1.7	1.5 ± 0.3	–	–	–	0
	Symphoricarpos rotundifolius leaf	2.4	0.7	0.8	1.3 ± 0.5	–	–	–	0
	TOTAL	55.2	51.5	34.2	46.9 ± 6.5A	7.4	4.3	10.7	7.5 ± 1.8B
Conifer	TOTAL	0.2	1.1	1.1	0.8 ± 0.3	0.1	0.4	0.5	0.3 ± 0.1
Forbs	*Achillea millefolium*	0.8	1.1	1.7	1.2 ± 0.3	0.8	1.8	1.8	1.5 ± 0.3
	Erigeron spp.	1.8	1.4	0.7	1.3 ± 0.3	–	–	–	0.1 ± 0.1
	Geranium spp.	1.7	1.0	0.5	1.0 ± 0.3	–	–	–	0.6 ± 0.4
	TOTAL	8.2	11.5	7.8	9.1 ± 1.2A	6.5	6.5	6.8	6.6 ± 0.1B
Grass	*Agropyron* spp.	1.6	0.7	1.8	1.4 ± 0.3	17.2	12.1	11.5	13.6 ± 1.8
	Agrostis spp.	–	–	–	0.7 ± 0.5	5.4	4.9	5.4	5.2 ± 0.2
	Bouteloua spp.	–	–	–	0	4.3	1.8	1.0	2.4 ± 1.0
	Bromus spp.	5.1	7.3	13.6	8.7 ± 2.5	5.6	8.4	6.9	6.9 ± 0.8
	Dactylis glomerata	–	-	-	0.8 ± 0.2	0.8	1.2	1.3	1.1 ± 0.2
	Agropyron trachycaulum	0.0	0.9	3.7	1.5 ± 1.1	–	–	–	0.7 ± 0.1
	Muhlenbergia spp.	–	–	–	0.9 ± 0.2	7.3	3.9	5.0	5.4 ± 1.0
	Oryzopsis spp.	–	–	–	0	2.2	0.5	0.7	1.1 ± 0.5
	Poa fendleriana	–	–	–	–	1.9	1.7	0.4	1.3 ± 0.5
	Poa pratensis	–	–	–	–	21.2	30.7	29.2	27.0 ± 3.0
	Poa spp.	20.4	16.7	24.1	20.4 ± 2.1	23.1	32.4	29.6	28.3 ± 2.8
	Stipa spp.	1.6	0.6	1.2	1.1 ± 0.3	5.6	7.5	7.0	6.7 ± 0.6
	TOTAL	32.2	29.8	49.5	37.2 ± 6.2A	75.0	77.1	72.2	74.8 ± 1.4B
Sedge/Rush	*Carex* spp.	4.3	4.9	5.7	5.0 ± 0.4	10.6	10.3	7.6	9.5 ± 0.9
	TOTAL	4.3	4.9	5.7	5.0 ± 0.4A	10.9	11.5	9.6	10.6 ± 0.6B

Also presented are species comprising > 1.0 % of either elk or cattle diets averaged over years
Class TOTALS not sharing a letter differ between species ($P > 0.90$)

impact browse resources, grazing on mountain meadows – the area of contention on LNF (Hurd 2002; Hurd et al. 2004; Halbritter and Bender 2011b) – indicated few impacts of herbivory. While relative habitat use of meadows by elk and cattle was generally equal to or higher than other habitat types, and conflicts can occur when wild and domestic ungulates concentrate their habitat use, significant variation existed in relative habitat use of individual meadows by elk and cattle (CV's > 80) on LNF. Additionally, no grass species was significantly overutilized based on minimum RSH recommendations (Allison et al. 2007). Further, RSH of *Poa pratensis* (and all grasses; Halbritter 2007) on meadows was not strongly correlated with relative habitat use of large herbivores. Relative habitat use by cattle was the primary large herbivore effect but accounted for ≤ 10 % of the variation in RSH of *Poa pratensis* among meadows despite the high prevalence (75 %) of grasses in general and *Poa pratensis* (55 %) in cattle diets. Thus, much of the variation in RSH of *Poa pratensis* was attributable to factors other than ungulate presence, including elevation, aspect, precipitation, location, size, and small herbivore use (Halbritter 2007; L. Bender, unpublished data).

Despite the conflict over perceived overutilization of meadows on LNF, our results illustrate that much of the variation in RSHs among meadows was associated with factors other than relative habitat use of meadows by large herbivores. Further, meadows were not grazed beyond sustainable levels. Although large herbivore presence is readily apparent to observers and thereby receives significant attention (Halbritter and Bender 2011b), other less obvious environmental factors affected RSHs on LNF. Management actions, such as selection of monitoring sites to assess degree of herbivore utilization of meadows vegetation, need to consider effects of other environmental variables (as well as the large variation in relative habitat use of meadows by large herbivores across LNF) or they risk erroneous conclusions as to the effect of large herbivores on mountain meadows. For example, while a limited number of meadows we surveyed could have indicated that grazing (or other factors) was substantially negatively affecting RSH of grasses, meadows across the LNF landscape showed no grazing-related impacts that affected sustainability. Selection and use of key sites as management indicators should represent landscape dynamics if such sites are used to drive decisions on management of landscapes, rather than to identify localized areas of conflict.

Relative habitat use by elk and cattle was generally equal to or higher in mountain meadows than other habitat types, and conflicts can occur when wild and domestic ungulates concentrate their use in vegetation types that represent a small proportion of the total available habitat. Because meadows comprised a small and declining proportion (< 5 %) of the total habitat available to ungulates in LNF, and relative habitat use (and, consequently, potential for herbivory) varied greatly, it is particularly important to identify those factors affecting use of meadows. On LNF, recommended management solutions to perceived overutilization of meadows (e.g., decreased stocking rates for cattle and reductions in elk populations throughout the game management unit; Wright 2000; Hurd 2002; Hurd et al. 2004) may have less than anticipated effects on RSH of meadows across the LNF landscape because of the weak correlations between RSH and relative habitat use of ungulates. Knowing which factors affect meadow use can aid management decisions by identifying those meadows where management practices would be most beneficial. For example, cattle primarily used meadows closer to water and with gentler slopes (Halbritter and Bender 2011b), so increasing the number or quality of these meadows could potentially alleviate use of some of the mountain meadows that receive relatively high use. Moreover, large differences in use of meadows, related to factors influencing distribution of elk and cattle (Halbritter 2007; Halbritter and Bender 2011b), further indicate that generalized reductions of elk and cattle numbers may have little effect alleviating perceived conflicts over use of meadows.

In contrast, utilization of browse on LNF was comparable to areas experiencing high herbivory (e.g., use of *Salix* spp. in Rocky Mountain National Park [70 %– > 90 %; Zeigenfuss et al. 1999] and use of *Populus tremuloides* in Yellowstone National Park [77–90 %; National Research Council 2002]. Relative habitat use by elk was correlated with the majority of the variation of browsing in all cover types (6–45 %; total combined use by cattle and deer accounted for only an additional 4–16 % of the variation in browsing intensity), reflecting the high prevalence of browse (34–55 %) in elk diets annually. The relatively high utilization suggested that elk might be exerting a substantial impact on browse species in LNF. Ungulate herbivory can have significant impacts on regenerating *Populus tremuloides* and other browse species which can influence the effectiveness of management practices such as prescribed burns and timber harvests (Cannon et al. 1987; Wisdom et al. 2006). Such effects may be magnified when browse is limited, such as on LNF (Table 6).

However, the relative scarcity of woody browse species in stands suggests that use rates might simply be high because little palatable browse was available in forested stands of LNF. Canopy cover in LNF negatively affected both herbaceous understory cover and woody stem densities, and canopy cover, particularly of conifer dominated stands, was high (71 %). Lack of browse (as well as herbaceous understory) may also contribute to elk use of mountain meadows, because forested stands, particularly conifer stands, provided little food for elk, deer, or cattle (Tables 3 and 6). Lack of browse (and forest encroachment of meadows) is a result of altered historical disturbance (especially fire) regimes and decreased timber harvest, which has increased stand densities and decreased the quantity of open habitats on LNF (Kaufmann et al. 1998; Frost et al. 2007). Increasing quantity (by overstory thinning or removal [McConnell and Smith 1970; Wallmo et al. 1972; Woods et al. 1982; Dahms and Geils 1997; Halbritter and Bender 2011a]) and quality (by prescribed burning in conjunction with harvesting [Covington et al. 1997; Dahms and Geils 1997; Halbritter and Bender 2011a]) of palatable browse could potentially alter elk distribution in LNF and lower use of other habitat types such as mountain meadows (Halbritter and Bender 2011a, b). The effect of overstory removal with or without burning was illustrated in the comparatively high relative habitat use of thinned and burned stands by all large herbivores (Table 1).

Competition results from shared use of limited resources resulting in decreased performance of one or more competitors (Holechek 1980; Vavra et al. 1989; Miller 2002). Quality of cattle diets on LNF were more than adequate for reproduction and growth (National Research Council 2000), and these levels of diet quality were consistent with variable but generally light overall use of forage and low population densities and stocking rates. Overall, elk diet quality was in the marginal to good range based on penned elk studies (Cook et al. 2004), which likely underestimate diet quality obtained by free-ranging elk because of the greater energetic costs faced by the latter. Although the lower quality seen in 2005 could result in reduced growth and productivity according to performance studies of penned elk (Cook et al. 2004), data from Halbritter and Bender (2011c) indicated no strong nutritional effects on elk survival or productivity. The only evident effect was on calf survival, which was mostly associated with above normal precipitation during parturition. Calf survival is more susceptible to slight changes in environmental variation than adult survival and thus is one of the first population parameters to show any evidence of resource limitations (Gaillard et al. 2000).

Thus, the few nutritional limitations seen on LNF appeared to result from density-independent effects, as 2005, the year of lowest fat levels in lactating elk (8.9 % versus 10.4–11.5 %), received only 82 % of normal precipitation (only 77 % of normal during the growing season). That precipitation rather than competition or density effects was likely the driving factor affecting these nutritional relationships was reflected in both plant cover and nutrient load (CP levels) showing significantly positive relationships with precipitation. Precipitation affected forage quality as reflected by lower CP levels of cattle diets in 2005. Further, an increase in precipitation in 2006 resulted in increased plant cover in thinned, burned, and mountain meadow cover types. However, despite lower forage quality in 2005, diet quality of both elk and cattle on LNF did not significantly limit individual performance (National Research Council 2000; Cook et al. 2004; Halbritter and Bender 2011c). Nor were diets of elk and cattle strongly similar (Schoener's index ≤ 0.57). Because the ultimate effect of competition is reduced fitness of one or all competitors due to reduced resource capture lowering individual and population performance (Holechek 1980; Vavra et al. 1989; Miller 2002), the relatively high diet quality achieved by cattle on LNF (and to a lesser extent elk), the lack of any significantly decreased performance parameter for elk, and low diet similarity all indicate that co-occurrence had little ultimate impact on either large herbivore.

Conclusions

Both elk and cattle acquired diets on LNF that generally did not limit individual or population performance, indicating that neither elk nor cattle populations were limited by condition of the extant plant community or by competition between the species. Nor was grazing by elk or cattle negatively impacting sustainability of meadows, the area of greatest contention on LNF. Diet composition and forage use also collectively indicated that at current stocking levels, no competition was occurring between elk and cattle on LNF. Thus, generalized reductions in stocking rates of cattle or elk populations were not warranted with respect to sustainability of meadows, and in fact increased numbers of both elk and cattle could be supported by current habitat conditions on LNF. However, with greater numbers of animals on the landscape comes a higher potential for impacts on plant communities unless large herbivore distribution is managed or available forage increased. Currently, only with browse utilization in burns was relative habitat use by large ungulates able to directly account for > 50 % of the variation in levels of browse utilization or RSH of *Poa pratensis* on LNF. Because use of palatable woody browse was high in all habitat types, likely due to low availability of browse, recruitment of palatable woody plants could potentially be reduced in LNF if saplings are too heavily browsed and woody browse remains relatively limited in forest cover types. Managers need to consider other factors in addition to large herbivore grazing if greater residual biomass is desired on meadows in LNF.

Elk used thinned conifer comparably to mountain meadows, despite the low occurrence of palatable browse species in thinned conifer stands. Burns were used by all large herbivores. Because understory herbaceous cover and woody densities were strongly negatively related to overstory canopy cover in LNF, management aimed at decreasing forest overstory to increase woody and herbaceous understories (forage quantity), preferably in conjunction with prescribed burns (to increase forage quality), could have strong effects in altering herbivore use of LNF landscapes. Rehabilitating encroached meadows by reestablishing historic disturbance regimes could also increase distribution, quantity, and quality of open habitats. Providing additional or higher quality foraging habitats could alter elk and cattle distribution away from meadows receiving relatively greater use, and thus alleviate perceived conflicts between elk and cattle, and herbivory and other land uses (Halbritter and Bender 2011b).

Competing interests

The authors declare that they have no competing interests.

Authors' contributions

All authors contributed equally to all aspects of this study and manuscript. Both authors read and approved the final manuscript.

Authors' information

HH is Wildlife Specialist and LCB is Senior Research Scientist with NMSU.

Acknowledgements

We thank the U.S. Forest Service, Rocky Mountain Elk Foundation, New Mexico Department of Game and Fish, and several private entities for funding this project. The New Mexico State University, Agricultural Experiment Station provided additional financial support. We thank B. Barrett, T. Kamienski, J. Piasecke, E. Watters, and M. Weisenberger for their contributions to this project. All activities were performed in accordance with NMSU Institutional Animal Use and Care Committee Permit No. 2003-023.

References

Allison CD, Holechek JL, Baker TT, Boren JC, Ashcroft N, Fowler JM (2007) Rapid assessment methodology for proactive rangeland management. Rangelands 29(2):45–50

Alstad KP, Welker JM, Williams SA, Trlica MJ (1999) Carbon and water relations of *Salix monticola* in response to winter browsing and changes in surface water hydrology: an isotopic study using $\delta^{13}C$ and $\delta^{18}O$. Oecologia 120:375–385

Bender LC (2007) Demographics, habitat use, and foraging of sympatric elk and cattle on Lincoln National Forest and Game Management Unit 34. Final Report, New Mexico Cooperative Fish and Wildlife and Research Unit, Las Cruces, NM

Cannon SK, Urness PJ, DeByle NV (1987) Habitat selection, foraging behavior, and dietary nutrition of elk in burned aspen forest. J Range Manage 40:433–438

Carpenter LH, Decker DJ, Lipscomb JF (2000) Stakeholder acceptance capacity in wildlife management. Hum Dim Wildl 5:5–19

Caughley G (1979) What is this thing called carrying capacity? In: Boyce MS, Hayden-Wing LD (eds) North American elk: ecology, behavior, and management. University of Wyoming Press, Laramie, WY, pp 2–8

Chadde SW, Kay CE (1991) Tall-willow communities on Yellowstone's northern range: a test of the "natural-regulation" paradigm. In: Keiter RB, Boyce MS (eds) The Greater Yellowstone ecosystem: redefining America's wilderness heritage. Yale University Press, New Haven, CT, pp 231–261

Collins WB, Urness PJ (1984) The pellet-group census technique as an indicator of relative habitat use: a response to Leopold et al. Wildl Soc Bull 12:327

Cook CW (1966) Factors affecting utilization of mountain slopes by cattle. J Range Manage 19:200–204

Cook RC, Cook JG, Murray DL, Zager P, Johnson BK, Gratson MW (2001) Development of predictive models of nutritional condition for Rocky Mountain elk. J Wildl Manage 65:988–997

Cook JG, Johnson BK, Cook RC, Riggs RA, Delcurto T, Bryant LD, Irwin LR (2004) Effects of summer-autumn nutrition and parturition date on reproduction and survival of elk. Wildl Monogr 155:1–61

Cottam G, Curtis JT (1956) The use of distance measures in phytosociological sampling. Ecology 37:451–460

Covington WW, Fule PZ, Moore MM, Hart SC, Kolb TE, Mast JN, Sackett SS, Wagner MR (1997) Restoring ecosystem health in ponderosa pine forests of the Southwest. J For 95:23–29

Dahms CW, Geils BW (1997) An assessment of forest ecosystem health in the Southwest, General Technical Report RM-GTR-295. U.S. Forest Service Rocky Mountain Research Station, Ft. Collins, CO

De Mazancourt C, Loreau M, Abbadie L (1998) Grazing optimization and nutrient cycling: when do herbivores enhance plant production? Ecology 79:2242–2252

Fleischner TL (1994) Ecological costs of livestock grazing in western North America. Cons Biol 8:629–644

Frost R, Roberts C, Hyatt G, Fowler J (2007) Mountain meadow and open area encroachment in the Lincoln Forest, Sacrament grazing allotment. Report 69. Range Improvement Task Force, New Mexico State University, Las Cruces, NM

Gaillard J-M, Festa-Bianchet M, Yoccoz NG, Loison A, Toïgo C (2000) Temporal variation in fitness components and population dynamics of large herbivores. Ann Rev Ecol System 31:367–393

Georgiadis NJ, Ruess RW, McNaughton SJ, Western D (1989) Ecological conditions that determine when grazing stimulated grass production. Oecologia 81:316–322

Halbritter H (2007) Demographics, habitat use, and foraging of sympatric elk and cattle on Lincoln National Forest. M.S. Thesis, New Mexico State University, Las Cruces, NM

Halbritter H, Bender LC (2011a) Elk habitat quality in the southern Sacramento Mountains, New Mexico. Southwest Nat 56: 1–8

Halbritter H, Bender LC (2011b) Contrasting observation and transect based models of cattle distribution on Lincoln National Forest, New Mexico. Range Ecol Manage 64: 514–520

Halbritter H, Bender LC (2011c) Condition, survival, and productivity of elk in the Sacramento Mountains of southern New Mexico. Southwest Nat 56: 305–314

Hamilton EW III, Giovannini MS, Moses SA, Coleman JS, McNaughton SJ (1998) Biomass and mineral element responses of a Serengeti short-grass species to nitrogen supply and defoliation: compensation requires a critical [N]. Oecologia 116:407–418

Holechek JL (1980) Concepts concerning forage allocation to livestock and big game. Rangelands 2:158–159

Hurd BJ (2002) Effects of big game and livestock on plant cover, composition, and herbaceous biomass in logged areas of the Sacramento Mountains. M.S. Thesis, New Mexico State University, Las Cruces, NM

Hurd BJ, Boren JC, Baker TT (2004) Foraging relationships of big game and livestock and effects on plant communities in logged areas of the Lincoln National Forest, Final Report, Resource joint venture Agreement RMRS-99150-RJVA. New Mexico State University, Range Improvement Task Force, Las Cruces, NM

Kaufmann MR, Huckaby LS, Regan CM, Popp J (1998) Forest reference conditions for ecosystem management in the Sacramento Mountains, General Technical Report RMRS-GTR-19. U.S. Forest Service Rocky Mountain Research Station, Ft. Collins, CO

Kay CE (1997) Viewpoint: Ungulate herbivory, willows, and political ecology in Yellowstone. J Range Manage 50:139–145

Kay CE, Bartos DL (2000) Ungulate herbivory on Utah aspen: assessment of long-term exclosures. J Range Manage 53:145–153

Lemmon RE (1956) A spherical densitometer for estimating forest overstory density. For Sci 2:314–320

Loft ER, Kie JG (1988) Comparison of pellet-group and radio triangulation methods for assessing deer habitat use. J Wildl Manage 52:524–527

Lyons RK, Stuth JW (1992) Fecal NIRS equations for predicting diet quality of free-ranging cattle. J Range Manage 45:238–244

McConnell BR, Smith JG (1970) Response of understory vegetation to ponderosa pine thinning in eastern Washington. J Range Manage 23:208–212

McNaughton SJ (1976) Serengeti migratory wildebeest: facilitation of energy flow by grazing. Science 191:92–94

Miller W (2002) Elk interactions with other ungulates. In: Toweill DE, Thomas JW (eds) North American elk; ecology and management. Smithsonian Institution Press, Washington D.C, pp 435–447

Murtaugh PA (2009) Performance of several variable-selection methods applied to real ecological data. Ecol Lett 12:1061–1068
National Research Council (2000) Nutrient requirements of beef cattle, 7th edn. National Academy Press, Washington D.C
National Research Council (2002) Ecological dynamics on Yellowstone's Northern Range. National Academy Press, Washington D.C
Neff DJ (1968) The pellet-group count technique for big game trend, census, and distribution: a review. J Wildl Manage 32:597–614
Peinetti HR, Kalkhan MA, Coughenour MB (2002) Long-term changes in willow spatial distribution on the elk winter range of Rocky Mountain National Park (USA). Land Ecol 17:341–354
Relva MA, Veblen TT (1998) Impacts of introduced large herbivores on *Austrocedrus chilensis* forests in northern Patagonia, Argentina. For Ecol Manage 108:27–40
Riggs RA, Cook JG, Irwin LL (2005) Management implications of ungulate herbivory in Northwest forest ecosystems. In: Wisdom MJ (ed) The Starkey Project: a synthesis of long-term studies of elk and mule deer. Alliance Communications Group, Lawrence, KS, pp 217–232
Ripple WJ, Larsen EJ (2000) Historic aspen recruitment, elk, and wolves in northern Yellowstone National Park, USA. Biol Cons 95:361–370
Ripple WJ, Larsen EJ, Renkin RA, Smith DW (2001) Trophic cascades among wolves, elk, and aspen on Yellowstone National Park's northern range. Biol Cons 102:227–234
Schoener TW (1970) Nonsynchronous spatial overlap of lizards in patchy habitats. Ecology 51:408–418
Severson KE, Medina AL (1983) Deer and elk habitat management in the Southwest. J Range Manage Monogr 2:1–64
Sparks DR, Malechek JC (1968) Estimating percentage dry weight in diets using a microscopic technique. J Range Manage 21:264–265
Stewart KM, Bowyer RT, Ruess RW, Dick BL, Kie JG (2006) Herbivore optimization by North American elk; consequences for theory and management. Wildl Monogr 167:1–24
Vavra M (2005) Livestock grazing and wildlife: developing compatibilities. Range Ecol Manage 58:128–134
Vavra M, McInnis M, Sheehy D (1989) Implications of dietary overlap to management of free-ranging large herbivores. Proc West Sect Amer Soc Anim Sci 40:489–495
Wallmo OC, Regelin WC, Reichert DW (1972) Forage use by mule deer relative to logging in Colorado. J Wildl Manage 36:1025–1033
West BT, Welch KB, Galecki AT (2014) Linear mixed models: a practical guide using statistical software, 2nd edn. CRC Press, Boca Raton, FL
Western Regional Climate Center (2007) Climate summary, Cloudcroft, NM. http://wrcc.dri.edu. Accessed 1 Jun 2007
Wisdom MJ, Vavra M, Boyd JM, Hemstrom MA, Ager AA, Johnson BK (2006) Understanding ungulate herbivory–episodic disturbance effects on vegetation dynamics: knowledge gaps and management needs. Wildl Soc Bull 34:283–292
Woods RF, Betters DR, Mogren EW (1982) Understory herbage production as a function of Rocky Mountain aspen stand density. J Range Manage 35:380–381
Wright BD (2000) Ungulate distribution and forage utilization in the Sacramento Mountains. Dissertation, New Mexico State University, Las Cruces, NM
Wyoming Game and Fish Department (1982) Handbook of biological techniques. Wyoming Game and Fish Department, Laramie, WY
Zar JH (1996) Biostatistical analysis, 3rd edn. Prentice Hall, Upper Saddle River, NJ
Zeigenfuss LC, Singer FJ, Bowden D (1999) Vegetation responses to natural regulation of elk in Rocky Mountain National Park, Biological Science Report USGS/BRD/BSR-1999-003. U.S. Government Printing Office, Denver, CO

6

The national forest inventory in China: history - results - international context

WeiSheng Zeng[1*], Erkki Tomppo[2], Sean P. Healey[3] and Klaus V. Gadow[4,5,6]

Abstract

Background: National forest resource assessments and monitoring, commonly known as National Forest Inventories (NFI's), constitute an important national information infrastructure in many countries.

Methods: This study presents details about developments of the NFI in China, including sampling and plot design, and the uses of alternative data sources, and specifically

- reviews the evolution of the national forest inventory in China through the 20th and 21st centuries, with some reference to Europe and the US;
- highlights the emergence of some common international themes: consistency of measurement; more efficient sampling designs; implementation of improved technology; expansion of the variables monitored; scientific transparency;
- presents an example of how China's expanding NFI exemplifies these global trends.

Results: Main results and important changes in China's NFI are documented, both to support continued trend analysis and to provide data users with historical perspective.

Conclusions: New technologies and data needs ensure that the Chinese NFI, like the national inventories in other countries, will continue to evolve. Within the context of historical change and current conditions, likely directions for this evolution are suggested.

Keywords: China; Europe; USA; National forest inventories; Forest inventory and analysis

Background

National forest resource assessment and monitoring, commonly known as National Forest Inventory (NFI), has become an important part of the national information infrastructure in many countries. NFI assessments provide an essential service, reconciling available resources with national priorities related to timeliness, precision, and forest values of interest. As a noun, the word 'inventory' refers to a detailed list of articles according to their properties with a commercial origin. As a verb, the word refers to the process of constructing that list. In a forest inventory, the tabulated information generally includes estimates for trees, tree properties and forests, often on the basis of areal units (Loetsch and Haller 1973; Davis et al. 2001; Tomppo et al. 2010b) and is regarded as reliable and adequate for its intended purposes.

Additional demands, including information on forest ecosystem health, are constantly emerging and new technology is applied to meet these challenges. Internationally, there is great diversity regarding definitions, sampling designs, reporting protocols and error estimation. Definitions and specifications may also change over time in one country. Substantial efforts have been made to meet the new challenges. Forest health monitoring has become an essential part of the inventories in many countries. Properties include symptom description, causal agent, and degree of damage. Inventories have also added measurements and assessments of biodiversity indicators, such as dead and decaying wood and key habitats (Tomppo et al. 2010a). Collaborative projects to reach agreement on common definitions and to make inventory results comparable highlighted differences in FAO definitions (UNFCCC LULUCF COST Action E43 2015; Tomppo et al. 2010a; Lanz et al. 2010). The involvement of experts from both organizations made it possible to make good progress towards common definitions, e.g., in land class and carbon pool definitions. The adoption of new

* Correspondence: zengweisheng@sohu.com
[1]Academy of Forest Inventory and Planning, State Forestry Administration, Beijing 100714, China
Full list of author information is available at the end of the article

definitions takes time and needs to be coordinated with a new inventory round. A constant awareness and work of the harmonisation of the concepts as well as the use of common definitions is therefore necessary. Before the new definitions are really in use, some kind of intermediate tools may be needed (Ståhl et al. 2012).

The loss of forest cover and forest degradation, especially in many developing countries, has become a problem of international concern, particularly in view of global warming and high CO_2 emissions (IPCC 2015). Inventories of land use and land use change, as well as biomass stock changes assessments, are integral tools in addressing these problems. The objective of this study is to describe the development of the national forest inventory in China, and to present the most recent estimates of the basic forest resource parameters, as well as the changes over time. Reference to forest resource inventories in some European countries, the FIA in the United States as well as the National Forest Assessment and Monitoring by FAO provides an international perspective.

Development of NFI's in Europe

The history of forest inventories dates back to the end of the Middle Ages when intensive use of forest resources first led to wood shortages which, in turn, forced users to assess timber resources, particularly near towns and mines (Loetsch and Haller 1973; Tomppo et al. 2010b). The first information collected for these purposes concerned forest area and crude estimates of growing stock.

The first inventories were often local with the aim of assessing the available timber resources for specific purposes and were often conducted by the timber users, e.g. commercial companies (Loetsch and Haller 1973; Davis et al. 2001). It soon became obvious that such inventories could not easily be used to compile national level forest information for purposes of formulating national forest policy; thus, NFIs were initiated.

Sample-based national forest inventories were initiated in the Nordic countries in the late 1910s and early 1920s, but were not introduced in other European countries until after World War II: in the late 1940s in the German Democratic Republic; in 1958 in France; in the 1960s in Austria; and in the 1980s in Switzerland. The early national inventories in the Nordic countries included not only information about areas, volume and increment of growing stock and the amount of timber, but also age, size and species structure of forests, silvicultural status of forests, accomplished and needed cutting and silvicultural regimes (Ilvessalo 1927). The purpose was to provide information for forest authorities, timber users, and planners who developed national forest policies. Some European countries have only recently introduced sample-based inventories. At the global level, national forest inventories are still lacking in many countries, although new inventories are being initiated all the time.

The sampling design in the first inventories in the three Nordic countries had some similarities and also some differences. All inventories used lines or strips passing through countries or country regions (Ilvessalo 1927; Tomppo et al. 2011). The intervals between the lines varied by regions depending on the variability of forests and land use. Norway and Sweden employed strip surveys: all trees over a certain diameter threshold were callipered on the strips with a width of 10 m (Tomter et al. 2010; Axelsson et al. 2010). The first NFI in Finland employed lines and visual assessment on the lines as well sample plots with exact mesurements and also with visual assessments on the plots, the latter one to calibrate the visual assessments (Ilvessalo 1927; Tomppo et al. 2011). All three countries changed the sampling designs later to utilize clusters of plots: Norway and Sweden in their inventories in 1957–1964 and 1952–1964, respectively, and Finland in the fifth inventory in 1964–1970. All inventories used only temporary plots first and later either a combination of permanent and temporary plots or only permanent plots. The Norwegian and Swedish NFIs have used fixed radius plots while the Finnish NFI angle count plots from 1964 until 2013. Concentric plots with two radii were adopted in the Finnish NFI in 2014.

Lawrence et al. (2010) gives a concise summary of the sampling designs in European countries and also in countries contributing the European Cost Action E43 and the country book written in the Action (Tomppo et al. 2011). Most of the countries use either detached field sample plots or clusters of plots. The forest area represented by one plot varied from 50 ha in Walloon region in Belgium and plantations in Iceland to 2400 ha in USA and 267,700 ha in Canada. There is quite a high diversity also in estimation methods, particularly in estimating the increment and the drain of trees. The country reports in the book by Tomppo et al. (2010a) presents more detailed descriptions of the inventory methods by countries and the changes in the designs.

New demands for forest inventories require timely and accurate spatially explicit information. Forest inventory groups have employed remotely sensed data for several decades to meet the requirements in a cost-effective way, first using aerial photographs and later also satellite images (Spurr 1960). The increasing availability of aerial photography in digital form and the ease of integration with auxiliary and other georeferenced data (GIS data) has greatly facilitated the use of aerial photographs (McRoberts and Tomppo 2007). NFI country reports (Tomppo et al. 2010a) show that aerial photography is still widely used by European NFIs. Koch (2013) gives reasons for continued use of aerial photography, such as long traditions, high spatial resolution, greater probability of acquiring cloud free data within a specific time window as well as smaller areas and fragmented land use in European countries compared to, e.g., larger non-European countries such as the USA, Canada, or countries in South America.

The use of satellite images, based on NFI field data dates back to 1980, first for forest mapping in stand-level inventories in an experimental way and later operationally at the national level in Finland and Sweden, and increasingly for spatialisation and modeling in the 1990s and 2000s (Reese et al. 2003; Tomppo et al. 2008a), as well as for regional inventories in the USA (McRoberts et al. 2002; McRoberts 2012).

The current trend is to use multi-resolution and multi-sensor data in the inventories. Statistically sound methods, the development of which requires resources and time, are a priority. Detailed and comprehensive field data remains always as the core information in the national forest inventories. Another trend in Europe and elsewhere is to establish statistically sound sampling based inventories. For the reasons, see Section Evolution of FIA in the United States. The tight connections among the inventory teams worldwide maintain the harmonization of the definitions and the methods.

Evolution of FIA in the United States

A national forest inventory was established by the US Congress, through the McSweeney-McNary Act of 1928, to assess issues related to national and regional timber supply. For several decades, periodic forest assessments were completed on a state-by-state basis with little effort to create or enforce methodological standardization (U.S. Department of Agriculture 1992). Labau et al. (1992) provide an account of the changing techniques used by this inventory, which was called the country's "Forest Survey" for most of its history and has been called the Forest Inventory and Analysis program (FIA) since approximately 1990 (earlier in some quarters).

Prior to World War II, strip sample methods adapted from European designs, such as that used in Finland's general survey (Ilvessalo 1927), were common across the country. Timber measurements were taken at fixed distances along transects that were typically established along regular, parallel lines. Following the war, advances in aerial photogrammetry made more efficient stratified designs feasible, and strip sampling became less common (Labau et al. 1992). Starting around 1950, plots began to be sampled using "variable radius" plot cruising, based upon sampling proportional to tree size. These techniques were adapted in the US from work in Austria by Bitterlich (Grosenbaugh 1958). From approximately 1960 through the mid-1980s, most inventory units used a 10-point cluster design within a 0.405 ha (1 acre) area. At the same time, programmable computers were making it possible to automate the calculation and tabling of volume estimates. This technology also supported compilation of data across many inventories, and in 1965, the USDA published what, according to Labau et al. (1992)), was the first truly nationwide forest survey report (U.S. Department of Agriculture 1965).

In addition to trends toward more efficient and more nationally consistent sampling and compilation, there were social and legislative developments that motivated the inventory to move beyond its roots in timber to assessment of a wider variety of forest resources. Public appreciation of non-timber forest resources such as water, habitat, and recreation use grew during the 1970s. The Forest and Rangeland Renewable Resources Planning Act of 1974 (RPA) broadened the mandate of the Forest Survey to include monitoring of a range of non-timber resources (U.S. Department of Agriculture 1992). Non-timber resource data from the inventory, such as habitat area estimates, informed several forest management debates, e.g., informed debates about management of public land (Bolsinger and Waddell 1993).

Several developments during the 1990s led to standardization of inventory techniques and analysis methods across the country. Beginning in 1996, all plots were required to use a four-point design that incorporated sampling proportional to plot area instead of tree size (Labau et al. 1992). Over the subsequent decade, all states migrated to a nationally consistent sample framework, in which one permanent plot was situated randomly within each 2430-ha (6000-acre) cell of a hexagonal tessellation of the country. This sample design is considered to be a "spatially balanced simple random sample," ensuring good geographic distribution of points while retaining a stochastic element (Reams et al. 2005). The sample can be intensified in areas of interest by further tesselating existing hexagons into finer spatial units, each with a randomly located sample. A national "core" set of variables is measured at each plot, and there is flexibility to add measurements to address regionally important monitoring questions. Until approximately 2010, additional measurements, focused primarily on forest health, were collected on 1/16 of all plots to allow broad-scale assessment of covered variables. Since then, a few of these variables have been moved to "core" status (measured on all plots), and the rest have been retired.

Plots are measured on a rotating basis in such a way that an approximately equal number of plots per state is measured each year. The re-measurement cycle is 5–10 years, depending upon the state, and because 10–20 % of the grid is visited each year, the system is known as an "annual inventory", as opposed to the earlier practice of periodically measuring all plots in the same year. As a result, instead of a single-date snapshot of conditions that becomes more dated until the next re-measurement, FIA estimates reflect a rolling average picture of conditions over a 5- to 10-year period.

A principle of the inventory of FIA (and other NFI's) is the focus on supporting a wide range of academic, industrial, governmental, and environmental clients. Reports on forest status and trends are published at both the state and national levels every 5 years. The national report (Smith et al. 2009) is considered the authoritative description of the US forestry sector. Beyond those reports, the public accesses the database itself approximately 100,000 times per year (U.S. Department of Agriculture 2015). While the exact coordinates of the plots are not publically available to protect the integrity of the sample (Healey et al. 2011), the local neighborhood of each plot is provided, as are all other elements of the plot measurements. This policy of availability is thought to maximize the return on the national investment in the inventory. NFIs vary in their willingness to share data beyond aggregate statistics. Outside of the free data access model pursued by the US and other countries, many NFIs share plot information only with select collaborators, while others, including China in the past, carefully restrict data access.

Today, sample-based inventories are conducted in most European and North American countries, although the tradition in Eastern Europe has been to aggregate data from stand-level inventories originally designed for management planning purposes. Stand level data are often assessed in different years and error estimation of such national aggregates are not possible. Consequently, many Eastern European countries have recently revised their systems in favour of statistical, sample-based NFIs. The main reason is to conduct and maintain forest inventories within a statistical framework presenting timely information and making it possible to estimate uncertainties of forest resource parameters. The Country reports in Tomppo et al. (2010a) give an overview of these changes by countries.

FAO NFMA

Since the early 2000s, the Forestry Department of the Food and Agriculture Organization of the United Nations (FAO) has invested substantial resources in developing a programme of support to national forest monitoring and assessment, commonly known as NFMA (Saket et al. 2010; FAO 2013; Tomppo et al. 2014). The NFMA operates mainly in developing countries, particularly in Tropical forests, and technical, financial, and institutional co-operation and support is often needed. Saket et al. (2010) characterized the NFMA approach as follows:

- Demand driven: countries request FAO support and define the assessment scopes as well as information requirements.
- Participatory: the involvement of a wide range of stakeholders is encouraged, including government institutions, the private sector and NGOs.
- Harmonized: the terms and definitions are consistent among national institutions and refer to internationally agreed terms and definitions. The harmonization allows comparison between countries and facilitates reporting to international reporting processes.

According to Saket et al. (2010), the objective of the NFMA is "to contribute to the sustainable management of forests and trees outside forests by providing national decision makers and stakeholders with the means of acquiring accurate, relevant and cost-effective information on the state, uses, management of the forestry resources and land use changes. Such information is particularly relevant for national and international dialogue on forestry related policy issues and socio-economic development". For more detailed objectives, see Saket et al. (2010). Assessing trees outside forests may develop into one of the most important challenges for the immediate future.

By 2013, NFMA had been completed in 10 countries and was in progress or anticipated in another 20 countries (FAO 2013). FAO's NFMA had employed a standard approach regarding sampling design and data collection until 2009. The major sampling unit was a 1 × 1 km square. Each unit contained a cluster of four plots with a size of 250 × 20 m, placed in perpendicular orientations. Small trees were measured on nested subplots. The details of the design are described by Saket et al. (2010) and FAO (2013). Due to the high workload for each cluster, the sampling intensity of the plots had been low. Estimates could have been computed on the national level only (Tomppo et al. 2014). However, when necessary, the sampling intensity was increased based on local information needs. FAO's NFMA has also launched specific studies to analyse and further develop the design based on local information needs (Tomppo and Katila 2008b; Tomppo et al. 2014). One aspect regarding inventories in Tropical countries isaccessibility,–the time needed to reach a field plot. Specific statistical methods are used to address this problem involving, for example, stratification. Considerations related especially to forest inventories in the Tropics are discussed by McRoberts et al. (2013).

Methods

Brief history of NFI in China

In China a national forest inventory was started relatively early. After the People's Republic of China was established in 1949, some national institutions for forest survey were set up at first in the northeastern region, and forest surveys were conducted in the Changbai Mountain and Xiaoxinganling forest areas. Subsequently, forest surveys were gradually extended all over the country. In 1962, the former Agriculture and Forestry Ministry (AFM) of China organized the provincial forestry departments to compile forest resource data from various inventories, such as reconnaissance surveys and forest management inventories conducted during the period 1950–1962 and prepared summaries for the whole country. For the first time in China forest resource data were gathered and reported for the country as a whole. This nation-wide assessment presented a general overview of forest resources in China. However, this first compilation of inventories only covered about 300 million ha, involving the main forest regions and State Forest Bureaus (Xiao 2005).

In 1973, the former AFM conducted two forest inventory pilot studies, one in northern China and another in southern China. Based on the findings, technical specifications on main methods in forestry surveys were issued. During 1973–1976, the first national forest inventory (NFI1), which is usually called the "4th 5-year Plan" Inventory, was implemented. Based on county-level inventories, the sampling designs and survey methods were not standardized for the whole country. In 1977, the former AFM conducted a pilot study involving a continuous forest inventory (CFI) in Jiangxi province. Based on findings of that pilot study, standard specifications on continuous forest inventory in China were issued for execution (AFM 1978).

During the period 1977–1981, the NFI2 was implemented, based on CFI principles with permanent plots and statistical sampling. The provinces, municipalities or autonomous regions, represented the target population. The basic sampling frame of the CFI system had become a good foundation for national forest inventory in China. Subsequently, a total of six NFIs (NFI3 to NFI8) were continuously conducted during the periods 1984–1988, 1989–1993, 1994–1998, 1999–2003, 2004–2008, and 2009–2013 (Xiao 2005; Lei et al. 2009; Lin et al. 2013). The key characteristics of the eight NFI's in China are listed in Table 1.

Following economic and technical advancement, the NFI system in China has been continually improved regarding sampling design, survey methods, and technical standards.

Sampling design

Sample grid

The history of NFI in China shows a progressive evolution towards statistical sampling techniques. All the provinces are now using a systematic sampling design and permanent field plots. For the purpose of efficiency, sampling designs should accommodate variation in forest attributes and land use structure, which suggests that the idea of a single design for the whole country may not be suitable. The most common NFI features include systematic sampling components based on two-dimensional grids, although the grid spacing varies considerably, from $1 * 2$ km^2 in Shanghai, $2 * 2$ km^2 in Beijing and Tianjin to $6 * 8$ km^2 in Yunnan and $8 * 8$ km^2 in Inner Mongolia (Table 2). A few provinces, such as Heilongjiang, Gansu, Qinghai and Xinjiang, divide the population into two or three strata according to eco-geographic regions. Each stratum is considered as an independent sub-population in which systematic sampling is applied. The inventory has a 5-year cycle, and about 1/5 provinces conduct inventories each year. The sampling population of the NFI in China is the total land area, including inland water. Plots are established on non-forest land for purposes of estimating the area of lands converted from non-forest to forest uses and vice versa.

For the estimation of inventory statistics, the estimators and error estimators were based on an asummption of simple random sampling in most provinces. In provinces with sub-populations, the estimators and error estimators of stratified sampling were used where each sub-population was considered as a stratum. Simple random sampling was assumed within each stratum. The national statistics were obtained from the summation of all provincial statistics completed in a 5-year cycle, and the errors were estimated using stratified sampling where each province was considered as a stratum (SFA 2011).

Plot configuration

The shape, size and number of sample plots vary greatly among provinces. Square plots are used in more than 90 % of the provinces. Circular plots are used only in Tibet, while rectangular plots are used in Inner Mongolia and in one sub-population of Heilongjiang. Point sampling or angle count sampling was used in Guangxi before the 6th NFI, but square plots have been used since 2005 (Cen et al. 2007). In the past, several provinces such as Henan and Hubei used to apply cluster sampling in plateau areas, where a cluster was composed of 4 or 5 plots (Xiao 2005).

Plot sizes vary between 0.06 and 0.10 ha, and most of them are 0.0667 ha (1 Chinese *mu*). Among the 31 provinces in China (not including Taiwan, Hongkong and Macao, the same hereafter), the smallest sample numbers are less than 3000 (Jiangxi 2608, Tianjin 2818, and Hainan 2829). The greatest number of samples exceeds 10,000 (Inner Mongolia 17,951, Hebei 11,709, Anhui 11,678, Henan 10,358, and Sichuan 10,007). For most of the provinces, the number of sample plots are between 3000 and 10,000. Several provinces had more sample plots before. For example, the number of sample plots in Guizhou was 11,017 before 1995, about twice the present number, while the sample in Jiangxi was 10,455 before 1996, about four times the current number of sample plots (Xiao 2005). Since the 6th NFI, the sampling design of each province did not change. The number, shape and size of sample plots for 31 provinces are summarized in the Table 2.

Table 1 Key characteristics of the eight NFIs in China

NFIs	Period	Key characteristics
NFI1	1973–1976	Based on county-level inventory, in most regions the inventory was not unified for the whole country
NFI2	1977–1981	The continuous forest inventory (CFI) method was applied, establishing an effective foundation for national monitoring
NFI3	1984–1988	The 1st re-inventory based on the CFI system, providing the changes on both quantity and quality of forest resources
NFI4	1989–1993	4 national forest monitoring centers were set up, which were responsible for quality check, statistical compilation, and output reporting for different regions
NFI5	1994–1998	The UNDP CPR 91/151 project was executed, and new technologies such as "3S" (RS-remote sensing, GPS-global positioning system, GIS-geographic information system) were started
NFI6	1999–2003	Remote sensing was widely applied, and full-coverage inventory for the mainland of China was achieved
NFI7	2004–2008	Several ecological variables were added, and forest ecological services in the whole country were evaluated
NFI8	2009–2013	Modeling of tree biomass equations for main tree species in China has been actively pursued

It should be pointed out that some provinces in western China base their inventories partially on interpretation plots using satellite imagery, such as LandSat, SPOT and RapidEye. For remote or inaccessible regions, interpretation of image plots represents the only information obtained. For example, about 30 % of forest plots had to be assessed through interpretation of image plots in Tibet.

Field survey methods

Field surveys represent the key activities of the NFI, relating directly to the subsequent statistical analysis and inventory outputs. A field survey usually involves three aspects: survey *items*, survey *procedures* and survey *methods*. Survey items (also called "variables" here) are defined by the technical specification. According to the newest *Technical Specifications on National Continuous Forest Inventory* issued for excution in 2014, the survey items include land use and cover, site and soil, forest stand characteristics, forest functions, ecological situations, species names and other items (SFA 2014a). Survey procedures generally involve plot location and establishment, survey of tree attibutes and plot attributes. In China's NFI, all plots are located on the kilometer grids of topographic maps 1:50,000 or 1:100,000, and should be permanently marked and periodically revisited. The survey of tree and plot attributes involves two different methods, measurement and judgement.

Quantitative measurement

The most important quantitative attibute is the diameter at breast height (dbh) of sample trees, which is measured using a diameter tape. The breast height is defined as 1.3 m, and the minimum dbh is 5.0 cm measured over bark. The maximum allowable measurement error (tolerance) for dbh measurement is 3 mm or 1.5 % (for trees with a dbh exceeding 20 cm).

The second most important quantitative tree attibute is height, which is measured using a traditional clinometer or ultrasonic hypsometer. The height of small trees can be measured using a wooden or bamboo stick, or even a fishing rod. The tolerance for tree height measurement is 3 % (for trees less than 10 m) or 5 % (for tree heights exceeding 10 m). In China's NFI, height measurements are only required for 3–5 average trees on each forest plot (SFA 2014a).

Besides dbh and tree height, other quantitative attibutes, including azimuth and horizontal distance from the plot center to the sample tree, are measured usually using a box compass. In addition, crown cover (or crown closure, canopy cover, canopy closure) of a forest stand need to be measured using transect sampling with two diagonal lines or cover area estimation based on mean crown width. Sometimes, crown cover is measured using visual estimation or subjective judgement, based on the experience of the cruising team.

Qualitative assessments

Most of tree and plot attributes are assessed using qualitative judgement. The qualitative attributes are usually classified into several classess, groups, or grades. Land type and forest category are two of the most important plot attributes. Land is classified into two types: forest land and non-forest land. Forest land and non-forest land are both divided into 5 classes (Table 3). Some of these can be divided further.

Forests are also classified into five categories related to administrative use: protection forest, special use forest, timber forest, fuel-wood forest, and economic forest. The first two categories are called "ecological forests", while the remaining three categories are known as "commercial

Table 2 Sampling designs for 31 provinces in China's NFI

Provinces	Sub-pop.	Area: km^2	number of Plots	Grid (Km)	Plot shape	Plot size (ha)	Year of last inventory
Beijing	/	16,410	4074	2*2	Square	0.0667	2011
Tianjin	/	11,305	2818	2*2	Square	0.0667	2012
Hebei	/	187,693	11,709	4*4	Square	0.06	2011
Shanxi	/	156,623	9915	4*4	Square	0.0667	2010
Inner Mongolia	/	1,183,000	17,951	8*8	Rectangular	0.06	2013
Liaoning	/	145,739	4613	4*8	Square	0.08	2010
Jilin[a]	/	189,193	8865	4*16/3	Square	0.06	2009
Heilongjiang	I	100,540	1571	8*8	Square	0.06	2010
	II	64,786	1013	8*8	Rectangular	0.06	2010
	III	289,282	9083	4*8	Square	0.06	2010
Shanghai	/	6341	3365	2*1	Square	0.0667	2009
Jiangsu	/	102,600	8536	4*3	Square	0.0667	2010
Zhejiang	/	101,800	4249	4*6	Square	0.08	2009
Anhui	/	138,165	11,678	4*3	Square	0.0667	2009
Fujian	/	121,501	5051	4*6	Square	0.0667	2013
Jiangxi	/	166,946	2608	8*8	Square	0.0667	2011
Shandong	/	152,221	9646	4*4	Square	0.0667	2012
Henan	/	167,000	10,358	4*4	Square	0.08	2013
Hubei	/	185,900	5820	4*8	Square	0.0667	2009
Hunan	/	211,835	6615	4*8	Square	0.0667	2009
Guangdong	/	176,769	3685	6*8	Square	0.0667	2012
Guangxi	/	237,600	4948	6*8	Square	0.0667	2010
Hainan	/	33,907	2829	4*3	Square	0.0667	2013
Chongqing	/	82,335	5133	4*4	Square	0.0667	2012
Sichuan	/	483,744	10,007	6*8	Square	0.0667	2012
Guizhou	/	176,167	5500	4*8	Square	0.0667	2010
Yunnan	/	382,644	7891	6*8	Square	0.0667	2012
Tibet	/	1,228,436	[a]5855	6*8	Circular	0.0667	2011
Shaanxi	/	205,977	6440	4*8	Square	0.08	2009
Gansu	I	449,734	2817	2*3	Square	0.08	2011
	II		4038	3*3	Square	0.08	2011
	III		10,846	4*8	Square	0.08	2011
Qinghai	I	721,514	76,616	2*2	Square	0.08	2013
	II		51,620	4*2	Square	0.08	2013
Ningxia	/	51,955	12,936	2*2	Square	0.06	2010
Xinjiang	I	164,700	16,474	3*4	Square	0.08	2011
	II		59,917	6*4	Square	0.08	2011

[a]The permanent plots in Jilin province were established through intensifying the 4*8 km grid by 50 %, that is, there are 3 plots for each 64 km^2 square. The number of plots (5855) in Tibet only includes the ground plots in the 1st sub-population (30 forestry counties). The number of plots in Gansu, Qinghai, Ningxia and Xinjiang include some RS-interpretation plots, as supplement to ground plots. Taiwan, Hongkong and Macao are not included in this table

forests". The definition of forest in China's NFI has changed to some extent during different periods. Before 1994, the minimum crown cover of a forest stand was 0.4 (or 35 %); but since the 5th NFI in 1994–1998, the minimum crown cover is defined as 20 %. In addition, since the 6th NFI, forests not only include arboreal forests and bamboo forests with a crown cover of more than 20 %, and economic shrubs with a crown cover of more than 30 % which provide mainly non-wood forest products and fruits, but also include other special shrubs, which are defined as shrub

Table 3 Land Classification System

Class name		Description
Forest land	Arboreal forest	Forest land with a minimum area of 667 m^2, a minimum width of 10 m, and a minimum crown cover of tree species of 20 %
	Bamboo forest	Forest land composed of bamboos with a minimum dbh of 2 cm
	Open forest land	Forest land with a crown cover of tree species of 10–19 %
	Shrub land	Forest land with a minimum crown cover of shrub species of 30 %, including two sub-classes: special shrubs and general shrubs
	Other forest land	Unclosed afforestation land, nursery land, clear-cut land, burned forest land, and planned forest land
Non-forest land	Cropland	Cultivated land, farmland
	Grazing land	Pasture, rangeland, grassland
	Inland water	Lakes, rivers, and other water bodies
	Built-up land	Industry and commerce, mining, traffic and transport, tourist facilities, dwellings and parking sites, gardens and parks
	Other land	Unused and unproductive non-forest land

lands located above the tree lines or in the area with less than 400 mm precipitation, or located in the karst region or dry-hot valley areas (SFA 2014a).

Other forest characteristics include origin, dominant species, age class/group, community structure, species composition, naturalness, disaster class, health class and so on. Forests are classified into two classes by origin: natural forests and plantations. They are also classified into five classes by age category: young, middle-aged, near-mature, mature, and post-mature forests. The details of definitions and specifications of all qualitative attributes can be found in SFA (2014a).

New technology

Remote sensing

The application of remote sensing (RS) in forest inventory can be traced back to early years of new China. The middle of the 1950s was the first time in China that aerial forest surveys and comprehensive ground investigations were implemented. Then, a technical system for forest inventory was established which was based on aerial photography and visual measurement. In the NFI1, remote sensing played an important role. At that time, the remotely sensed data were aerial photos. In the NFI2, remote sensing was applied to forest inventory of Tibet in 1977. Black-and-white MSS images of 1:500,000 scale and aerial photos of small scale were used to determine forest area. In the NFI3, the application of remote sensing was less important. In inaccessible regions in some provinces in western China, remotely sensed data was used as supplement of field survey. In the NFI4, remote sensing was applied to forest inventory of Tibet again in 1991, where Landsat TM images were used to interpret forest area, and field plots were measured to obtain forest volume per hectare. In NFI5, combining with the implementation of UNDP CPR 91/151 project, double sampling for stratification based on remote sensing interpretation and ground survey was tested in the Jiangxi pilot in 1996, which settled a technical foundation for wide application of remote sensing. In NFI6, remote sensing was widely used in all of the 31 provinces. Several provinces in western China, such as Tibet, Xinjiang, Gansu, Qinghai and Sichuan, conducted firstly their full-coverage inventories based on remote sensing (Zeng 2004; Wang et al. 2005; Zhang and Wang 2007). In the 7th and 8th NFIs, remote sensing continued to be appllied as an effective supplement to ground plots and an important information source for forest mapping. During this period, integrated application of multi-scale remotely sensed data has been taken into consideration.

GPS and other technologies

Global positioning systems (GPS) have been applied to forest inventory for less than 20 years in China. In the NFI5, combining with the implementation of UNDP CPR 91/151 project, the application of GPS in NFI was first tested in the Jiangxi pilot in 1996. In the NFI6, GPS was gradually applied to field survey in several provinces (Luo et al. 2002). Since the 7th and 8th NFIs, GPS techniques have been widely used in field work of NFI in all 31 provinces. Using the functions of positioning and navigation decreases the time for locating plots, increases the ratio of re-measured plots, and provides assurance in some extent to increase the quality of inventory data (Luo et al. 2002; Wu and Jiang 2006).

Besides RS and GPS, additional new technologies include advanced database applications and modeling, and the use of portable digital assistants (PDA/i-pad) and geographical information systems (GIS). Statistics and calculations mainly depended on manual work, or were supported to some extent by calculator and simple computer before the NFI4. During the NFI4 in 1989–1993, normative database files based on dBase and BASIC started to be applied in the office work. In the NFI5, mini-type of database based on Foxplus was used. Since theNFI6, large database based on Oracle has been applied, which can ensure security and stability of NFI data, and increase the efficiency of data processing and information management.

Modeling techniques were only used for establishment of tree volume equations between NFI1 and NFI4. In NFI5 (1994–1998), the application of growth and yield models was used in the Jiangxi pilot, with the implementation of UNDP CPR 91/151 project. Since NFI6, modeling techniques have been widely used for estimation of growth and drain (including cut and dead volume) (SFA 2011). During NFI8, specific modeling techniques were been applied to develop biomass equations for the main tree species in China (Zeng 2011, 2014, 2015).

The first pilot of PDA application was made in Tianjin in 2002. Since NFI7, PDA techniques have been used in several provinces, especially in north-eastern China (Wang 2006). In recent years, data collection system based on i-pad has started to be applied in NFI (Xiao et al. 2013). Application of PDA/i-pad system in NFI can achieve paper-free data collection, increase the accuracy of data, and strengthen the pertinence of quality control (Zhang and Zhang 2008). However, because of the disadvantages of poor waterproof capability, less battery life, small memory space, and limited CPU processing ability, PDA is still not applied to NFI in most provinces, especially in southern China.

The application of GIS (geographical information system) in NFI was relatively late. In NFI5, combining with the implementation of UNDP CPR 91/151 project, the application of GIS in was first tested in the Jiangxi pilot, which laid a foundation for wide application of GIS in NFI. Since NFI6, GIS techniques have been widely used in NFI, providing a technical platform for management of inventory data, querying and analysis of spatial data, and increasing the quality and efficiency of forest mapping (Wang and Li 2010; Shu 2014).

Results and Discussion

Forest area and growing stock volumes

According to the results of NFI8 in 2009–2013, the total forest area in China is 207.69 million hectares, the relative forest cover is estimated at 21.63 %, and the total growing stock volume at 15.14 billion m^3 (SFA 2014b). Estimates of forest area, forest cover percent and growing stock volume from NFI1 to NFI8 are listed in Table 4. Forest area, forest cover percent and growing stock volume show continuous increase, except from NFI1 to NFI2, due to large scale afforestation and strict natural forest protection.

As mentioned before, the original definition of forest in China's NFI has changed somewhat during different inventory periods. Thus, the values of the forest areas in Table 4 are not completely comparable. Figure 1 shows the resulting differences of forest areas, using three independent time series and the published total.

In addition, there are other factors affecting the comparability between different measurement occasions, such as improved sampling and progress in remote sensing. For example, during the three decades from NFI1 to NFI6, only three inventories were conducted in Tibet, and the results were not completely comparable. Tibet is an autonomous region with extensive forest areas, the highest growing stock volumes and a low population density. Thus, the forest area and growing stock volumes are relatively stable. But during three inventories in 1977, 1991 and 2001, the forest area has increased from 6.32, 7.17 to 8.45 million ha, and the growing stock volume increased from 1.40 to 2.05 and 2.27 billion m^3. The increase of forest area and growing stock volume in Tibet amounts to 0.85 million ha and 0.65 billion m^3 respectively from NFI3 to NFI4, representing 9 and 65 % of the total increase of forest area and volume in the whole country. The specific inventory results of other provinces, such as Sichuan, also influenced the comparability of national statistics between different inventory periods.

Forest types, origins and age classes

Forest types

Forests include three types: arboreal forests, bamboo forests, and special shrubs. Among the total forest area, arboreal forests are 175.34 million ha (not including 2.13 million ha forests in Taiwan, Hongkong and Macao), bamboo forests are 6.01 million ha, and special shrubs are 24.21 million ha. Forests are further classified into 5 categories, i.e., protective forest, special-use forest, timber forest, fuelwood forest, and economic forest, which are respectively 99.67, 16.31, 67.24, 20.57, 1.77 million ha (SFA 2014b). The areas and proportions of different forest categories from NFI1 to NFI8 in China are listed in Table 5 (Lei 2005; SFA 2009, 2014b).

Table 5 shows that the proportions of protective forest and special-use forest are increasing gradually from NFI1 to NFI8 while those of timber forest and fuelwood forest are decreasing. The changes from NFI5 to NFI7 are especially obvious, reflecting the strategic shift in China's forest policy from timber production to ecological conservation at the turn of the century. The proportion of econmic forest increases gradually from NFI1 to NFI5 but decreases slightly from NFI5 to NFI8 which is the result of the market share of fruits and other non-timber forest products reaching close to saturation levels.

Table 4 Forest area, forest cover percent and growing stock volume of the eight NFIs in China

NFIs	Forest area (million ha)	Forest cover percent (%)	Growing stock volume (billion ha)
NFI1	121.86	12.69	8.66
NFI2	115.28	12.01	9.03
NFI3	124.65	12.98	9.14
NFI4	133.70	13.92	10.14
NFI5	158.94	16.55	11.27
NFI6	174.91	18.21	12.46
NFI7	195.45	20.36	13.72
NFI8	207.69	21.63	15.14

Forest origins

Forests are classified into two main types: natural forests and plantations. In the traditional concepts in China's NFI, natural forests and plantations do not include other special shrubs except economic shrubs. According to the results of NFI8, the area and growing stock volume of natural forests amount to 121.84 million ha and 12.30 billion m^3 respectively, while those of plantations are 69.33 million ha and 2.48 billion m^3 (SFA 2014b).

Most of the forests in China are natural forests occupying 64 % of the total forest area and 83 % of the total growing stock volume respectively. Looking at the results of previous NFIs, we find that the plantation area and volume proportions gradually increase from 20 and 2 % in NFI1 to 36 and 17 % in NFI8 (Table 6). This increase is mainly due to the extensive afforestations during the past decades.

Forest age classes

Forests, excepting bamboo and economic forests, are classified into five age classes: young, middle-aged, near-mature, mature, and over-mature forests. According to the results of NFI8, the area and volume of young forests are 53.32 million ha and 1.63 billion m^3 respectively. Those of middle-aged forests are 53.11 million ha and 4.11 billion m^3, and those of near-mature, mature, and over-mature forests are respectively 58.17 million ha and 9.04 billion m^3 (SFA 2014b).

The majority of the forests in China are young and middle-aged, occupying 65 % of the total forest area. The results of previous NFIs show that the proportions of area and volume of forests in two age classes are relatively constant (Table 7). The young and middle-aged forests occupy about 2/3 of the area and less than 40 % of the total volume, and the mature forests occupy about 1/3 of the area and more than 60 % of the volume.

Increment and drain

The increment estimated by China's NFI is defined as the volume increment of survivor trees between two inventory periods plus half of the volume increment of harvested or dead trees and the volume of ingrowth trees that exceed the dbh-threshold of 5.0 cm between the two inventories (SFA 2011). According to the results of NFI8, the annual increment of forests is 0.76 billion m^3 where the increments of survivor trees, harvested or dead trees, and ingrowth trees are 0.52, 0.07, and 0.17 billion m^3, respectively. Looking at the results of previous NFIs, we find that the annual increment has gradually increased from 0.19 billion m^3 in NFI1 to 0.76 billion m^3 in NFI8. The main reason for this increase is probably the rapid development of plantations in China.

The drain estimated by China's NFI is defined as the volume of trees that were found to be harvested or dead between two inventory periods. The volume of the

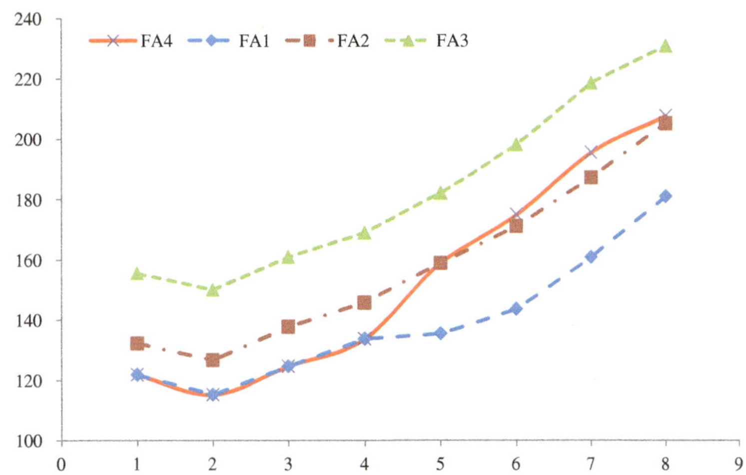

Fig. 1 Development of Forest Areas of the eight NFIs in China. (FA1—Forest areas only inclcuding ≥0.4 (or 35 %) closed cover forests and economic shrubs; FA2—Forest areas also including 0.2–0.3 (or 20–35 %) cover forests and economic shrubs; FA3—Forest areas including both ≥0.2 (or 20 %) cover forests and all special shrubs (Wang and Hou 2014); FA4—Forest areas published in eight NFIs

Table 5 The areas and proportions of forest categories from eight NFIs in China (Note: The forests in Taiwan, Hongkong and Macao are not included in this table)

NFIs	Total		Protection Forest		Special-use forest		Timber forest		Fuelwood forest		Economic forest	
	Area	%	Area	%	Area	%	Area	%	Area	%	Area	%
NFI1	119.78	100.00	7.85	6.55	0.67	0.56	99.07	82.71	3.67	3.07	8.52	7.11
NFI2	113.42	100.00	13.21	11.65	1.30	1.14	83.94	74.01	3.69	3.25	11.28	9.95
NFI3	122.80	100.00	17.76	14.47	3.12	2.54	83.73	68.18	4.44	3.62	13.74	11.19
NFI4	131.85	100.00	19.28	14.62	3.35	2.54	88.83	67.38	4.29	3.25	16.10	12.21
NFI5	156.84	100.00	24.59	15.68	3.97	2.53	103.61	66.06	4.45	2.84	20.22	12.89
NFI6	172.79	100.00	58.51	33.86	6.38	3.69	83.47	48.31	3.04	1.76	21.39	12.38
NFI7	193.33	100.00	93.79	48.51	13.22	6.84	64.16	33.19	1.75	0.90	20.41	10.56
NFI8	205.56	100.00	99.67	48.49	16.31	7.93	67.24	32.71	1.77	0.86	20.57	10.01

sample trees that have been cut or died between successive measurement occasions is calculated by using the tree measurements at the first occasion and adding half of estimated growth between measurements (SFA 2011). According to the results of NFI8, the annual drain is 0.51 billion m^3 with the drains of harvested trees and dead trees being 0.39 and 0.12 billion m^3. Comparing the results of previous NFIs, the annual drain gradually increased from 0.34 billion m^3 in NFI3 (having no data about drain in NFI1 and NFI2) to 0.51 billion m^3 in NFI8.

Geographical distribution

The forests in China are mainly distributed in the north-eastern, south-western and southern regions. About one half of the forest area is located in the 5 provinces Inner Mongolia, Heilongjiang, Yunnan, Sichuan, and Tibet (Fig. 2).

Less than 10 % of the forest areas are distributed in the following 16 provinces (municipality, or autonomous region): Beijing, Tianjin, Hebei, Shanxi, Shanghai, Jiangsu, Anhui, Shandong, Henan, Haihan, Chongqing, Gansu, Qinghai, Ningxia, Honghong, and Macao. To simplify the analysis of change and distribution patterns, six regions have been identified for the whole country: the Northern, North-Eastern, Eastern, South-Central, South-Western and North-Western region. Table 8 presents the forest areas and volumes of the six regions in eight NFIs.

Table 8 shows that the proportions of forest areas in the six regions are not very different except the lower proportion in north-western region. The developments of forest areas in the six regions are very similar corresponding to overall changes in the whole country, but the trends of forest area proportions are variable. The proportion of forest area in the north-eastern region decreased from 21 % in the NFI1 to 16 % in the NFI8, whereas the proportion in the north-wesertn region increased in recent decades from 6.6 % in the NFI5 to 9.5 % in the NFI8.

Similarly, Table 8 shows that the proportions of forest volumes in the six regions differ greatly. More than 40 % of the total volume is found in the south-western region, with only about 1/3 in the northern, eastern, south-central and north-western regions. The proportion of forest volume in the north-eastern region decreased continuously from 25 % in the NFI1 to 19 % in the NFI8. The development of volumes in south-western region is similar to that in the north-eastern region. In contrast, the proportions of volumes in the eastern and south-central regions increased from 6 and 7 % in the

Table 6 Areas and growing stock volumes of natural forests and plantations from eight NFIs in China

NFIs	Natural forests				Plantations			
	Area (million ha)	%	Volume (billion m^3)	%	Area (million ha)	%	Volume (billion m^3)	%
NFI1	96.09	80.22	8.31	98.11	23.69	19.78	0.16	1.89
NFI2	91.12	80.42	8.53	96.93	22.19	19.58	0.27	3.07
NFI3	91.67	74.72	8.38	94.05	31.01	25.28	0.53	5.95
NFI4	97.48	74.00	9.20	92.84	34.25	26.00	0.71	7.16
NFI5	110.17	70.24	9.90	90.74	46.67	29.76	1.01	9.26
NFI6	115.76	68.49	10.59	87.52	53.26	31.51	1.51	12.48
NFI7	119.69	65.99	11.40	85.33	61.69	34.01	1.96	14.67
NFI8	121.84	63.73	12.30	83.20	69.33	36.37	2.48	16.80

Table 7 Areas and volumes of young and middle-aged forests and mature (including near-mature and over-mature) forests from eight NFIs in China

NFIs	Young and middle-aged forests				Mature forests			
	Area (million ha)	%	Volume (billion m^3)	%	Area (million ha)	%	Volume (billion m^3)	%
NFI1	70.68	64.14	2.49	28.75	39.51	35.86	6.17	71.25
NFI2	68.19	69.00	3.39	38.52	30.64	31.00	5.41	61.48
NFI3	72.17	68.48	3.36	37.71	33.22	31.52	5.55	62.29
NFI4	77.84	69.60	3.71	37.44	34.00	30.40	6.20	62.56
NFI5	92.27	69.69	4.20	38.50	40.14	30.31	6.71	61.50
NFI6	96.88	67.85	4.71	38.93	45.91	32.15	7.39	61.07
NFI7	104.63	67.25	5.35	40.04	50.96	32.75	8.01	59.96
NFI8	106.43	64.66	5.74	38.84	58.17	35.34	9.04	61.16

NFI1 to 11 and 12 % in the NFI8 respectively while the volume proportions in the northern and north-western regions show little change.

Discussion

In reviewing the evolution of forest inventories in China, Europe, and the US through the 20th and 21st centuries, common themes emerge. Each of these inventories has shown movement toward: nationally consistent measurement; more efficient sample designs; implementation of improving technology; expansion of the variables monitored; and scientific transparency. Our focus below is how China's expanding NFI exemplifies these global trends.

Nationally consistent measurement

Initial inventory efforts are almost necessarily local, using methods appropriate for meeting specific information needs. As national-scale resource assessment has

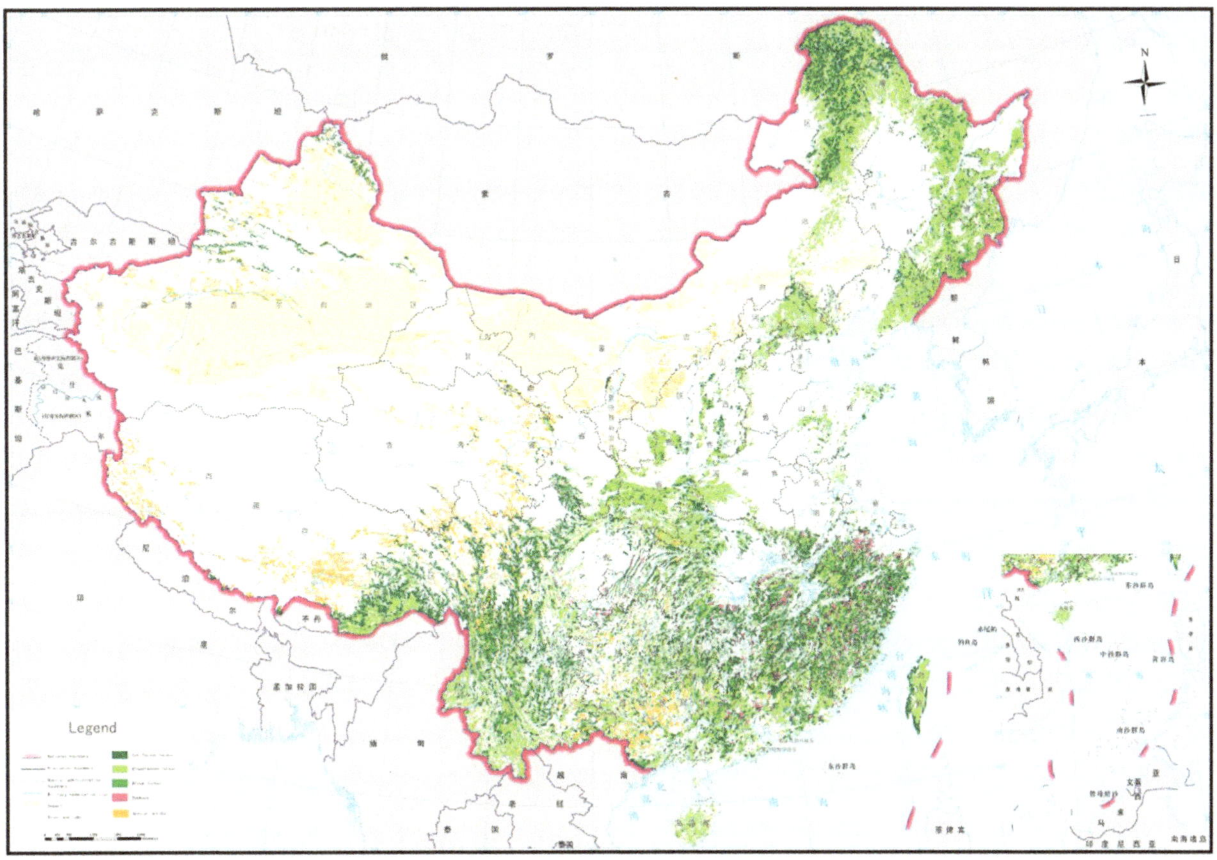

Fig. 2 Current Forest Map of China. Source: SFA (2014b)

Table 8 Forest areas and volumes of the six regions in eight NFIs

Regions	NFI1	NFI2	NFI3	NFI4	NFI5	NFI6	NFI7	NFI8
Forest areas (million ha)								
Total	119.78	113.31	122.68	131.73	156.84	172.79	193.33	205.56
NO	15.43	16.40	17.12	18.17	20.37	23.29	27.61	29.74
NE	25.16	25.02	25.79	26.43	29.18	29.85	31.63	32.71
EA	19.39	16.42	19.29	21.56	26.99	28.65	30.51	32.07
SC	25.72	23.18	23.14	26.28	32.83	36.04	40.03	42.46
SW	25.90	24.64	28.74	30.71	37.14	42.39	46.31	49.06
NW	8.18	7.65	8.62	8.58	10.33	12.57	17.24	19.52
Forest volumes (billion m^3)								
Total	84.70	88.01	89.14	99.10	109.08	120.98	133.63	147.79
NO	8.75	9.09	9.51	10.00	11.06	12.39	13.50	15.68
NE	21.33	21.94	21.48	22.41	23.59	23.66	25.68	28.18
EA	5.39	6.74	6.10	6.83	8.06	10.43	12.89	15.70
SC	5.94	7.15	6.78	7.52	9.24	12.24	15.31	17.44
SW	36.78	36.73	38.78	45.59	49.46	54.23	57.48	60.82
NW	6.51	6.36	6.49	6.75	7.67	8.03	8.77	9.97

NO northern region, including Beijing, Tianjin, Hebei, Shanxi, and Inner Mongolia, *NE* north-eastern region, including Liaoning, Jilin, and Heilongjiang, *EA* eastern region, including Shanghai, Jiangsu, Zhejiang, Anhui, Fujian, Jangxi, and Shandong, *SC* south-central region, including Henan, Hubei, Hunan, Guangdong, Guangxi, and Hainan, *SW* south-western region, including Chongqing, Sichuan, Guizhou, Yunan, and Tibet, *NW* north-western region, including Shaanxi, Gansu, Qinghai, Ningxia, and Xinjiang. The forests in Taiwan, Honghong and Macao are not included in this table

become a priority in the countries we have surveyed, the need for consistent measurement standards has become apparent. Standards often originate from influential pilot studies or through lessons learned by sub-national or international inventories. In China, the continuous forest inventory (CFI) pilot in Jiangxi province strongly influenced standards, beginning with the second NFI (1977–1981). Sampling strategies were tested and validated through the Jiangxi pilot application. Later, remote sensing and GPS techniques now used in China's NFI were also first applied in Jiangxi. International collaboration affected the volution of forest inventory strategies. One example is the use of permanent sample plots in a statistically sound way (e.g., Matérn 1984).

Regional differences in inventory methods still do exist. For example, plot design varies widely within the country. However, the NFI performs the central function of integrating regional inventory methods by, for example, developing standard classification methods that are compatible with the measurements made in constituent regions. China's capacity to coherently report national forest resources across a wide range of ecosystems has depended upon convergence toward consistent methodologies, and this trend will likely continue. Note also that a cost-efficient inventory may require different sampling intensities in different ecozones. The most important aspect is that the estimates within a country and between countries are comparable.

More efficient sample designs

National forest inventories are charged with monitoring large areas with limited budgets. Our survey of international inventories suggests that improvements to plot and sample designs are made in the interests of obtaining better estimate precision without requiring additional field costs. Plot designs evolve as more efficient measurement configurations are shown to adequately capture local conditions, while new sample designs are developed to improve variance estimates by controlling how plots are distributed or analyzed. A "statistical" design based framework has been used in most inventories. Systematic or stratified systematic sampling designs are often used and three-stage design in US. Arranging plots into clusters is also common particularly in Europe, a cluster aiming to be a one-day workload on average. The FIA in USA uses detached plots. A circular plot is the most common plot design, with either one radius or a few radii. Angle count plot is used by some countries.

Through the 1950s and 60s, China's developing capacity to carry out forest inventory was dependent upon methods that were heterogeneous across counties. Pilot studies in the 1970s led to establishment of continuous forest inventory principles that have since influenced sample design parameters nationally. While plot size, orientation, and remeasurement cycle varies by province, permanent plots are now standard across the country. The concept of unbiased plot distribution is standard, too, with simple random designs used in some provinces and stratified designs used in others. With more and more application of high resolution remote sensing, it is possible to improve the sampling design of the Chinese NFI, for example with reference to the annual inventory system in the US, allowing more immediate response of inventory results to broad-scale disturbances or trends.

Implementation of improving technology

New technology has affected how national forest inventories get to, measure, store, and analyze plot information. Just as basal area prisms enabled advances in decades past to efficiency of plot data collection, new technologies are influencing how China assembles its inventory estimates. Remote sensing has been used as the primary measurement of forest area in inaccessible regions, and is currently used to develop estimation strata in some provinces. Global positioning greatly facilitates the navigation needed to re-measure the permanent plots now used throughout the country, and use of PDAs has reduced time and human error associated with data input.

New modeling techniques have improved tree-level estimation of variables such as volume and biomass, and have more recently been used to estimate growth, mortality, biomass and carbon storage (Ge et al. 2004; Li and Lei 2010). Finally, the use of GIS has improved how inventory data are stored, queried, and applied to the production of resource maps (Wang and Li 2010).

Expansion of variables studied

China's early inventory efforts, like those throughout the world at the time, focused on the timber supply. In China, as elsewhere, a growing appreciation of the ecosystem services and non-timber forest products provided by forests has created a need for broader assessment capacities. Beginning with NFI7 (2004–2008), a suite of environmental variables has been added to the inventory to support assessment of ecosystem services (Table 1). This has enabled new assessments of both ecological services (Dai et al. 2009) and forest health (Yang et al. 2015). Like the rest of the world, China continues to develop better biomass equations so that climate change mitigation function of forest carbon storage may be better understood (Zeng 2014).

Scientific transparency

As national forest inventory data are being used in a wider array of scientific inquiries, users need to understand the properties and limitations of the sample. As inventories have matured, methodology increasingly is both influenced and vetted by the international scientific community, a process which ensures rigour and provides documentation for the user community. This paper and other recent works (Yan et al. 2011; Zeng 2013) contribute to peer understanding of Chinese methods. FAO and other projects, such COST Action E43 have done great work in adopting common definitions worldwide, a process which relies upon clear methodological description from cooperating countries (COST Action E43 2015).

Recent increases in Chinese forest area and growing stock have resulted from huge afforestation and reforestation efforts. It is important both within and outside of the country to understand methodological details used to verify these increases. It is also essential to see the extensive efforts, detailed here (Fig. 1), that have gone into reconciling changes to the inventory in the interests of making estimates that are consistent over time. A next step in the process of transparency might be the public access to the plot-level observations themselves. In addition to providing users more context about how estimates are made, public access would likely stimulate important ties to remotely sensed data as well as new analyses at more local levels.

Conclusions

This paper documents important changes in the Chinese NFI both to support continued trend analysis and to apprise data users of important changes to the inventory over the years. Critical details about sample design and plot design are likewise given, and uses of other data sources are described. New technologies and data needs ensure that the Chinese inventory, like the inventories in other countries, will continue to evolve. With historical changes and current conditions as context, we have suggested likely directions for this evolution.

Competing interests

The authors declare that they have no competing interests.

Authors' contributions

WeiSheng Zeng carried out the second and third sections about the NFI in China and the results. Erkki Tomppo carried out the first section National Forest Inventories and participated in the fourth section Discussion. Sean Healty carried out the Discussion and participated in the first section. Klaus Gadow drafted the outline of the manuscript, and participated in the first and Discussion sections. All authors read and approved the final manuscript.

Acknowledgements

The authors thank Ms Yang XueYun of the Academy of Forest Inventory and Planning, State Forestry Administration (Beijing), for compiling the main results of previous NFIs in China.

Author details

[1]Academy of Forest Inventory and Planning, State Forestry Administration, Beijing 100714, China. [2]Natural Resources Institute Finland, Helsinki, Finland. [3]Forest Service, United States Department of Agriculture, Ogden, UT, USA. [4]University of Göttingen, Göttingen, Germany. [5]Beijing Forestry University, Beijing, China. [6]Department of Forestry and Wood Technology, University of Stellenbosch, Stellenbosch, South Africa.

References

AFM (1978) Specifications on continuous forest inventory in China. Agriculture and Forestry Ministry, People's Republic of China, p 18

Axelsson A-L, Ståhl G, Söderberg U, Petersson H, Fridman J, Lundström A (2010) Sweden. In: Tomppo E, Gschwantner TH, Lawrence M, McRoberts RE (eds) National forest inventories-pathways for common reporting. Springer., pp 19–32, ISBN 978-90-481-3232-4

Bolsinger CL, Waddell KL (1993) Area of old-growth forests in California, Oregon, and Washington. U.F.S.P.N.R. Station, Portland, OR, p 26

Cen JY, Li QY, Zeng WS, Li XM (2007) Assessment of continuous forest inventory system in Guangxi. Central South Forest Inventory and Planning 26(3):8–13

COST Action E43 (2015) Harmonisation of national forest inventories in Europe: techniques for common reporting., http://www.metla.fi/eu/cost/e43/. Last accessed. May 18, 2015

Dai L, Zhao F, Shao G, Zhou L, Tang L (2009) China's classification-based forest management: procedures, problems, and prospects. Environ Manag 43:1162–1173

Davis LS, Johnson KN, Bettinger P, Howard TE (2001) Forest management to sustain ecological, economic, and social values, 4th edn. Waveland Press, Inc, Long Grove, IL, ISBN 1-57766-436-1

FAO (2013) Support to forest monitoring and assessment, country projects [online]., Available from www.fao.org/forestry/fma/73410/en/ [accessed 26 May2015]

Ge HL, Meng XY, Tang XM (2004) Growth model system for continuous forest inventory. For Res 17(4):413–419 (In Chinese)

Grosenbaugh LR (1958) Point-sampling and line-sampling: probability theory, geometric implications, synthesis. In: F.S. U.S. Department of Agriculture (ed) Southern forest experimental station., p 34

Healey SP, Lapoint E, Moisen GG, Powell SL (2011) Maintaining the confidentiality of plot locations by exploiting the low sensitivity of forest structure models to different spectral extraction kernels. Int J Remote Sens 32:287–297

Ilvessalo Y (1927) The forests of Suomi (Finland). Results of the general survey of the forests of the country carried out during the years 1921–1924. Communicationes Ex Instituto Quaestionum Forestalium Finlandiae. Editae 11

IPCC (2015) Fifth Assessment Report (AR5)., http://www.ipcc.ch/pdf/assessment-report/ar5/syr/AR5_SYR_FINAL_SPM.pdf (retrieved 30th May, 2015)

Koch B (2013) Remote sensing supporting national forest inventories. In: FAO knowledge reference for national forest assessments., http://forestportal.efi.int/lists.php?pl=01&sf=1

Labau VJ, Bones JT, Kingsley NP, Lund HG, Smith WB (1992) A history of the forest survey in the United States: 1830–2004. F.S. U.S. Department of Agriculture, Washington DC, p 82

Lanz A, Alberdi I, Barbati A, Barsoum N, Brändli U-B, Chirici G, Cienciala E, Condés S, Di Cosmo L, Freudenschuss A, Gabler K, Gschwantner T, Hylen G, Ilvesniemi H, Kusar G, Kändler G, Lawrence M, McRoberts RE, Nabuurs G-J, Petersson H, Priwitzer T, Robert N, Rondeux J, Schadauer K, Ståhl G, Tomter S, Tomppo E, Tosi V, Vidal C, Weiss P, Winter S (2010) Harmonisation of national forest inventories in Europe. Reference definitions. In: Tomppo E, Gschwantner TH, Lawrence M, McRoberts RE (eds) National forest inventories-pathways for common reporting. Springer., pp 1–18, ISBN 978-90-481-3232-4

Lawrence M, McRoberts RE, Tomppo E, Gschwantner Th, Gabler K (2010) Comparisons of National Forest Inventories. In: Tomppo E, Gschwantner Th, Lawrence M, McRoberts RE (eds.). National Forest Inventories - Pathways for common reporting. Springer, p. 19–32. ISBN 978-90-481-3232-4.

Lei JF (2005) Forest resources in China. Chinese Forestry Press, Beijing, p 408

Lei XD, Tang MP, Lu YC, Hong LX, Tian DL (2009) Forest inventory in China: status and challenges. Int For Rev 11(1):52–63

Li HK, Lei YC (2010) Estimation of forest biomass carbon storage in China. Chinese Standard Press, Beijing, p 60

Lin GZ, Wen XR, Zhou CG, She GH (2013) Review and progress of China's continuous forest inventory system. Open J For 3(1):17–22

Loetsch F, Haller KE (1973) Forest inventory. Vol I. BLV Verlagsgesellschaft mbH, München

Luo HB, Chen XQ, Zhou GH, Yu S, Zeng WS (2002) Benefit analysis to application of GPS positioning technique in continuous forest inventory of Tibet. Central South Forest Inventory and Planning 21(1):25–26

Matérn B (1984) Four lectures on forest biometry. Report 23. Swedish University of Agricultural Sciences. Section of Forest Biometry. Umeå. Sweden., ISBN 91-576-2001-6

McRoberts RE (2012) Estimating forest attribute parameters for small areas using nearest neighbors techniques. For Ecol Manag 272:3–12

McRoberts RE, Tomppo EO (2007) Remote sensing support for national forest inventories. Remote Sens Environ 110:412–419

McRoberts RE, Nelson MD, Wendt DG (2002) Stratified estimation of forest area using satellite imagery, inventory data, and the k-Nearest Neighbors technique. Remote Sens Environ 82:457–468

McRoberts RE, Tomppo EO, Vibrans AC, de Freitas JV (2013) Design considerations for tropical forest inventories. Pesquisa Florestal Brasileira-Brazilian Journal of Forestry Research 33(74):189–202

Reams GA, Smith WD, Hansen MH, Bechtold WA, Roesch FA, Moisen GG (2005) The forest inventory and analysis sampling frame. In: Bechtold WA, Patterson PL (eds) The enhanced forest inventory and analysis program-national sampling design and estimation procedures. US Forest Service Southern Research Station, Asheville, NC

Reese H, Nilsson M, Pahén TG, Hagner O, Joyce S, Tingelöf U, Egberth M, Olsson H (2003) Countrywide estimates of forest variables using satellite data and field data from the National Forest Inventory. Ambio 32:542–548

Saket M, Branthomme A, Piazza M (2010) FAO NFMA—support to developing countries on national forest monitoring and assessment. In: Tomppo E, Gschwantner TH, Lawrence M, McRoberts RE (eds) National forest inventories—pathways for common reporting. Edited by Springer., pp 583–594

SFA (2009) Report of forest resources in China (2004–2008). China Forestry Press, Beijing, p 68

SFA (2011) Specification for data processing and statistic in national forest inventory (LY/T 1957). Chinese Standard Press, Beijing, p 50

SFA (2014a) Technical specifications on national continuous forest inventory

SFA (2014b) Report of forest resources in China (2009–2013). China Forestry Press, Beijing, p 86

Shu J (2014) Application of geographical information system (GIS) in China's forestry. Inner Mongolia Forestry Investigation and Design 37(3):59–61

Smith WB, Miles PD, Perry CH, Pugh SA (2009) Forest resources of the United States, 2007. F.S. U.S. Department of Agriculture, Washington DC, p 336

Spurr SH (1960) Photogrammetry and photo-interpretation, 2nd edn. The Ronald Press Company, New York

Ståhl G, Cienciala E, Chinci G, Lanz A, Vidal C, Winter S, McRoberts RE, Rondeux J, Schadauer K, Tomppo E (2012) Bridging national and reference definitions for harmonizing forest statistics. For Sci 58(3):214–223

Tomppo E, Katila M (2008) Comparing alternative sampling designs for national and regional forest monitoring. In: Tomppo E, Andersson K (eds) Technical review of FAO's approach and methods for National Forest Monitoring and Assessment (NFMA), FAO, NFMA working paper 38., pp 55–86, Available from www.fao.org/docrep/016/ap153e/ap153e.pdf [accessed 26 May 2015]

Tomppo E, Haakana M, Katila M, Peräsaari J (2008) Multi-source national forest inventory-methods and applications. Managing forest ecosystems 18. Springer., p 374, ISBN 978-1-4020-8712.-7

Tomppo E, Gschwantner T, Lawrence M, McRoberts RE (eds) (2010a) National forest inventories: pathways for common reporting. Springer, New York, USA, pp 113–130

Tomppo E, Schadauer K, McRoberts RE, Gschwantner TH, Gabler K, Ståhl G (2010b) Introduction. In: Tomppo E, Gschwantner TH, Lawrence M, McRoberts RE (eds) National forest inventories-pathways for common reporting. Springer., pp 1–18, ISBN 978-90-481-3232-4

Tomppo E, Heikkinen J, Henttonen HM, Ihalainen A, Katila M, Mäkelä H, Tuomainen T, Vainikainen N (2011) Designing and conducting a forest inventory-case: 9th national forest inventory of Finland. Springer, Managing forest ecosystems 21., p 270, ISBN 978-94-007-1651-3

Tomppo E, Malimbwi R, Katila M, Mäkisara K, Henttonen HM, Chamuya N, Zahabu E, Otieno J (2014) A sampling design for a large area forest inventory: case Tanzania. Can J For Res 44(2014):931–948

Tomter ST, Hylen G, Nilsen J-E (2010) Norway. In: Tomppo E, Gschwantner TH, Lawrence M, McRoberts RE (eds) National forest inventories-pathways for common reporting. Springer., pp 19–32, ISBN 978-90-481-3232-4

U.S. Department of Agriculture, F.S. (1965) Timber trends in the United States. F.S. U.S. Department of Agriculture, Washington Office, p 713

U.S. Department of Agriculture, F.S. (1992) Forest service resource inventories: an overview. USDA, Washington, DC, p 39

U.S. Department of Agriculture, F.S. (2015) The forest inventory and analysis strategic plan. USDA, Washington, DC, p 40

Wang XK (2006) Application of PDA in forest resources inventory. Shanxi Forestry Science & Technology 2:34–36

Wang XK, Hou XW (2014) The special shrub resources in China. For Res Manage S0:13–16

Wang LR, Li CG (2010) Developing of a mobile GIS system of digital forest resource continuous inventory. Scientia Silvae Sinicae 46(8):174–175

Wang XJ, Han AH, Huang GS, Zhang YX, Ding XQ (2005) Establishment of a national RS-based forest resources monitoring operational system. For Res Manage 2:70–74

Wu HG, Jiang LY (2006) Some measures to improve the precision and application level of GPS in forestry. For Res Manage 2:46–50
Xiao XW (2005) National forest inventory of China. China Forestry Press, Beijing, p 282
Xiao ZH, Xue CQ, Xiong ZP, Liu KC (2013) National integrated monitoring for forest resources and ecological situations in 2012: a pilot study in Guangdong. China Forestry Press, Beijing, p 130
Yan HW, Huang GS, Zeng WS, Nie XY, Gao XL (2011) Considerations on construction of the integrated forest resource monitoring system in China. For Res Manage 6:6–11
Yang J, Dai GH, Wang SR (2015) China's national monitoring program on ecological functions of forests: an analysis of the protocol and initial results. Forests 6:809–826
Zeng WS (2004) Discussion on application of remote sensing in forest inventories. Central South Forest Inventory and Planning 23(1):47–49
Zeng WS (2011) Methodology on modeling of single-tree biomass equations for national biomass estimation in China. Dissertation. Chinese Academy of Forestry, Beijing
Zeng WS (2013) Discussion on methods for producing annual national estimates of forest resources. For Res Manage 1:26–31
Zeng WS (2014) Development of monitoring and assessment of forest biomass and carbon storage in China. Forest Ecosystems 1:20
Zeng WS (2015) Using nonlinear mixed model and dummy variable model approaches to construct origin-based single tree biomass equations. Trees 29(1):275–283
Zhang YX, Wang ZX (2007) Research on application of remote sensing in forest inventory. China Forestry Press, Beijing, p 170
Zhang HJ, Zhang J (2008) Using PDA and "3S" tecnniques to achieve paper-free data collection in forest inventory. Forestry Engineering 24(3):39–40

7

Comparison of carbon-stock changes, eddy-covariance carbon fluxes and model estimates in coastal Douglas-fir stands in British Columbia

Colin J Ferster[1*], JA (Tony) Trofymow[2,4], Nicholas C Coops[1], Baozhang Chen[1] and Thomas Andrew Black[3]

Abstract

Background: The global network of eddy-covariance (EC) flux-towers has improved the understanding of the terrestrial carbon (C) cycle, however, the network has a relatively limited spatial extent compared to forest inventory data and plots. Developing methods to use inventory-based and EC flux measurements together with modeling approaches is necessary evaluate forest C dynamics across broad spatial extents.

Methods: Changes in C stock change (ΔC) were computed based on repeated measurements of forest inventory plots and compared with separate measurements of cumulative net ecosystem productivity (ΣNEP) over four years (2003 – 2006) for Douglas-fir (*Pseudotsuga menziesii var menziesii*) dominated regeneration (HDF00), juvenile (HDF88 and HDF90) and near-rotation (DF49) aged stands (6, 18, 20, 57 years old in 2006, respectively) in coastal British Columbia. ΔC was determined from forest inventory plot data alone, and in a hybrid approach using inventory data along with litter fall data and published decay equations to determine the change in detrital pools. These ΔC-based estimates were then compared with ΣNEP measured at an eddy-covariance flux-tower (EC-flux) and modelled by the Carbon Budget Model - Canadian Forest Sector (CBM-CFS3) using historic forest inventory and forest disturbance data. Footprint analysis was used with remote sensing, soils and topography data to evaluate how well the inventory plots represented the range of stand conditions within the area of the flux-tower footprint and to spatially scale the plot data to the area of the EC-flux and model based estimates.

Results: The closest convergence among methods was for the juvenile stands while the largest divergences were for the regenerating clearcut, followed by the near-rotation stand. At the regenerating clearcut, footprint weighting of CBM-CFS3 ΣNEP increased convergence with EC flux ΣNEP, but not for ΔC. While spatial scaling and footprint weighting did not increase convergence for ΔC, they did provide confidence that the sample plots represented site conditions as measured by the EC tower.

Conclusions: Methods to use inventory and EC flux measurements together with modeling approaches are necessary to understand forest C dynamics across broad spatial extents. Each approach has advantages and limitations that need to be considered for investigations at varying spatial and temporal scales.

Keywords: Forest carbon; Micrometeorology; Biometry; Remote sensing; Geographic information systems

* Correspondence: colin.ferster@ubc.ca
[1]Department of Forest Resources Management, University of British Columbia, 2424 Main Mall, Vancouver, BC V6T 1Z4, Canada
Full list of author information is available at the end of the article

Background

Forests are a large component of the global carbon (C) stocks, containing an estimated 1146 Pg C (Dixon et al. 1994). Forest processes, which may be influenced by forest management, can therefore have a large impact on the global C budget, either by storing or releasing C. It is therefore critical to understand forest C dynamics, including forest C stock components and transfer mechanisms in order to develop accurate forest C models such as the Carbon Budget Model of the Canadian Forest Sector (CBM-CFS3) (Kurz et al. 2009), 3PG (Landsberg and Waring 1997), and Ecosys (Grant et al. 2007), amongst others, as well as to inform forest management policy and for national and international reporting (Kurz et al. 2002). Factors affecting forest C dynamics include natural (e.g. fire, insect outbreaks) and anthropogenic disturbances from land use management and change (e.g. harvest, reforestation, deforestation) (Kurz et al. 2002) as well as weather (Morgenstern et al. 2004), fertilization (Jassal et al. 2010), and stand age and species composition, which can be related to disturbance history, site edaphic characteristics, or management practices (e.g. silvicuture) (Humphreys et al. 2006; Krishnan et al. 2009).

To measure the net exchange of C between land ecosystems and the atmosphere (net ecosystem exchange, NEE, with -NEE referred to as net ecosystem productivity, NEP), a global network of over 400 eddy-covariance (EC) flux stations has been established across a range of ecosystems, building an extensive data record of NEP, in some cases spanning up to two decades. The majority of these towers are located on sites not undergoing major disturbances so the fluxes measured reflect the interaction of weather, vegetation composition, stand age, and seasonal phenology. EC flux-towers use micrometeorological equipment to take measurements at the canopy scale of the exchanges of gasses including CO_2, water vapor, sensible heat, and some are equipped to measure other trace gases (Baldocchi 2008). The source area contributing to measurements made by the instruments mounted on the EC flux-tower, the flux-tower footprint, is variable in size and shape depending on height of measurement, surface roughness length, wind speed and direction, and atmospheric stability (Leclerc and Thurtell 1990; Schmid 2002), and proper interpretation of EC flux-tower based measurements depends on the flux-footprint over which the fluxes are sampled (Chen et al. 2008). Early estimates modelled flux-footprints as simple ovals. Recently, estimates of flux-footprint climatology may demonstrate more complex geometries and continuous probability density surfaces to quantify the upwind distribution of weighting factors over long time periods (Chen et al. 2009).

Forest measurements can be taken to quantify forest C stocks (Dixon et al. 1994; Clark and Brown 2001; NFI 2008), and by measuring forest C stocks at two times with a sufficient interval between measurements, the C stock change (ΔC) can be determined and used to estimate the cumulative net uptake of C from the atmosphere to the forest for the period of measurement (ΣNEP) (Clark and Brown 2001). Measurements are typically made of individual plants, for example, all trees within a plot are measured for diameter, species, and height, then allometric relationships are used to determine tree mass (Ter-Mikaelian 1997). Similarly, woody debris pieces are measured along transects using line intersect sampling and allometric relationships are applied to find biomass (Brown 1971). Other components are made by direct measurements, for example, understory vegetation can be sampled, dried, and weighed (Bailey et al. 1998). To measure soil C, field samples are typically dried and weighed for calculation of bulk density, and sent for laboratory analysis of C using dry combustion (Janzen 2005). Soil C measurements are important because the underground processes driving soil respiration form a large component of ecosystem gas exchanges, but these are less well-understood and measured compared to above ground stocks and processes (Ehman et al. 2002). Changes in soil C storage over time have implications as a source or sink for the global C cycle, and as an indicator of environmental function and health (Janzen 2005).

One application of forest measurement data to provide accounting of forest management actions and subsequently inform forest management decisions is as an input to C budget models such as CBM-CFS3. CBM-CFS3 is a forest C budget model that utilizes growth and yield information for biomass and generates explicit simulation of dead organic matter dynamics (Kurz et al. 2009). In contrast, forest process models use approaches based on the understanding of processes such as photosynthesis; these process-based models have more potential to model changing conditions because they do not rely on historic growth and yield and therefore are better suited to simulate forest conditions under global change situations (Landsberg and Waring 1997). CBM-CFS3 was designed to meet reporting requirements for forest management, provide policy support, and a function as a C budgeting tool for operating foresters, thus functioning at a range of spatial scales, and estimating records of C stocks, transfers between pools, and emissions (Kurz et al. 2002; Kurz et al. 2009).

Comparing EC flux-tower cumulative NEP (ΣNEP) and inventory ground plot measurements of C stock changes (ΔC) over the same period can help inform forest C processes for several reasons. First, ground plot measurements can serve as an independent validation of ΣNEP measurements made at EC flux-towers. Second, inventory plot data can provide more detailed information about the stand structural changes that may be driving variation stand-level EC C exchange. Third, broad stand inventory measurement datasets are available for a greater spatial extent than the global EC flux-tower network. For example, the Canadian National Forest Inventory (NFI) samples approximately 22,000 photo plots locations across Canada with detailed ground plots measured at over 1000 locations (Gillis et al. 2005). However, comparing inventory plot changes and EC flux-tower measurements requires comparisons to be made across different spatial and temporal scales. Therefore, such comparisons pose methodological challenges beyond individual inventory plot and EC flux-tower measurements.

Comparisons between EC tower measurements of ΣNEP and inventory plot measurements of ΔC stocks have been completed at a number of sites globally (Schulze et al. 2000; Granier et al. 2000; Law et al. 2001; Barford et al. 2001; Ehman et al. 2002; Curtis et al. 2002; Miller et al. 2004; Black et al. 2005; Gough et al. 2008; Kominami et al. 2008; Yashiro et al. 2010; Gielen et al. 2013; Babst et al. 2014), and at two sites of the Canadian Carbon Program (CCP), Boreal Ecosystems Research and Monitoring Sites (BERMS) station, located in the boreal forest of northern Saskatchewan (Theede 2007). Over a ten-year interval (1994 – 2004), Theede (2007) found a convergence of 15.6 ± 4.0 MgC ha^{-1} 10^{-1} years (inventory plots) and 18.2 ± 8.09 MgC ha^{-1} 10^{-1} years (EC tower) at the Old Aspen site and 5.8 ± 2.0 MgC ha^{-1} 10^{-1} years (inventory plots) and 6.9 ± 1.6 MgC ha^{-1} 10^{-1} years (EC tower) at the Old Jack Pine site. However, the homogeneous stand structure contributing to canopy-level EC flux measurements and large fetch in the boreal forest reduced the need for consideration of spatial vegetation structure and footprint distributions.

These previous studies have contributed considerably to the understanding of forest C dynamics; however, at sites with more complex vegetation structure and topography, such as the forests found in coastal British Columbia, the spatial distribution of vegetation structure and the footprint distribution are important considerations for accurate comparisons of EC flux-tower and inventory plot measurements (Schmid and Lloyd 1999). More complex ecosystems, therefore, require more complex models and methods developed for accurate comparisons of EC flux-tower measurements and inventory-based measurements. For example, a limited number of studies, including Ehman et al. (2002), Chen et al. (2009), Ferster et al. (2011), and Gielen et al. (2013) applied weighting by EC flux-tower footprint climatology to inventory-based measurements, where areas of the footprint that contribute more to tower-measured flux are weighted more heavily than other areas, resulting in an overall improvement when compared to simple flux footprint geometries.

In this paper, the difference between 2002 and 2006 inventory plot measurements of ecosystem C stocks was determined and the ΔC stocks compared to CBM-CFS3 modelled and EC-flux-tower estimates of ΣNEP at the CCP coastal British Columbia Flux Station. Through a combination of variable topography and a complex disturbance history (Trofymow et al. 2008), the EC flux-tower sites possess fine-scale spatial variation in forest structure, thus presenting a challenge to the interpretation of EC flux data (Schmid and Lloyd 1999; Göckede et al. 2004). To allow the comparison between measurements from the inventory plots, C budget models, and EC flux-towers, our approach consisted of four steps: first, we found the change in C based on forest inventory ground plot measurements made in 2002 and 2006 (four year period); second, we developed an additional approach that utilized litterfall data and published decay equations, for a second inventory-based estimate of C stocks change; third, we defined stand attributes at a fine spatial resolution, stratified the site based on stand attributes, and weighted the two inventory plot estimates; and fourth, calculated ΔC. These were compared with CBM-CFS3 model-based estimates of ΣNEP and EC flux-tower-measured ΣNEP for four years (2003, 2004, 2005 and 2006) in the same period. Finally, we evaluated how the forest age and conditions at the different sites accounted for convergence or lack of convergence in estimates for the different methods, and discussed the relative advantages and constraints of each measurement method.

Methods

Study area

The coastal British Columbia Flux Station sites are located on the leeward central east coast of Vancouver Island (Figure 1), in the Coastal Western Hemlock biogeoclimatic zone (Green and Klinka 1994) with a mean annual air temperature of 10° Celsius. Douglas-fir (*Pseudotsuga menziesii var menziesii* (Mirb.) Franco) is the dominant species, with lesser amounts of western hemlock (*Tsuga heterophylla* (Raf.) Sarg.), western redcedar (*Thuja plicata* Donn *ex* D. Donn), and red alder (*Alnus rubra* Bong.). Three eddy-covariance and meteorological towers were placed in stands of different ages (Table 1).

EC flux-tower footprints, which delineate the spatial distribution of upwind source-weighting factors, were calculated by Chen et al. (2009) to represent the size, shape, direction, and magnitude of the flux as a function of wind speed and direction, measurement height, and surface roughness. Chen et al. (2009) calculated hourly footprints for 10 m by 10 m cells, weighted by NEP, and averaged over the 2002 – 2006 measurement period to determine the footprint climatology for the sites. In this study, the flux-footprint climatology 85% cumulative flux probability boundary was selected as the maximum extent of analysis, since the majority of the probability

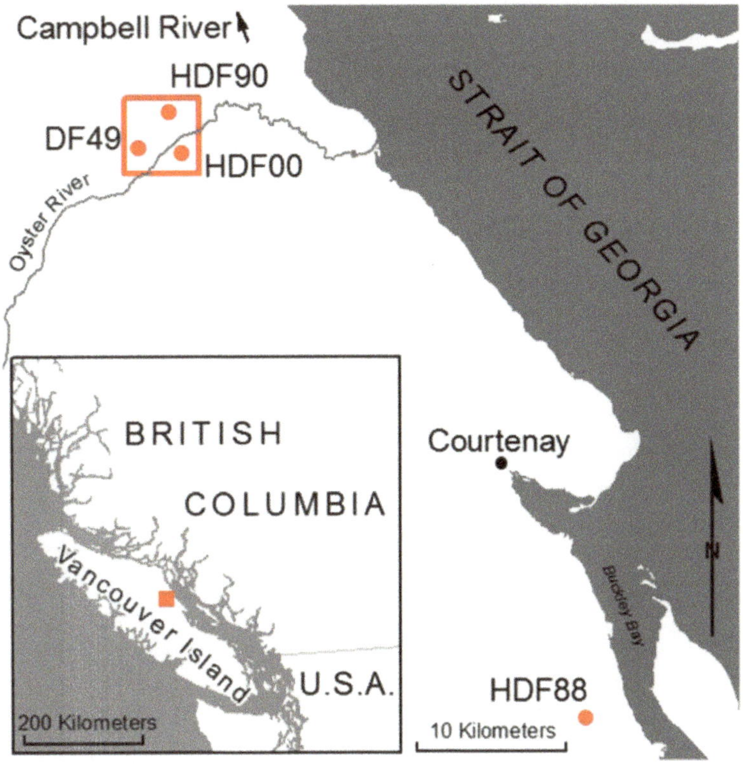

Figure 1 The Canadian Carbon Program Coastal British Columbia Flux Station consists of three sites with flux towers (DF49, HDF00, HDF88) as well as ancillary inventory ground plots (HDF90) on the central east coast of Vancouver Island. A 5 x 5 km area (red square) spanning the Oyster River area has been used for spatial modelling studies.

density is concentrated within this area and the remainder extends across a large area (Figure 2).

Inventory ground plots

All ground plots were established and measured by the Canadian Forest Service at the four sites following Canadian National Forest Inventory (NFI) guidelines and protocols (NFI 2008) though they are not part of the primary NFI ground plot network. Plot locations were chosen to represent each site series class (based on air photo interpretation and field transects) within the preliminary flux footprint boundary estimated in 2002 (Figure 2). Establishment and initial measurements were made in September 2002 and re-measurements were made in September 2006, constituting a four-year measurement interval. NFI data-compilation software used to determine plot-level values through application of allometric equations to overstory trees and density values for woody debris NFI (2008). Since live overstory vegetation roots were not measured in 2002 or 2006, coarse and fine root C mass was estimated from overstory biomass using equations by Li et al. (2003).

Live biomass

Live trees greater than 1.3 m tall were measured in an 11.28 m radius circular plot. Where trees were very numerous (for example, at HDF88), a half plot or 5 m radius circular plot was measured following NFI guidelines (NFI 2008). Allometric equations are used in the NFI compiler to estimate stem, bark, branch, and foliar biomass based on work by Lambert et al. (2005). These biomass values were converted to C content by assuming 52, 56, 52, and 52% dry C concentration, respectively (Matthews 1993).

Understory vegetation from four 1 m × 1 m microplots was destructively sampled and sent to the laboratory for drying and weighing for determination of mass which was converted to C content assuming 50% dry C concentration.

Detritus

Woody debris, fallen woody material > 1 cm diameter, was measured for diameter, species, and decay class along four 15 m transects at each plot. Decay class was determined by field crews in the five-class ordinal rating system

Table 1 Study site locations and characteristics

Site	Location (latitude and longitude)	Site description
DF49	49.868797°,	• Near-rotation stand established in 1949.
(Figure 2a)	- 125.33515°	• 162 ha flux tower footprint.
		• 12 ground plots.
		• 260 to 470 m elevation.
		• Harvested of old growth timber in 1937, 1938, and 1943.
		• Broadcast burned in 1938 and 1943 following harvest.
		• Planted in 1949.
HDF88	49.536655°,	• Juvenile pole-sapling stand established in 1988.
(Figure 2b)	- 124.90146°	• 35 ha flux tower footprint.
		• 6 ground plots.
		• 150 to 220 m elevation.
		• Harvested of second-growth timber in 1987.
		• Broadcast burned following harvest.
		• Planted in 1988.
HDF90	49.893666°,	• Juvenile pole-sapling stand established in 1990.
	- 125.304415°	• No flux tower.
		• 6 ground plots.
		• 175 m elevation.
		• Harvested of second-growth timber in 1990.
		• Broadcast burned following harvest.
		• Planted in 1990.
		• Similar in age and composition to HDF88.
HDF00	49.872177°,	• Regenerating clearcut stand established in 2000.
(Figure 2c)	- 125.29235°	• 14 ha flux tower footprint.
		• 9 ground plots.
		• 160 to 190 m elevation.
		• Harvested from a second-growth stand in 2000.
		• Logging debris were piled and burned.
		• Planted in 2000.

utilized by the NFI, from intact (Class 1) to highly decomposed (Class 5) (NFI 2008). NFI compilation routines utilize algorithms for line intercept sampling to determine plot level volume and a lookup table of woody debris density values by species and decay class to determine mass. C content was determined assuming 50% dry C concentration of woody debris mass. Fine woody debris, fallen woody material 2 mm to <1 cm diameter, was sampled from within the four 1 × 1 m understory microplots in each sample plot and sent for laboratory drying, weighing, and determination of mass, and C content assuming dry mass is composed of 50% C. Dead standing trees were measured for diameter, height, and species and decay class was observed and recorded by field crews. Stumps were measured for diameter and decay class in 2002 but were not remeasured in 2006.

Surface substrates were measured along four 15 m long transects to determine the average depth and percent area surface coverage in the entire plot. The organic litter, fibric, and humus layer (forest floor) was sampled and average depths measured from within four 20 × 20 cm templates (one at each microplot). These destructive samples were collected, sieved, dried, and weighed to determine bulk density and subsamples sent for laboratory analysis of C concentration.

Mineral soil

In 2002, <2 mm mineral soil was sampled for bulk density and C concentrations from 10 – 12 cm diameter excavated holes at three depth intervals 0–15 cm (4 samples), 15–35 cm (2 samples), and 35–55 cm (one sample) and % volumetric coarse fragments determined from one 55-cm-deep soil pit per ground plot. Data were scaled to the entire groundplot discounting for the area without mineral soil (*i.e.* exposed bedrock). In 2006, the 0–15 cm layer was re-measured for C content at four locations in each plot using a 2 cm diameter soil corer. Samples were stored in plastic bags at 2°C prior to laboratory processing. For a detailed description of soil sampling methods see NFI (2008). For 2002–2006 soil C stock changes, only the 0–15 cm layer was considered, soil C at deeper depths was assumed unchanged.

Changes in inventory ground plot C stocks

Changes in C stocks over the four-year interval were calculated from the difference between C stocks in 2002 and 2006. The total change in C stocks was calculated as:

$$\Delta C = \Delta C_L + \Delta C_D + \Delta C_S$$

where ΔC_L is the change in total live biomass C stocks including live trees, and shrubs, herbs, and bryophytes; ΔC_D is the total change in detrital C stocks including dead standing trees, woody debris, and the forest floor; and ΔC_S is the change in mineral soil C stocks (0–15 cm).

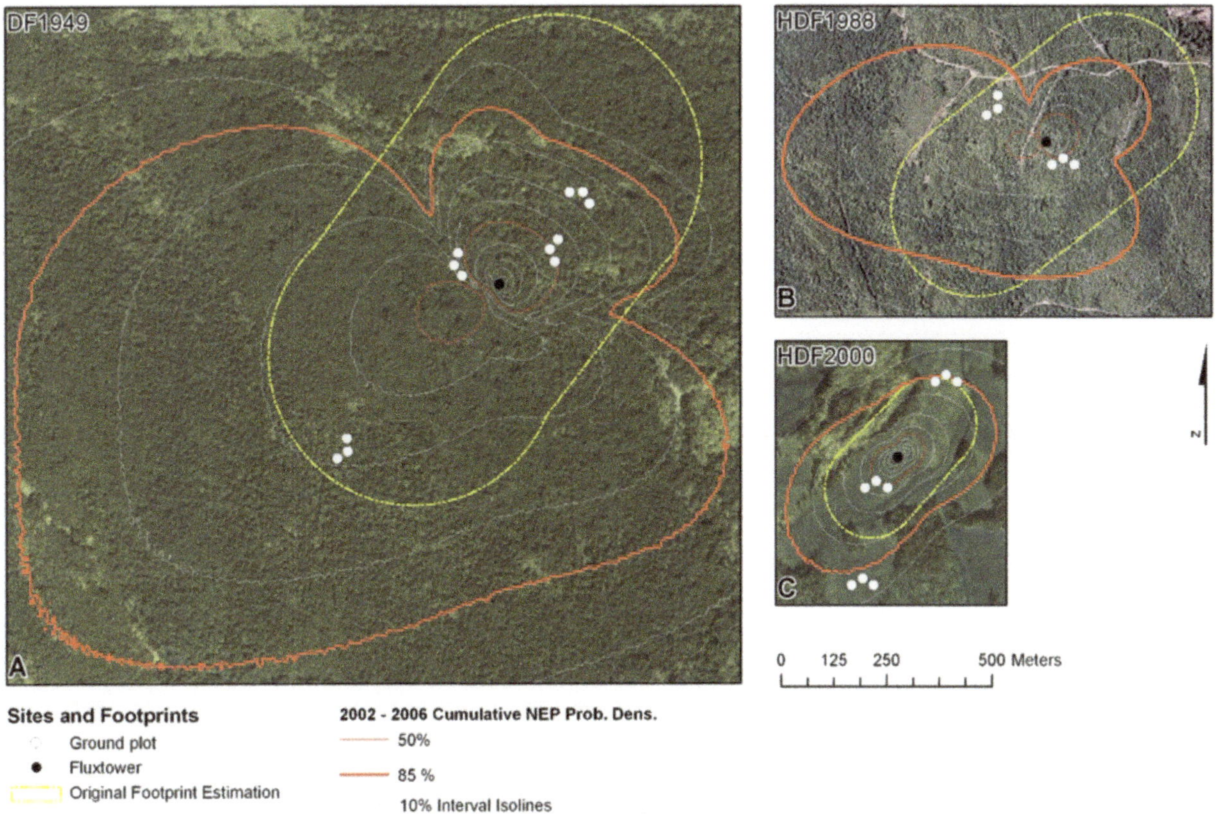

Figure 2 Flux footprints and inventory plot locations for the **A)** DF49, **B)** HDF88, **C)** HDF00 sites. Background imagery is pan-sharpened 2004 Quickbird for **A** and **C**, and true-colour orthophoto for **B**.

ΔC_{D2} Detritus annual accounting method

Preliminary examination of the ΔC_D values showed larger-than-expected changes in the measured C stocks over the four-year interval, especially for the forest floor component. To evaluate these measured changes, ΔC_D was estimated using a second method (referred to as ΔC_{D2}) that accounted for the annual inputs into the detrital pools from litterfall and mortality (*i.e.* the transfers from live biomass to forest floor and soil , and live trees to dead trees respectively) and annual losses and transfers from decomposition estimated using published equations and parameters (Smyth et al. 2010; Kurz et al. 2009).

Inputs to detrital C pools from 2002 – 2005 included overstory fine litterfall (needles, leaves, cones, twigs) which was collected quarterly in 3 0.189 m^2 conical-mesh litterfall traps located in each ground plot at DF49, HDF88 and HDF90. Annual litter fall masses were calculated and converted to C assuming 50% dry C content. Herbaceous material, which was not tall enough to be sampled using litterfall traps especially at the HDF00 site, was collected and measured at the sample plots in 2002 and 2006, linearly interpolated between measurement dates, and assumed to turn over at a rate of 80% annually. Shrubby litterfall material was divided into stem and branch and foliage components, and assumed to turn over at a rate of 3% and 95%, respectively similar to the fine branch wood and foliage for deciduous trees (Kurz et al. 2009). When trees died within the measurement interval, the biomass was transferred to the detrital pools half way through the measurement interval. Finally, the biomass of any trees (live or dead) that were measured in 2002 and not recorded by the field crew in 2006 was transferred to the detrital pools half-way through the measurement interval under the assumption that the tree fell during the measurement interval.

Losses through decomposition and transfer to the soil organic C of each detrital pool was calculated at an annual time step with an annual average temperature of 10°C using coefficients that were developed and validated nationally (Smyth et al. 2010) for the CBM-CFS3 model (Kurz et al. 2009). The transfers and coefficients used are presented in Figure 3. Soil C stock changes were assumed negligible, and ΔC_2 was calculated as: $\Delta C_2 = \Delta C_L + \Delta C_{D2}$.

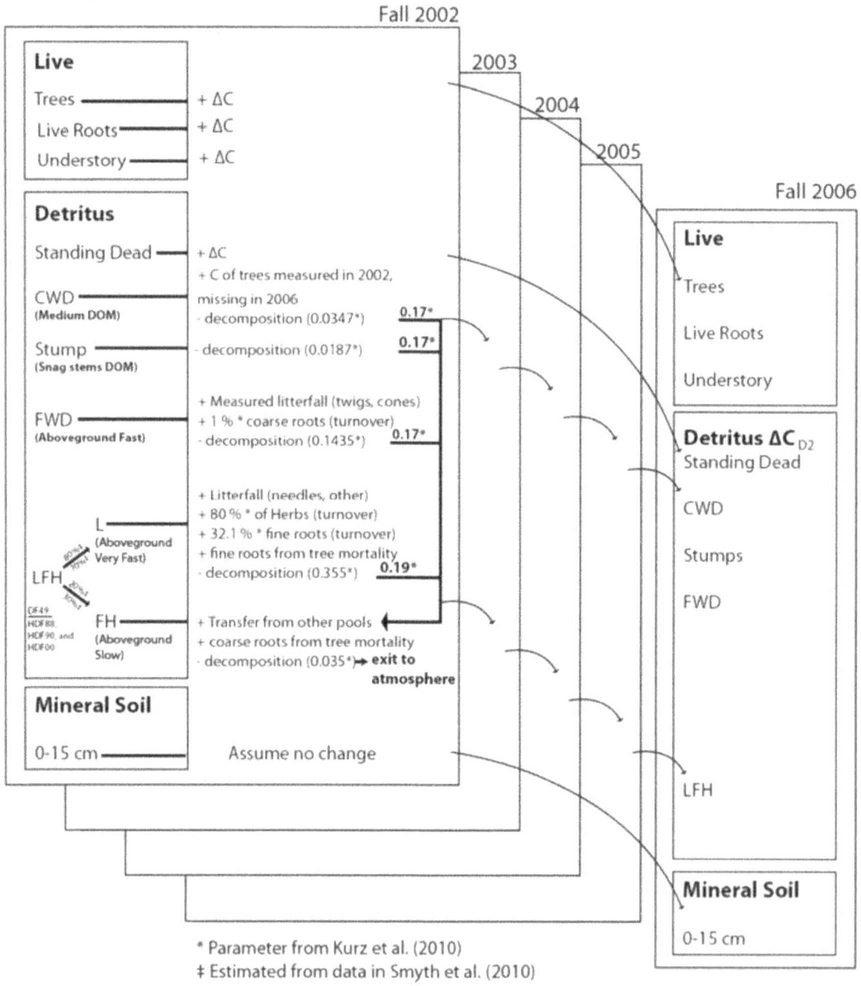

Figure 3 ΔC_2 stocks inputs (+) and outputs (–). Detrital pools decomposition and transfer coefficients from Kurz et al. (2009) and Smyth et al. (2010).

Spatial distribution of C stocks and ΔC at the stand scale

To account for spatial heterogeneity due to stand structure and topography, the methodology developed by Ferster et al. (2011) was applied to the ground plot data at the flux tower sites. Following this approach, predictor variables from GIS, topography, and remote sensing data were used to impute measurements from the inventory ground plots across the forest site based on the three-most-similar-neighbours (K-MSN with K = 3) (Crookston and Finley 2008) using the Mahalanobis distance as a measure of similarity (Mahalanobis 1936). For each footprint cell, the detailed inventory target measurements were estimated as the weighted mean of the three nearest neighbours (inversely weighted by the Mahalanobis distance) and used for calculation of site-level means of ΔC. To assess the error of the model, the root mean squared difference (RMSD) was calculated based on leave-one-out cross validation (Stage and Crookston 2007). The procedure was completed using R version 2.11 (R Development Core Team 2011) and yaImpute version 1.0-10 (Crookston and Finley 2008). Predictor variables were selected if correlations with inventory-based C stocks in 2002 and 2006 were significant at the 95% confidence level and there was no significant co-linearity with other predictor variables. Following the calculation, the 2002 – 2006 cumulative NEP footprint probability density surface for each site (calculated by Chen et al. 2008 and Chen et al. 2009) was used to weight each 10 m by 10 m footprint cell for the calculation of the site level mean value.

Site-level estimates of ΔC and ΔC_2 stocks and were calculated three ways. First, inventory- based site estimates ΔC and ΔC_2 stocks were calculated as an arithmetic mean of ground plots at each site. Second, ΔC_2 stocks were calculated with the K-MSN prediction. Third, and finally, ΔC_2 stocks were calculated with the K-MSN prediction weighted by flux footprint probability density.

Flux tower ΣNEP

Measurement and calculations of flux tower annual NEP published by Black et al. (2008) and Krishnan et al. (2009) were summed to find the cumulative ΣNEP (2004–2006) used for this study. These values were gap filled for conditions at night when friction velocity was low (i.e. inadequate turbulent mixing) and corrected for energy-balance closure. Annual NEP values for 2003, 2004, 2005, and 2006 were summed for each site to estimate the ΣNEP total for the four-year period. Morgenstern et al. (2004) reported that uncertainty in the annual NEP measured using EC at the near-rotation stand may be as much as 0.9 MgC ha^{-1} $year^{-1}$ (due to systematic error in the EC measurements). This suggests that the uncertainty in the estimate at the near-rotation stand may be ±3.6 MgC ha^{-1} over the full four-year measurement interval. Detailed estimates of the uncertainty that propagates through the calculation based on the friction velocity thresholds (e.g. following Richardson and Hollinger 2007) was beyond the scope of this paper.

CBM-CFS3 ΣNEP

CBM-CFS3 is a forest C accounting model used for national reporting of annual C inventories for Canada's managed forests (Kurz et al. 2009). This model uses growth and yield equations and allometric equations to estimate tree growth and stand net primary production (NPP). The C mass of overstory trees and overstory tree roots is use to estimate litterfall and root mortality transfers to the aboveground and belowground detritus C pools, respectively. A soil submodel estimates decomposition and heterotrophic respiration (Rh) based on the size of various detrital pools and mean annual temperature. The model tracks all major C pools to ensure closure and estimates annual NEP from NPP - Rh. Wang et al. (2011) modeled forest processes using the CBM-CFS3 model for the 5 × 5 km area spanning the Oyster River and encompassing the DF49, HDF90, and HDF00 sites (Figure 1). Model runs for the area were performed on 1 hectare grid cells of forest disturbance history data, forest cover data, growth and yield equations, and disturbance transition matrices from Trofymow et al. (2008). The modeled ΣNEP values for the DF49, HDF00 sites (from Wang et al. 2011) were calculated as site-level means unweighted and weighted by the flux probability distribution for the model cells in the footprint area; and at HDF90 as an unweighted mean of model cells over the spatial extent of the ground plots. Annual model values of NEP for 2003, 2004, 2005 and 2006 were summed to calculate the ΣNEP and mean annual NEP for the four year period (Figure 4).

Comparing convergence of ΔC, EC tower ΣNEP, and CBM-CFS3 ΣNEP

Comparisons were made among the estimates by first evaluating how each method ranked the stand ages. Second, estimates of EC tower ΣNEP and CBM-CFS3 ΣNEP were compared. Third, at each stand age from youngest to oldest, the estimates of ΔC and tower and CBM-CFS3 ΣNEP were compared for convergence by comparing the means, variance around the means indicated by the standard deviation, and spatial variance indicated by the RMSD. Comparisons of convergence among methods were made for the following estimates: ΔC_2 as an arithmetic mean of plots (ΔC_2 unweighted), ΔC_2 with K-MSN classification (ΔC_2 K-MSN), ΔC_2 with K-MSN classification and footprint weighting (ΔC_2 K-MSN FPW), CBM-CFS3 ΣNEP as an arithmetic mean of cells within the tower footprint, CBM-CFS3 ΣNEP as a footprint weighted mean (ΣNEP CBM-CFS3 FPW), and EC tower ΣNEP.

Results

Live biomass ΔC_L

For all sites, there was a positive change in the live C mass of overstory trees (Table 2). For the near-rotation stand and juvenile stands, the increase was larger than at the regenerating site. The increase in live biomass C was nearly equal at the near-rotation and juvenile stands. An increase in shrub, herb, and bryophyte understory ΔC was observed at the near rotation and regenerating stands, while a small decrease in understory ΔC stocks occured at the juvenile stands, possibly related to an expected increase in canopy closure as the overstory trees matured. Comparing the two juvenile stands (HDF88 and HDF90), HDF88 had higher live biomass C than HDF90 (for example, live stem C and total live C were more than 1 standard deviation larger).

Detritus ΔC_{D1}

Dead standing tree C decreased in the near-rotation stand due to dead standing trees falling during the measurement interval, and there was a small increase at the other sites due to tree mortality (Table 2). Large woody debris increased in the near-rotation stand, and decreased at the other sites. There was considerable variability among plots, indicated by the high standard deviations compared to the means. Fine woody debris decreased at all sites, except one of the juvenile sites (where there was a small increase). The large decrease in fine woody debris at the regenerating clearcut was likely due to decomposition of debris from the recent harvest. Finally, forest floor material increased at all sites, with the smallest increase at the regenerating clearcut, and the largest increase at the juvenile stands.

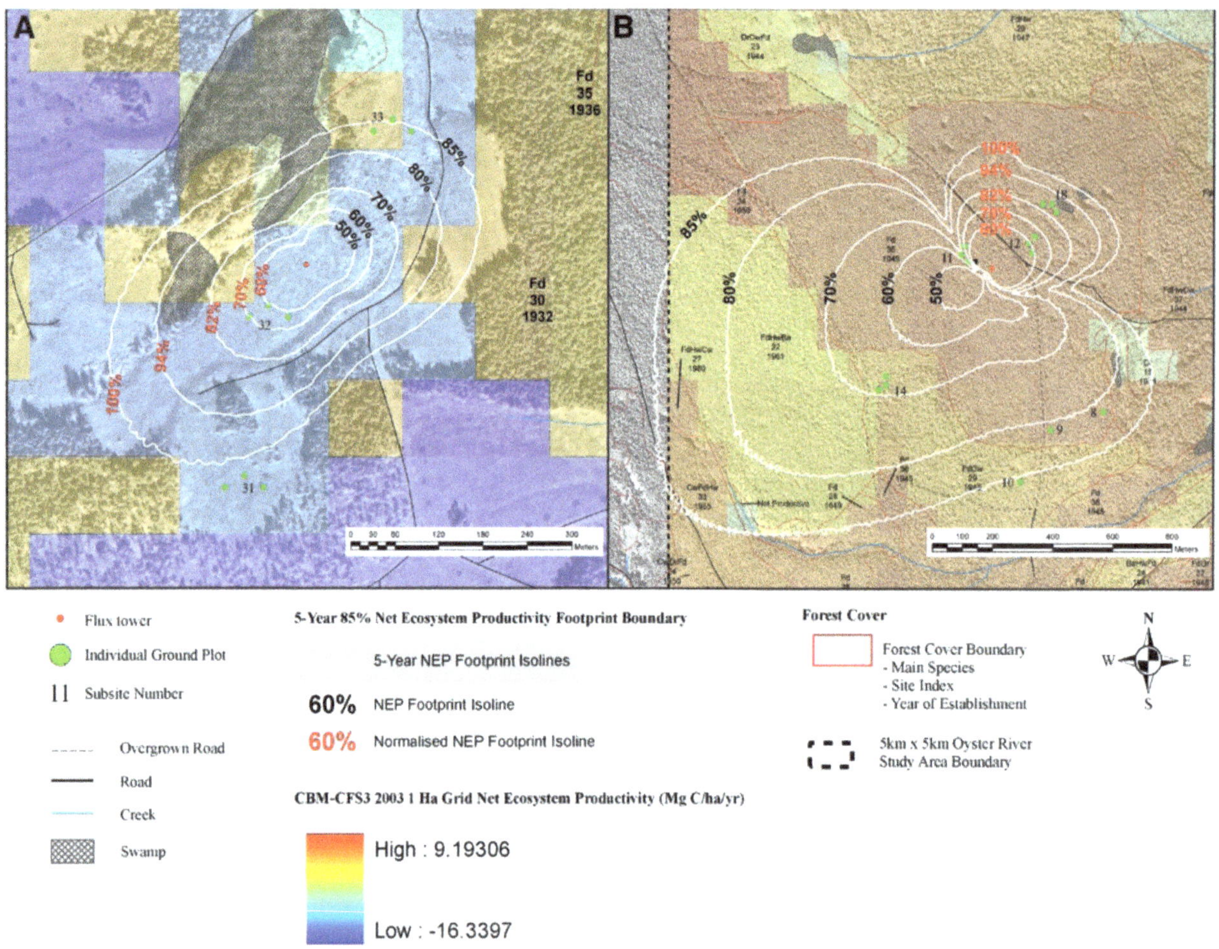

Figure 4 Flux tower footprint isolines and CBM-CFS3 model grids for **a)** the regenerating clearcut (HDF00) **b)** the near-rotation stand (DF49). For main species, Fd = Douglas-fir (*Pseudotsuga menziesii*), Hw = western hemlock (*Tsuga heterophylla*), Cw = western redcedar (*Thuja plicata*), Dr = red alder (*Alnus rubra*), Ba = Amabilis Fir (*Abies amabilis*). Site index is an estimation of height of typical dominant and co-dominant trees in even-aged and undisturbed sites at 50 years age to indicate productivity.

Mineral soil ΔC_s

Soil C (0–15 cm) stocks was highest at the near-rotation site, followed by the two juvenile stands, and lowest at the regenerating clearcut (Table 2). Over the measurement interval, large increases in mineral soil (0–15 cm) ΔC_s were measured at all sites, with the highest at one of the juvenile stands (HDF88).

Detritus annual accounting method ΔC_{D2}

Measured litterfall was highest at the near-rotation stand, followed by the juvenile stands, and the regenerating stand had the lowest amount (Figure 5). The majority of litterfall at the near rotation stand and juvenile stands was needles followed by twigs. At the near-rotation stand, litterfall was highest in 2001–2002, decreased through 2004–2005, and slightly increased in 2005–2006. The other sites were relatively constant through time.

Results from the annual accounting method for detritus, showed a net gain for the sum of components at the near-rotation stand, and a decrease at the other sites (Table 3). The increase in C at the near-rotation stand was slightly larger using ΔC_{D2} than the direct measurement method; however, the biggest difference was at the juvenile stand, which showed a net loss using ΔC_{D2}, and a large positive balance using the direct measurement method. For stumps, this method indicated that decomposition may have been greater than was initially expected (stumps were not re-measured due to an expectation of minimal decomposition), especially at the more recently disturbed juvenile and regenerating clearcut sites. Large woody debris increased at the near-rotation stand, and decreased at the other sites, due to less dead standing trees to fall and less branchfall. The increase in fine woody debris was estimated to be much lower using ΔC_{D2}, with the near-rotation stand having a net-accumulation of fine

Table 2 Inventory-based site C stocks and ΔC estimates using arithmetic plot means (standard deviation)

	DF49	*HDF90*	*HDF88*	*HDF00*
Live				
Foliage C 2002	12.7 (1.8)	2.7 (0.9)	2.6 (0.8)	0 (0.1)
Foliage ΔC	−0.1 (0.7)	1.2 (0.7)	1.5 (0.4)	0.4 (0.5)
Branch C 2002	22.1 (4)	2.5 (1)	1.5 (0.4)	0 (0.1)
Branch ΔC	0.7 (1.2)	1.3 (0.9)	1.8 (0.5)	0.1 (0.2)
Bark C 2002	17.6 (3.8)	0.7 (0.1)	1.1 (0.4)	0 (0)
Bark ΔC	0.9 (0.9)	0.7 (0.3)	1 (0.3)	0.1 (0.2)
Stem C 2002	90.6 (23.4)	2.4 (0.4)	4.7 (2)	0.1 (0.3)
Stem ΔC	5.7 (4.7)	3.3 (1.1)	4.4 (1.3)	0.6 (1.1)
Roots C 2002	32.6 (8.3)	1.8 (0.3)	2.8 (1.3)	0.1 (0.1)
Roots ΔC	1.6 (1.6)	1.8 (0.6)	3.9 (0.9)	0.7 (1.5)
Understory C 2002	1.3 (0.6)	8.3 (1.9)	11.2 (4.7)	4.9 (1.8)
Understory ΔC	1.3 (1.8)	−0.2 (2.8)	−2.6 (4.5)	1.7 (1.9)
Total Live C 2002	177 (39.9)	18.5 (3.7)	24 (4.7)	5.2 (2)
Total Live ΔC	10.2 (8.7)	8.1 (6)	10 (5.9)	3.7 (4.4)
Detritus				
Standing Dead C 2002	21.8 (13.9)	0.2 (0.2)	0 (0.1)	0 (0.2)
Standing Dead ΔC	−3.4 (11.1)	0.5 (0.6)	0.1 (0.2)	0 (0.2)
Stumps C 2002	6.4 (3.2)	5.7 (2.8)	11.7 (6.1)	11.9 (10.6)
Stumps ΔC	0 (0)	0 (0)	0 (0)	0 (0)
Large Woody Debris C 2002	34.9 (20.5)	12.3 (8.7)	43.3 (25.3)	14.4 (11.1)
Large Woody Debris ΔC	1.6 (12.4)	−3.5 (9.7)	−1.5 (18.8)	−0.9 (6.3)
Fine Woody Debris C 2002	3.1 (1)	1.3 (0.7)	0.5 (0.6)	6.1 (2.9)
Fine Woody Debris ΔC	−0.9 (1)	−0.7 (0.8)	0.8 (0.8)	−4.5 (2.6)
Forest Floor C 2002	13.8 (7.3)	12.1 (7.2)	18.8 (13.1)	18.2 (11.6)
Forest Floor ΔC	11 (6.5)	1.9 (11.2)	24.8 (19.7)	0.1 (12.3)
Total Detritus C 2002	79.9 (32.7)	31.6 (8.4)	74.4 (38.7)	50.6 (16.8)
Total Detritus ΔC	8.2 (21)	−1.9 (8.1)	24.2 (30.3)	−5.4 (12.6)
Soil				
Mineral Soil 0–15 cm C 2002	38.6 (21.9)	29 (8.9)	34.6 (16.1)	23.6 (11.3)
Mineral Soil 0–15 cm ΔC	18.8 (24.1)	27.8 (27.7)	10.7 (32.4)	24.2 (15.2)
Totals				
Total Live ΔC + Total Detritus ΔC	18.3 (23.4)	6.2 (10.5)	34.2 (30.6)	−1.7 (14.9)
Total Live ΔC + Total Detritus ΔC + Mineral Soil 0 - 15 cm ΔC	37.2 (34.7)	34 (29.8)	44.9 (41.3)	22.5 (14.2)

All units are Mg C ha^{-1}.

woody debris, the juvenile sites had a small loss, and the regeneration site experienced the largest decrease. This indicates that, given the available inputs to litterfall and expected rates of decomposition for fine woody debris, the direct measured stock changes in fine woody debris was very likely unrealistically high. Finally, for the forest floor layer, the near-rotation stand was estimated to have a large increase, while the other sites decreased, with the greatest decrease at the regenerating clearcut.

Since the plot values for ΔC_{D2} were judged more reliable, all subsequent analyses to spatially scale the plot data to the site were made using ΔC_2.

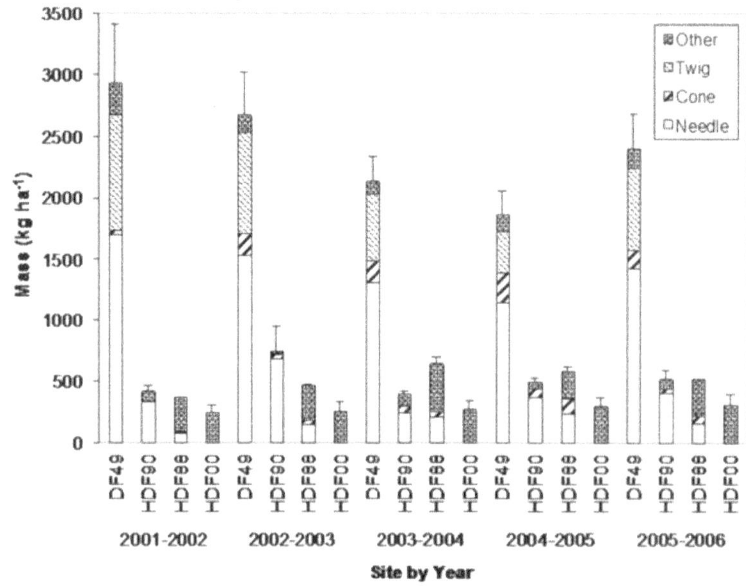

Figure 5 Measured annual litterfall 2001–2006. Needles include green and senescent foliage, cones include male and female cones, twigs include woody material such as twigs or small branches, and other includes miscellaneous pieces such as leaves, mosses, and lichens.

Spatial distribution of C stocks and ΔC_2 stocks

Predictor variables (Table 4) were selected and used for the K-MSN procedure at the sites with flux towers (Figure 6). Slope was an important predictor for all sites as it was strongly correlated with soil C stocks (with less sloping plots having higher soil (0 – 15 cm) C stocks), and at the regenerating clearcut, slope was also correlated with living tree C stocks. At the near rotation site, forest-inventory mapping was a significant predictor variable for overstory C stocks. Due to the smaller footprints at the other sites, forest-inventory mapping varied little within the site; however, at all sites the other spatial predictor variables showed finer scale variation. At the regenerating clearcut, NDVI was a significant predictor for woody debris, forest floor, and mineral soil C stocks.

The Mahalanobis distance (Figure 7) demonstrates how comprehensively the inventory ground plots represent stand conditions across the footprint. Lower percentiles represent small distances that were relatively well represented by the ground plots. Areas with the smallest Mahalanobis distances and best ground plot representation were close to the towers in the areas contributing most to NEP measurements (within the 50%

Table 3 Inventory-based site detrital C in 2002 and ΔC_{D2} (2002–2006) using plot mean (standard deviations) litterfall, mortality and decomposition values

	DF49	*HDF90*	*HDF88*	*HDF00*
Detritus				
C_{2002} Standing Dead	21.8 (13.9)	0.2 (0.2)	0 (0.1)	0 (0.2)
ΔC_{D2} Standing Dead	−3.4 (11.1)	0.5 (0.6)	0.1 (0.2)	0 (0.2)
C_{2002} Stumps	6.4 (3.2)	5.7 (2.8)	11.7 (6.1)	11.9 (10.6)
ΔC_{D2} Stumps ΔC	−0.5 (0.2)	−0.4 (0.2)	−0.9 (0.4)	−0.9 (0.8)
C_{2002} Large Woody Debris C	34.9 (20.5)	12.3 (8.7)	43.3 (25.3)	14.4 (11.1)
ΔC_{D2} Large Woody Debris	3.9 (9.6)	−1 (1)	−4.3 (2.6)	−1.4 (1)
C_{2002} Fine Woody Debris	3.1 (1)	1.3 (0.7)	0.5 (0.6)	6.1 (2.9)
ΔC_{D2} Fine Woody Debris	1 (0.8)	−0.4 (0.3)	−0.1 (0.2)	−2.3 (1.1)
C_{2002} Forest Floor	13.8 (7.3)	12.1 (7.2)	18.8 (13.1)	18.2 (11.6)
ΔC_{D2} Forest Floor	8.2 (4.1)	−4.8 (4.2)	−7.4 (7.2)	−9.3 (6.7)
C_{2002} Total Detritus	79.9 (32.7)	31.6 (8.4)	74.4 (38.7)	50.6 (16.8)
ΔC_{D2} Total Detritus	9.2 (8.7)	−6.1 (4.2)	−12.5 (9.4)	−13.8 (6.5)

All units are Mg C ha^{-1}.

Table 4 Environmental predictor variables used to determine stratification units at DF49, HDF88, and HDF00

Coverage	Variable	Units	Selection		
			DF49	*HDF88*	*HDF00*
Forest-Inventory	Site Index[1]	m	selected		
	Top Height	m			
Disturbance History (Trofymow et al. 2008)	1st Harvest	year			
	2nd Harvest	year			
	1st Fire	year			
	2nd Fire	year			
	Date Est. 1953	year			
	Date Est. 2003	year			
	1st Fertilization	year			
	2nd Fertilization	year			
	Fire Cause 1	nominal			
	Fire Cause 2	nominal			
Topography[3]	aspect	azimuth	selected	selected	selected
	Elevation	m asl			selected
	Slope	degrees	selected	selected	selected
	SCOSA[2]			selected	selected
	SSINA[2]			selected	selected
	TSRAI[2]		selected	selected	selected
1999 Orthophoto (Gougeon 1995)	Dominant Canopy Tree Density	stems ha^{-1}	selected	selected	
2004 Multispectral[4]	NDVI	NDVI	selected		selected
Forest-Inventory (Trofymow et al. 2008)	Cover Species				
	Site Species				
Soil Survey of Canada (Jungen 1985)	Most Common Soil Association				
CFS and Forest Companies (Trofymow et al. 2008)	Site Series		selected	selected	

[1]Site Index: Tree height at 50 years age at breast height (1.3 m).
[2]Topographic Variables: SCOSA = Slope × cos (Aspect), SSINA = Slope × sin (Aspect); TSRAI (Topographic Solar Radiation Aspect Index) = (1-cos((π/180°)(Aspect-30°)))/2 (Roberts and Cooper 1989).
[3]Topography from 2004 LiDAR survey at DF49 and HDF00 (Coops et al. 2007). Topography from 1:50,000 National Topographic Series map at HDF88.
[4]Multispectral data from 2004 Quickbird survey at DF49 and HDF00. Multispectral data from 2004 Landsat scene at HDF88.

cumulative flux probability density isoline). Notable areas with large Mahalanobis distances included areas with different forest cover types (e.g., patches of hardwoods at the near-rotation stand), very few overstory trees (at the juvenile stand), and short steep slopes (between bench terraces at the regenerating clearcut).

For site level estimations of ΔC_L at the regenerating clearcut, K-MSN was more than 1 standard deviation lower than the unweighted mean, and K-MSN-FPW was 1 standard deviation lower than K-MSN. For ΔC_{D2}, K-MSN and K-MSN FPW were less than 1 standard deviation higher than the unweighted mean (Tables 3 and 5). The totals for ΔC_2 K-MSN and K-MSN FPW were less than 1 standard deviation higher than the arithmetic mean (Table 6).

For the juvenile stands, K-MSN and K-MSN FPW ΔC_L was less than 1 standard deviation higher than the unweighted mean. The K-MSN estimate was higher than the K-MSN FPW estimate. For ΔC_{D2}, K-MSN and K-MSN FPW estimates were slightly higher than the mean (less than 1 standard deviation). For the total ΔC_2, the estimates were less than 1 standard deviation different, and the standard deviations were very large compared to the means, indicating a large amount of variability at the site.

At the near-rotation stand, ΔC_L K-MSN and K-MSN FPW were slightly higher than the arithmetic mean (less than 1 standard deviation). The estimates using KMSN and KMSN-FPW were identical. For ΔC_{D2}, the unweighted mean was higher than K-MSN and K-MSN

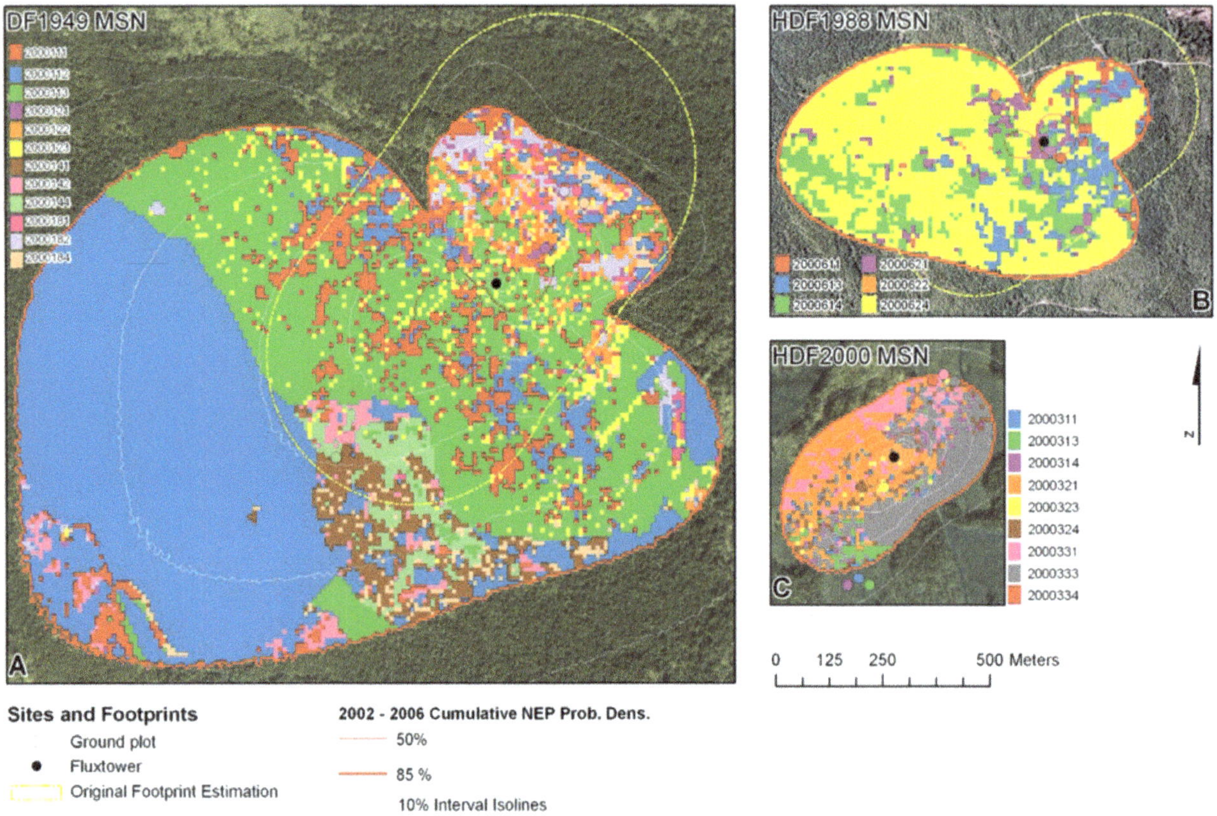

Figure 6 First most similar neighbour (MSN) sample plots demonstrate the patterns of spatial variability for **A)** the near rotation stand, **B)** juvenile stand, **C)** recent clearcut. Background imagery is pan-sharpened 2004 Quickbird for **A** and **C**, and true-colour orthophoto for **B**.

FPW, but less than 1 standard deviation different. For ΔC_2 total, all estimates were within 1 standard deviation.

Comparing the ΔC_2 and ΣNEP methods at the juvenile stands, HDF88 was compared with measurements from the flux tower installed at the site, and HDF90 was compared with CBM-CFS3 model output. While the two stands were similar in seral stage, species composition, and stand history, there were differences in stand composition. For example, HDF88 had higher live C stocks, in particular, live stem C, understory stem C, and total stem C. In addition, HDF88 had larger detritus C stocks including stumps C, large woody debris C, and total detritus C. The stand variability at HDF88 was also greater than at HDF90 indicated by the large standard deviations compared to the means. Given the large amount of spatial variability at HDF88, spatial scaling, and footprint weighting had a notable effect on the amount and sign at this site.

Comparing inventory ΔC_2, EC tower ΣNEP and CBM-CFS3 ΣNEP

All methods ranked the near-rotation stand as a sink for C (net C uptake), the juvenile stands as a weak sink or weak source (small C uptake or release), and the regenerating clearcut as a source (net C release). While the ranking was consistent for all methods, there were differences in the magnitude of values for each stand (Table 6).

At the regenerating clearcut, the CBM-CFS3 estimate of ΣNEP with footprint weighting converged more closely with the EC tower estimate of ΣNEP than the unweighted estimate, and both were within 1 standard deviation. Evaluating the yearly means (Figure 8a), demonstrated that footprint weighting improved convergence between the CBM-CFS3 estimate of ΣNEP and the EC flux ΣNEP. This was primarily due to less weighting for patches of uncut forest in the periphery of the flux footprint (Figure 4).

At the juvenile sites, CBM-CFS3 ΣNEP indicated that HDF90 was a weak sink (net C uptake) and EC tower ΣNEP indicated that HDF88 was a weak source (net C release) (Tables 3 and 6). At the near rotation stand, the estimates of ΣNEP from CBM-CFS3 were higher than the EC flux tower ΣNEP, with the footprint weighted estimate being the highest. Considering the means of ΣNEP from CBM-CFS3 on a yearly basis, footprint weighting decreased convergence with the EC flux tower measurement of ΣNEP (Figure 8b).

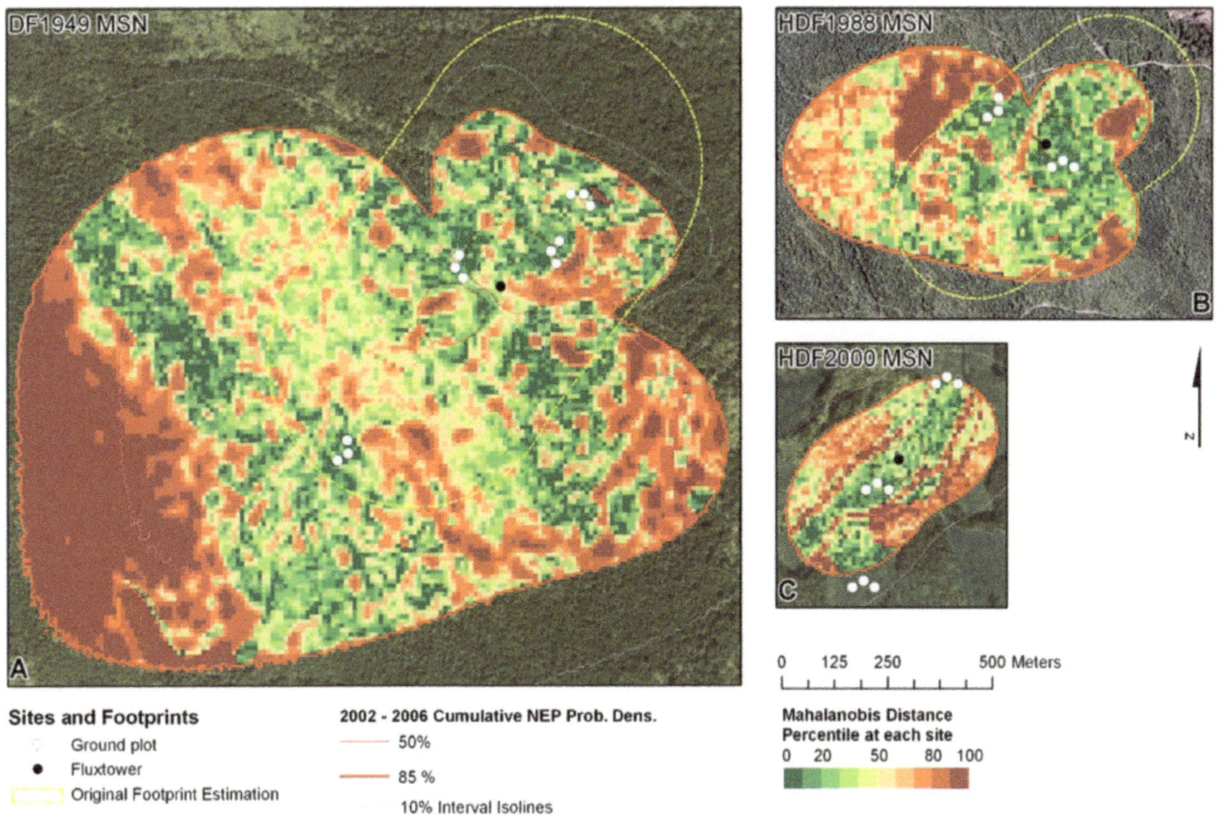

Figure 7 Total Mahalanobis distance for K-MSN stratification units for **A)** the near rotation stand, **B)** juvenile stand, **C)** recent clearcut. Mahalanobis distance percentiles are for each site. Footprint cells with green tones are below the median Mahalanobis distance at site and are relatively well represented by inventory-based sample plots, while footprint cells with red tones are above the median Mahalanobis distance for each site and are less well represented. Background imagery is pan-sharpened true-colour 2004 Quickbird for **A** and **C**, and true-colour orthophoto for **B**.

Table 5 Site mean (± RMSD) changes in live biomass (ΔC_L) and detrital (ΔC_{D2}) C stocks calculated using three most similar neighbours classification (K-MSN), and K-MSN with footprint weighting means (K-MSN FPW) of plots for sites with flux-towers

		DF49		*HDF88*		*HDF00*	
Pool	**Component**	**K-MSN**	**K-MSN FPW**	**K-MSN**	**K-MSN FPW**	**K-MSN**	**K-MSN FPW**
Live	ΔC_L Foliage	0.1 ± 1.7	0.2 (0.4) ± 1.7	1.3 ± 2	1.3 (0.3) ± 2	0.2 ± 0.9	0.1 ± 0.9
	ΔC_L Branches	0.9 ± 1.6	1 (0.6) ± 1.6	1.9 ± 1.3	1.8 (0.6) ± 1.3	0.1 ± 1.1	0 ± 1.1
	ΔC_L Bark	1.1 ± 1.5	1.1 (0.4) ± 1.5	1 ± 1.7	1 (0.3) ± 1.7	0.1 ± 0.8	0 ± 0.8
	ΔC_L Stem	6.7 ± 1.4	6.1 (2.2) ± 1.4	4.4 ± 1.7	4.2 (1.2) ± 1.7	0.4 ± 0.9	0.1 ± 0.9
	ΔC_L Understory	1.8 ± 1	2.5 (2.3) ± 1	−0.9 ± 0.8	−1.7 (5.2) ± 0.8	0.8 ± 1.3	0.6 ± 1.3
	ΔC_L Roots	2 ± 1.5	1.9 (0.8) ± 1.5	4.3 ± 2	4.4 (0.6) ± 2	0.3 ± 1.1	0.1 ± 1.1
	ΔC_{LTotal}	12.8 ± 1.6	12.8 (3) ± 1.6	11.9 ± 0.8	10.9 (7.2) ± 0.8	1.8 ± 0.8	1 ± 0.8
Detritus	ΔC_{D2} Standing Dead	−5.7 ± 1.2	−8.1 (16.1) ± 1.2	0.1 ± 1.3	−0.7 (0.4) ± 1.2	0 ± 1.4	0 ± 1.4
	ΔC_{D2} Stumps	−0.5 ± 1.1	−0.5 (0.1) ± 1.1	−0.6 ± 1.2	−3.5 (1.1) ± 1.9	−0.7 ± 1.3	−0.8 ± 1.3
	ΔC_{D2} Large Woody Debris	3.5 ± 0.9	6.3 (13.7) ± 0.9	−3.5 ± 1.9	−0.1 (0.2) ± 1.1	−1.5 ± 0.3	−1.5 ± 0.3
	ΔC_{D2} Find Woody Debris	0.6 ± 1.4	0.2 (0.8) ± 1.4	−0.1 ± 1.1	−6.9 (4.3) ± 1.5	−2 ± 1.3	−1.6 ± 1.3
	ΔC_{D2} Litter	7.1 ± 0.9	8.7 (2.5) ± 0.9	−6.2 ± 1.5	−0.2 (11.2) ± 1.6	−6.1 ± 1.3	−4.8 ± 1.3
	$\Delta C_{D2Total}$	5 ± 1.5		−10.3 ± 1.6	1.3 (0.3) ± 2	−10.4 ± 1.3	−8.7 ± 1.3
Total	$\Delta C_2 = \Delta C_{LTotal} + \Delta C_{D2Total}$	17.8 ± 1.2	19.5 (5.9) ± 1.5	1.6 ± 1.2	−0.2 ± 1.2	−8.6 ± 1.3	−7.7 ± 1.3

The Root Mean Square Difference (RMSD) is presented as an estimate of spatial estimation error.

Table 6 Comparisons of inventory-based site C stock changes (ΔC_2) determined from plot means (live biomass and detrital fluxes), K-MSN scaling mean (standard deviation), and K-MSN with footprint weighting (K-MSN FPW) mean (standard deviation) to estimates of ΣNEP from flux towers and ΣNEP from CBM-CFS3 for the (a) 4-year period 2003–2006 Mg C ha^{-1} and the (b) mean annual value for each method in Mg C ha^{-1} yr^{-1}

a)	Site			
Method	***DF49***	***HDF90***	***HDF88***	***HDF00***
ΔC_2 Mean	19.4 (7.2)	2.0 (8.5)	−2.5 (13.3)	−10.1 (8.2)
ΔC_2 K-MSN	17.8 (5.8) ± 1.2	na/	1.6 (11.7) ± 1.2	−8.6 (5.0) ± 1.3
ΔC_2 K-MSN FPW	19.5 (5.9) ± 1.2	na/	−0.2 (11.2) ± 1.2	−7.7 (3.5) ± 1.3
ΣNEP Flux Tower	13.6 ± 3.6	na/	−1.9	−20.1
ΣNEP CBM-CFS3*	18.9	2.0*	na/	−11.5
ΣNEP CBM-CFS3 FPW	24.1	na/	na/	−17.9
b)	**Site**			
Method	***DF49***	***HDF90***	***HDF88***	***HDF00***
ΔC_2 Mean	4.9 (1.8)	0.5 (2.1)	−0.63 (3.3)	−2.5 (2.0)
ΔC_2 K-MSN	4.5 (1.5) ± 1.2	na/	0.4 (2.9) ± 1.2	−2.2 (1.2) ± 1.3
ΔC_2 K-MSN FPW	4.9 (1.5) ± 1.2	na/	−0.1 (2.8) ± 1.2	−1.9 (1.9) ± 1.3
ΣNEP Flux Tower	3.4 ± 0.9	na/	−0.5	−5.0
ΣNEP CBM-CFS3*	4.7	0.5*	na/	−2.9
ΣNEP CBM-CFS3 FPW	6.0	na/	na/	−4.47

*average of all grid cells encompassing the six plots 2003 - 2006.
na - not available or not applicable.

At the regenerating clearcut, unweighted ΔC_2 was within 1 standard deviation of the CBM-CFS3 estimates of ΣNEP, and closest to unweighted CBM-CFS3 ΣNEP. The unweighted mean of ΔC_2, ΔC_2 K-MSN, and ΔC_2 K-MSN FPW were greater than 1 standard deviation higher than the EC flux tower estimate of ΣNEP. ΔC_2 K-MSN was within 1 standard deviation of the CBM-CFS3 unweighted ΣNEP. Both ΔC_2 K-MSN and ΔC_2 K-MSN FPW were greater than 1 standard deviation higher than ΣNEP CBM-CFS 3 FPW (Table 6).

At the juvenile stands, all estimates ΔC_2 were within 1 standard deviation of the estimates of ΣNEP. At HDF90, the unweighted estimate of ΔC_2 matched ΣNEP CBM-CFS3. At HDF88, the unweighted mean of ground plots was the closest to the EC tower measurements of ΣNEP, while the ΔC_2 K-MSN had a different sign (indicating a small source), and the K-MSN ΔC_2 FPW had the same sign as ΔC_2 unweighted (Table 6).

At the near-rotation stand, all estimates of ΔC_2 were within 1 standard deviation of all estimates of ΣNEP. The estimates of ΔC_2 (and CBM-CFS3 ΣNEP) higher than the EC flux tower measurements of ΣNEP (Table 6).

Discussion

Several approaches can be used to compare forest inventory data with EC ΣNEP. For example, several authors have used measurements or estimates of soil and detritus respiration in combination with inventory measurements of overstory productivity (Law et al. 2001; Ehman et al. 2002; Kolari et al. 2004; Black et al. 2005). In the present study, comparisons were based on measurements of ΔC which were higher than EC ΣNEP. Black et al. (2005) also found a divergence between inventory measurements of ΔC (including ΔC_S measurements) and EC ΣNEP, with ΔC higher than EC ΣNEP; however, they also calculated ΣNEP using measurements of heterotrophic respiration combined with biometric estimates of NPP, and found that the bias was not observed. They concluded that ΔC systematically overestimated NEP due to unaccounted decomposition processes and uncertainties in ΔC_S. At DF49, Jassal et al. (2010) measured heterotrophic soil respiration in 2007, which was a major portion of total soil respiration. Including measurements of soil heterotrophic respiration with net primary production estimates from forest inventory may improve convergence with EC-flux tower measurements. However, a limitation of taking measurements of soil respiration is that the measurements require long-term collection using special equipment and are not typically collected in traditional forest inventory.

Kolari et al. (2004) found that management practices, such as clearcut harvesting had a large impact on the soil C balance, with clearcut sites soil C sources, while

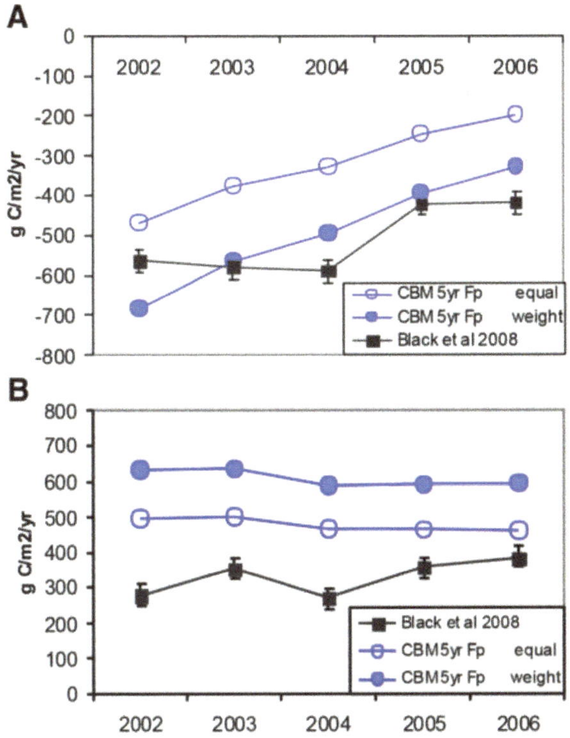

Figure 8 Yearly NEP estimates for CBM-CFS3 and EC flux tower (Black et al. 2008) for **(A)** the regenerating clearcut and **(B)** the near-rotation stand. CBM-CFS3 is shown as yearly mean of footprint cells with equal weighting (Fp equal) and weighting by NEP flux probability density distribution (Fp weight).

non-disturbed sites had more stable soil C balances. In this study, a trend in soil C stocks, with the highest soil C stocks at the oldest site, and the lowest at the recently disturbed site, was consistent with the trend by seral stage reported by Kolari et al. (2004). For the two similar-aged juvenile stands, since there was no flux tower installed at HDF90 and CBM-CFS3 model runs results were not available at HDF88, a direct comparison was not possible. In addition to the differences in measurement methods, the differences in ΣNEP between the two juvenile stands could also be partially attributed to site conditions. Notably, HDF88 had more detrital C stocks, likely leading to larger releases of C due to decomposition. This was reflected in the more negative ΔC_{D2} at HDF88 compared to HDF90.

For the forest inventory measurement of ΔC and its components ΔC_L, ΔC_D, and ΔC_S, differences in procedure, measurement error by field and lab crews, and sample design may introduce error into inventory measurements that may be larger than the magnitude of change in C stocks over short measurement intervals, such as the 4 year period used in this study. Overall, ΔC_{D1} and ΔC_S were much larger than reported in previous studies (e.g. Law et al. 2001), and was much larger than expected given the measured litterfall inputs and estimated decomposition rates. The NFI (2008) calls for 10-year remeasurement intervals, which may be appropriate for trees and detritus, but may be too short for ΔC_S. In addition, direct use of forest inventory data relies on subjective classification of decay class that can differ from one-field-crew to the next, so utilizing decay and transfer algorithms may reduce the potential effect of these classifications. Developing site-specific allometric equations is labour intensive, destructive, and costly, and therefore uncommon in most studies; however, use of existing relationships can lead to errors due to differences in tree architecture, and wood density. Further, the use of inappropriate allometric equations can be a significant source of error in forest productivity studies (Clark and Brown 2001). The NFI data compilation procedure provided regionally applicable allometric equations; however, the errors in this large pool of equations have not yet been estimated and therefore is a limitation.

The measured changes in ΔC_S were much larger than was expected for a four year period. For example, Law et al. (2001) estimated soil sequestration of 0.7 – 0.8 MgC ha^{-1} year^{-1} in a Douglas-fir stand and Gielen et al. (2013) found no significant change over an eight-year interval in a Scots pine forest. In other studies by Ehman et al. (2002), Kolari et al. (2004), Miller et al. (2004), Ohtsuka et al. (2007), and Granier et al. (2008) ΔC_S was assumed to be zero over the measurement interval. The large value for ΔC_S in this study may be explained by several factors related to measurement methodology. First, mineral soil bulk densities in the 0–15 cm layer were assumed not to change and thus the BD values measured in 2002 were also used in 2006. If the bulk density decreased then the C_S change would be overestimated. Second, at the regenerating clearcut and juvenile stands, an increase in fine root mass from 2002 to 2006 included in the <2 mm soil fraction could also have accounted for some of the increase in soil C. Third, samples in 2002 were taken by 10–12 cm diameter hole excavations, while samples in 2006 were collected using a 2 cm diameter soil corer, which, due to compression at the opening of the sample corer, may be biased to soil from shallower depths where C content is likely higher. Therefore, in this study, measurements of ΔC_S were deemed unreliable, due to the large positive changes in ΔC_S, and assumed to be negligible over the four year period and thus not used for subsequent calculations of site level ΔC and further comparisons. In future work, efforts to reduce any chances of variation in the way the samples are collected and analysed may result in more reliable measurements of C stock changes. For example, collecting and processing the samples using identical methods at nearly the same locations, or collecting a much larger number of samples to capture a wider range of spatial variation could reduce the effect of spatial variability on measurements of ΔC_S.

Several authors have applied footprint analysis in an effort to correct for non-homogenous environments, and horizontal advection, for example due to daily upslope or downslope winds, which may introduce bias into the eddy-covariance measurements (Baldocchi 2008). For example, Ehman et al. (2002) used footprints to calculate an estimate of sensor bias given using a vegetation index, and this gave confidence that the sensor measurements were representative of the inventory measurements. Gielen et al. (2013) used a more complex footprint model applied to the processing of NEP data, were fluxes originating outside of the area where inventory measurements were sampled were removed from the NEP estimates. In this study, footprint weighting by ecological attributes and footprint probability density was evaluated where it facilitated comparison between inventory-based measurements of ecosystem ΔC stocks and tower ΣNEP by accounting for fine scale spatial variability in forest structure. The effect of K-MSN classification and footprint weighting had an effect on the estimated site-level values; however, this effect was smaller than differences due to measurement methodology. Notably, applying footprint weighting to the estimation of CBM-CFS3 ΣNEP at the regenerating clearcut increased convergence with EC tower ΣNEP; however, footprint weighting did not have the same effect for ΔC_2. The largest differences observed among methods were seen at the regenerating clearcut, where the inventory approach indicated the site was a much lower source than the ΣNEP methods. A possible cause is that the inventory methods did not properly account for C losses from decomposition of coarse roots from stumps and large woody debris at recently harvested sites, which can be a substantial source of respiration (Janisch et al. 2005). Therefore, if an estimate of stump coarse root decomposition were included in ΔCD_2 it may have increased convergence with the ΣNEP estimates. Additionally, the original sample locations were designed to capture the range of conditions in near proximity of the tower (based on interpretation of aerial photographs), and the relationship was similar when scaled up to the site level. Future studies may seek to derive accurate footprint estimates prior to establishing sample plots to ensure they are representative of the broader site conditions. Another more costly alternative is to establish a much greater number of plots in a systematic basis around the tower, to ensure the spatial variation within the tower footprint is adequately captured.

This research highlights how different measurements and approaches for C accounting include different constraints, and as a result, computations using different data sources may not converge. Each type of measurement has "different strengths and weaknesses, but the combination of multiple measurements and modeling has the potential for refining estimates of C stocks and fluxes" (Law et al. 2004). For example, eddy-covariance data are temporally detailed, and provide information about daily and seasonal processes. However, the data are less spatially extensive and detailed than the forest inventory data, for example, only three sites were available in this study each covering an area with diverse vegetation within the tower footprint. Measurement error was estimated to be relatively low, on the order of 0.9 Mg C ha^{-1} year $^{-1}$ (Morgenstern et al. 2004). In contrast, forest inventory data are more spatially extensive and include detailed measurements of forest attributes. The main constraint for forest inventory data is that they are less temporally detailed (for example, measurements were available at only two points in time), and the error in these measurements may be greater than the expected rate of change over the measurement interval. The inventory plots in this study sampled a broad range of vegetation conditions and provided insight into the representativeness of flux-tower footprints. Estimates of error of 1.2 to 1.3 Mg C ha^{-1} were identified relating to spatial representativeness of the inventory plots within the tower footprints. Finally, C budget models are important for understanding landscape scale C processes. Since these models depend on numerous coefficients, algorithms and assumptions, it is valuable to compare model outputs with EC flux and inventory measurements to assess their performance and evaluate their behaviour. Considering all of these approaches together, the relative strengths and limitation of each approach can be taken into account in evaluating the suitability of each type of measurement at varying spatial and temporal scales.

Conclusions

The comparison of ΔC from inventory measurements with ΣNEP from CBM-CFS3 and EC flux-tower measurements demonstrated an agreement among the methods in trends across stand seral stage; however, due to differences in the measurement approaches, there was some divergence in the results. Convergence amongst methods was closest for the juvenile sites (HDF90 and HDF88), which were in transition from C source- to sink. At the most recently harvested site (HDF00), the EC flux ΣNEP and CBM CFS3 ΣNEP indicated that the site was a greater source of C over the time period than ΔC from inventory methods. At the near-rotation site the inventory ΔC_2 method, unweighted CBM CFS3 ΣNEP and EC flux ΣNEP also converged.

Each of the measurement methods had advantages and limitations. For example, while EC flux towers provide temporally rich data, they had limited spatial coverage. Obtaining ΔC using forest inventory measurements included several challenges such as the consistency of data collected over long time intervals, small changes in ΔC compared to uncertainty in the measurements, and insufficient measurement of processes such as decomposition of all detrital pools. Forest inventory data are available over broad areas; therefore methods that combine measurements with estimates of decomposition are needed to provide information over a regional scale. Footprint analysis using the spatial predictors from remote sensing and topography can give confidence that sample locations represent broader area site conditions. Considering the advantages and constraints of each type of measurement is helpful in choosing appropriate measurement methods at varying spatial and temporal scales.

Competing interests
The authors declare that they have no competing interests.

Authors' contributions
CJF set research objectives, devised and undertook analysis, and composed the manuscript. JAT set research objectives, directed the collection and synthesis of field data, summarized output from CBM-CFS3 runs, advised analysis, assisted with manuscript composition, and provided extensive editorial work on the manuscript. NCC advised on research objectives, advised analysis, and assisted with manuscript composition. BC advised the footprint analysis, and provided commentary on manuscript. TAB advised on research objectives, advised the analysis, and provide extensive review and commentary throughout the work. All authors read and approved the final manuscript.

Acknowledgements
We thank Bob Ferris, Frank Eichel, and Glenda Russo of the Canadian Forest Service and staff of B.A. Blackwell and Associates for their help processing and collecting National Forest-inventory-style ground plot data. We also thank François Gougeon, CFS, for determining canopy tree density values for the sites and to Graham Stinson, CFS, for the CBM-CFS3 output files used in Wang et al. (2011). We thank the UBC Land and Food Systems Biometeorology and Soil Physics Group, in particular Paul Jassal, Praveena Krishnan, Kai Morgenstern, and Elyn Humphreys for their work processing EC flux data, and Zoran Nesic, Dominic Lessard, Andrew Sauter, and Andrew Hum for their work running and maintaining the EC flux tower sites. We also thank Mark Johnson for his comments and suggestions. Funding for this study was provided by the Canadian Forest Service Pacific Forestry Centre Graduate Student Award, a CFCAS grant to the Canadian Carbon Program (CCP), and Natural Sciences and Engineering Research Council of Canada (NSERC) Discovery Grant to NCC. The LiDAR data were acquired by Benoit St-Onge of the University of Quebec at Montreal as part of an ongoing collaborative project with funds provided by NSERC and BIOCAP.

Author details
[1]Department of Forest Resources Management, University of British Columbia, 2424 Main Mall, Vancouver, BC V6T 1Z4, Canada. [2]Canadian Forest Service (Pacific Forestry Centre), Natural Resources Canada, 506 Burnside Road West, Victoria, BC V8Z 1M5, Canada. [3]Faculty of Land and Food Systems, University of British Columbia, 2357 Main Mall, Vancouver, BC V6T 1Z4, Canada. [4]Department of Biology, University of Victoria, PO Box 1700, Station CSC, Victoria, BC V8W 2Y2, Canada.

References
Babst F, Bouriaud O, Papale D, Gielen B, Janssens I, Nikinmaa E, Ibrom A, Wu J, Bernhofer C, Köstner B, Grünwald T, Seufert G, Ciais P, Frank D (2014) Above-ground woody carbon sequestration measured from tree rings is coherent with net ecosystem productivity at five eddy-covariance sites. New Phytol 201:1289–1303, doi: 10.1111/nph.12589

Bailey JD, Mayrsohn C, Doescher PS, St Pierre E, Tappeiner JC (1998) Understory vegetation in old and young Douglas-fir forests of western Oregon. For Ecol Manage 112:289–302

Baldocchi DD (2008) "Breathing" of the terrestrial biosphere: lessons learned from a global network of carbon dioxide flux measurement systems. Aust J Bot 56:1, doi:10.1071/BT07151

Barford CC, Wofsy SC, Goulden ML, Munger JW, Pyle EH, Urbanski SP, Hutyra L, Saleska SR, Fitzjarrald D, Moore K (2001) Factors controlling long- and short-term sequestration of atmospheric CO2 in a mid-latitude forest. Science 294:1688–1691, doi:10.1126/science.1062962

Black K, Bolger T, Davis P, Nieuwenhuis M, Reidy B, Saiz G, Tobin B, Osborne B (2005) Inventory and eddy covariance-based estimates of annual carbon sequestration in a Sitka spruce (Picea sitchensis (Bong.) Carr.) forest ecosystem. Eur J For Res 126:167–178, doi:10.1007/s10342-005-0092-4

Black TA, Fredeen A, Jassal R (2008) Carbon Sequestration in British Columbia's Forests and Management Options. University of Victoria, Victoria, BC, White paper. Pacific Institute for Climate Solutions

Brown JK (1971) A planar intersect method for sampling fuel volume and surface area. For Sci 17:96–102

Chen B, Chen JM, Mo G, Black TA, Worthy DEJ (2008) Comparison of regional carbon flux estimates from CO 2 concentration measurements and remote sensing based footprint integration. Global Biogeochem Cycles. doi:10.1029/2007GB003024

Chen BB, Black TA, Coops NC, Hilker T, Trofymow JA, Morgenstern K, Black AT (2009) Assessing tower flux footprint climatology and scaling between remotely sensed and eddy covariance measurements. Boundary-Layer Meteorol 130:137–167, doi:10.1007/s10546-008-9339-1

Clark D, Brown S (2001) Measuring net primary production in forests: concepts and field methods. Ecol Appl 11:356–370

Coops NC, Hilker T, Wulder MA, St-Onge B, Newnham GJ, Siggins A, Trofymow JA (2007) Estimating canopy structure of Douglas-fir forest stands from discrete-return LiDAR. Trees-Structure Funct 21:295–310. doi:10.1007/s00468-006-0119-6

Crookston NL, Finley AO (2008) yaImpute: an R package for kNN imputation. J Stat Softw 23(10):1–16

Curtis P, Hanson P, Bolstad P, Barford C, Randolph J, Schmid H, Wilson K (2002) Biometric and eddy-covariance based estimates of annual carbon storage in five eastern North American deciduous forests. Agric For Meteorol 113:3–19, doi:10.1016/S0168-1923(02)00099-0

Dixon RK, Solomon AM, Brown S, Houghton RA, Trexier MC, Wisniewski J (1994) Carbon pools and flux of global forest ecosystems. Science 263:185–190, doi:10.1126/science.263.5144.185

Ehman JL, Schmid HP, Grimmond CSB, Randolph JC, Hanson PJ, Wayson CA, Cropley FD (2002) An initial intercomparison of micrometeorological and ecological inventory estimates of carbon exchange in a mid-latitude deciduous forest. Glob Chang Biol 8:575–589, doi:10.1046/j.1365-2486.2002.00492.x

Ferster CJ, Trofymow JA, Coops NC, Chen B, Black TA, Gougeon FA (2011) Determination of ecosystem carbonstock distributions in the flux footprint of an eddy-covariance tower in a coastal forest in British Columbia. Can J For Res 41:1380–1393, doi:10.1139/X11-055

Gielen B, De Vos B, Campioli M, Neirynck J, Papale D, Verstraeten A, Ceulemans R, Janssens IA (2013) Biometric and eddy covariance-based assessment of decadal carbon sequestration of a temperate Scots pine forest. Agric For Meteorol 174:135–143, doi:10.1016/j.agrformet.2013.02.008

Gillis M, Omule A, Brierley T (2005) Monitoring Canada's forests: the national forest inventory. For Chron 81:214–221

Göckede M, Rebmann C, Foken T (2004) A combination of quality assessment tools for eddy covariance measurements with footprint modelling for the characterisation of complex sites. Agric For Meteorol 127:175–188, doi:10.1016/j.agrformet.2004.07.012

Gougeon F (1995) A crown-following approach to the automatic delineation of individual tree crowns in high spatial resolution aerial images. Can J Remote Sens 21:274–284

Gough C, Vogel C, Schmid H, Su H, Curtis P (2008) Multi-year convergence of biometric and meteorological estimates of forest carbon storage. Agric For Meteorol 148:158–170, doi:10.1016/j.agrformet.2007.08.004

Granier A, Ceschia E, Damesin C, Dufrene E, Epron D, Gross P, Lebaube S, Le Dantec V, Le Goff N, Lemoine D, Lucot E, Ottorini JM, Pontailler JY, Saugier B (2000) The carbon balance of a young Beech forest. Funct Ecol 14:312–325, doi:10.1046/j.1365-2435.2000.00434.x

Granier A, Bréda N, Longdoz B, Gross P, Ngao J (2008) Ten years of fluxes and stand growth in a young beech forest at Hesse, North-eastern France. Ann For Sci 65:704–704, doi:10.1051/forest

Grant RF, Black TA, Humphreys ER, Morgenstern K (2007) Changes in net ecosystem productivity with forest age following clearcutting of a coastal Douglas-fir forest: testing a mathematical model with eddy covariance measurements along a forest chronosequence. Tree Physiol 27:115–131

Green RN, Klinka K (1994) A Field Guide to Site Identification and Interpretation for the Vancouver Forest Region. Land Management Handbook Number 28. British Columbia Ministry of Forests, Research Program, Victoria, BC

Humphreys ER, Black TA, Morgenstern K, Cai T, Drewitt GB, Nesic Z, Trofymow JA (2006) Carbon dioxide fluxes in coastal Douglas-fir stands at different stages of development after clearcut harvesting. Agric For Meteorol 140:6–22, doi:10.1016/j.agrformet.2006.03.018

Janisch JE, Harmon ME, Chen H, Fasth B, Sexton J (2005) Decomposition of coarse woody debris originating by clearcutting of an old-growth conifer forest. Ecoscience 12:151–160, doi:10.2980/i1195-6860-12-2-151.1

Janzen HH (2005) Soil carbon: a measure of ecosystem response in a changing world? Can J Soil Sci 85:467–480

Jassal RS, Black TA, Trofymow JA, Roy R, Nesic Z (2010) Soil CO_2 and N_2O flux dynamics in a nitrogen-fertilized Pacific Northwest Douglas-fir stand. Geoderma 157:118–125, doi:10.1016/j.geoderma.2010.04.002

Jungen J (1985) Soils of Southern Vancouver Island. Victoria, BC

Kolari P, Jukka P, Rannik U, Hannu I, Pertti H, Berninger F (2004) Carbon balance of different aged Scots pine forests in Southern Finland. Glob Chang Biol 10:1106–1119, doi:10.1111/j.1365-2486.2004.00797.x

Kominami Y, Jomura M, Dannoura M, Goto Y, Tamai K, Miyama T, Kanazawa Y, Kaneko S, Okumura M, Misawa N, Hamada S, Sasaki T, Kimura H, Ohtani Y (2008) Biometric and eddy-covariance-based estimates of carbon balance for a warm-temperate mixed forest in Japan. Agric For Meteorol 148:723–737, doi:10.1016/j.agrformet.2008.01.017

Krishnan P, Black TA, Jassal RS, Chen B, Nesic Z (2009) Interannual variability of the carbon balance of three different-aged Douglas-fir stands in the Pacific Northwest. J Geophys Res 114:G04011, doi:10.1029/2008JG000912

Kurz W, Apps M, Banfield E, Stinson G (2002) Forest carbon accounting at the operational scale. For Chron 78:672–679

Kurz WA, Dymond CC, White TM, Stinson G, Shaw CH, Rampley GJ, Smyth C, Simpson BN, Neilson ET, Trofymow JA, Metsaranta J, Apps MJ (2009) CBM-CFS3: a model of carbon-dynamics in forestry and land-use change implementing IPCC standards. Ecol Modell 220:480–504, doi:10.1016/j.ecolmodel.2008.10.018

Lambert M-C, Ung C-H, Raulier F (2005) Canadian national tree aboveground biomass equations. Can J For Res 35:1996–2018, doi:10.1139/x05-112

Landsberg JJ, Waring RH (1997) A generalised model of forest productivity using simplified concepts of radiation-use efficiency, carbon balance and partitioning. For Ecol Manage 95:209–228, doi:10.1016/S0378-1127(97)00026-1

Law B, Thornton P, Irvine J (2001) Carbon storage and fluxes in ponderosa pine forests at different developmental stages. Glob Chang Biol 7:755–777

Law BE, Turner D, Campbell J, Sun OJ, Van Tuyl S, Ritts WD, Cohen WB (2004) Disturbance and climate effects on carbon stocks and fluxes across Western Oregon USA. Glob Chang Biol 10:1429–1444, doi:10.1111/j.1365-2486.2004.00822.x

Leclerc M, Thurtell G (1990) Footprint prediction of scalar fluxes using a Markovian analysis. Boundary-Layer Meteorol 52(3):247–258

Li Z, Kurz WA, Apps MJ, Beukema SJ (2003) Belowground biomass dynamics in the Carbon Budget Model of the Canadian Forest Sector: recent improvements and implications for the estimation of NPP and NEP. Can J For Res 33:126–136, doi:10.1139/x02-165

Mahalanobis P (1936) On the generalized distance in statistics. Proc Natl Inst Sci 12:49–55

Matthews G (1993) The carbon content of trees. Forestry Commission Technical Paper, vol. 4. Forestry Commission, London

Miller SD, Goulden ML, Menton MC, da Rocha HR, de Freitas HC, Figueira AMES, de Sousa CA D (2004) Biometric and micrometeorological measurements of tropical forest carbon balance. Ecol Appl 14:114–126, doi:10.1890/02-6005

Morgenstern K, Andrew Black T, Humphreys ER, Griffis TJ, Drewitt GB, Cai T, Nesic Z, Spittlehouse DL, Livingston NJ (2004) Sensitivity and uncertainty of the carbon balance of a Pacific Northwest Douglas-fir forest during an El Niño/La Niña cycle. Agric For Meteorol 123:201–219, doi:10.1016/j.agrformet.2003.12.003

NFI (2008) Canada's National Forest Inventory - Ground Sampling Guidelines. Natural Resources Canada, Canadian Forest Service, Victoria, BC, [Electronic Resource]. Available: https://nfi.nfis.org/documentation/ground_plot/Gp_guidelines_v5.0.pdf. Accessed 13 April 2015

Ohtsuka T, Mo W, Satomura T, Inatomi M, Koizumi H (2007) Biometric based carbon flux measurements and net ecosystem production (NEP) in a temperate deciduous broad-leaved forest beneath a flux tower. Ecosystems 10:324–334, doi:10.1007/s10021-007-9017-z

R Development Core Team (2011) R: a language and environment for statistical computing. R Foundation for Statistical Computing, Vienna, Austria, URL http://www.R-project.org/

Richardson AD, Hollinger DY (2007) A method to estimate the additional uncertainty in gap-filled NEE resulting from long gaps in the CO2 flux record. Agric For Meteorol 147:199–208, doi:10.1016/j.agrformet.2007.06.004

Roberts D, Cooper S (1989) Concepts and techniques of vegetation mapping. In: Ferguson D, Morgan P, Johnson FD (eds) Land classifications based on vegetation: applications for resource management. USDA Forest Service General Technical Report: INT-257; USDA Forest Service Intermountain Forest and Range Experiment Station, Ogden, UT, USA, pp 90–96

Schmid H (2002) Footprint modeling for vegetation atmosphere exchange studies: a review and perspective. Agric For Meteorol 113:159–183, doi:10.1016/S0168-1923(02)00107-7

Schmid HP, Lloyd CR (1999) Spatial representativeness and the location bias of flux footprints over inhomogeneous areas. Agric For Meteorol 93:195–209, doi:10.1016/S0168-1923(98)00119-1

Schulze ED, Hogberg P, Van Oene H, Persson T, Harrison A, Read D, Kjoller A, Matteucci G (2000) Interactions Between the Carbon and Nitrogen Cycles and the Role of Biodiversity: A Synopsis of a Study Along a North-south Transect Through Europe. In: Schulze ED (ed) Carbon and Nitrogen Cycling in European Forest Ecosystems. Springer, Heidelberg, pp 468–491

Smyth C, Trofymow JA, Kurz W (2010) Decreasing Uncertainty in CBM-CFS3 Estimates of Forest Soil Carbon Sources and Sinks Through use of Long-Term Data from the Canadian Intersite Decomposition Experiment. Information Report BC-X-422. Natural Resources Canada, Canadian Forest Service, Pacific Forestry Centre, Victoria, BC

Stage AR, Crookston NL (2007) Partitioning error components for accuracy-assessment of near-neighbor methods of imputation. For Sci 53:62–72

Ter-Mikaelian M (1997) Biomass equations for sixty-five North American tree species. For Ecol Manage 97:1–24, doi:10.1016/S0378-1127(97)00019-4

Theede AD (2007) Biometric and Eddy-Covariance Estimates of Ecosystem Carbon Storage at Two Boreal Forest Stands in Saskatchewan: 1994–2004. University of Saskatchewan, MSc thesis

Trofymow JA, Stinson G, Kurz WA (2008) Derivation of a spatially explicit 86-year retrospective carbon budget for a landscape undergoing conversion from old-growth to managed forests on Vancouver Island, BC. For Ecol Manage 256:1677–1691, doi:10.1016/j.foreco.2008.02.056

Wang Z, Grant RF, Arain MA, Chen BN, Coops N, Hember R, Kurz WA, Price DT, Stinson G, Trofymow JA, Yeluripati J, Chen Z (2011) Evaluating weather effects on interannual variation in net ecosystem productivity of a coastal temperate forest landscape: a model intercomparison. Ecol Modell 222:3236–3249, doi:10.1016/j.ecolmodel.2011.06.005

Yashiro Y, Lee N-YM, Ohtsuka T, Shizu Y, Saitoh TM, Koizumi H (2010) Biometric-based estimation of net ecosystem production in a mature Japanese cedar (Cryptomeria japonica) plantation beneath a flux tower. J Plant Res 123:463–472, doi:10.1007/s10265-010-0323-8

8

Inventory of aspen trees in spruce dominated stands in conservation area

Matti Maltamo[1*], Annukka Pesonen[2], Lauri Korhonen[1], Jari Kouki[1], Mikko Vehmas[3] and Kalle Eerikäinen[4]

Abstract

Background: The occurrence of aspen trees increases the conservation value of mature conifer dominated forests. Aspens typically occur as scattered individuals among major tree species, and therefore the inventory of aspens is challenging.

Methods: We characterized aspen populations in a boreal nature reserve using diameter distribution, spatial pattern, and forest attributes: volume, number of aspens, number of large aspen stems and basal area median diameter. The data were collected from three separate forest stands in Koli National Park, eastern Finland. At each site, we measured breast height diameter and coordinates of each aspen. The comparison of inventory methods of aspens within the three stands was based on simulations with mapped field data. We mimicked stand level inventory by locating varying numbers of fixed area circular plots both systematically and randomly within the stands. Additionally, we also tested if the use of airborne laser scanning (ALS) data as auxiliary information would improve the accuracy of the stand level inventory by applying the probability proportional to size sampling to assist the selection of field plot locations.

Results: The results showed that aspens were always clustered, and the diameter distributions indicated different stand structures in the three investigated forest stands. The reliability of the volume and number of large aspen trees varied from relative root mean square error figures above 50% with fewer sample plots (5–10) to values of 25%–50% with 10 or more sample plots. Stand level inventory estimates were also able to detect spatial pattern and the shape of the diameter distribution. In addition, ALS-based auxiliary information could be useful in guiding the inventories, but caution should be used when applying the ALS-supported inventory technique.

Conclusions: This study characterized European aspen populations for the purposes of monitoring and management of boreal conservation areas. Our results suggest that if the number of sample plots is adequate, i.e. 10 or more stand level inventory will provide accurate enough forest attributes estimates in conservation areas (minimum accuracy requirement of RMSE% is 20%–50%). Even for the more ecologically valuable attributes, such as diameter distribution, spatial pattern and large aspens, the estimates are acceptable for conservation purposes.

Keywords: Diameter distribution; Historical continuity; Inventory; LiDAR; *Populus tremula* L; Simulation; Spatial arrangement; Stand characteristics

Background

One of the most interesting minor tree species in boreal forests of northern Europe is the European aspen (*Populus tremula* L.). The importance of aspen is closely related to its biodiversity values because it hosts particularly diverse groups of associated species, many of which are threatened in Fennoscandia (Esseen et al. 1992; Kouki et al. 2004). In addition, large-sized aspens have generally disappeared from managed forests because they have low economic value and are intermediate hosts of the pine rust fungus (*Melampsora pinitorqua* [Braun] Rostr.) that causes serious damage to young pine stands (Kurkela 1973; Heliövaara and Väisänen 1984).

Although aspen is a typical species in post-disturbance, early successional stages, recent studies have indicated that aspen can maintain its populations in natural old-growth coniferous forests for up to several hundred years,

* Correspondence: matti.maltamo@uef.fi
[1]University of Eastern Finland, School of Forest Science, P.O. Box 111, FI-80101 Joensuu, Finland
Full list of author information is available at the end of the article

even though they may slowly decline in abundance (Lilja et al. 2006; Vehmas et al. 2009b). In particular, old and large-sized aspen trees, which are most valuable for biodiversity, are mostly found in mature and old-growth mixed forests where they grow in small groups or as scattered individuals (Tikka 1954; Syrjänen et al. 1994). Because the spatiotemporal continuity of ecologically important characteristics is regarded as important for conservation purposes (Stokland et al. 2002; Kouki et al. 2004) the ability to inventory and monitor aspen trees is essential for the management of conservation areas.

Several variables can be used to describe aspens in stand-level forest inventories. First, the existence of aspen can be recorded. Secondly, detail on the amount and size of aspen trees is of interest; in stand-level inventories, they are usually described using basal area, mean diameter, and mean height (Koivuniemi and Korhonen 2006). Thirdly, from a biodiversity point of a view, information on size variation and spatial distribution is highly relevant (Kouki et al. 2004). The determination of spatial distribution requires that the trees are individually mapped, which is usually practically impossible to conduct in field surveys, except for research purposes. Correspondingly, the tree height distributions are usually not assessed due to the laborious field measurements, whereas diameter distributions can be obtained. With this information, some indicators of the naturalness of the forest structure, such as the shape of the diameter distribution, of the given aspen population can be assessed. Furthermore, it is easy to define the proportion of large aspens when their diameters at breast height (*dbh*) are, for instance, greater than 25 cm.

The problem related to the assessment of aspen at stand-level inventories is that the low density of the aspen trees results in high estimates of sampling errors. It is also possible that aspens are not separated from other economically less-important deciduous species in tree stock descriptions for forest management. In validation studies of the inventories by compartments, the root mean square errors (RMSEs) obtained for the total growing stock volume have ranged from 15% – 38% (Poso 1983; Haara and Korhonen 2004). However, species-specific errors are considerably higher, being 29%, 43% and 65% for Scots pine (*Pinus sylvestris* L.), Norway spruce (*Picea abies* L.), and the group consisting of silver birch (*Betula pendula* Roth) and downy birch (*B. pubescens* Ehrh), respectively (Haara and Korhonen 2004). While the errors are usually acceptable for the dominant coniferous species, minor deciduous tree species are described too inaccurately for many purposes. For European aspen, the relative RMSE can be several hundreds of percent (Arto Haara, personal comm.).

Airborne laser scanning (ALS) -based technology has been successfully applied to stand-level inventories during recent years (Næsset 2007; Maltamo and Packalen 2014). Forest characteristics are usually estimated with 100% coverage for the inventory area by utilising the area-based approach (ABA), i.e. statistical relationships between forest attributes and ALS metrics at the plot level. The first applications estimated forest characteristics as a whole, but the inventory system by Packalén and Maltamo (2007), which also utilizes aerial photographs and relies on non-parametric imputation, was the first species-specific estimator for forest attributes. However, deciduous tree species are usually pooled into one single group (Packalén and Maltamo 2007). Thus, the previously developed ALS based inventory approaches are not appropriate for providing species-specific information on aspen. In studies by Breidenbach et al. (2010) and Pippuri et al. (2013) species-specific ALS inventory has also been applied to identify aspens, but the RMSE values have been over 100%.

ALS can also provide information about individual trees, which can be aggregated to the stand level. In a study by Säynäjoki et al. (2008), single aspen trees were detected from dense ALS data. The inventory system was, however, rather complex, including, for instance, visual interpretations by aerial images to separate coniferous trees from deciduous trees. As a result, the classification accuracy of large ($dbh > 25$ cm) aspen trees was 78.6%. In addition, discrimination of aspen can be difficult because the ALS intensity metrics that are important in species detection overlap with spruce and birch (Ørka HO et al. 2007; Korpela et al. 2010).

Despite the recent advances in tree-level identification, it is challenging to obtain stand-level information on aspen in remote sensing-based forest inventories. Correspondingly, the accuracy estimates have been rather low for aspen, or it has been completely ignored in traditional field inventories. Since the trend in forest inventories is toward remote sensing applications, rare and scattered tree species, such as aspen, could become neglected in inventories. On the other hand, there is an increasing need to have forest inventory information on aspen, especially in conservation areas where the occurrence and long-term persistence of scattered aspen trees may be crucial for many conservation-dependent species.

The goal of this study was to characterise aspen populations in a boreal nature reserve. The study data are based on mapped individual aspens in three separate spruce dominated forest stands. In this unique data set the aspen populations have developed without the effects of active forest silviculture during recent decades. We characterised aspen using diameter distribution, spatial pattern of trees and forest attributes volume (V, $m^3 \cdot ha^{-1}$), number of stems (N, ha^{-1}), number of stems of large aspens ($N_{dbh > 25\,cm}$, ha^{-1}) and basal area median diameter (D_gM, cm). Furthermore, we applied stand level inventory simulations to examine the accuracy of the estimates of these characteristics. Finally, probability proportional to size (PPS) sampling using ALS metrics as auxiliary information was evaluated as a method to improve inventory estimates.

Methods

Study area and field measurements

The study area was located in Koli National Park (NP) in eastern Finland (29°50′E, 63°5′N). The area is characterised as a highly variable boreal landscape, where the altitude varies from 94 – 347 m above sea level (Lyytikäinen 1991; Kärkkäinen 1994). The area lies in the transitional area between the southern and middle boreal vegetation zones (Kalliola 1973). Most forests in the area are dominated by Norway spruce (*Picea abies* L. Karst.) and Scots pine (*Pinus sylvestris* L.) with a highly variable admixture of silver birch, downy birch, European aspen, and grey alder (*Alnus incana* [L.] Moench) (Lyytikäinen 1991; Grönlund and Hakalisto 1998).

In a study by Vehmas et al. (2009b), the historical continuity of aspen was studied based on inventory registers, and some areas where large aspens have survived from 1910 were found within the current Koli NP. Three of the largest of these stands were selected for this study (Figures 1 and 2). Other stands were very small sized, included only a few aspens or had highly irregular shape. The total area of forest stand 1 was 8.05 ha, whereas stands 2 and 3 covered 5.96 and 12.93 ha of forest, respectively (Table 1). Within the three stands, both *dbh* and GPS position were recorded for all living aspen trees having a *dbh* larger than 5 cm in 2006. Stem volumes of

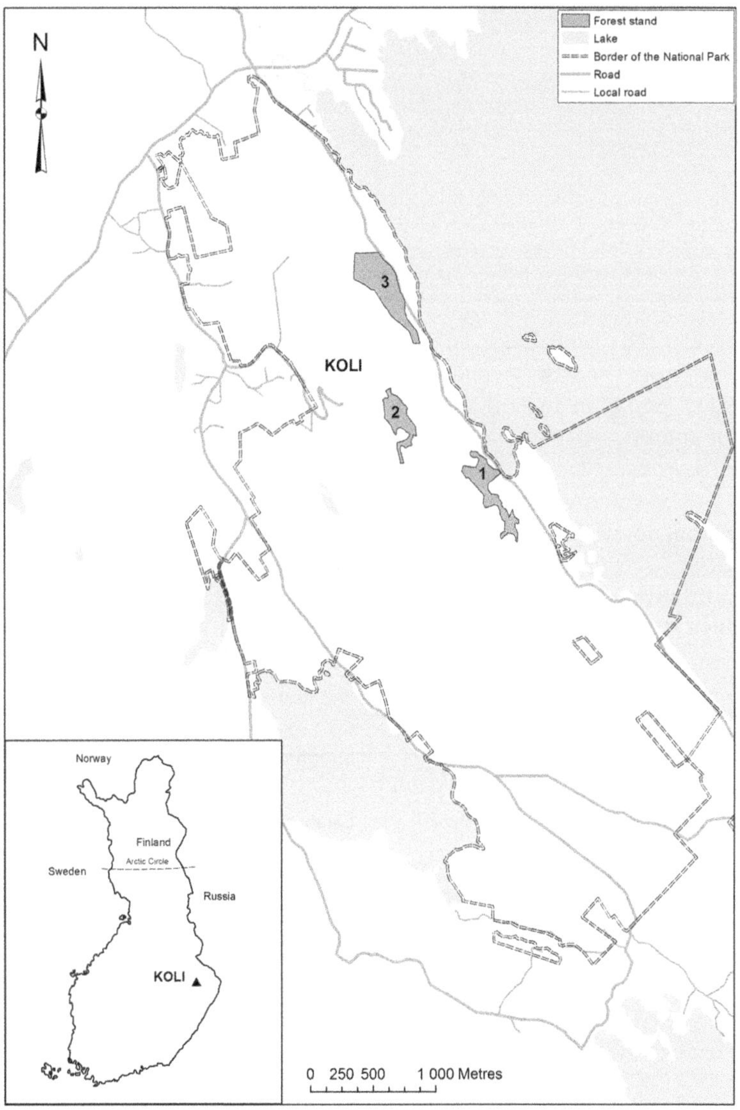

Figure 1 The location of the study area.

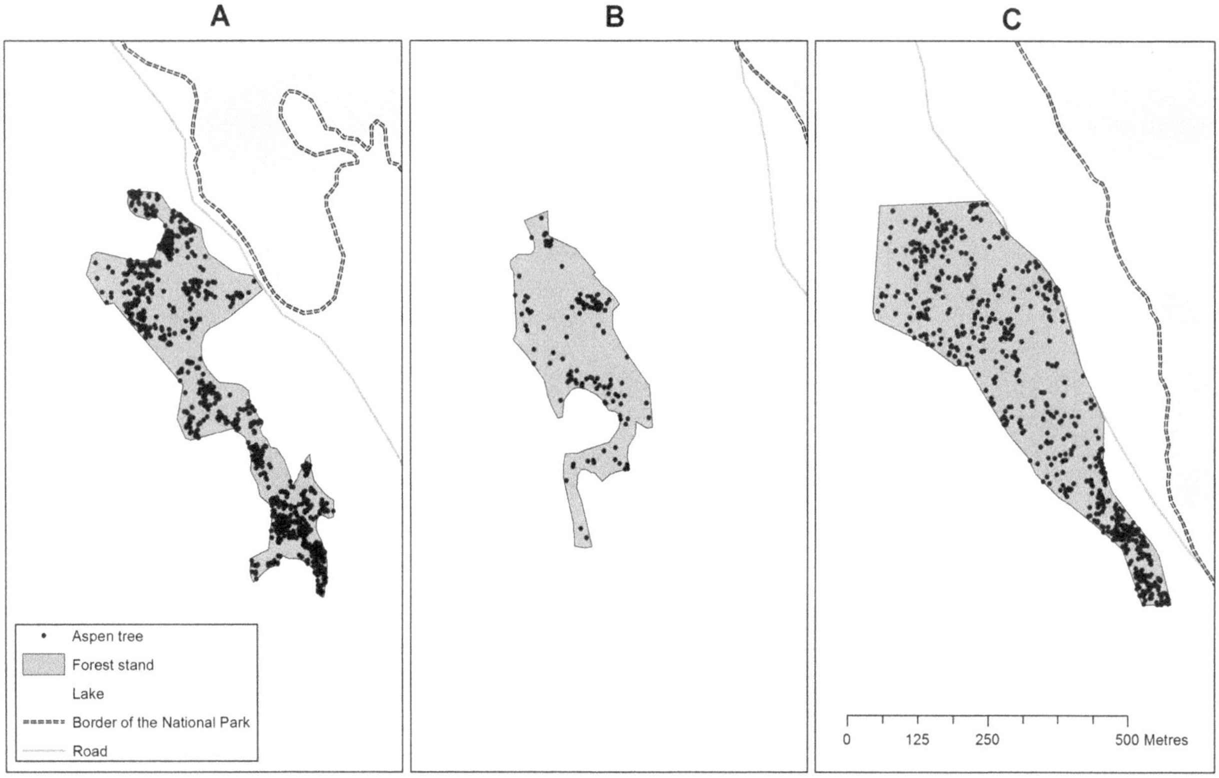

Figure 2 The study stands (**A**= stand 1, **B** = stand 2, **C** =stand 3) with aspen tree locations.

the tallied aspens were calculated using Laasasenaho's (1982) volume function for Scots pine, since published equations were not available for European aspen (Kinnunen et al. 2007). The choice of volume model was based on expert opinion. Sum characteristics were converted to per-hectare levels, and D_gM was calculated for the three stands (Table 1). In addition, total stand volume and basal area were taken from the existing stand register data and updated into aspen measurement date (see Vehmas et al. 2009b).

Table 1 Areas and forest attributes for European aspen and stand totals for the three study stands

Attribute	Stand		
	1	2	3
Area (ha)	8.05	5.96	12.94
D_gM (cm)	22.4	39.1	49.7
G (m^2·ha^{-1})	3.8	1.9	6.0
V (m^3·ha^{-1})	22.2	12.4	35.7
N (ha^{-1})	148.6	21.3	54.1
$N_{dbh>25cm}$ (ha^{-1})	16.0	15.3	32.9
G_{total} (m^2·ha^{-1})	23.2	34.3	36.8
V_{total} (m^3·ha^{-1})	168.9	363.1	422.2

Area denotes the size of the stand, D_gM, cm denotes basal area median diameter of aspen, G, m^2·ha^{-1} denotes basal area of aspen, V, m^3·ha^{-1} denotes volume of aspen, N, ha^{-1} denotes number of stems of aspen, $N_{dbh>25cm}$ ha^{-1} denotes number of stems of large aspens, G_{total}, m^2·ha^{-1} denotes basal area of the total growing stock and V_{total}, m^3·ha^{-1} denotes volume of the total growing stock.

Additionally, 15 rectangular sample plots located in Koli NP that did not overlap with the three stands previously described were used to find the best-correlating ALS metric with aspen volume. These data were used in PPS sampling. These plots were originally established to examine single-tree detection of aspen from remote sensing data and included at least one aspen tree (Säynäjoki et al. 2008). The *dbh* was measured and stem volumes calculated for all trees. More detailed information of the data obtained from the 15 sample plots can be found in Säynäjoki et al. (2008).

Laser data

The geo-referenced ALS point cloud data from Koli NP were collected on 13 July 2005 using an Optech ALTM 3100 scanner operating at a mean altitude of 900 m above ground level (a.g.l), which resulted in a nominal sampling density of ca. 4 measurements·m^{-2}. Both the first and last pulse data were recorded, and the last pulse data were employed to generate a digital terrain model (DTM) by the method explained in Axelsson (2000) using a grid cell size of 1 m. Above ground heights (i.e., canopy heights) for the laser points were obtained by subtracting the DTM at the corresponding location. In this study, the pulse data obtained with the ALS sensor was reclassified to "first echo" or "last echo". It is worth noting here that the original single echoes were duplicated to both first and last echo classes, whereas the intermediate echoes were completely ignored. For more details on the original ALS data, see Vehmas et al. (2009a).

The height distribution of the first and last pulse canopy height hits was used to calculate plot-wise percentiles for 0, 1, 5, 10, 20, ..., 90, 95, 99, and 100% heights (h_0, h_1, ..., h_{100}) (Næsset 2004), and cumulative proportional crown densities (p_0, p_1, ..., p_{100}) were calculated for the respective quantiles. The height distributions contained only those laser points that were classified as above-ground hits; a threshold value of 0.1 m was used. The h_5, for example, denotes the height at which the accumulation of laser hit heights in the vegetation was 5%, and, correspondingly, p_5 denotes the proportion of laser hits that accumulated at the 5% height. In addition, the following variables were calculated by sample plots: the laser pulse intensities accumulating in percentiles (i_{10}, i_{30},..., i_{90}), the average intensity value of above-ground hits, the proportion of ground hits versus canopy hits using a threshold value of 0.1 m (*veg*), and the average height (h_{mean}) and standard deviation of the above-ground hits (h_{sd}). The intensity values were used as outputted by the sensor without calibration. All metrics were calculated separately for the first and the last pulse data.

Stand level inventory

The methods for aspen inventory were studied based on simulations using field data from the three stands where all aspens were mapped. We simulated stand level inventory by placing circular plots of size 400 m^2 (radius 11.28 m) into the stands both systematically and randomly. The size of the plot was chosen to correspond to the grid cell size in PPS sampling (see methodology below). Five, ten, fifteen, or twenty plots were located in each study stand for different sampling intensities. All sampling alternatives were repeated 2500 times. In the simulations, plots were only included if the centre point of the plot was within the study stand. For plots located at the edge of the stand, an edge correction was applied by multiplying the attribute value of an edge plot by its expansion factor (Beers 1966):

$$\text{Attribute} = \text{Attribute}_{\text{Edge plot}} \times \frac{\text{Plot size}}{\text{Edge plot size}} \qquad (1)$$

where, attribute is attribute value after correction, attribute$_{\text{Edge plot}}$ is attribute value of the edge plot, plot size is size of the sample plot, i.e., 400 m^2, and edge plot size is the size of the edge plot within the stand.

This correction was made for sum attributes *V*, *N*, and $N_{dbh>25cm}$ but not for D_gM. This edge correction is slightly biased but leads to considerably more accurate results than without applying any correction (Schreuder et al. 1993). Finally, the estimates of forest attributes were calculated as sample means for each sample.

Furthermore, we also tested if the use of ALS data as auxiliary information would improve the accuracy of the stand level inventory by applying PPS sampling. The basic idea of this approach is to use the ALS metric to guide the selection of field plot locations (Pesonen et al. 2010a, b). We applied the same number of sample plots as in the case of systematic and random sampling, but sampling probabilities varied according to ALS information. This was done to choose the most promising plot locations for aspen plots. First, probability layers were produced, i.e., the auxiliary data values were directly calculated for the whole stand that was divided into a grid of 20 m × 20 m sample units. When applying PPS sampling, the sample units were square and are referred to as grid cells. ALS based auxiliary data values were calculated for each sample unit ($i = 1,..., N_{grid}$, where N_{grid} is the total number of sample units in a stand), and the probabilities of each unit *i* being selected were determined. The selection probabilities for the sample units (p_i) were calculated by dividing the auxiliary data value x_i for the sample unit *i* by the sum of the auxiliary data values over the whole area of the probability layer ($p_i = x_i/\Sigma x_i$). These selection probabilities were finally utilised in sampling 5, 10, 15 or 20 sample units. The calculations were repeated 2500 times.

Shape of the diameter distribution

In stands 1 and 3 the shape of the diameter distribution estimated using the simulated fixed-radius, plot-based inventory approach was compared with the actual empirical distribution according to the developed rules. The unimodal form of diameter distribution (stand 2) was not considered. In the comparison of measured and estimated diameter distributions, the goal was to examine if the sampled distributions followed the underlying actual size distribution of aspen. The sampled diameter distributions were determined in 5- (bimodal stand) or 10-cm (descending stand) diameter classes within the range from 10 – 95 cm (See Figure 3 for actual distributions). For descending diameter distributions, the following rule was applied:

Number of stems in 10–20-cm *dbh* class > number of stems in 20–30-cm *dbh* class > number of stems in 30–40-cm *dbh* class.

If this rule was fulfilled by the estimate, it was classified as a realistic estimate for the underlying empirical

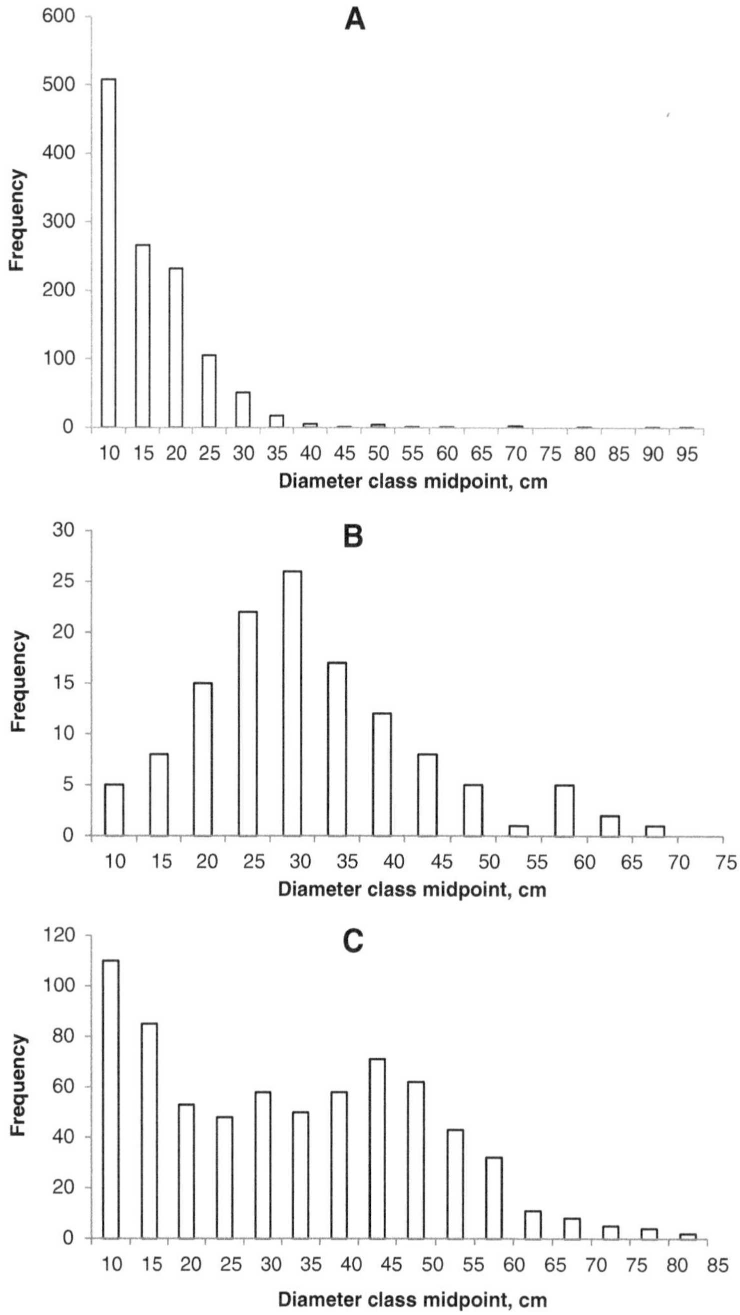

Figure 3 Diameter distribution of stands 1–3 (**A** = stand 1, **B**= stand 2, **C** =stand 3).

distribution. Correspondingly, in the case of bimodal distribution, the following rule was applied:

The first mode in the distribution is within *dbh* classes from 10 – 20 cm, and the second mode in the distribution is after the 25–30-cm *dbh* class.

Spatial pattern of aspens

The spatial pattern of the aspens within the three study stands was determined by applying Ripley's K(t) function (Ripley 1981). It describes the expected number of trees at distance t from a randomly selected tree. If the value of the function is larger than what would be expected based on random spacing, the spatial pattern is clustered; if smaller, it is systematic. We applied the library *spatstat* (Baddeley and Turner 2005) in statistical software R to calculate the K(t) values for each of the three stands. Isotropic correction was applied to minimize edge effects in the calculation (Ripley 1988).

The spatial patterns derived for the entire stands were compared with estimates obtained from the simulated samples. Therefore, a Fisher index (I) was calculated for each stand-wise simulation obtained using the following equation:

$$I = \frac{s_n^2}{\bar{n}}, \tag{2}$$

where s_n^2 is the variance of the plot-wise numbers of aspens in the sample of 20 plots and $\bar{n}$ is the mean of the plot-wise numbers of aspens. The I values greater than 1 indicate clustered spatial patterns.

Reliability characteristics

The simulation results were validated in term of relative RMSE.

$$RMSE\% = \frac{100x\sqrt{\frac{\sum_{i=1}^{N}\left(y-\frac{\sum_{i=1}^{n}\hat{y}_i}{n}\right)^2}{N}}}{y} \tag{3}$$

and bias

$$bias\% = \frac{100x\frac{\sum_{i=1}^{N}\left(y-\frac{\sum_{i=1}^{n}\hat{y}_i}{n}\right)}{N}}{y} \tag{4}$$

where N is the number of simulations, y is the observed value for the stand, $\hat{y}_i$ is the predicted value for sample plot simulation i, and n is number of sample plots in one sample.

Finally, in the case of the ALS-guided inventory, the relative improvement in volume estimate compared to selecting sample units of 20 m × 20 m with equal probabilities was calculated.

Results

Reliability figures for attributes of simulated stand level inventories

In general, the results are more accurate by means of RMSE% when the number of sample plots increases (Table 2). An exception is stand 3 with systematic placement of plots where accuracy decreased in the case of 20 plots. This is related to the shape of the stand and, thus, to the decreased possibilities to locate the systematic sample plot network to the narrow, densely stocked southern part of the stand in simulations. In general, the results are also slightly more accurate for systematic than random plot locations especially in the case of stands 1 and 2. The biases are in most cases below 2% and there are only a few cases where the values are over 5%.

In the case of RMSE% of V, which is usually regarded as the most important stand attribute, the figures are rather high with smaller number of sample plots and still remain approximately at the level of 25%–40% even with 20 sample plots (Table 2). In stand 2 the RMSE% values were larger for V and also for N compared to stands 1 and 3. This outcome may be related to the smaller quantities of aspen in stand 2 (see Table 1). From the ecological point of a view, $N_{dbh>25\text{cm}}$ is the most important forest attribute. Especially in stand 1, but also in stand 2, most of the aspens had smaller *dbh* values, less than 25 cm (Table 1, Figure 3) and correspondingly the RMSE% figures are high. On the other hand, the RMSE% figures are lower in stand 3 where the diameter distribution (Figure 3C.) shows that a remarkable proportion of aspens that reside in the group of trees is in the *dbh*-class larger than 25 cm. Finally, in general the results are most accurate for D_gM.

In the case of PPS sampling, the chosen auxiliary information metric from ALS was h^2_{mean}, which is based on the correlation estimate (0.76) between aspen V and this ALS metric in 15 sample plots of Koli NP. Corresponding correlations between this ALS metric and grid cell level values in the study area were also calculated, and the effect of the PPS sampling on the reliability of V estimates in general is presented in Table 3. As shown, the correlation was close to zero in stand 2 and the effect of PPS sampling is negative in this case. Regarding the two other stands the correlations between ALS metric and volume were greater than 0.3, and the improvements in volume estimates were more than 10% and 3%, respectively. The minor improvement in stand 3 may be related to the existence of very large aspens.

Shape of the diameter distribution estimate

The shape of the diameter distribution of aspen was unimodal and skewed to the right in stand 2, descending in stand 1, and bimodal in stand 3 (Figure 3A–C). The shape of sample plot based diameter distribution estimates obtained from the simulations was examined in stands 1 and 3, including ecologically interesting descending and bimodal distributions, respectively. Examination was implemented by classifying the diameter distribution estimate of each simulation according to the rules presented in the methods. In stand 1 the proportion of fixed-radius plot estimates, which correctly classified the descending structure, ranged from 50% – 80% for 5 – 20 plots (Table 4). This was the case both for systematically and randomly located plots. For stand 3 with a bimodal structure, the proportion of correctly classified plots with different number of sample plots corresponded to those of stand 1, but the success rates were lower.

Table 2 Relative RMSE and bias (in brackets) values of the forest attributes in three study stands

Stand	Sampling rate (%)	Sampling alternative	Number of plots	RMSE (%) (bias %)			
				V	*N*	D_gM	$N_{dbh>25cm}$
1	2.48	Random	5	62.0 (−3.4)	64.3 (−5.0)	56.9 (−10.6)	81.8 (−2.4)
	4.97	Random	10	43.8 (−3.3)	45.6 (−2.9)	35.2 (−6.0)	58.2 (−3.5)
	7.45	Random	15	35.2 (0.6)	36.2 (0.2)	23.5 (−3.6)	47.7 (−0.5)
	9.94	Random	20	30.5 (−1.4)	31.0 (−1.2)	15.5 (−2.1))	41.8 (−2.1)
		Systematic	5	50.8 (−3.0)	56.0 (−4.4)	48.2 (−6.8)	74.3 (−1.2)
		Systematic	10	34.5 (−0.1)	40.1 (−1.4)	35.8 (−6.1)	48.5 (0.1)
		Systematic	15	25.9 (−2.3)	21.5 (−1.5)	17.8 (−1.8)	49.7 (−4.2)
		Systematic	20	24.5 (−1.8)	25.5 (−1.2)	13.5 (−2.5)	38.5 (−2.8)
2	3.36	Random	5	85.7 (−0.2)	83.8 (0)	29.4 (1.3)	87.1 (−0.8)
	6.72	Random	10	59.8 (1.2)	59.7 (0.6)	24.5 (−1.4)	60.6 (0.2)
	10.08	Random	15	50.0 (−0.2)	49.2 (0)	21.1 (−1.3)	50.1 (−0.2)
	13.45	Random	20	41.8 (0.7)	42.7 (−0.6)	18.7 (−0.4)	42.8 (−0.1)
		Systematic	5	84.4 (−3.8)	98.2 (−8.2)	28.6 (0.9)	91.9 (−5.9)
		Systematic	10	53.2 (0)	47.6 (−2.4)	23.6 (−0.6)	52.2 (−1.4)
		Systematic	15	37.6 (1.4)	38.3 (−0.5)	18.5 (−0.3)	42.2 (0.5)
		Systematic	20	30.6 (0.1)	31.1 (−0.6)	16.9 (−0.2)	34.6 (−0.3)
3	1.55	Random	5	51.1 (0.2)	64.5 (−0.5)	21.0 (0.5)	52.8 (−0.1)
	3.09	Random	10	36.0 (−0.2)	46.3 (−1.8)	13.1 (0.5)	37.7 (−0.7)
	4.64	Random	15	29.0 (−0.5)	37.3 (−1.0)	10.1 (−0.1)	30.1 (−0.5)
	6.18	Random	20	25.2 (0.1)	31.7 (−0.8)	8.5 (0.1)	25.6 (0.2)
		Systematic	5	53.6 (−1.2)	58.4 (2.9)	18.5 (0.5)	52.9 (−0.5)
		Systematic	10	33.6 (−0.4)	35.5 (−0.7)	12.8 (−1.6)	36.4 (−0.6)
		Systematic	15	26.7 (0)	29.8 (−0.7)	7.7 (−0.2)	24.2 (0)
		Systematic	20	33.8 (0.1)	35.7 (0.4)	7.2 (−0.5)	31.9 (−0.2)

Spatial pattern of trees

The analysis based on mapped aspen data with Ripleys K-function showed that in all three stands the aspens are clustered, because the expected value of other trees close to each tree is larger than a Poisson distribution would suggest (Figure 4A–C). Correspondingly, the analysis based on sampling simulations and Fisher's index showed that in each case the average value showed that spatial pattern was clustered (Table 5). Also the proportion of simulations showing clustered spatial pattern was always over 50%, even with just five sample plots.

Table 3 Correlation and the improvement in the RMSE of *V* (%) due to the use of ALS auxiliary information in PPS sampling

Statistical variable	Stand		
	1	2	3
Correlation	0.35	0.04	0.33
Improvement in the RMSE of *V* (%)	10.6	−7.0	3.2

Discussion

This study considered stand level aspen populations in a boreal nature reserve. The analysis was based on diameter distribution, spatial pattern of aspen trees and reliability figures of forest attribute estimates of stand level inventory. Our unique data included mapped aspen trees

Table 4 Proportion (%) of correctly classified diameter distribution types in 2500 simulations

Stand	Number of plots	Inventory method	
		Systematic	Random
1	5	51.2	54.9
	10	65.4	66.9
	15	76.0	75.0
	20	78.8	78.6
3	5	35.3	36.7
	10	52.2	49.9
	15	55.2	54.7
	20	65.1	61.7

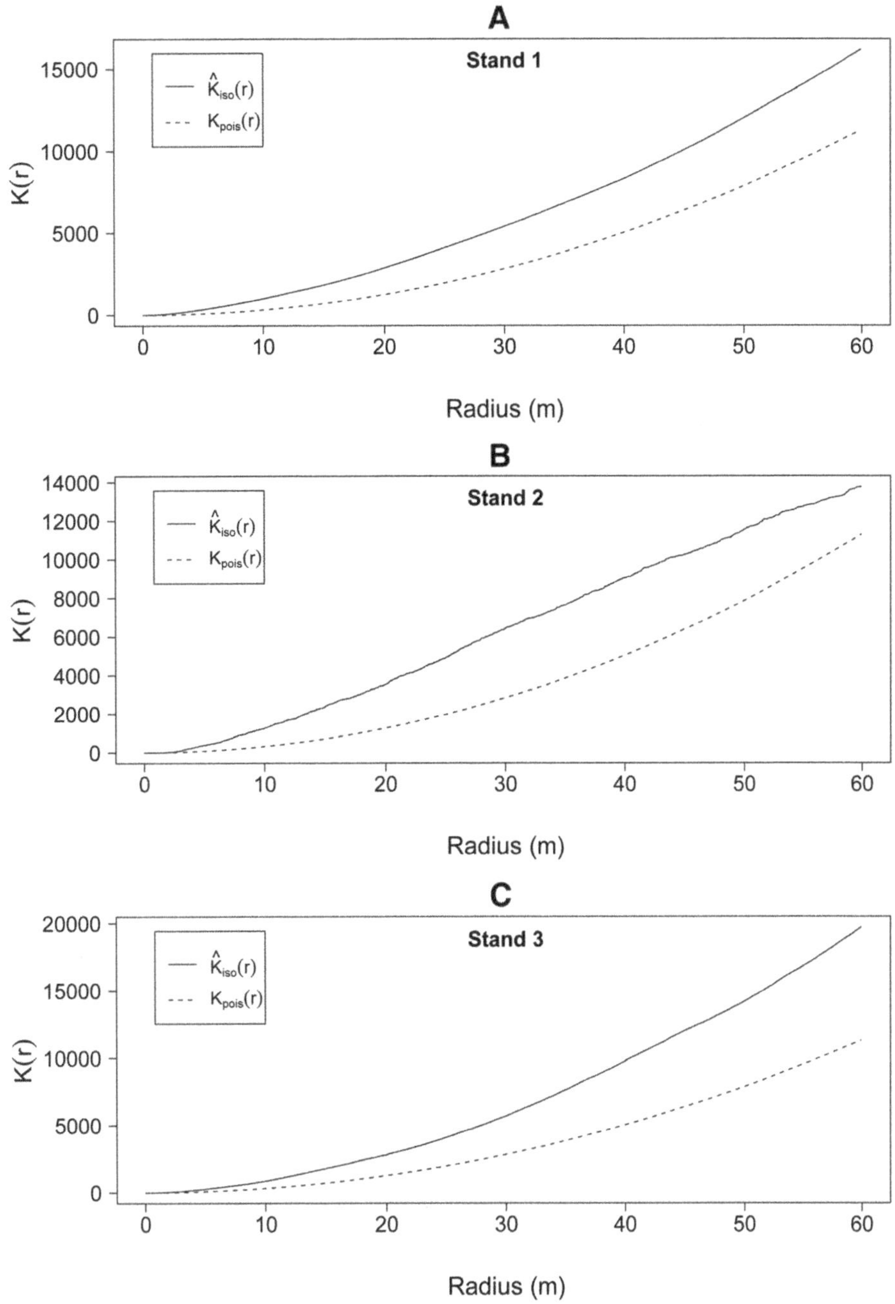

Figure 4 Ripley's K function for stands 1–3 (**A**= stand 1, **B** = stand 2, **C** =stand 3). The dashed line describes the expected value of trees based on Poisson distribution with radius *r*, and solid line the estimate obtained using Ripley's K function and isotropic edge correction.

in three spruce dominated forest stands where local aspen populations have survived during the past 100 years (Vehmas et al. 2009b). These stands also represent favourable growing environments of aspen. The proportion of aspen in these stands is considerably higher than the average value of 1.5% in Finland (Tomppo et al. 2001) being 16.3%, 5.5% and 16.2% of basal area in stands 1, 2 and 3, respectively. These statistics are for the forest area in general, but in conservation areas the proportion of aspen is usually considerably larger. It also should be noted that our study stands were rather large in size compared to the average stand size, which is ca. 2 ha in southern Finland. However, with respect to the state-owned forests and conservation areas where aspen is common, the stand sizes used in this study were broadly similar.

The acceptable level of forest attribute results is, of course, dependent on the need for information, but according to inventory by compartments, the RMSE

Table 5 Average values of Fisher index and proportion of clustered spatial patterns of aspen trees in 2500 simulations

Stand	Number of plots	Random		Systematic	
		Average	Proportion of clustered samples	Average	Proportion of clustered samples
1	5	8.6	94.1	9.5	96.1
	10	9.9	99.6	10.0	99.2
	15	10.0	100.0	11.4	100.0
	20	10.6	100.0	10.6	100.0
2	5	1.9	56.1	2.0	60.4
	10	2.3	73.4	2.6	84.1
	15	2.6	84.5	2.7	90.9
	20	2.7	90.2	2.8	95.6
3	5	3.0	94.1	3.1	96.1
	10	3.5	95.6	3.7	89.8
	15	3.7	95.6	4.1	98.7
	20	3.9	98.0	4.0	97.0

values of V for deciduous tree species in mixed stands should be between 20% – 50% in Finnish conditions (Uuttera et al. 2002). However, in the previous studies the RMSE% figures have been higher (65%) for deciduous tree species (birches) (Haara and Korhonen 2004). Our results showed, in general, that with the lower number of sample plots the RMSE figures are over 50% but the requirement set by Uuttera et al. (2002) is possible to obtain by measuring 10 or more plots. Between the three stands, the differences in accuracy were caused by the amount of aspen growing stocks, the sizes of the stands (i.e., sampling intensity) and small differences in the spatial patterns of trees. The effect of these aspects is to some extent inversely related. For example, in stand 2, the sampling intensity was highest, but also the RMSE figures were still the highest. This is most likely due to the low amount of aspen in the growing stock.

We also guided random sample plot placement by applying auxiliary ALS information with PPS sampling. This kind of approach previously has been applied in studies by Pesonen et al. (2010a, b) in the estimation of quantities of coarse woody debris (CWD). The chosen ALS metric, the square of the mean height of laser echoes emphasized the grid cells with the tallest trees. In our case, this technique decreased sampling efficiency in stand 2. This was due to the negative correlation between aspen volume and the square of the mean height of laser, i.e. the tallest trees in stand 2 are not aspens. This can be considered a drawback of the approach. If pre-information concerning the chosen variable does not hold true, the benefit is completely lost. In our case, the correlation between aspen volume and ALS metric-derived mean height was very strong in the 15 large sized fixed-area aspen sample plots, which were earlier used in single tree-based aspen detection from ALS data in the same Koli area (Säynäjoki et al. 2008), but obviously this kind of information cannot be generalised without risks associated with the extrapolation.

With respect to aspen populations in conservation areas, the detailed information on diameter distribution (e.g., the number of large aspens and the shape of distribution) is also of primary interest because many aspen-associated species are highly specialised to specific tree properties (Kouki et al. 2004; Sahlin and Ranius 2009). While descending diameter distribution shapes are interpreted as indicators of uneven-aged stand structure and, thus, may reflect the continuity of aspen populations, bimodal distributions reveal the existence of more than one aspen layer, which is usually also strongly related to the stand age structure. Information on both of these distribution types can be utilised in the management of conservation areas, and without this information, the management lacks primary attributes characterising stands. Regarding our results on mimicking distribution types with 10 fixed-radius plots, the proportions of correctly described diameter distributions were about 65% and 50% when obtained for the descending and bimodal diameter distributions, respectively. These proportions can be further increased more than 10 percentage units by increasing the number of sample plots. The same trend is also true for $N_{dbh>25cm}$, 10 or more measured sample plots may be required to reach the 25%–50% level of RMSE%.

During the last fifteen years, numerous field-based sampling methods for assessing different sparse populations have been presented (e.g. Holopainen et al. 2006; Ringvall et al. 2007; Gove et al. 2013). Although aspen populations are sparse in general this is not the case in our study data. The abundance of aspen in our stands is comparable to the abundance of birch in Finland which constitutes 17% of growing stock and is the third most frequent tree species of the country (Metsätilastollinen vuosikirja 2013). In such conditions the use of sparse population inventory methods may lead to high cost and time-consuming fieldwork. Since there are numerous sampling methods for sparse populations, the suitability of some of these, such as parallel strips suggested by Marquardt et al. (2012), could be investigated for our data that is, in any case, a topic for future studies.

According to the analysis of spatial patterns, the aspens were strongly clustered in all three stands. This is in line with previous findings (e.g. Syrjänen et al. 1994). In general clustered spatial patterns of trees make inventory more challenging (Pippuri et al. 2012) which is consistent with the RMSE% levels of our study. On the other hand, it is worth noting that the clustered spatial pattern of aspens was also successfully identified from sample plots without the information of tree location. This is an important outcome for planning aspen inventories.

In our study, remote sensing was only applied as auxiliary information in PPS sampling. However, in earlier studies, individual aspens have been detected from ALS data or they have been part of tree stock descriptions in area-based approaches. The problems related to single-tree detection include the typically very low general detection rate and overlapping of aspen intensity values with other tree species, such as birch and pine. Also, the vast size of the crown of mature aspen trees would eventually cause difficulties for interpretation when crown sizes of other trees were considerably smaller. On the other hand, when successful, single-tree detection would reveal unforeseen information on aspen crowns. Here, we did not apply single-tree detection, since the technique was already tested with data from Koli NP by Säynäjoki et al. (2008). In the case of ABA, earlier studies have shown poor accuracy estimates obtained for aspen. In our case, this approach was inapplicable, since the number of measured training plots available in Koli NP was not adequate and the forest vertical structure and tree species constitution outside Koli NP is considerably different.

Conclusions

This study characterized European aspen populations for the purposes of monitoring and management of boreal conservation areas. Our results suggest that if the number of sample plots is adequate, i.e. 10 or more using plot size 400 m^2, stand level inventory will provide accurate enough forest attributes estimates in conservation areas (minimum accuracy requirement of RMSE% is 20%–50%). Even for the more ecologically valuable attributes, such as diameter distribution, spatial pattern and large aspens, the estimates are acceptable for conservation purposes. Between the three stands, the differences in accuracy were caused by the amount of aspen growing stocks, the sizes of the stands and small differences in the spatial patterns of trees. ALS-based auxiliary information might also be useful in guiding the inventory. However, there is still the major risk that relying on ALS may decrease accuracy. Completely remote sensing-based inventory applications for such detailed attributes obtainable for aspens must still await further development of sensors and algorithms, such as multispectral ALS or combination of ALS and hyper spectral data.

Competing interests

The authors declare that they have no competing interests.

Authors' contributions

MM participated to all phases of the study. AP calculated the results concerning sampling simulations and participated to writing of corresponding parts of the study. LK calculated the results concerning spatial pattern of trees and participated to writing of corresponding parts of the study. JK was responsible of the ecological part of the Background and Discussion. MV conducted fieldwork and ALS metrics analysis. KE participated planning and writing of the study. All authors have read and commented the manuscript. All authors read and approved the final manuscript.

Acknowledgements

This work was supported by by the strategic funding of the University of Eastern Finland. We thank Ms Anne Nylander for her help with the compilation of the figures.

Author details

[1]University of Eastern Finland, School of Forest Science, P.O. Box 111, FI-80101 Joensuu, Finland. [2]Blom Kartta Oy, Kauppakatu 15, 80100 Joensuu, Finland. [3]City of Joensuu, 80100 Joensuu, Finland. [4]Natural Resources Institute Finland, Joensuu Unit, P.O. Box 68, FI-80101 Joensuu, Finland.

References

Axelsson P (2000) DEM generation from laser scanner data using TIN models. In: The International Archives of the Photogrammetry and Remote Sensing, vol 33, Part B4/1, Amsterdam, pp 110 – 117

Baddeley A, Turner R (2005) Spatstat: an R package for analyzing spatial point patterns. J Stat Soft 12:1–42

Beers TW (1966) The direct correction for boundary-line slopover in horizontal point sampling. Research Progress Report 224, Purdue University, Agricultural Experiment Station, Lafayette, Indiana, p 8

Breidenbach J, Næsset E, Lien V, Gobakken T, Solberg S (2010) Prediction of species-specific forest inventory attributes using a nonparametric semi-individual tree crown approach based on fused airborne laser scanning and multispectral data. Remote Sens Environ 114:911–924

Esseen PA, Ehnström B, Ericson L, Sjöberg K (1992) Boreal forests—the focal habitats of Scandinavia. In: Hansson L (ed) Ecological Principles of Nature Conservation. Elsevier Applied Science, London

Gove JH, Ducey MJ, Valentine HT, Williams MS (2013) A comprehensive comparison of perpendicular distance sampling methods for sampling downed coarse woody debris. Forestry 86:129–143

Grönlund A, Hakalisto S (1998) Management of traditional rural landscapes in Koli National Park. Separate plan of Koli National Park. North Karelia Regional Environment Centre, Joensuu, Regional environmental publications 104, pp 81

Haara A, Korhonen KT (2004) Kuvioittaisen arvioinnin luotettavuus. Metsätieteen aikakauskirja 4(2004):489–508

Heliövaara K, Väisänen R (1984) Effects of modern forestry on northwestern European forest invertebrates: a synthesis. Acta For Fenn 189:1–32

Holopainen M, Leino O, Kämäri H, Talvitie M (2006) Drought damage in the park forests of the city of Helsinki. Urban For Urban Gree 4:75–83

Kalliola R (1973) Suomen kasvimaantiede. Werner Söderström Osakeyhtiö, Porvoo

Kärkkäinen S (1994) Herb-rich forest vegetation of the Koli area. Publication of the Water and Environment Administration: series A 172, pp 51

Kinnunen J, Maltamo M, Päivinen R (2007) Standing-volume estimates of forests in Russia: how accurate is the published data? Forestry 80:53–64

Koivuniemi J, Korhonen KT (2006) Inventory by compartments. In: Kangas A., Maltamo M. (eds) Forest Inventory. Methodology and Applications. Managing Forest Ecosystems, vol 10, Springer, Dordrecht, pp 271–278

Korpela I, Ørka H-O, Maltamo M, Tokola T, Hyyppä J (2010) Tree species classification in airborne LiDAR data: influence of stand and tree factors, intensity normalisation, and sensor type. Silva Fenn 44:319–339

Kouki J, Arnold K, Martikainen P (2004) Long-term persistence of aspen, a key host for many threatened species, is endangered in old-growth conservation areas in Finland. J Nat Conserv 12:41–52

Kurkela T (1973) Epiphytology of *Melampsora* rusts of Scots pine (*Pinus sylvestris* L.) and aspen *Populus tremula* L. The Finnish Forest Research Institute Research Report 79, pp 68

Laasasenaho J (1982) Taper curve and volume functions for pine, spruce, and birch. Commun Inst For Fenn 108:74

Lilja S, Wallenius T, Kuuluvainen T (2006) Structural characteristics and dynamics of old *Picea abies* forests in northern boreal Fennoscandia. EcoScience 13:181–192

Lyytikäinen A (1991) Kolin luonto, maisema ja kulttuurihistoria. Kolin luonnonsuojelututkimukset. Vesi- ja ympäristöhallituksen monistesarja 308

Maltamo M, Packalen P (2014) Species-specific management inventory in Finland. In: Maltamo M, Naesset E, Vauhkonen J (eds) Forestry Applications of Airborne Laser Scanning: Concepts and Case Studies. Managing Forest Ecosystems, vol. 27th edn. Springer, Dordrecht, pp 241–252
Marquardt T, Temesgen H, Eskelson BNI, Anderson P (2012) Evaluation of sampling methods to quantify abundance of hardwoods and snags within conifer dominated riparian zones. Ann For Sci 69:821–828
Metsätilastollinen vuosikirja (2013) http://www.metla.fi/julkaisut/metsatilas tollinenvsk. Accessed 20 Dec 2013
Næsset E (2004) Practical large-scale forest stand inventory using a small-footprint airborne scanning laser. Scand J For Res 19:164–179
Næsset E (2007) Airborne laser scanning as a method in operational forest inventory: status of accuracy assessments accomplished in Scandinavia. Scand J For Res 22:433–442
Ørka HO, Næsset E, Bollandsås OM (2007) Utilising airborne laser intensity for tree species classification. In: International Archives of the Photogrammetry, Remote Sensing, and Spatial Information Sciences, vol 36, Part 3/W52, pp 300–304
Packalén P, Maltamo M (2007) The k-MSN method in the prediction of species-specific stand attributes using airborne laser scanning and aerial photographs. Remote Sens Environ 109:328–341
Pesonen A, Kangas A, Maltamo M, Packalén P (2010a) Different sources of auxiliary information in coarse woody debris inventory. Forest Ecol Manag 259:1890–1899
Pesonen A, Maltamo M, Kangas A (2010b) The comparison of airborne laser scanning-based probability layers as auxiliary information for assessing coarse woody debris. Int J Remote Sens 31:1245–1259
Pippuri I, Kotamaa E, Maltamo M, Peltola H, Packalén P (2012) Exploring horizontal area-based metrics to discriminate the spatial pattern of trees and need for first thinning using airborne laser scanning. Forestry 85:305–314
Pippuri I, Maltamo M, Packalen P, Mäkitalo J (2013) Predicting species-specific basal areas in urban forests using airborne laser scanning data and existing stand register data. Eur J For Res 132:999–1012
Poso S (1983) Basic features of forest inventory by compartments. Silva Fenn 17:313–349
Ringvall A, Snäll T, Ekström M, Ståhl G (2007) Unrestricted guided transect sampling for surveying sparse species. Can J For Res 37:2575–2586
Ripley BD (1981) Spatial statistics. John Wiley & Sons, New York
Ripley BD (1988) Statistical inference for spatial processes. Cambridge University Press, Cambridge
Sahlin E, Ranius T (2009) Habitat availability in forests and clearcuts for saproxylic beetles associated with aspen. Biodivers Conserv 18:621–638
Säynäjoki R, Packalén P, Maltamo M, Vehmas M, Eerikäinen K (2008) Detection of aspens using high-resolution aerial laser scanning data and digital aerial images. Sensors 8:5038–5055
Schreuder HT, Gregoire TG, Wood GB (1993) Sampling Methods for Multiresource Forest Inventory. John Wiley & Sons, New York
Stokland JN, Holien H, Gaarder G (2002) Arealtall for boreal regnskog i Norge 2002. NIJOS-rapport 2:1–20
Syrjänen K, Kalliola R, Puolasmaa A, Mattson J (1994) Landscape structure and forest dynamics in sub-continental Russian European taiga. Ann Zoo Fenn 31:19–34
Tikka PS (1954) Structure and quality of aspen stands. I. Structure. Commun Inst For Fenn 44:1–33
Tomppo E, Henttonen H, Tuomainen T (2001) Valtakunnan metsien 8. inventoinnin menetelmä ja tulokset Metsäkeskuksittain Pohjois-Suomessa 1992–94 sekä tulokset Etelä-Suomessa 1986–92 ja koko maassa 1986–94. Metsätieteen aikakauskirja B/2001: 99–248
Uuttera J, Hiltunen J, Rissanen P, Anttila P, Hyvönen P (2002) Uudet kuvioittaisen arvioinnin menetelmät- Arvio soveltuvuudesta yksityismaiden metsäsuunnittelu. Metsätieteen Aikakauskirja 3(2002):523–531
Vehmas M, Eerikäinen K, Peuhkurinen J, Packalén P, Maltamo M (2009a) Airborne laser scanning-based identification of herb-rich mature forests in the Koli National Park, eastern Finland. Forest Ecol Manag 257:46–53
Vehmas M, Kouki J, Eerikäinen K (2009b) Long-term spatiotemporal dynamics and historical continuity of European aspen (*Populus tremula* L.) stands in Koli National Park, eastern Finland. Forestry 82:135–148

9

Land use change detection in Solan Forest Division, Himachal Pradesh, India

Shipra Shah[1*] and DP Sharma[2]

Abstract

Background: Monitoring the changing pattern of vegetation across diverse landscapes through remote sensing is instrumental in understanding the interactions of human activities and the ecological environment. Land use pattern in the state of Himachal Pradesh in the Indian Western Himalayas has been undergoing rapid modifications due to changing cropping patterns, rising anthropogenic pressure on forests and government policies. We studied land use change in Solan Forest Division of Himachal Pradesh to assess species wise area changes in the forests of the region.

Methods: The supervised classification (Maximum likelihood) on two dates of IRS (LISS III) satellite data was performed to assess land use change over the period 1998–2010.

Results: Seven land use categories were identified namely, chir pine (*Pinus roxburghii*) forest, broadleaved forest, bamboo (*Dendrocalamus strictus*) forest, ban oak (*Quercus leucotrichophora*) forest, khair (*Acacia catechu*) forest, culturable blank and cultivation. The area under chir pine, cultivation and khair forests increased by 191 ha (4.55 %), 129 ha (13.81 %) and 77 ha (23.40 %), whereas the area under ban oak, broadleaved, culturable blank and bamboo decreased by 181 ha (16.58 %), 152 ha (6.30 %), 71 ha (2.72 %) and 7 ha (0.47 %), respectively.

Conclusions: The study revealed a decrease in the area under forest and culturable blank categories and a simultaneous increase in the area under cultivation primarily due to the large scale introduction of horticultural cash crops in the state. The composition of forests also exhibited some major changes, with an increase in the area of commercially important monoculture plantation species such as pine and khair, and a decline in the area of oak, broadleaved and bamboo which are facing a high anthropogenic pressure in meeting the livelihood demands of forest dependent communities. In time deforestation, forest degradation and ecological imbalances due to the changing forest species composition may inflict irreversible damages upon unstable and fragile mountain zones such as the Indian Himalayas. The associated common property externalities involved at local, regional and global scales, necessitate the monitoring of land use dynamics across forested landscapes in developing future strategies and policies concerning agricultural diversification, natural forest conservation and monoculture tree plantations.

Keywords: Land use; Solan Forest Division; Supervised Classification; Maximum likelihood

* Correspondence: drshiprashah1984@gmail.com
[1]Department of Forestry, College of Agriculture, Fisheries and Forestry, Fiji National University, NasinuP.O. Box 7222, Fiji Islands
Full list of author information is available at the end of the article

Background

Very few landscapes remain on Earth that have not been significantly altered by the human race in one way or another (Yang 2001). Land use land cover (LULC) change is the primary modifier of landscapes, affecting a wide range of socioeconomic, biological, climatic, and hydrological systems (Sohl and Sohl 2012). Understanding the spatial extent and distribution of LULC change is vital to the study of environmental changes at various levels (Ojima et al. 1994). LULC change is one of the key drivers of local and regional climate change (Chase et al. 1999), biodiversity decline (Sala et al. 2000), soil degradation (Tolba and El-Kholy 1992) and the loss of ecosystem services, thereby affecting the ability of ecosystems to support human needs (Vitousek et al. 1997). Population growth, rapid economic development and poverty have been identified as the underlying causes of land use change resulting in deforestation and land degradation (Giri et al. 2003; Bolland et al. 2007).

During the 1960s the concept of vegetation mapping gained momentum, resulting in an increase in studies of LULC change worldwide (Yang 2001). It is universally acknowledged that declining forest cover triggers ecological problems like changes in global climate, habitat degradation and unprecedented species extinction (Goldsmith 1998). Atmospheric carbon dioxide concentration is now higher than at any time in the past 10 million years (Kennedy and Hanson 2006). It is estimated that forests (including associated soils, peat deposits and lake sediments) hold 62–78 % of the world's terrestrial biospheric carbon (IPCC 2001). Forest clearing is one of the major drivers of global warming and climate change. The world's forests particularly those in the tropics are shrinking at a dramatic rate (Apan 1999). Anthropogenic activities involving forests, such as land use change can alter the amount and temporal distribution of C storage (Adams et al. 1993; Haynes et al. 1994). According to Achard et al. (2014), the gross loss of tropical forests was 8.0 million ha yr^{-1} in the 1990s and 7.6 million ha yr^{-1} in the 2000s, which resulted in carbon losses of 887 MtC yr^{-1} and 880 MtC yr^{-1} respectively during the same period. It is imperative that regular periodic assessments of forest cover change in tropical regions are carried out to recognize previous patterns, assist proper planning and predict future trends. The use of remote sensing data has been instrumental in monitoring the changing pattern of vegetation across diverse landscapes. The capacity for large geographical coverage, high temporal frequency and wide selection of spatial and spectral resolution options, further enhances the use of remotely sensed imagery for LULC change detection.

LULC change is primarily a local event, since the characteristics of such change can vary dramatically from one region to another. In order to capture the uniqueness of each region it is necessary that LULC studies are conducted at local to regional scales to better understand the cumulative impacts at multiple scales (Sohl et al. 2004). Over the years land use and cropping patterns in the western Himalayan state of Himachal Pradesh have been changing. The expansion of cultivation at the expense of forests has been relatively more profound in Himachal Pradesh than other parts of India (Melkania and Melkania 1987). The introduction of horticultural cash crops and the commercialization of agriculture have placed greater demands on the forests (Gouri et al. 2004) of the state which have not only been undergoing a change in area but also a change in composition primarily due to human interference (Deshingkar et al. 1997). To arrest forest degradation the state government has been involved in raising monoculture tree plantations of species such as pine which offer relatively less ecological and economic services to forest dependent communities as compared to mixed broadleaved forests (Baland et al. 2008). Another grievous cause of concern is the replacement of oak by pine in the Indian Western Himalayas. Thus, the forested landscapes in the state are undergoing major transformations. Some LULC change studies have been conducted in Himachal Pradesh (Sharma et al. 2014, Singh et al. 2014, Chand and Chatranta 2014, Pareta 2014, Pareta and Pareta 2014, Ramachandra et al. 2012) but none have specifically focused on the change in the forests of the region to highlight specific trends in species area and composition. Therefore, the present study is an attempt to study land use change in Solan Forest Division, to highlight species wise area changes which can play an indispensible role in developing future strategies and policies concerning natural forest conservation and monoculture tree plantations.

Methods

Study area

The population of Himachal Pradesh as per the Census of India, 2011 is 6.86 million, with a decadal growth rate of 12.81 %. Around 90 % of the total population lives in rural areas and consequently people are highly dependent upon forests for their livelihoods (Gouri et al. 2004). Local communities depend on forests for fuelwood, fodder, grazing, non-timber forest products and construction timber through Timber Distribution (TD) rights. There has been an ever increasing pressure on the forest resources due to the rising population on one hand as well as limited and fragmented agricultural land on the other.

The state has a predominantly agro-horti-pastoral economy. Apart from forests around 87 % of the population also depends on agriculture (Gouri et al. 2004).

While in the lower valleys agriculture and animal husbandry form the backbone of the economy, at higher altitudes, agro-pastoral systems predominate. Due to a high population density per unit area of agricultural land, and small size of landholdings– 64 % are less than 1 ha (Gouri et al. 2004), agriculture which was primarily for subsistence and considered an uneconomic vocation was diversified into horticulture which has relatively higher yields per unit area in terms of money than other agricultural crops (Ahluwalia 1998). Temperate fruits like apple, pear, stone fruits, nuts; subtropical fruits like mango, litchi, guava, citrus fruits; and new fruits crops like kiwi, strawberry, olive and hazelnuts have been introduced in recent years and have since become a major source of the export earnings of the state. Himachal Pradesh is now known as the fruit bowl of India and today the agriculture sector provides direct employment to around 71 % of the total working population (Kumar and Prashar 2012).

The present study was carried out in Solan Forest Division of Himachal Pradesh, which is located between 30° 45' 00" to 31° 10'00" N latitude and 76° 55'00" to 77° 15'00' E longitude. The area falls under subtropical region where climate varies from extreme hot in the lower elevations to extreme cold in higher elevations. Temperature in lower areas ranges between 15 and 36 °C and in higher areas it varies from 0 to 24 °C. Precipitation is in the form of rains mainly during rainy season and the area receives on an average 1000 mm annual rainfall.

The forests of the division are pure and mixed stands of chir pine and mostly conform to Champion and Seth (1968) $9C_1a$ - lower Shiwalik chir pine forests, lying between an elevation of 900–2100 m a.s.l. The area above 1800 m a.s.l is inhabited by oak forests. Low lying areas show widespread population of bamboos and are also a transitional zone for *Acacia catechu*.

Methodology

Cloud free IRS 1D (LISS-III) satellite data were acquired for winter months, 26 December, 1998 and 17 January, 2010 (Fig. 1), from the National Remote Sensing Agency (NRSA), Government of India, Hyderabad. Obtaining images at near anniversary dates is considered important for change detection studies (Jensen, 2007). The stock map and inventory reports available with Solan Forest Division were used to derive information on forest ranges, forest compartments and land uses. These were further confirmed by ground truthing. The stock map was digitized for forest compartments and land use types distributed in the different ranges of the division. Forest compartment coverage was then imported into IDRISI Taiga (Clark lab, Clark University, Worcester MA, USA) for further processing. Toposheets (1: 25000 scale) pertaining to Solan Forest Division were procured from FSI (Forest Survey of India) and used to digitise contours at 20 m interval using CARTALINX 1.2 data builder. The Digital Elevation Model (DEM) for the division was prepared by processing these cartographic features using IDRISI Taiga.

Preprocessing of the images was done through atmospheric and geometric rectifications. The LISS III images were corrected for atmospheric path radiance using dark object subtraction method (Chavez 1988), wherein the pixel (associated with the dark object) having minimum brightness value in the near infrared band was detected and the corresponding pixel values in all other bands were subtracted from the specific raw bands. This resulted in an image corrected for atmospheric scattering. The images were re-sampled for all bands using the nearest neighbor method, so that the original brightness values of pixels were kept unchanged. The resultant root mean squared error was found to be less than 1 pixel for both the images, 0.001465 pixels for the 1998 image and 0.000201 pixels for the 2010 image. Image enhancement

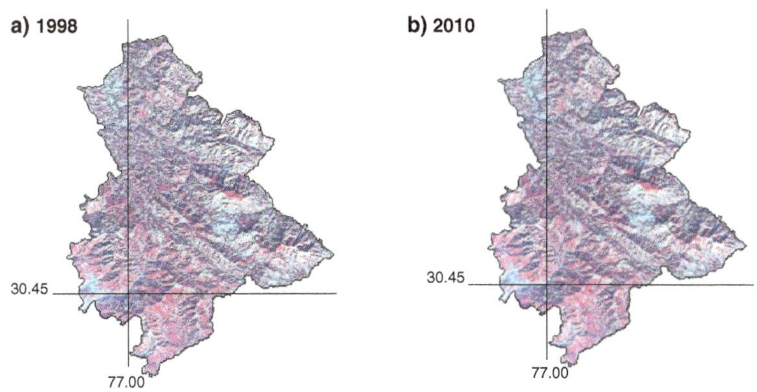

Fig. 1 IRS 1D (LISS-III) satellite images **a**) 1998 and **b**) 2010

techniques such as histogram stretching, median filtering and compositing were also applied to aid visual interpretation. While histogram stretching is a linear stretch technique used for adjusting image intensities to enhance contrast; median filtering aids in random noise removal by creating an output image in which each pixel's value is based on its value and those of its immediate neighbors in an input image. Colour composite images allow us to view the reflectance information from three separate bands in a single image, and were created by using the bands which had the greatest amount of information, least redundancy and low relativity.

The training sites were determined and a supervised classification was performed on both images using Maximum Likelihood algorithm. The supervised classification technique is preferred, because the data of the study area is available and the authors have a prior knowledge of the area. The Maximum Likelihood decision rule is one of the most widely used supervised classification algorithms (Wu and Shao 2002; McIver and Friedl 2002). Land use classification was first done for the whole area of Solan Forest Division, 57,158 ha (recorded forest area + non-forest area) and then from these classified images, the forest compartments 13,067 ha (recorded forest area) within the division were extracted, to assess the land use change in the forests of the division. As a result the land use maps depicting seven different land uses were derived. The flowchart elucidating the development of land use maps is given in Fig. 2.

A complete accuracy assessment was performed on classified images of the two dates generated during this study. A sample of ground truth points were distributed randomly on the classified image. Total 450 points were overlaid over the classified image to assign to each point an identifier of the land cover type. The ground truth data (reference data) used were collected from field surveys and existing stock map that had been field checked. An error matrix was established to evaluate the accuracy of the classification. This is a very effective way to represent the accuracy of the classification results, as the accuracy of each category is clearly described (Fan et al. 2007). Overall accuracy, user and producer accuracies and the Kappa statistics, were derived from the error matrices. The overall accuracy was defined as the total number of correctly classified pixels divided by the total number of reference pixels (total number of sample points) (Rogan et al. 2002). The user accuracy was defined as the proportion of the correctly classified pixels in a class to the total pixels that were classified in that class. It indicates the probability that a classified pixel actually represents that category in reality. The producer accuracy was calculated by dividing the total number of correctly classified pixels in a class by the total number of reference measurements of that class and it is the probability that a class identified from the reference data is correctly classified on the map. The KHAT statistics (an estimate of Kappa), which provides a measure of how many more pixels were correctly classified than expected by chance (Congalton and Green 1999), was computed through the following formula:

$$k = \frac{0\text{-}E}{1\text{-}E}$$

where: k represents the Kappa Index of Agreement; O is the observed accuracy or proportion of matching values (the matrix diagonal) i.e. the sum of the diagonal elements divided by the overall total; and E is the expected proportion of matches in this diagonal i.e. the sum for each diagonal cell of the row total times the column total divided by the overall total squared.

Results

Cartographic features obtained from the stockmap of Solan Forest Division revealed that it covers five forest ranges namely Dharampur, Parwanoo, Solan, Kandaghat and Subathu (Fig. 3). The area under these ranges was in the order, Subathu (14198 ha) > Dharampur (11564 ha) > Solan (11027 ha) > Parwanoo (10688 ha) > Kandaghat (9681 ha). Within these five forest ranges are distributed around 256 forest compartments (Fig. 4). Dharampur is the largest range having the maximum number of forest compartments (94), followed by Parwanoo (75), Solan (37), Kandaghat (35) and Subathu range (15). In these forest compartments, seven types of land uses were identified. These are: 1) chir pine (*Pinus roxburghii*) forests, 2) ban oak (*Quercus leucotricophora*) forests 3) khair (*Acacia catechu*) forests 4) bamboo (*Dendrocalamus strictus*) forests 5) broad leaved forests 6) cultivation and 7) culturable blank.

Accuracy assessment

The user accuracy ranged between 74–93 % in 1998 and, 85–92 % in 2010 (Table 1). Producer accuracy ranged between 79–96 % in 1998 and, 86 and 92 % in 2010 (Table 1). Classification accuracy as a result of maximum likelihood classification resulted in an overall accuracy of 87 and 89 % and KHAT accuracy of 85 and 87 % in 1998 and 2010, respectively (Table 2).

Changes in Land use (1998–2010)

Land use classification was first done for the whole area of Solan Forest Division which is spread over 57,158 ha and includes both recorded forest area and non-forest area (Fig. 5) and then from these classified images, the forest compartments spread over 13,067 ha within the division (recorded forest area) were extracted, to assess the land use change in the forests of the division (Fig. 6).

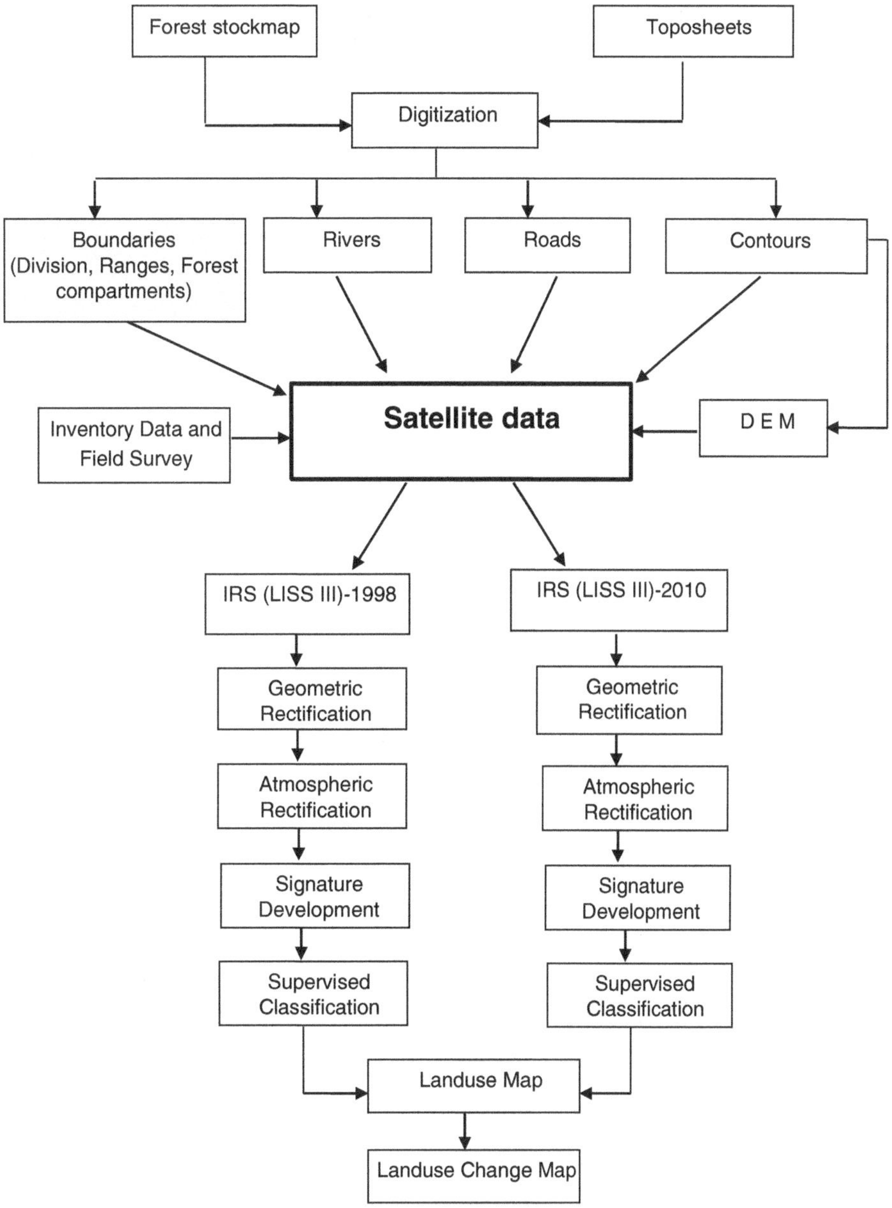

Fig. 2 Flow chart of the procedure followed to develop land use maps for Solan Forest Division

In 1998 the distribution of the various land use categories i.e. chirpine, culturable blank, broadleaved, bamboo, ban oak, cultivation and khair was over 4200, 2608, 2413, 1491, 1092, 934 and 329 ha area which accounted for 32.14, 19.96, 18.47, 11.41, 8.36, 7.15, and 2.52 % of the total area, respectively (Table 3). In 2010 chirpine, culturable blank, broadleaved, bamboo, cultivation, ban oak and khair land use categories covered 4391, 2537, 2261, 1498, 1063, 911 and 406 ha which accounted for 33.60, 19.42, 17.30, 11.46, 8.13, 6.97 and 3.11 % of the total area, respectively. Chirpine, cultivation and khair increased in area by 191 ha (4.55 %), 129 ha (13.81 %) and 77 ha (23.40 %) respectively from 1998 to 2010. Ban oak, broadleaved, culturable blank and bamboo decreased in area by 181 ha (16.58 %), 152 ha (6.30 %), 71 ha (2.72 %) and 7 ha (0.47 %) respectively (Table 3). The area under forest and culturable blank decreased by 72 and 71 ha respectively while that under cultivation increased by 129 ha.

Discussion

The computed accuracies of the classification process were reasonable which can be explained by the fact that the total number of correctly classified pixels was high. User accuracy as a result of maximum likelihood classification ranged between 74–93 % in 1998 and 85–92 % in

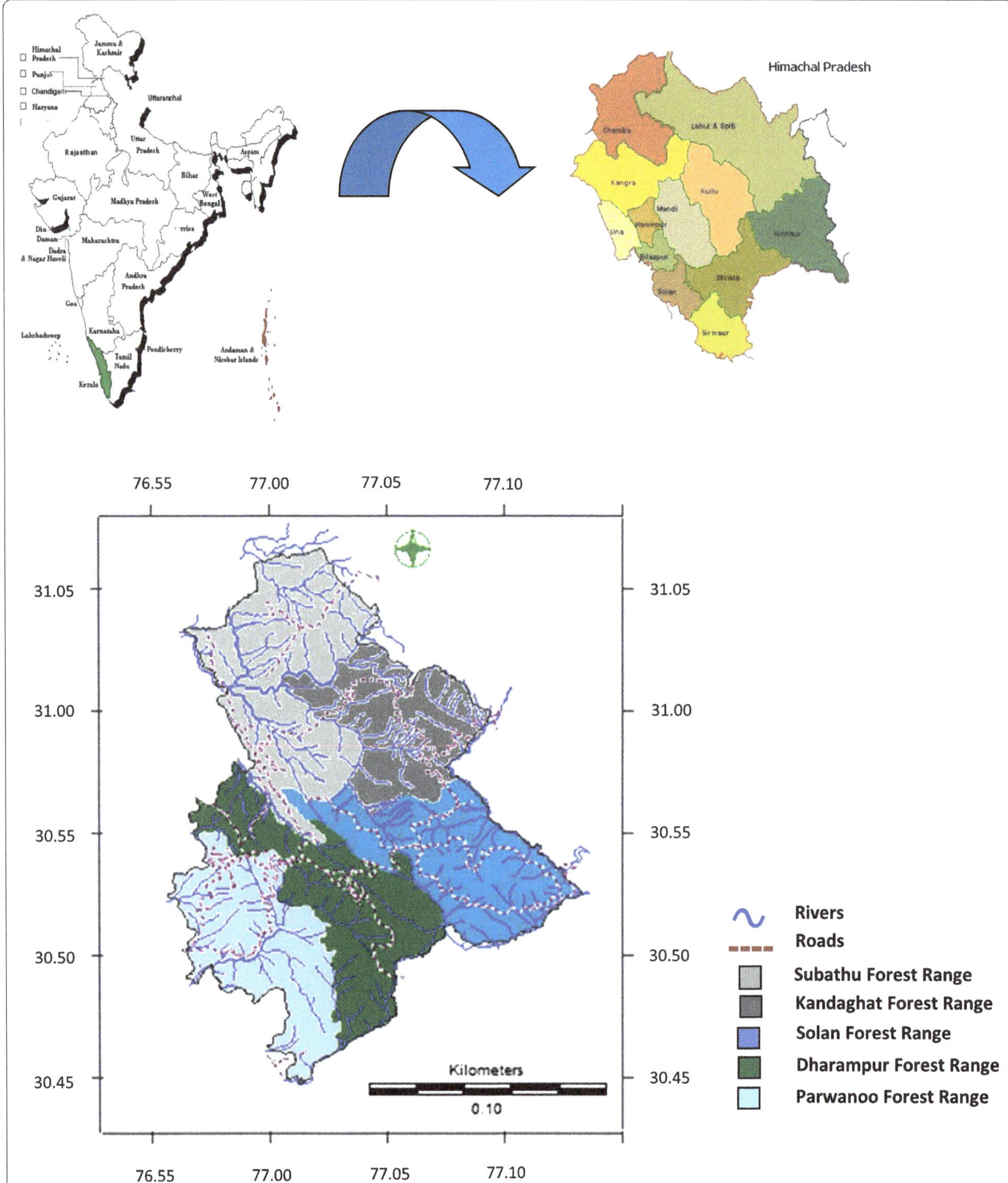

Fig. 3 Basemap of Solan Forest Division showing the five Forest Ranges

2010; while producer accuracy ranged between 79–96 % in 1998 and 86–92 % in 2010. Pareta (2014) in a study on land use and land cover change in Baddi district of Solan, has reported user accuracy for various lands cover classes between 93–100 % while producer accuracy between 83–100 %. Sharma and Leon (2005)) while working on land cover classification of Solan district using IRS (LISS III) data of different seasons reported a user accuracy ranging between 64–89 %, 32 and 68 %, 23–62 %; and producer accuracy between 46–100 %, 33

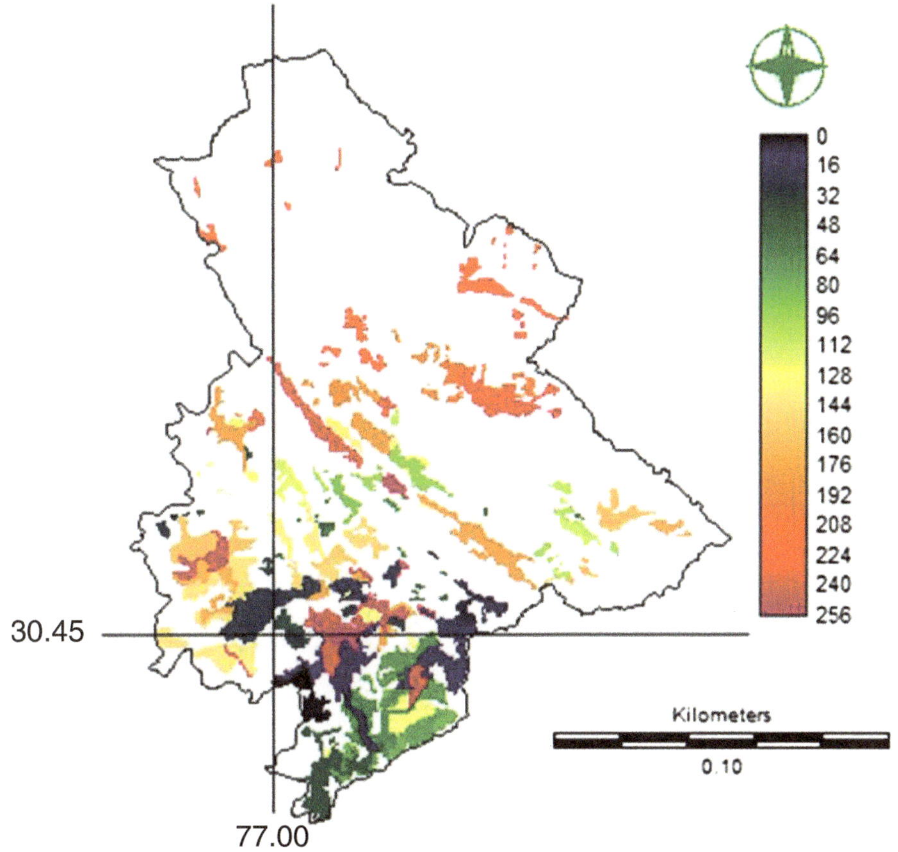

Fig. 4 Data Layer of Forest Compartments in Solan Forest Division

and 74 %, 28–80 %, in summer, winter and spring datasets respectively, as a result of maximum likelihood classification. Overall accuracy in the present study for the different land use categories was 87–89 % in 1998 and 2010, respectively. Overall accuracies of 76 % in summer dataset as compared to 49 % and 46 % in winter and spring datasets respectively, have been reported by Sharma and Leon (2005). Ramachandra et al. (2012) while studying landscape dynamics of Mandhala watershed in Solan district of Himachal Pradesh using IRS LISS III and Landsat data reported overall accuracies ranging between 78.52–86.52 %. KHAT accuracies of 85 % and 87 % in 1998 and 2010 land use classifications respectively were computed in the present study. Comparatively lower accuracy as per 1998 classification was due to relatively higher mixing of spectral signatures between different land use categories in 1998 than in 2010. Sharma and Leon (2005) have reported KHAT accuracies of 71 %, 40 % and 38 % for summer, winter and spring datasets respectively.

The land use classification showed a decrease in the area under both forest and culturable blank, and an increase in the area under cultivation. Similar results have been reported by Ramachandra et al. (2012), in Mandhala watershed, district Solan, where the area under forest decreased by 537.04 ha and the area under cultivation increased by 33.93 ha over the period 1972–2007. Pareta (2014) reported a decrease in the forest cover by 54.95 km^2 and an increase in cropland by 1.66 km^2 in Baddi town of district Solan over 1990–2013. Pareta and Pareta (2014) reported a decline in agricultural land and dense forest by 41.42 km^2 and 112 km^2 respectively over 2001–2013 in Chamba district of Himachal Pradesh.

Land use pattern in Himachal Pradesh has undergone tremendous transformation due to changes in agriculture cropping pattern, urbanization and industrialization. Crop diversification gained momentum in the nineties and now covers several new areas in low and mid hill districts of the state (Sharma, 2011). With the integration of horticultural cash crops, changing land use and cropping patterns have evolved (Sharma et al. 2007; Singh 1999). The area under horticulture in the state registered an enormous increase from 86230 ha in 1980–1981 to 186900 ha in 2005–2006. In Solan the per cent share of area under non-foodgrain crops increased from 8.25 % in 1982–1983 to over 12.96 % in 2004–2005 (Sharma, 2011). While only about 11 % of the total geographical area of the state is available for cultivation

Table 1 Error matrices as a result of land use classification for 1998 and 2010

Reference data 1998									
Classified data	Ban oak	Chirpine	Culturable Blank	Broadleaved	Bamboo	Cultivation	Khair	Total	User accuracy
Ban oak	46	3	1	0	0	1	0	51	90
Chirpine	1	64	2	4	0	0	0	71	90
Culturable Blank	1	6	65	0	0	3	0	75	87
Broadleaved	1	0	3	59	2	0	3	68	87
Bamboo	1	0	0	9	48	5	2	65	74
Cultivation	0	1	4	0	0	66	0	71	93
Khair	0	1	0	3	0	0	45	49	92
Total	50	75	75	75	50	75	50	450	
Producer accuracy	92	85	87	79	96	88	90		87
Reference data 2010									
Ban oak	43	3	1	0	0	0	0	47	91
Chirpine	4	67	1	3	0	1	0	76	88
Culturable Blank	0	1	68	3	0	3	0	75	91
Broadleaved	1	2	1	65	2	1	2	74	88
Bamboo	0	1	0	3	46	1	3	54	85
Cultivation	2	1	4	0	0	68	0	75	91
Khair	0	0	0	1	2	1	45	49	92
Total	50	75	75	75	50	75	50	450	
Producer accuracy	86	89	91	87	92	91	90		89

(Sharma, 2011), over the years growing area under horticultural cash crops has put increasing pressure on forest resources through land clearing (Sharma et al. 2010). In a study on forest based livelihoods in Himachal Pradesh, Gouri et al. (2004) reported that in two of the villages where apple cultivation was adopted 35 years ago, encroachment in forest areas for fruit production has reduced the forest cover to just 5 % in Kiari and 3 % in Dhadi Rawat village. In Kullu district of Himachal, encroachment of apple orchards into un-demarcated protected forest and demarcated protected forest is a common practice (ODA 1993). The use of wood for fruit packing cases also resulted in extensive deforestation in the state (Singh 1992). Up to mid-1980s, 0.2 million m^3 of wood was extracted to make 20 million boxes annually for packaging of fruits and vegetables (Stokes, 1983). Alternative packaging materials like corrugated cardboard are being experimented with to deal with the problem of forest exploitation; however fruit growers have issues with the quality of the boxes.

Table 2 Comparison of two classification accuracy measures for two dates

Classification	Overall accuracy (%)	KHAT accuracy (%)
1998	87	85
2010	89	87

Rising population pressure and inappropriate policies of the government have further threatened forest sustainability and are contributing to deforestation and forest degradation. Nautor is an ancient right under which landless people are permitted to break fresh agricultural land in common land areas by village elders (ODA 1993). In 1968 the Himachal Pradesh Nautor Land Rules came into force whereby the government started granting nautor land (redistributed land) upto one acre to landless and other eligible people for agriculture and horticulture (Chowdhry, 2008). The un-demarcated forest is the land that was designated for allocation under nautor rights. This practice of giving away un-demarcated forest land to landless cultivators under the provisions of the said rules has resulted in deforestation in the Himalayan state (Gupta 2007). Another policy in the history of forest management in Himachal is the Timber Distribution (TD) system under which landowners or right holders claim rights to timber, primarily to meet house construction or repair needs. This policy while meeting the basic needs of the local population has also been the single largest reason for timber harvest in the western Himalayan region (Vasan 1998) due to the rampant misuse of these rights. Although the system was directed towards villagers, TD rights extended in urban areas as well. While previously the right to timber was unlimited it was later restricted to one or two trees once every five years. However,

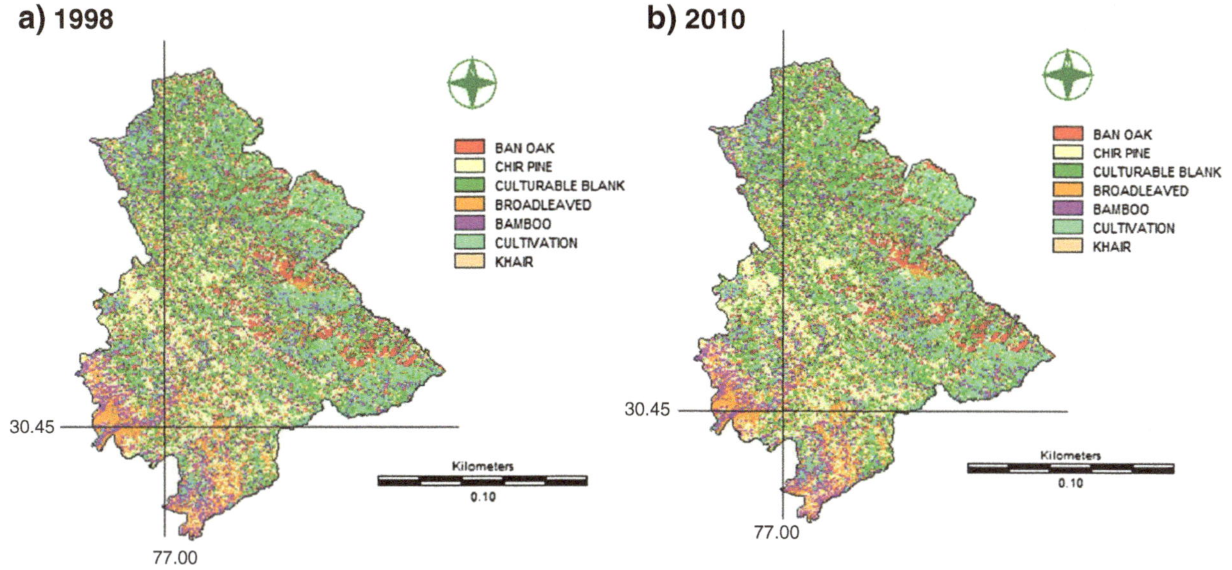

Fig. 5 Land use classification based on Minimum distance classifier for the whole area of Solan Forest Division – recorded forest area plus non-forest area **a**) 1998 and **b**) 2010

people were still paying the same nominal amount fixed in the last century which was about 20 % of the market rate at the time of the initial forest settlements (Vasan 1998) i.e. a right holder could buy a tree for a subsidized rate as low as Rs 3 to 5. In 1992–1993 the subsidy provided by the forest department under TD was to the tune of Rs.795,600,000 (DFFC 1994). In time, the traditional timber distribution system was overrun by a timber mafia with approximately 150,000 m^3 of timber (against a total growing stock of 96.8 million m^3) carrying a market value of around Rs 750 million, extracted annually in the state under the TD rights, being siphoned off to timber industries (Gouri et al. 2004). This policy was later reviewed in light of a strong opposition, and the Himachal Pradesh Forest (Timber Distribution to the Right Holders) Rules, 2013 were notified on 26 December, 2013, which enhanced both the rates and periodicity for grant of TD rights. The government has also been funding massive monoculture plantation drives across the state to arrest forest degradation which however is able to generate relatively less ecological and economic services to forest dependent communities when

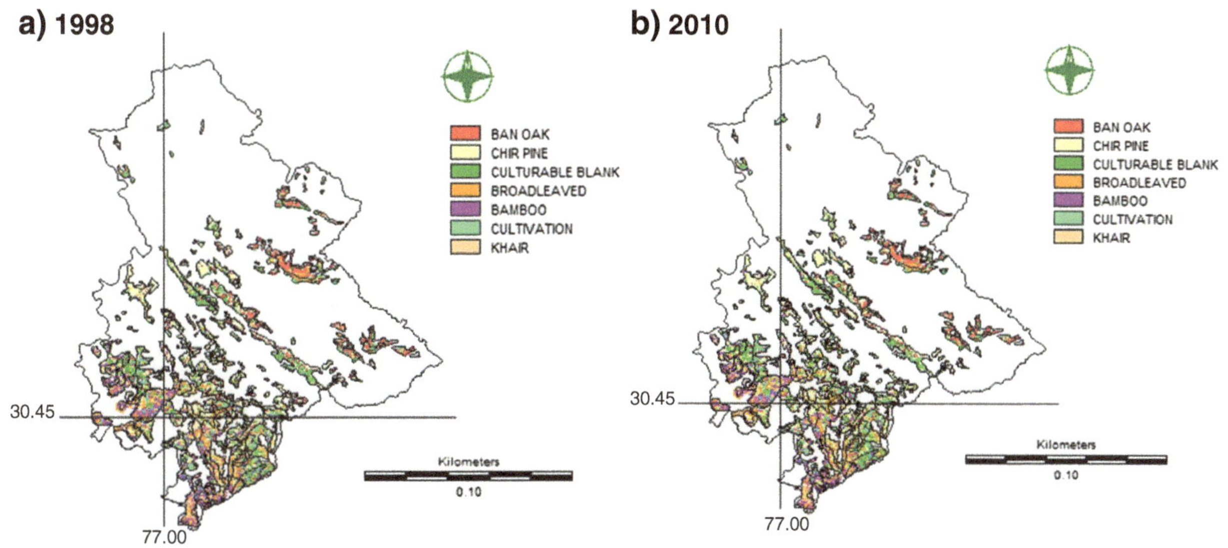

Fig. 6 Land use classification based on Minimum distance classifier for forest compartments extracted from Solan Forest Division – recorded forest area **a**) 1998 and **b**) 2010

Table 3 Territorial (compartment wise) Land use change in Solan Forest Division

Category	Area under Land Use (ha)		% Change in Land use Area
	1998	2010	
Cultivation	934	1063	129 (+)
	(7.15)	(8.13)	(13.81)
Culturable Blank	2608	2537	71(–)
	(19.96)	(19.42)	(2.72)
Forest Land use			
Chir Pine	4200	4391	191 (+)
	(32.14)	(33.6)	(4.55)
Ban oak	1092	911	181 (–)
	(8.36)	(6.97)	(16.58)
Broad Leaved	2413	2261	152 (–)
	(18.47)	(17.30)	(6.30)
Khair	329	406	77 (+)
	(2.52)	(3.11)	(23.40)
Bamboo	1491	1498	7 (–)
	(11.41)	(11.46)	(0.47)
Total	13,067	13,067	

Figures in parenthesis are in percent

compared to mixed broadleaved forests and so the vicious cycle of forest degradation has remained unchecked. Such policies have a profound impact on forested landscapes and play a pivotal role in change of land use.

Within the forest land use category, chirpine and khair showed an increase in area, while ban oak, broadleaved and bamboo decreased in area over 1998–2010 period. Both chir pine and khair have been the primary plantation species of the state, where the focus has revolved more around commercially important species rather than ecologically valuable ones. An increase in the area of chir pine over the period 1998–2010 may be attributed to an increase in the area under plantations of chir pine. This species is being tapped for resin and is the largest species raised under plantations in Himachal Pradesh constituting about 30.60 % of the total area under plantations and it registered an increase in area of 8265 ha over 1998–2001 (Planning Commission 2005). Also profuse regeneration under chir pine forests was observed during ground surveys of the study area. An increase in the area of khair may also be attributable to an increase in the area under plantations of khair which are a commercial source of kattha and cutch. While kattha is used in betel-vine preparations normally referred to as paan, cutch is a by-product of the kattha industry used for dyeing, colouring pulp in paper industry, water softening and in deep oil drilling operations. Khair is the second single largest species being raised under plantations in the state after chirpine constituting 17.64 % of the total area under plantations, and over 1998–2001 period, there has been reported an increase by over 7497 ha in the area under this species (Planning Commission 2005).

Conversely, a decrease in the area of ban oak may be attributed to heavy anthropogenic pressure on oak forests in hill states which results in indiscriminate lopping of the species. Ban oak is a multipurpose species having a good fodder and fuel quality; hence, it is under a high biotic stress in Himalayan forests (Saxena et al. 1978; Tiwari and Singh 1982; Bankoti et al. 1986; Joshi and Tiwari, 2011). Gouri et al. (2004), reported scarcity of fodder in Himachal Pradesh due to factors like forest fire, decrease in grasslands and mounting land utilisation pressure due to the adoption of horticulture. This has resulted in people keeping fewer cattle thereby affecting the availability of cow dung and natural manure. As a result the fuel requirements which were to some extent met through cow dung cakes, are now solely dependent on forests (Gouri et al. 2004) such as those of oak which are facing the problem of severe degradation. The total annual consumption of rural domestic fuel in the state is around 2.5 to 3.2 million tones, half of which is extracted from public forests (Singh and Sikka 1992). Decrease in the area of ban oak may also be due to poor regeneration observed in these forests compared to forests of chir pine in ground surveys of the study area. According to Bhandari et al. (1997), the regeneration potential of *Pinus roxburghii* is greater than that of *Quercus* species.

In the present study it is pertinent to observe that the increase in the area of pine namely 191 ha, is comparable with the decline in the area of oak namely 181 ha. This may be due to the replacement of oak by pine which has become an ever increasing phenomenon in the Western Himalayas (Singh et al. 1984). The exploitive management practices exerted by hill population in oak forests have encouraged pine (Saxena et al.1984). Forests in the Indian Himalayas are burnt periodically by the local communities to encourage the growth of grasses. This increases the preponderance of fire-resistant species such as chir pine. The aggressiveness of chir pine and its capacity to colonize disturbed sites have enabled it to spread at the expense of ban oak forests which are under immense biotic pressure (Singh and Singh 1984).

A decrease in the area of broadleaved forests which includes species such as *Terminalia belerica, T. chebula, Dalbergia sissoo, Pyrus pashia, Albizzia chinensis, Juglans regia* and *Celtis australis,* has also been observed in the present study, primarily due to the anthropogenic pressure of the forest fringe communities. According to Baland *et al.*, 2008, broad-leaved forests tend to be more useful and consequently more degraded than coniferous forests. The former are of greater utility to adjacent villages due to the superior quality of firewood and fodder, and are also a source of several non-timber forest products. The decrease in the area under bamboo forests may be attributed to pressure on bamboo for its commercial purposes. In Himachal Pradesh, bamboo forests are considered economically very important and support local livelihoods. Several villagers generate income from them by making use of bamboo in basket making (Gouri et al. 2004).

Conclusion

Land use change reflects the role of human activities on natural resources and the environment. Analysis of the spatial and temporal pattern of land use and assessment of the key driving factors behind the associated changes is imperative for sustainable use of land and its resources. The present study characterized the land use change of Solan Forest Division in Himachal Pradesh where the area under forests and culturable blank has been decreasing while that under cultivation shows an increasing trend. The forests of the division also exhibit some major changes in terms of species composition and area, specifically, an increase in the area of commercially important monoculture plantation species such as pine and khair, and a simultaneous decrease in the area of ecologically important oak and broadleaved forests which are under a high biotic stress. Human driving forces emerge as the key factor responsible for land use change in the region. This study highlights how the commercialization of agriculture, anthropogenic pressure on forests and government policies have, the potential to significantly alter land use pattern across mountain ecosystems. There is an immediate need to periodically monitor LULC change across forested areas to assist decision making and future policies concerning agriculture and forestry, both of which are a way of life for the people in the hills of India and nevertheless play a pivotal role in landscape modification. The cumulative effects of deforestation, forest degradation and changes in the species composition of forests in the long run may have far reaching adverse impacts on people, livelihoods, sustainability and the forest ecosystem services flowing from the Indian Himalayan Region.

Abbreviations

LULC: Land use land cover; IRS: Indian Remote Sensing; LISS III: Linear Imaging Self-Scanning Sensor 3; NRSA: National Remote Sensing Agency; DEM: Digital Elevation Model; TD: Timber Distribution.

Competing interests

The author(s) declare that they have no competing interests.

Authors' contributions

SS carried out the image analysis, designed and drafted the manuscript. DP was involved in the acquisition of the satellite imageries, interpretation of the data and in revising the manuscript. Both authors read and approved the final manuscript,

Authors' information

1) SS participated in this research as a doctoral student. Currently she is working in the Department of Forestry, College of Agriculture, Fisheries and Forestry, Fiji National University, Nasinu, Fiji Islands.
2) DP is a professor of Silviculture at Dr. Y.S. Parmar University of Horiculture and Forestry, Solan, Himachal Pradesh, India.

Acknowledgements

Authors are grateful to Forest Division Solan, Himachal Pradesh for providing necessary help and guidance.

Author details

[1]Department of Forestry, College of Agriculture, Fisheries and Forestry, Fiji National University, NasinuP.O. Box 7222, Fiji Islands. [2]Department of Silviculture and Agroforestry, Dr. Y.S. Parmar University of Horiculture and Forestry, Solan, Himachal Pradesh, India.

References

Achard F, Beuchle R, Mayaux P, Stibig HJ, Bodart C, Brink A, Carboni S, Desclee B, Donnay F, Eva HD, Lupi A, Rasi R, Seliger R, Simonetti D (2014) Determination of tropical deforestation rates and related carbon losses from 1990 to 2010. Glob Chang Biol 20(8):2540–2554

Adams R, Adams D, Callaway J, Chang C, McCarl B (1993) Sequestering carbon on agricultural land: social costs and impacts on timber markets. Contemp Pol Issues 11:76–87

Ahluwalia MS (1998) Social, Cultural, and Economic History of Himachal Pradesh. Indus Publishing, New Delhi

Apan AA (1999) GIS applications in tropical forestry. Faculty of engineering and surveying. University of Southern Queensland, Toowoomba, Queensland, Australia

Baland JM, Bardhan P, Das S, Mookherjee D (2008) Forests to the People: Decentralization and Forest Degradation in the Indian Himalayas. World Dev 38(11):1642–1656

Bankoti TNS, Melkania U, Saxena AK (1986) Vegetation analysis along an altitudinal gradient in Kumaun Himalaya. Indian J Ecol 13:211–221

Bhandari BS, Mehta JP, Nautiyal BP, Tiwari SC (1997) Structure of Chir Pine (Pinus roxburghii Sarg.) community along an altitudinal gradient in Garhwal Himalaya. Int J Ecol Environ Sci 23:67–74.

Bolland LP, Ellis EA, Gholz HL (2007) Land use dynamics and landscape history in La Montana, Campeche, Mexico. Landsc Urban Plan 82:198–207

Champion HG, Seth SK (1968) A revised survey of the forest types of India. Government of India Publications Division, New Delhi

Chand K, Chatranta M (2014) GIS & remote sensing technique to examine Landuse / Landcover Changes: A Case Study of emerging industrial hub of District Solan, Himachal Pradesh. Indian J Appl Res 4(6):218–220

Chase TN, Pielke RA, Kittel TGF, Nemani RR, Running SW (1999) Simulated impacts of historical land cover changes on global climate in northern winter. Clim Dyn 16:93–105

Chavez PS (1988) An improved dark object subtraction technique for atmospheric correction of multispectral data. Remote Sens Environ 24:459–479

Chowdhry P (2008) Gender Discrimination in Land Ownership. SAGE Publications, India

Congalton RG, Green K (1999) Assessing the Accuracy of Remotely Sensed Data: Principles and Practices. Lewis Publications, Boca Raton, USA

Department of Forest Farming and Conservation (1994) Annual Administrative Report. Himachal Pradesh

Deshingkar P, Bradley PN, Chadwick MJ, Leach G (1997) Adapting to climate change in a forest-based land use system: a case study of Himachal Pradesh, India. Stockholm Environment Institute, Stockholm

Fan F, Weng Q, Wang Y (2007) Land use and land cover change in Guangzhou, China, from 1998 to 2003, based on Landsat TM/ETM+ imagery. Sensors 7:1323–1342

Giri C, Defourny P, Shrestha S (2003) Land cover characterization and mapping of continental Southeast Asia using multi-resolution satellite sensor data. Int J Remote Sensing 24(21):4181–4196

Goldsmith FB (1998) Tropical rain forest. Chapman and hall, London

Gouri MS, Morrison E, Mayers J (2004) Policy influences on forest-based livelihoods in Himachal Pradesh, India. International Institute for Environment and Development, London
Gupta HK (2007) Deforestation and Forest Cover Changes in the Himachal Himalaya, India. Int J Ecol Environ Sci 33(2–3):207–218
Haynes R, Alig R, Moore E (1994) Alternative Simulations of Forestry Scenarios Involving Carbon Sequestration Options. USDA Forest Service Gen. Tech. Rep. PNW-GTR-335, Pacific Northwest Research Station, Portland, Oregon.
International Panel on Climate Change (IPCC) (2001) Climate Change: The Scientific Basis. Contribution of Working Group I to the Third Assessment Report of the Intergovernmental Panel on Climate Change. Cambridge University Press, Cambridge, UK
Jensen JR (2007) Introductory Digital Image Processing: a remote sensing perspective. Science Press and Pearson Education Asia Ltd, Beijing
Joshi NR, Tiwari A (2011) Regeneration status and phytosociology in *Quercus leucotrichophora* (A. Camus) and *Pinus roxburghii* (Sarg.) mixed forests in two different aspects influenced by forest fires in community managed forests of Kumaun Central Himalaya, India. Nature and Science 9(9):160–166
Kennedy D, Hanson B (2006) Ice and history. Science 311:1673
Kumar S, Prashar D (2012) An analysis on changing trends of foodgrains in Himachal Pradesh. Int J Pharm Life Sci 3(6):1739–1742
McIver DK, Friedl MA (2002) Using prior probabilities in decision-tree classification of remotely sensed data. Remote Sens Environ 81:253–261
Melkania NP, Melkania U (1987) Man and environment in Himachal Pradesh, India. In: Chadha SK (ed) Himachal Himalayas: ecology and environment. Today and Tomorrow's Printers and Publishers, New Delhi, pp 81–99
Ojima DS, Kalvin KA, Turner BL (1994) The global impact of land use change. BioScience 44(5):291–356
Overseas Development Administration Report (1993) Himachal Pradesh Forestry Project. Forestry Development in India and the Project Area. British High Commission, New Delhi
Pareta K (2014) Land Use and Land Cover Changes Detection using Multi Temporal Satellite Data. Int J Manag Soc Sci Res 3(7):10–17
Pareta K, Pareta U (2014) Climate Change Impact on Land and Natural Resource in Chamba Tehsil of Himachal Pradesh State, India. Int J Sc Tech 2(4):38–48
Planning Commission, Government of India (2005) Himachal Pradesh Development Report. State Plan Division, Academic Foundation
Ramachandra TV, Kumar U, Joshi NV (2012) Landscape Dynamics in Western Himalaya – Mandhala Watershed, Himachal Pradesh, India. A J G 12(1):1–8
Rogan J, Franklin J, Roberts DA (2002) A comparison of methods for monitoring multitemporal vegetation change using Thematic Mapper imagery. Remote Sens Environ 80(1):143–156
Sala OE, Chapin FS, Armesto JJ, Berlow E, Bloomfield J, Dirzo R, Huber-Sanwald E, Huenneke LF, Jackson RB, Kinzig A, Leemans R, Lodge DM, Mooney HA, Oesterheld M, Poff NL, Sykes MT, Walker BH, Walker M, Wall DH (2000) Biodiversity: global biodiversity scenarios for the year 2100. Science 287:1770–1774
Saxena AK, Pandey U, Singh JS (1978). On the ecology of oak forest in Nainital Hills, Kumoun Himalaya. In: Singh JS, Gopal B (eds) Glimpses of Ecology: Prof. R. Misra Commemoration Volume. Jaipur International Scientific Publication, Jaipur, India, pp 167–180
Saxena AK, Singh SP, Singh JS (1984) Population structure of forests of Kumaun Himalaya. Implications for management. J Environ Manage 19:307–324
Sharma H (2011) Crop Diversification in Himachal Pradesh: Patterns, Determinants and Challenges. Ind Jn of Agri Econ 66(1):97–114
Sharma DP, Bren L (2005) Effect of seasonal spectral variations on land cover classification. J Indian Soc Remote 33(2):203–209
Sharma RK, Sankhayan PL, Hofstad O, Singh R (2007) Land use changes in the Western Himalayan region - a study at watershed level in the state of Himachal Pradesh, India. Int J Ecol Environ Sci 33:197–206
Sharma RK, Sankhayan PL, Singh R (2010) Land Use Changes and Forest Degradation in a Himalayan Watershed: Analysis and Policy Alternatives. J Nat Res Pol Res 2(3):263–280
Sharma R, Rishi MS, Ahluwalia AS, Lata R (2014) Comparative change in landuse/landcover in the buffer zone of Kashlog limestone mines, Darlaghat, Himachal Pradesh, India using remote sensing and GIS tools. Int J Remote Sens Geosci 3(6):27–30
Singh RB (1992) Dynamics of Mountain Geosystems. Ashish Pub, New Delhi
Singh RB (1999) Land use change, diversification of agriculture and agroforestry in North-west India. In: Haque T (ed) Land Use Planning in India. NCAEPR, New Delhi, pp 122–130
Singh DV, Sikka BK (1992) Wood balance study of Himachal Pradesh. Himachal Pradesh University, Shimla, Agro Economic Research Centre
Singh JS, Singh SP (1984) An integrated Ecological study of Eastern Himalaya with Emphasis on Natural Resources. Kumaun University, Nainital
Singh JS, Rawat YS, Chaturvedi OP (1984) Replacement of oak forest with pine in the Himalaya affects the nitrogen cycle. Nature 311:54–56
Singh SK, Chandel V, Kumar H, Gupta H (2014) RS & GIS based urban land use change and site suitability analysis for future urban expansion of Parwanoo planning area, Solan, Himachal Pradesh (India). Intl J Dev Res 4(8):1491–1503
Sohl TL, Sohl LB (2012) Land-Use Change in the Atlantic Coastal Pine Barrens Ecoregion. Geogr Rev 102(2):180–201
Sohl TL, Gallant AL, Loveland TR (2004) The Characteristics and Interpretability of Land Surface Change and Implications for Project Design. Photogramm Eng Rem S 70(4):439–448
Stokes V (1983) Need for Substitute Packaging for Environmental Harmony. In: National Workshop on substituting forest timber for packaging of horticultural produce, tea, tobacco and textiles, Vigyan Bhawan, New Delhi, 30–31 July 1983.
Tiwari JC, Singh SP (1982) Vegetation analysis of a forest lying in transition zone between lower and upper Himalayan moist temperate forest. In: Paliwal GS (ed) The Vegetational Wealth of Himalayas. Puja Publishers, New Delhi, pp 104–119
Tolba MK, El-Kholy OA (1992) The World Environment 1972–1992: Two Decades of Challenge. Chapman & Hall, London
Vasan S (1998) Political Ecology of Timber Rights in the Western Himalayas. In: Proceedings of the IGES International Workshop on Forest Conservation Strategies for the Asia and Pacific Region, Institute for Global Environmental Strategies, Kanagawa, Japan, 21–23 July, 1998
Vitousek PM, Mooney HA, Lubchenco J, Melillo JM (1997) Human domination of earth's ecosystems. Science 277:494–499
Wu W, Shao G (2002) Optimal. Combinations of Data, Classifiers, and sampling methods for Accurate Characterizations of deforestation" Can J Remote Sens 28(4):601–609
Yang X (2001) Change Detection Based on Remote Sensing Information Model and its Application on Coastal Line of Yellow River Delta. Earth Observation Research Center, Tokyo, China

10

Spatial pattern of tree diversity and evenness across forest types in Majella National Park, Italy

Mohammad Redowan

Abstract

Background: Estimation of tree diversity at broader scale is important for conservation planning. Tree diversity should be measured and understood in terms of diversity and evenness, two integral components to describe the structure of a biological community. Variation of the tree diversity and evenness with elevation, topographic relief, aspect, terrain shape, slope, soil nutrient, solar radiation etc. are well documented.

Methods: Present study explores the variation of tree diversity (measured as Shannon diversity and evenness indices) of Majella National Park, Italy with five available forest types namely evergreen oak woods, deciduous oak woods, black/aleppo pine stands, hop-hornbeam forest and beech forest, using satellite, environmental and field data.

Results: Hop-hornbeam forest was found to be most diverse and even while evergreen Oak woods was the lowest diverse and even. Diversity and evenness of forest types were concurrent to each other i.e. forest type which was more diverse was also more even. As a broad pattern, majority portion of the study area belonged to medium diversity and high evenness class.

Conclusions: Satellite images and other GIS data proved useful tools in monitoring variation of tree diversity and evenness across various forest types. Present study findings may have implications in prioritizing conservation zones of high tree diversity at Majella.

Keywords: Tree diversity; Tree evenness; Forest type; Shannon diversity index; Shannon evenness index; Neural network; Kappa statistic

Background

Understanding relationship between environmental factors and plant diversity is a core research area in ecological studies (Zhang and Zhang 2011). Variations in the spatial patterns of plant diversity with various ecological factors e.g. elevation (Wallace 1878; Wang et al. 2003), topographic relief (Richerson and Lum 1980; O'Brien et al. 2000), terrain shape (Tivy 1993; Bolstad et al. 1998), aspect, slope and soil nutrients (Zhang and Zhang 2011), energy (Currie and Paquin 1987) solar radiation and potential evapotranspiration (Currie 1991; Pierce et al. 2005), etc. have been documented over centuries. However, studying the variations of the plant diversity with forest types is important for adopting appropriate management and conservation strategies. Majella national park in Italy is a protected area of high ecological importance. With more than 2000 floristic species, the Park hosts 65 % flora of Abruzzo region, 37 % of Italy and 22 % of the European species (Majella National Park 2011). Vegetation in Majella is divided into several distinct forest types. Each forest type has characteristic tree species composition and is named after the dominant tree species. Present study aimed to investigate pattern of variation of tree diversity across various forest types in Majella.

The concept of species compositional variation of plant communities (or 'diversity') should be expressed in two terms; diversity and evenness (Whittaker 1965; Gillespie et al. 2008), although numerous publications defined biodiversity in terms of species richness (number of species per unit area) only. There are two principal considerations/terms to understand in studying biological diversity which are a) number of different species in a community and b) number of individuals within each of these species in that community (Hamilton 2005). Species diversity is the number of different species in a particular area weighted by some measure of abundance such as number of individuals or biomass, while evenness is the relative abundances of individuals within each species. Both diversity and evenness are

Correspondence: redowanforestry@yahoo.com
Department of Forestry and Environmental Science, Shahjalal University of Science and Technology, Sylhet 3114, Bangladesh

generally indicated by several indices e.g. Shannon index, Simpson index, Gini-Simpson index, Berger-parker index etc. However, Shannon indices (both diversity and evenness) (Shannon 1948) are the most commonly used and acceptable indicator of biodiversity (Kent and Coker 1992; Mouillot and Lepretre 1999) and have been used in this research to indicate the level of tree diversity.

At present, satellite data are potential source of information of tree diversity due to large scale complete spatial coverage at reasonable cost (Duro et al. 2007). Images are in many cases free to obtain (e.g. Landsat Thematic Mapper) and can be effectively used for studying tree diversity at broader scale with high accuracy for feasible spatial and spectral resolution. In this study, the tree diversity and evenness maps of Majella, produced by classifying Landsat Thematic Mapper (TM) image in an artificial neural network classifier were overlapped with five forest-type thematic maps available from the park office with an aim to explore the variation of diversity and evenness across forest types.

Methods

The study was conducted in the Morrone Region (42°4'30"–42°14'40" N; 13°49'55"–14°1'51" E) of Majella National Park located in central Italy. Morrone is the most forest-type rich area of the park at close proximity to each other (Fig. 1). Morrone mountain range (peak mount Morrone, 2061 m above mean sea level) passes through the western part of the park where this study site largely engulfs part of the range. Vegetation in Morrone grades with altitude. There are four distinct phytoclimatic belts located at various altitudinal range; Mediterranean (below 400/500 m), sub-Mediterranean (500–900/1000 m), temperate Montane (900–1600/1700 m) and subalpine (above 1700 m). Field data of present study were collected from the sub-Mediterranean and Montane zones. Sub-Mediterranean belt consist mainly of Pubescent Oak (*Quercus pubescens*), Turkey Oak (*Quercus cerris*), different Mapples (*Acer* spp.), Ash species (*Fraxinus* spp.), Hop-hornbeam (*Ostrya carpinifolia*), some conifers and farmlands (Fig. 2). Montane is deciduous, mesophilic forest where European Beech (*Fagus sylvatica*) is the most dominant species. Sub-Mediterranean belt possess warm, dry climate with frequent annual summer draught due to hot spell (July-August). Montane belt posses low temperature, harsh winter conditions, cooler summer and high evenly-distributed rainfall than contiguous lowlands. Tree vegetation of Morrone is broadly classified into five prominent forest types and named after the dominant species located sparsely or contiguously in different altitudinal levels. The forest types are evergreen oak woods (comprising 1 % of the area), deciduous oak woods (31 %), black/aleppo pine stands (26 %), hop-hornbeam forest (11 %) and beech forest (31 %).

For capturing field information regarding the tree diversity and evenness, 100 sample plots of 30 m square size were surveyed from 7th September to 30th October in 2009. Stratified adaptive sampling technique with strata based on forest type was adopted in the field. Tree species count in each sample plot was done by line transact method (detailed in Runkle 1982). Nomenclature of the trees was done following Conti (1998). Location of the sample plots was recorded using handheld iPAQ Global Positioning System (GPS).

For each of the plots, Shannon diversity and evenness indices were calculated using the following formula:

$$\text{Shannon diversity index(SDI)} = -\sum_{1}^{s} Pi \ln Pi$$

$$\text{Shannon evenness index (SEI)} = H / \log(S).$$

Here P_i is the proportion of total number of species made up of the ith species in the plot, and S refers to the total number of species or species richness (also called α diversity) of that particular plot (Whittaker 1972; Pielou 1975).

Calculated diversity and evenness indices values for 100 plots were then arranged in ascending order and divided into 5 equal-spaced discrete diversity and evenness classes namely very low, low, medium, high and very high (Table 1). Each diversity or evenness class thus contained values of 20 field plots. The lowest value of both SDI and SEI was 0 while the highest value was 1.818 for SDI and 0.985 for SEI (Table 1). Each of the 5 diversity and evenness classes however contained a range of diversity or evenness values (Table 1). Zero (0) value of SDI and SEI indicates there is only one species present in that plot, while the highest values of SDI and SEI indicate maximum number of species found in that particular plot.

A cloud and defect free Landsat TM image captured on 27 September 2003 and four ancillary GIS layers namely elevation, slope, aspect and solar radiation were used in a back-propagation artificial neural network (NN) classifier. NN is a promising non-parametric image classifier successfully used by some researchers in similar works (e.g. Skidmore et al. 1997; Mutanga and Kumar 2007). NN shows better performance in common landcover classification than some other methods e.g. maximum likelihood (Civco 1993). One big advantage of NN classifier is, image and other ancillary spatial data layers can be jointly put into the classifier to produce required number of discrete landcover classes. Thirty meter resolution ASTER Digital Elevation Model (DEM) of the study site was acquired from the archive of National Aeronautic and Space Administration (NASA) via ITC (Faculty of Geoinformation Science and Earth Observation, Univesity of Twente, The Netherlands). Slope, aspect and solar radiation data were created from DEM using surface analysis tool in ArcGIS 9.3. Field sampled data of all diversity and evenness classes were converted to

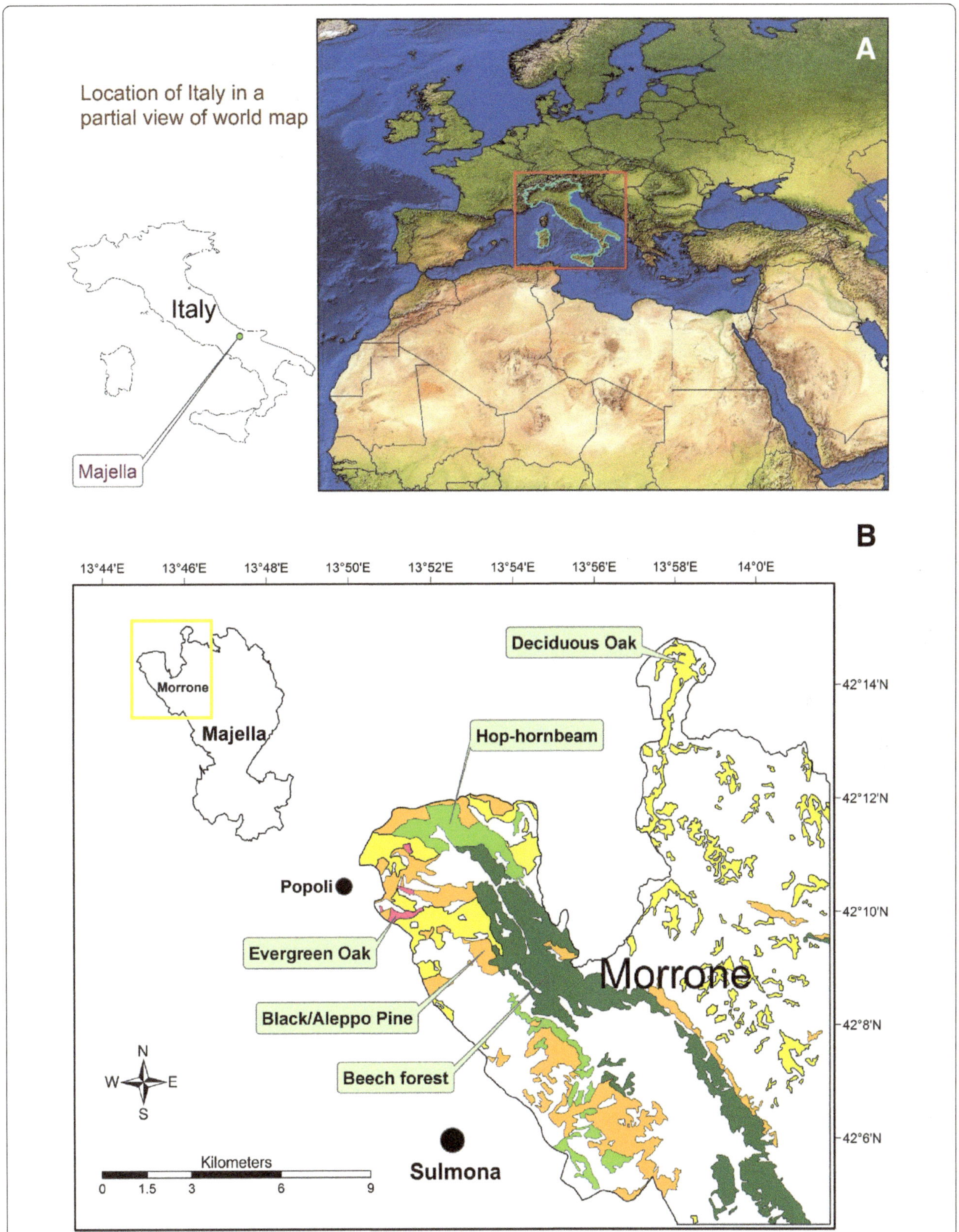

Fig. 1 Location of Italy in partial world map (**A**) and five forest types in Morrone Region of Majella National Park, Italy (**B**). Source of the park boundary shapefile: (Parco Nationale della Majella 1999)

Fig. 2 Glimpses of Majella National Park, Italy. Landscape view (**A**), mixed stand (**B**), pure stand of Beech (**C**) and Pine (**D**). Picture courtesy: Author

area of interest (AOI) point shape file in ENVI 4.3. Seventy percent of the field data were used for training the image and thirty percent for validation of classification result, which is a standard proportion (e.g. Mutanga and Rugege 2006). TM image and GIS layers were layer-stacked and fed to the NN classifier along with the training data set in ENVI 4.3. The outputs from the NN were diversity and evenness thematic maps of the study site with five classes in each. The NN classifier was run for the diversity and evenness maps independently. Map validation was done by generating error matrix using the training data set with overall accuracy and kappa statistic (Cohen 1960; Congalton and Green 1999). Kappa statistics is an index of inter-rater reliability that is commonly used to measure the level of agreement between two sets of dichotomous ratings or scores (Manel et al. 2001; Guisan et al. 2002). Kappa value of 1 indicates perfect agreement, >0.75 good classification, >0.80 excellent agreement and 0.50 a performance not better than chance. Forest type thematic map (shape file) of the study area was obtained from the park office. Diversity and evenness maps of Morrone produced using TM image and GIS layers were overlapped with forest type shape files separately using ArcGIS to explore the variation of diversity and evenness with forest types. Average value of Shannon diversity (SDI) and evenness (SEI) indices obtained from the field samples falling in each forest types was compared with the remotely sensed diversity and evenness levels/classes under each of forest type.

Table 1 Shannon diversity and evenness classes

Shannon diversity/ Evenness classes	Shannon Diversity Index (SDI)		Shannon Evenness Index (SEI)	
	Lower limit	Upper limit	Lower limit	Upper limit
Very Low	0	0.311	0	0.404
Low	0.325	0.637	0.428	0.660
Medium	0.643	0.939	0.663	0.754
High	0.960	1.136	0.755	0.868
Very High	1.170	1.818	0.869	0.985

For assessing the overall quantitative structure of the tree communities in all forest types, five parameters namely density (the numerical strength of a species per unit area), relative density (the proportion of a species in a stand as a whole), frequency (the degree of dispersion in terms of percentage occurrence), relative frequency (the dispersion of a species in relation to that of all the species) and abundance (the number of individuals of different species in the community per unit area) of individual trees were calculated using the following formula (Shukla and Chandel 2000).

$$\text{Density of a species per unit area} = \frac{\text{Total number of individuals of a species in all the sample plots}}{\text{Total number of sample plots studied}}$$

$$\text{Relative density of a species} = \frac{\text{Total number of individuals of a species}}{\text{Total number of individuals of all species}} \times 100$$

$$\text{Frequency of a species} = \frac{\text{Total number of quadrates in which the species occured}}{\text{Total number of quadrates studied}} \times 100$$

$$\text{Relative frequency of a species} = \frac{\text{Frequency of the species}}{\text{Sum of the frequencies for all species}} \times 100$$

$$\text{Abundance of a species} = \frac{\text{Total number of individuals of a species in all plots}}{\text{Total number of plots in which the species occured}}$$

Results and Discussion

A total of 24 tree species belonging to ten families were recorded in all available forest types in Morrone region of Majella National Park. A list of such species along with structural information of tree communities namely density, relative density, frequency, relative frequency and abundance has been presented in Table 2. Majority of the trees found in Morrone belonged mainly to Fagaceae, Sapindaceae and Pinaceae families. Among the tree species, some were most frequently found in the sample plots while few others were rare. Most frequently found species were Mona Ash/Fraxinus (F 51, RF 15.64), Hophornbeam (F 49, RF 15.03), Black Pine (F 49, RF 15.03), Beech (F 30, RF 9.02), Pubescent Oak (F 25, RF 7.67), Maple (F 24, RF 7.36) and Aleppo Pine (F 16, RF 4.91). However in this case, the frequencies and relative frequencies of Mona Ash, Hophornbeam and Beech are nearly the same i.e. these species were equally frequent (Fig. 3, Table 2). Species like Wild Cherry (F 2, RF 0.61), Salix/White Willow (F 2, RF 0.61), Common Hazel (F 1, RF 0.31), Norway spruce (F 1, RF 0.31) and European Ash (F 1, RF 0.31) were equally infrequent or rare. Species with highest relative density (RD) include Beech (18.41), Black Pine (15.95), Hophornbeam (15.19), Mona Ash (9.60), Aleppo Pine (7.34) and Evergreen Oak (4.22). And species with lowest relative density include Common Hazel (0.03), European

Table 2 List of main tree species found in the forest types in the study site with several measures of their quantitative structure

No.	English name	Scientific name	Family	Density (D)	Relative density (RD %)	Frequency (F)	Relative frequency (RF %)	Abundance
1	Ailanthus	*Ailanthus altissima*	Simaroubaceae	0.11	0.37	4	1.23	2.75
2	Aleppo Pine	*Pinus alepensis*	Pinaceae	2.21	7.34	16	4.91	13.81
3	Beech	*Fagus sylvatica*	Fagaceae	5.54	18.41	30	9.20	18.47
4	Black Pine	*Pinusnigra*	Pinaceae	4.80	15.95	49	15.03	9.80
5	Common Hazel	*Corylu savellana*	Betulaceae	0.01	0.03	1	0.31	1.00
6	Common Laburnum	*Laburnum anagyroides*	Fabaceae	0.19	0.63	7	2.15	2.71
7	European Ash	*Fraxinus excelsior*	Oleaceae	0.01	0.03	1	0.31	1.00
8	Evergreen/White Oak	*Quercus spp*	Fagaceae	1.27	4.22	11	3.37	11.55
9	Hophornbeam	*Ostria carpinifolia*	Butalaceae	4.57	15.19	49	15.03	9.33
10	Italian Alder	*Alnus cordata*	Betulaceae	0.09	0.30	4	1.23	2.25
11	Maple	*Acer pseudoplatanus*	Sapindaceae	0.72	2.39	24	7.36	3.00
12	Mona Ash/Fraxinus	*Fraxinus ornus*	Oleaceae	2.89	9.60	51	15.64	5.67
13	Montpellier Maple	*Acer monspessulanum*	Sapindaceae	0.10	0.33	7	2.15	1.43
14	Norway Mapple	*Acer platanoides*	Sapindaceae	0.11	0.37	3	0.92	3.67
15	Norway spruce	*Picea abies*	Pinaceae	0.01	0.03	1	0.31	1.00
16	Port-Orford-Cedar	*Chamaecyparus cawsoniana*	Cupressaceae	0.22	0.73	5	1.53	4.40
17	Pubescent Oak	*Quercus pubescens*	Fagaceae	4.66	15.49	25	7.67	18.64
18	Robinia	*Robinia pseudoacacia*	Fabaceae	0.15	0.50	5	1.53	3.00
19	Salix/White Willow	*Salix alba*	Salicaceae	0.04	0.13	2	0.61	2.00
20	Small leaved Maple/Field Maple	*Acer campestre*	Sapindaceae	0.49	1.63	7	2.15	7.00
21	Stone/Holly Oak	*Quercus ilex*	Fagaceae	1.16	3.86	14	4.29	8.29
22	Tilia/Linden/Lime	*Tilia cordata*	Malvaceae	0.40	1.33	7	2.15	5.71
23	Turkey Oak	*Quercus cerris*	Fagaceae	0.19	0.63	2	0.61	9.50
24	Wild Cherry	*Prunus avium*	Rosaceae	0.17	0.56	2	0.61	8.50

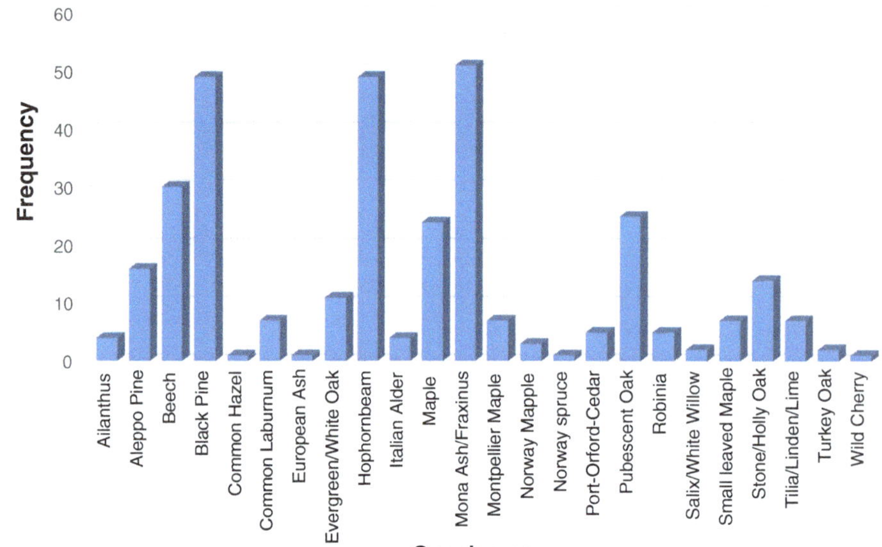

Fig. 3 Tree species found in Morrone region of Majella with frequency of occurrence

Ash (0.03), Norway spruce (0.03), Salix/White Willow (0.13) and Montpellier Maple (0.33).

The tree diversity and evenness maps of Morrone resulted from the classification of satellite image with other ancillary data in neural network with highest accuracies are given in Fig. 4. When only the image was used in NN, overall accuracies of the resulted tree diversity map was 53 % and tree evenness map was 57 %. Map accuracies significantly increased when ancillary data layers were jointly used with the satellite image in the NN classifier. In that case, the highest overall accuracy of the tree diversity map was 91 % (with kappa value 0.9) (Table 3) and the tree evenness map was 83 % (with kappa value 0.79) (Table 4). Obtained aaccuracies of the diversity and evenness maps were very high. Use of only image could not have produced such high accuracy maps unless other four ancillary environmental variables (elevation, aspect, slope and solar radiation) had been jointly used with the TM image in the NN classifier. As environmental variables have profound bearing on regulating plants' life, diversity and distribution, the use of such variables in NN classifier has increased classification accuracies significantly. The effect of ancillary data layers in increasing image classification accuracies in NN has been explained in details in other articles of this author (Redowan 2013; Redowan and Riadh 2013).

The diversity and evenness maps were overlapped with the forest type map and the results of overlapping with area (%) under each class are presented in Table 5 and Table 6 respectively.

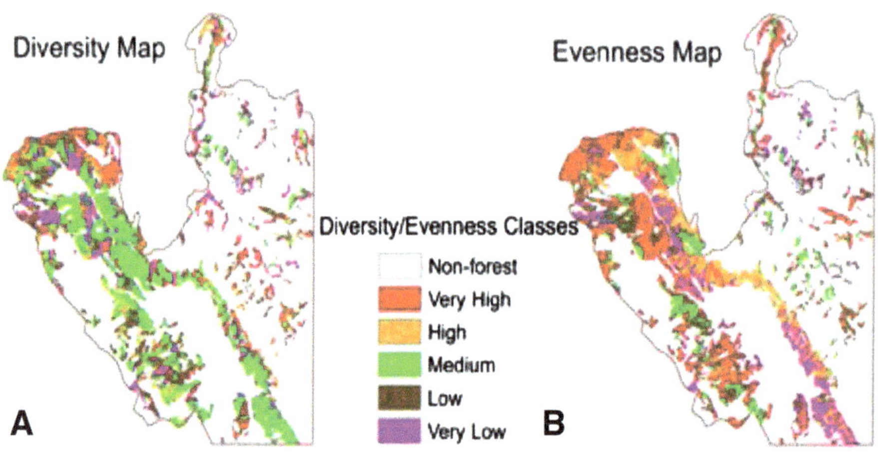

Fig. 4 Tree diversity (**A**) and evenness (**B**) maps resulted from classification of Landsat TM imagery

Table 3 Contingency table of accuracy assessment of tree diversity map with Kappa Statistic, Overall Accuracy, Producer's Accuracy (PA) and User's Accuracy (UA)

		Validation data							
		High	Low	Medium	Very high	Very low	Row total	UA%	PA%
Image data	High	5	0	0	0	0	5	100	83
	Low	0	5	0	0	0	5	100	83
	Medium	1	0	6	1	0	8	75	100
	Very high	0	1	0	5	0	6	83	83
	Very low	0	0	0	0	6	6	100	100
	Column total	6	6	6	6	6	30		

Overall accuracy 91 %, Kappa Statistic 0.90

Good agreement exists between average value of SDI and SEI in each forest type with the majority portion of that forest type belonging to the respective or contiguous diversity and evenness classes in the classified maps. As for example, average value of Shannon diversity index of ground survey in Beech forest is 0.65 which belongs to medium diversity category, while majority area of that forest type (34 %) belongs to medium diversity class as well, a good agreement between remotely sensed data with ground information (Table 5). Similar trend of diversity is noticed in other forest types as well.

For Shannon evenness index map, majority area of an evenness class corresponds to the respective ground information. As for example, majority area (36 %) of Black/Aleppo Pine forest belongs to medium evenness class of the map where average value of Shannon evenness index of the field samples is also of medium level (0.67). Similar trend is true for other forest types (Table 6).

Hop-hornbeam forest is the most diverse and even (average SDI 1.14 and SEI 0.82) followed by Black/Aleppo Pine forest, deciduous Oak woods, Beech forest and evergreen Oak woods (Fig. 5). Evergreen Oak woods is the least diverse and even (average SDI and SEI are 0.52 and 0.34 respectively). The diversity and evenness follow similar trend i.e. the forest type which is more diverse is also more even (Fig. 5).

As a broad pattern, 37 % of the study area belongs to medium diversity class, 16 % to low diversity class, 16 % to very high diversity classes, 18 % to very low diversity class and 12 % to high diversity class (Fig. 6). On the other hand, majority portion (27 %) of the study area belongs to high evenness class, nearly equal portion (21 % and 22 %) to low and very high evenness classes, 17 % to medium evenness class and 13 % to very low evenness classes (Fig. 6).

The study area is a mountainous landscape with mount Morrone (second highest peak of Majella) passing through the centre of it. Altitude is a principal regulatory factor governing the distribution and diversity of vegetation in the region (which is also documented in other studies e.g. Stanisci et al. 2005). Change in diversity is less prominent at lower altitude than at higher altitude. Vertical gradation of vegetation with altitude was very evident and noticed in the spatial extension and arrangement of typical forest types as a broad pattern; so were the species diversity and evenness (Table 5). As for example, beech forest was found confined to the specific phytoclimatic belt Montane (900/1000-1600/1700 m) with average SDI 0.65, belonging to *Medium* diversity class. Deciduous oak woods were mainly found below 900 m altitude in the sub-Mediterranean belt with average SDI 0.72, belonging to *High* diversity class. Diversity and evenness classes vary widely across the forest

Table 4 Contingency table of accuracy assessment of tree evenness map with Kappa Statistic, Overall Accuracy, Producer's Accuracy (PA) and User's Accuracy (UA)

		Validation data							
		High	Low	Medium	Very high	Very low	Row total	UA%	PA%
Image data	High	5	1	0	0	0	6	83	83
	Low	0	4	0	1	0	5	80	67
	Medium	1	0	5	0	0	6	83	83
	Very high	0	0	1	5	0	6	83	83
	Very low	0	1	0	0	6	7	86	100
	Column total	6	6	6	6	6	30		

Overall accuracy 83 %, Kappa Statistic 0.79

Table 5 Statistics of overlapping TM diversity map with forest types map along with average values of Shannon Diversity Index (SDI) in each forest type

Forest types		Area of each diversity class (%) that falls under a forest type					SDI* (avg)
	Total area (ha)	Very high	High	Medium	Low	Very low	
Beech	1842	20	26	**34**	11	9	0.65^{M}
Black/Aleppo Pine	1507	8	22	**32**	29	9	0.82^{M}
Evergreen Oak	42	27	9	4	**39**	21	0.52^{L}
Hop-hornbeam	631	**32**	14	17	29	8	1.14^{VH}
Deciduous Oak	1820	**35**	16	14	17	18	0.72^{H}

Major diversity class in each forest type has been boldfaced
SEI* = Average value of Shannon Evenness Index in each forest types; Diversity classes: M = Medium; L = Low; H = High; VH = Very high

types due mainly to other associated species in each type beyond the dominant one along with altitude. As for example, highest average SDI (1.14) and SEI (0.82) were found in the hop-hornbeam forest situated at a wide altitudinal range (500–1200 m) while lowest average SDI (0.52) and SEI (0.34) were found in the evergreen oak woods situated in the lower altitude (up to 500 m). Of course it does not necessarily mean that tree diversity is high at the higher altitude and low at the lower altitude. Rather the opposite trend i.e. decreases of species diversity with altitude was observed. Indeed, generally the Mediterranean (<500 m) and sub-Mediterranean (500–900 m) phytoclimatic belts are the most species rich and even with assemblages of various species while Montane belt (900–1600 m) is the least species rich and even which is solely dominated by Beech (*Fagus sylvatica*). The forest types are basically broad categorization of the woody species assemblage of the park, each of them occupying a broad altitudinal ranges, and scale of that variation as 'forest stands' is wide too. For example "beech" is short for broad-leaved mesophilic deciduous forest mostly beech, but at lower altitudes (e.g. 800–1000 m) it is mixed with e.g. maples (*Acer* spp.), pubescent Oak (*Quercus pubescens*) and Linden/Lime (*Tilia cordata*). As the altitudinal range of occurrence of some of the forest type overlaps with each other, the variation in average SDI or SEI values of each of them does not narrowly attribute to altitude but very broadly. But variation of the diversity and evenness of trees across the forest types is distinct (Jiang et al. 2007). Of all forest types, hop-hornbeam forest is the most diverse and even while evergreen oak woods are the least diverse and even.

This study revealed a good agreement between the produced diversity and evenness maps when they were overlapped with the forest type map of Majella. Broadly, areas with high tree diversity also have high species evenness. This phenomenon is of course very usual. Plots or areas with more number of species will definitely count for higher diversity and evenness as diversity and evenness in most cases are concurrent to each other.

The level of diversity and evenness that is found in Majella is not high in comparison to some other similar studies. Jiang et al. (2007) reported average Shannon-Weiner index value as 2.03 in deciduous broadleaved forest and 1.90 in temperate evergreen coniferous and deciduous broadleaved mixed forest in northwestern China. Jing et al. (2004) documented SDI ranging from 0.92–1.96 and SEI from 0.15–0.46 at different altitudinal belts (1100–1700 m) in a coniferous forest at Changbai mountainous landscape. Marsden and Pilgrim (2003) studied tree diversity of two lowland rainforest sites on New Britain in Papua New Guinea and reported SDI value at different habitats ranging from 1.63 to 2.38. In comparison to these study areas, it can be opined that diversity in Morrone is comparatively neither high nor low, rather of intermediate level.

Table 6 Statistics of overlapping TM evenness map with forest types map with forest type map along with average values of Shannon Evenness Index (SEI) in each forest types

Forest types		Area of each diversity class (%) that falls under a forest type					SEI* (avg)
	Total area (ha)	Very high	High	Medium	Low	Very low	
Beech	1842	23	31	3	7	**37**	0.40^{VL}
Black/Aleppo Pine	1507	22	20	**36**	16	7	0.67^{M}
Evergreen Oak	42	17	11	12	12	**48**	0.34^{VL}
Hop-hornbeam	631	24	**35**	16	16	9	0.82^{H}
Deciduous Oak	1820	**32**	12	25	15	18	0.76^{H}

Major evenness class in each forest type has been boldfaced
SEI* = Average value of Shannon Evenness Index in each forest types; Diversity classes: M = Medium; L = Low; H = High; VH = Very high

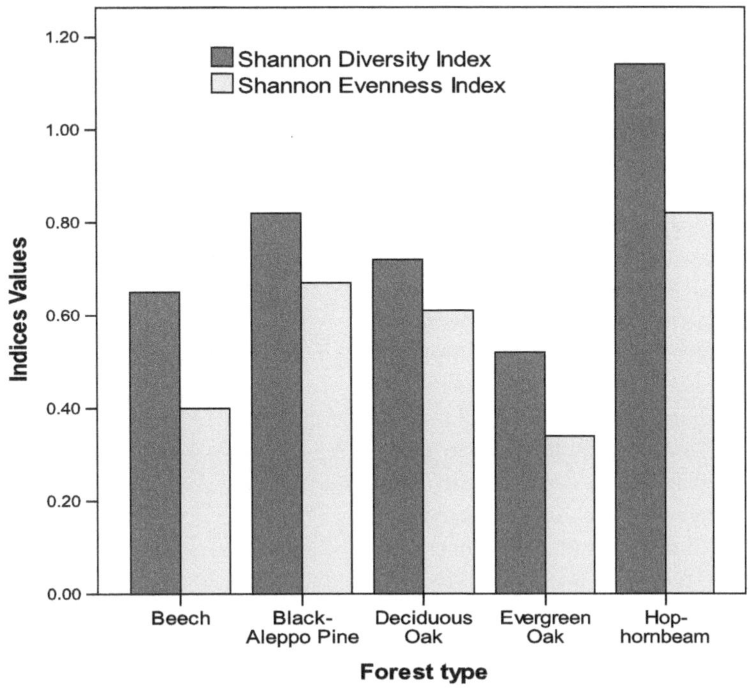

Fig. 5 Average value of diversity and evenness in each forest type

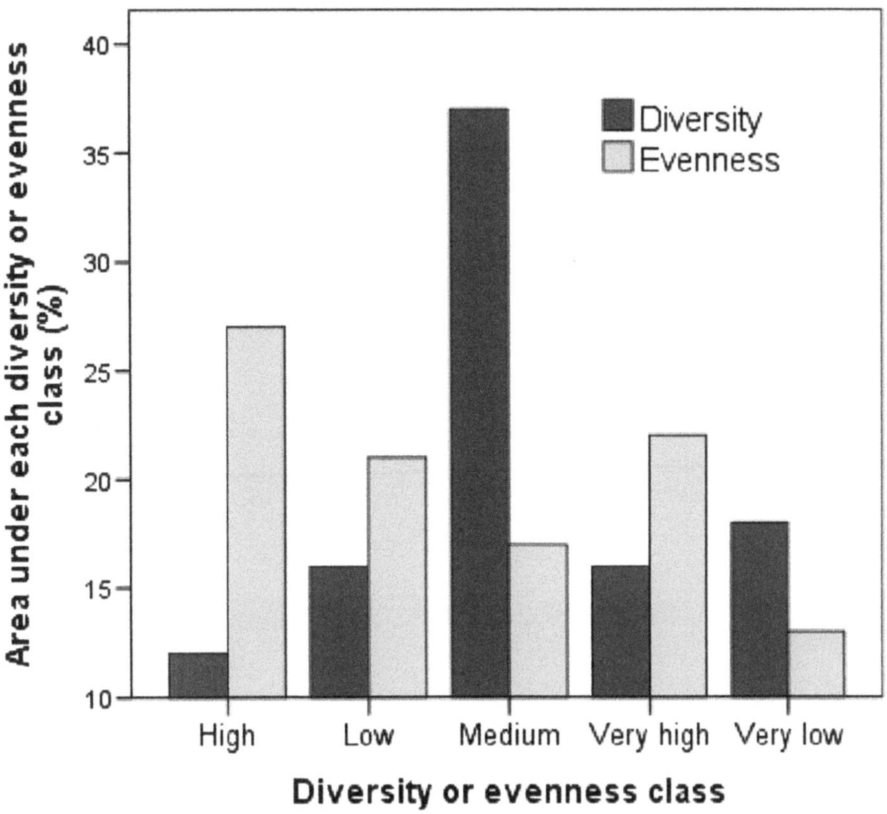

Fig. 6 Area falling under a particular diversity or evenness classes

Conclusion

Diversity and evenness are two integral components to describe, measure and compare biological diversity of forest communities. This study demonstrated how tree diversity and evenness of an Italian forest site varied with forest types and could be mapped and compared against each other using a combination of satellite image, GIS and ground survey data. The diversity and evenness went parallel to each other i.e. the forest type which was more diverse was also more even. As a broad pattern, majority portion of the study area belonged to medium diversity but high evenness classes. In comparison to the other similar studies, diversity and evenness of Morrone were of medium level. Findings of the present study might have significant implications in prioritizing biodiversity conservation areas of different forest types/ecoregions, undertaking silvicultural operations and selecting sites or enrichment planting. Further research can investigate the variation in mapping accuracy of diversity and evenness using high-resolution image and with less number of field samples in other areas of the park.

Competing interests
The author declares that he has no competing interests.

Acknowledgement
The article is a partial output of a MSc research done in the Natural Resources (NRs) department of ITC, University of Twente, The Netherlands under Erasmus Mundus scholarship scheme of the European Union. Majella national park authority provided the shape files of forest types, accommodation and species identification supports during the field campaign.

References

Bolstad PV, Swank W, Vose J (1998) Predicting Southern Appalachian over story vegetation with digital terrain data. Landsc Ecol 13:271–283
Civco DL (1993) Artificial neural networks for land-cover classification and map. Int J Geogr Inf Syst 7:173–186
Cohen J (1960) A coefficient of agreement for nominal scales. Educ Psychol Meas 20:37–46
Congalton RG, Green K (1999) Assessing the Accuracy of Remotely Sensed Data: Principles and Practices, Mapping Sciences Series. Lewis Publishers, Boca Raton, Florida
Conti F (1998) Flora D'Abruzzo: An annotated check-list of the flora of the Abruzzo. Herbarium Mediterraneum Panormitanum, Palermo
Currie DJ (1991) Energy and large-scale patterns of animal and plant species richness. Am Nat 137(1):2–24
Currie DJ, Paquin V (1987) Large scale bio-geographical pattern of species richness of trees. Nature 329:326–331
Duro DC, Coops NC, Wulder MA, Han T (2007) Development of a large area biodiversity monitoring system driven by remote sensing. Prog Phys Geogr 31:235–60
Gillespie TW, Foody GM, Rocchini D, Giorgi AP, Sassan Saatchi S (2008) Measuring and modelling biodiversity from space. Prog Phys Geogr 32(2):203–221
Guisan A, Edwards TC, Hastie T (2002) Generalized Linear and generalized additive models in studies of species distributions: setting the scene. Ecol Model 157:89–100
Hamilton AJ (2005) Species diversity or biodiversity? J Environ Manag 75:89–92
Jiang Y, Kang M, Zhu Y, Xu G (2007) Plant biodiversity patterns on Helan Mountain, China. Acta Oecologica 32:125–133
Jing Wei J, Deng H, Wu G, Zhao J (2004) Biological diversity in the coniferous forest on the northern slope of Changbai Mountain, northeast China. Int J Sustain Dev World Ecol 11:181–190
Kent M, Coker P (1992). Vegetation description and analysis: A practical approach, John Wiley and Sons, INC
Parco Nationale della Majella (1999). Realizzazione degli studi preliminarie e dell'elaborato technico del piano del parco e del regolamento. Volume 2
Majella National Park (Parco Nazionale della Majella) (2011). Information from the official website, Available at http://www.parcomajella.it, accessed on 25 May, 2011
Manel S, Williams HC, Ormerod SJ (2001) Evaluating presence-absence models in ecology: the need to account for prevalence. J Appl Ecol 38:921–931
Marsden SJ, Pilgrim JD (2003) Diversity and abundance of fruiting trees in primary forest, selectively logged forest, and gardens on New Britain, Papua New Guinea. Trop Biodiv 8(1):15–29
Mouillot D, Leprêtre A (1999) A comparison of species diversity estimators. Res Popul Ecol 41:203–215
Mutanga O, Kumar L (2007) Estimating and mapping grass phosphorus concentration in an African savannah using hyper spectral image data. Int J Remote Sens 28(21):4897–4911
Mutanga O, Rugege D (2006) Integrating remote sensing and spatial statistics to model herbaceous biomass distribution in a tropical savannah. Int J Remote Sens 27(16):3499–3514
O'Brien EM, Field R, Whittaker RJ (2000) Climatic gradients in woody plant (tree and shrub) diversity: water energy dynamics, residual variation and topography. Oikos 89(3):588–600
Pielou EC (1975). Ecological diversity. John Wiley and Sons, INC
Pierce KB, Lookingbill T, Urban DA (2005) Simple method for estimating potential relative radiation (PRR) for landscape-scale vegetation analysis. Landsc Ecol 20:137–147
Redowan M (2013) Tree diversity detection with mid-resolution images and environmental data in a neural network. J Indian Soc Remote Sens 41(3):567–576
Redowan M, Riadh SM (2013) Effect of image band number and band multicollinearity on mapping accuracy of tree diversity and evenness. J Remote Sens GIS 4(2):14–22
Richerson PJ, Pierce Lum KL (1980) Patterns of plant species diversity in California: relation to weather and topography. Am Nat 116:504–536
Runkle JR (1982) Patterns of Disturbance in Some Old-Growth Mesic Forests of Eastern North America. Ecology 63:1533–1546
Shannon CE (1948) A mathematical theory of communication. Bell Sys Tech J 27:379–423
Shukla RS, Chandel PS (2000). Plant Ecology and Soil Science. 9th Edn. S. Chand & Company Limited, Ramgor, New Delhi-110055, India. pp 121–376
Skidmore AK, Turner BJ, Brinkh W, Pierce Knowles E (1997) Performance of a neural network: mapping forests using GIS and remotely sensed data. Photogram Engineer Remote Sens 63:501–514
Stanisci A, Pelino G, Pierce Blasi C (2005) Vascular plant diversity and climate change in the alpine belt of the central Apennines (Italy). Biodiv Conserv 14:1301–1318
Tivy J (1993). Biogeography - A Study of Plants in the Ecosphere. Longman, Harlow
Wallace AR (1878) Tropical Nature and Other Essays. Macmillan, New York
Wang G, Zhou G, Yang L, Li Z (2003) Distribution, species diversity and life-form spectra of plant communities along an altitudinal gradient in the northern slopes of Qilianshan Mountains, Gansu, China. Plant Ecol 1652:169–181
Whittaker RH (1965) Dominance and Diversity in Land Plant Communities. Science 14:250–260
Whittaker RH (1972) Evolution and measurement of species diversity. Taxon 21(2):213–251
Zhang J, Zhang F (2011) Ecological relations between forest communities and environmental variables in the Lishan Mountain Nature Reserve, China. Afr J Agric Res 6(2):248–259

11

Negative consequences of positive feedbacks in US wildfire management

David E Calkin[1*], Matthew P Thompson[1] and Mark A Finney[2]

Abstract

Over the last two decades wildfire activity, damage, and management cost within the US have increased substantially. These increases have been associated with a number of factors including climate change and fuel accumulation due to a century of active fire suppression. The increased fire activity has occurred during a time of significant ex-urban development of the Wildland Urban Interface (WUI) along with increased demand on water resources originating on forested landscapes. These increased demands have put substantial pressure on federal agencies charged with wildfire management to continue and expand the century old policy of aggressive wildfire suppression. However, aggressive wildfire suppression is one of the major factors that drive the increased extent, intensity, and damage associated with the small number of large wildfires that are unable to be suppressed. In this paper we discuss the positive feedback loops that lead to demands for increasing suppression response while simultaneously increasing wildfire risk in the future. Despite a wealth of scientific research that demonstrates the limitations of the current management paradigm pressure to maintain the existing system are well entrenched and driven by the existing social systems that have evolved under our current management practice. Interestingly, US federal wildland fire policy provides considerable discretion for managers to pursue a range of management objectives; however, societal expectations and existing management incentive structures result in policy implementation that is straining the resilience of fire adapted ecosystems and the communities that reside in and adjacent to them.

Keywords: Wildfire suppression; Wildfire paradox; Wildland urban interface; Resilience

Review

In the past half century, wildland fire managers in the US has developed an increasingly sophisticated suppression response organization, including large aircraft, smoke jumpers, infrared mapping, satellite detection, computer dispatching, fire simulation, and nationwide coordination. Between 2011 and 2014, federal expenditures on suppressing large wildfires have exceeded $1.5 each year (National Interagency Fire Center 2014), but wildfires are becoming larger and more expensive and damages to watersheds and communities are still rising. Do these outcomes evince success of a suppression-centric strategy? Are practical alternatives available? In this article we review the structure of US wildfire management organizations. Attention is given to the incentives for agencies as well as the public. Evaluation of incentives and consequences within the framework of actuarial risk assessment reveals that modern wildfire problems derive from the self-reinforcing cycle of counter-effective actions.

* Correspondence: decalkin@fs.fed.us
[1]Forestry Sciences Laboratory, US Forest Service Rocky Mountain Research Station, 800 East Beckwith, Missoula, MT 59801, USA
Full list of author information is available at the end of the article

Historical context

Wildfire has been long recognized as an essential and perpetuating process in the ecology of most North American forests and rangelands (Wright 1982). In some places today, such as the southern states fire has an accepted presence and is seen as a vital management tool (Fowler and Konopik 2007). By contrast, the western US contributes most of the suppression costs and damages and policies emphasizing fire exclusion have come to be regarded as both feasible and desirable. The evolution of current policy is well documented by historians (see for example Pyne 1982). Consequences of fire exclusion can be generalized from detailed ecological research that shows low and mid-elevation forests with relatively frequent fires have become denser and spatially continuous and support large crown fires (Hessburg et al. 2005).

Forests with long-interval fire regimes show changes in landscape patterns and proportions of age and structure (Keane et al. 2002). Grassland and shrubland ecosystems have also experienced changes in fire regime in the past century, with some losing diversity without fire (Brockway et al. 2002) and some because of increased fire frequency after invasion by exotic annual grasses that increase continuity and flammability under a wide range of weather conditions (Knapp 1996).

Perhaps the strongest evidence of changes to fire regimes in forested ecosystems caused by attempted fire exclusion comes from comparisons between contemporary management and direct accounts of management at the close of the 19th century. The series of Annual Reports to the Department of Interior regarding the forest reserves from 1897–1905 offers a broad set of descriptions of forests and fire activities throughout the west and depicts startling contrasts with contemporary conditions. In California, for example, the 257,314 acre Rim Fire in 2013 (Lydersen et al. 2014) and the 97,717 acre King fire in 2014 burned as extensive crown fires through multiple elevation zones in the west-central portion of the Sierra Nevada mountains. Little more than a century before, however, Sudworth (1900) described a universal surface fire regime in forests of the Stanislaus and Tahoe Forest Reserves:

> *"The fires of the present time are peculiarly of a surface nature, and with rare exception there is no reason to believe that any other type of fire has occurred here. Parts of the older forests may have had a deep humus, which, being burned, would have destroyed timber by deep burning at the roots. But there being no humus at the present time, deep burning is impossible. The tree roots are for the most part buried deep in the crevices of bare rock, in gravel, sand, or shale, over which surface fires run annually without the slightest direct injury to the roots. Barring the debris left from timbercutting, the only food for these fires is the scanty fall of pine and fir needles, irregular patches of low conifer seedlings, and chaparral. In general, these materials limit the fires to surface burning" (Sudworth* 1900*).*

The wholesale conversion of California forests from a surface fire regime to the modern infrequent crown fire (Show and Kotok 1924) occurred rapidly because fires were easily suppressed in those early fuel conditions and the high productivity of the California environment generated biomass quickly. The same trends are documented in low- and mid-elevation forests throughout the west (Keane et al. 2002; Hessburg et al. 2005; Pechony and Shindall 2010) but more slowly where productivity is lower (e.g. Colorado). In California, both factors encouraged foresters to think that fire could be excluded to grow more trees. Several decades later, Show and Kotok (1924) recorded effects of pursuing total fire exclusion:

> *"The establishment of the national forests in California, beginning as early as 1891, thus found forest burning an established practice. The idea that fires could be excluded entirely from millions of acres was generally regarded as preposterous and the most gloomy pictures were drawn of any such attempt. It was claimed that the uncontrollable crown fire was to be expected as the inevitable consequence of allowing ground cover and litter to accumulate. Thus, in the early years of protection of the national forests, the forests were still open as a result of the repeated fires of the past. The great outbreak of incendiarism and agitation for light burning did not come until later. As fire protection became an accomplished fact and the young growth began to fill up the open forest, the amount of inflammable material in the forests increased greatly."*

Even with the advantage of witnessing firsthand the fuel and fire transformations, advocates for prescribed fire ("light burning") in the early 1900s had political difficulties reversing the trend because of the primacy of timber-management perspectives. Show and Kotok (1924) detail considerable evidence against early attempts to reintroduce frequent surface burning resulting in damage to timber and killing of tree reproduction. Nowadays, forest scientists and ecologists find these historical reports convincing evidence that attempted "protection" paradoxically caused forest destruction. Protection really was not "an accomplished fact" because wildfire could not be permanently deferred – only changed in character when it inevitably occurred. However, the existing incentives and organizational structure maintain this pattern of wildfire suppression response.

Modern context

Increasing wildfire activity, damage, and associated management cost within the US (Williams 2013) is driven by a complex web of social and ecological factors (Spies et al. 2014). Although disturbance is a critical characteristic of many ecological systems that have evolved with wildfire and human communities over the past several millennia (Stewart 2002), fire control remains the dominant management paradigm in the western US beginning in the 20th century (Pyne 1982). Now, after more than a century of aggressive suppression the wildfire paradox (Arno and Brown 1991) is fully realized in most western forests. Reduced wildfire on the landscape has led to increased fuel loading and continuity on most forested landscapes in the western US. Under extreme but not infrequent conditions suppression is less successful particularly where fuels have accumulated (Arno and Brown 1991; Stephens and Ruth 2005) and impacts to natural and developed resources are greater. Essentially, through our management efforts we have changed the distribution of fire behavior to only the most extreme. Public land management agencies have responded to this increase in wildfire activity with increased effort to remove wildfire from the landscape through aggressive initial attack (IA) and extensive wildfire suppression efforts aimed at minimizing the size of and/or damage from those few fires that escape IA. Although IA rates appear to be relatively stable (averaging between 97 and 99 percent successful, Calkin et al. 2005), the amount of area burned from large fires has been increasing over the last several decades (see Figure 1) further prompting increased wildfire suppression effort.

A report by USDA Forest Service (2014) documents the financial scale of wildfire management activities including large fire suppression and associated consequences to the Agency's budget. Large wildfire suppression activities by US Federal agencies have cost nearly $24 billion between 2000 and 2013 (adjusted to 2013 US$). This does not include state and local government costs nor does it include federal cost for pre-suppression activities (planning, equipment and labor acquisition), wildfire initial attack, or hazardous fuels management. The USDA Forest Service maintains the largest federal wildfire organization representing approximately 70 percent of all federal expenditures on wildfire management. In 1995 wildfire related expenses represented 17 percent of the US Forest Service's appropriated funds, by 2014, 51 percent of the appropriated funds were related to wildfire management. This increased demand for wildfire suppression has created a budgetary cycle where ever increasing demands for wildfire suppression funding come at the cost of other public land management programs, some of which are intended to directly reduce future wildfire damage. During this period of time, the Agency's budget has not increased commensurately with the rise in wildfire management expenditures. Thus other major budget items have experienced significant reduction; for example vegetation and watershed management (22 percent reduction), facilities (67 percent reduction), roads (46 percent reduction), and deferred maintenance (95 percent reduction) (USDA Forest Service 2014). Increasing wildfire damage to human and ecological systems compromise public land management agencies such as the USDA Forest Service's ability to maintain key ecological function and the provisioning of ecosystem services.

Drivers of wildfire response

Although a number of recent articles have focused on the application of active forest management of fire adapted ecosystems to improve social and ecological conditions (see for example Franklin et al. 2014; Spies et al. 2014) we examine the drivers of wildfire management response and

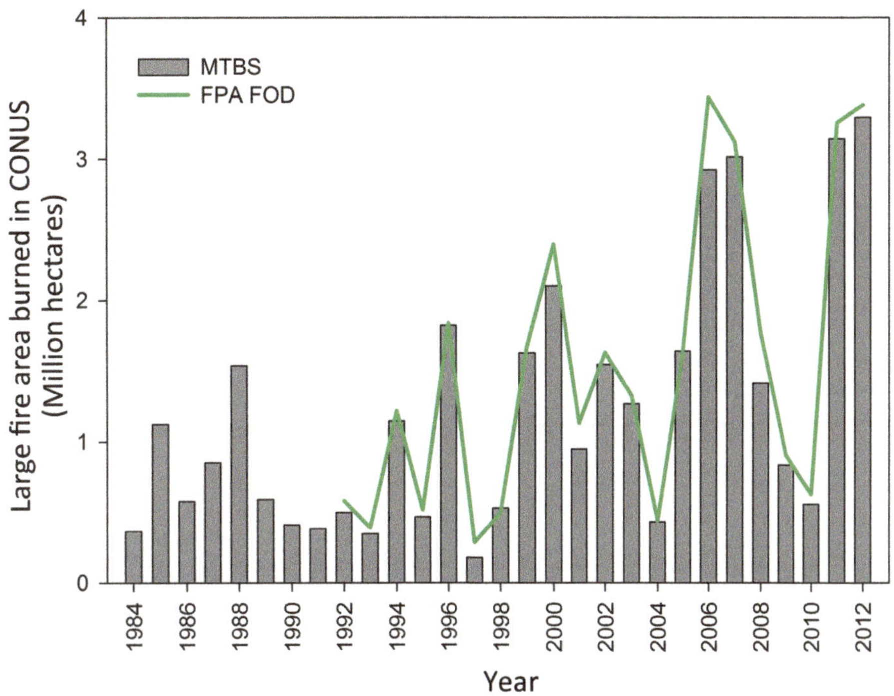

Figure 1 Estimates of area burned by large wildfires in the contiguous United States (CONUS), 1984–2012. A large wildfire is defined here as ≥ 405 ha in the Western US, ≥ 203 ha in the Eastern US. Estimates for 1984–2012 (bars) are based on large perimeters mapped as part of the Monitoring Trends in Burn Severity (MTBS) project (Eidenshink et al. 2007). An additional set of estimates for 1992–2012 (line) is based on records of wildfires > 405 ha included in the Fire Program Analysis Fire-Occurrence Database (Short 2014).

implications to the implementation of wildfire policy and associated consequences. Figure 2 demonstrates primary managerial, social, and ecological drivers of wildfire management response in both the fire season that is currently being managed and how these factors affect the trajectory of wildfire risk and associated management response into the future. In this simplified framework, wildfire risk is jointly determined by the landscape hazard and the susceptibility of values-at-risk (Finney 2005). Hazard reflects forest and fuel conditions, and is typically quantified in terms of the likelihood and intensity of wildfire (Scott et al. 2013). The susceptibility of resources (e.g., habitat) and assets (e.g., homes) is often framed as potential loss, however for some resources fire can lead to improved ecological condition (Scott et al. 2014). The critical point of this figure is that all of the primary drivers of wildfire suppression response, both within the current season and the trajectory for future response, drive an increasing suppression response with consequences to the trajectory of future wildfire risk and associated loss.

Wildfire risk drives managers to attempt to suppress wildfires to reduce potential resource loss from the current event. Economic efficiency suggests that the cost of the last resource assigned to manage a wildfire should be equal to the reduction in net loss from the assignment of that resource. Although US federal wildland fire policy does not specifically direct managers to seek economically efficient strategies it does dictate that the cost of the management response should be commensurate with values to be protected (Interagency Working Group 2001). Despite guidance on the application of economic principles to wildfire management there are many reasons to believe that current suppression response is excessively risk averse. Specifically, the loss of value to resources due to wildfire may be less than the cost of the suppression response developed to protect those values. In a choice survey of federal wildfire managers Calkin et al. (2013) demonstrated that, all else equal, mangers were more likely to select wildfire suppression strategies with higher suppression costs after accounting for potential risk to homes, ecological values, and firefighter exposure.

Wildfire management response that promotes increased application of wildfire may be hindered by societal expectations of the suppression only orientation of past management practices (Steelman and McCaffrey 2013). Socio-political influence from politicians, landowners and the affected public on fire managers increases the level and cost of suppression response (Canton-Thompson et al. 2008). For example Donovan et al. (2011) demonstrated that per area cost of

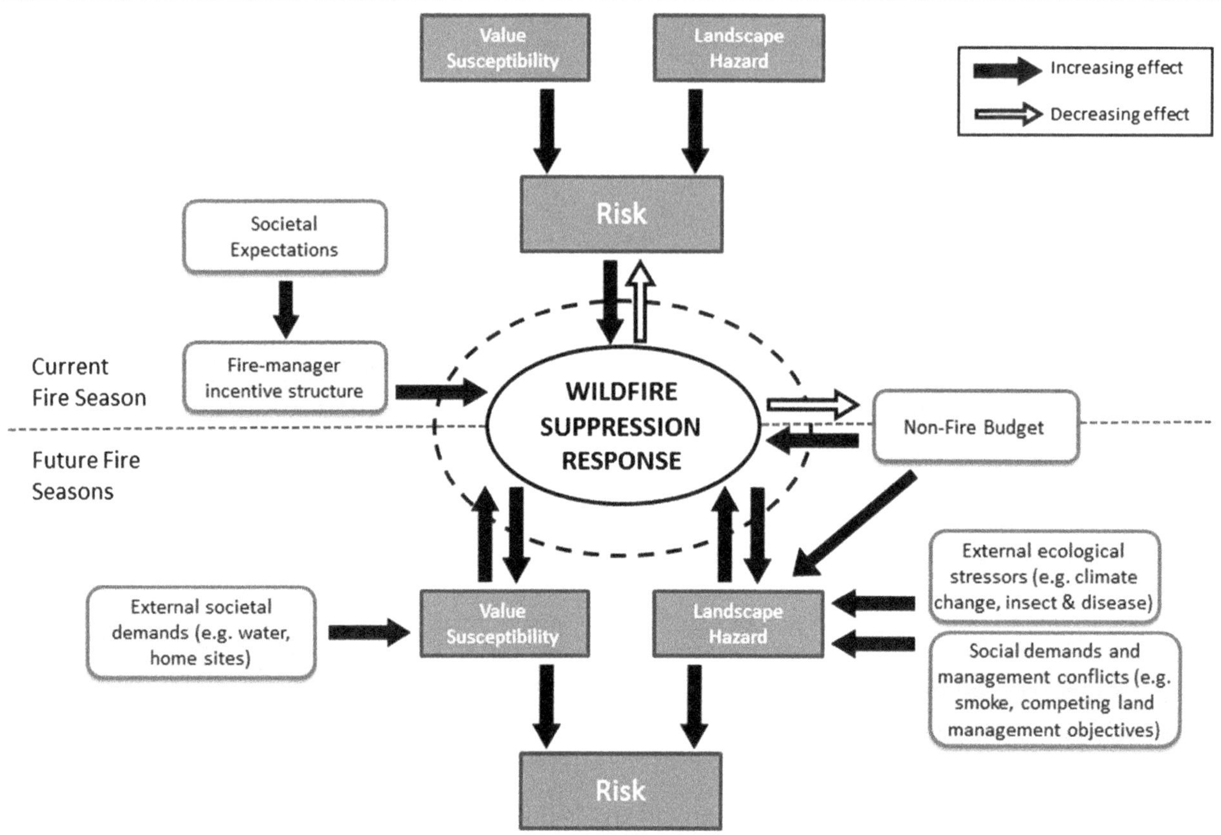

Figure 2 Conceptual model of managerial, social, and ecological drivers of near-term and future wildfire management response.

managing a wildfire was positively associated with the seniority of the congressional representative of the district in which the fire occurred and the amount of media coverage. In addition to the positive socio-political influence on the level of suppression response, the existing incentive structure and budget process provides only weak feedback in terms of internalizing the impact of wildfire suppression spending on decision makers charged with managing wildland fires (Donovan and Brown 2005; Thompson et al. 2013).

Further, federal fire managers exhibit decision biases and heuristics that are common to environments characterized by significant complexity and uncertainty, which can lead to suboptimal decisions and outcomes (Thompson 2014). When selecting strategies, fire managers have displayed the discounting bias (overemphasizing short-term risk reduction over longer-term considerations), the status quo bias (reverting to suppressing all fires relative to allowing fires to burn for ecological considerations), and the loss aversion bias (preferring safe options when consequences are framed as potential gains) (Wilson et al. 2011). Fire managers can also exhibit systematic errors in estimating and interpreting probabilistic information (such as underestimating likely outcomes while overestimating rare events), which can lead to poor risk management (Donovan and Noordijk 2005; Maguire and Albright 2005; Wibbenmeyer et al. 2013). Knowledge gaps and limited understanding of the socioeconomic and ecological consequences of fire engender further uncertainty when evaluating alternative courses of action (Thompson and Calkin 2011; Venn and Calkin 2011; Hyde et al. 2013) which may lead to managers pursuing aggressive suppression strategies under conditions with low levels of risk to valued assets (Wibbenmeyer et al. 2013).

Budgetary implications of spending on wildfire suppression have both immediate and long term negative consequences. Over recent years, spending on current wildfire seasons have frequently resulted in transferring budget away from other land management activities including those that could reduce landscape hazard such as fuel reduction treatments. However, it has become increasingly clear that the impact of increasing fire management costs on non-fire related expenditures may have more significant long term implications to the agencies' ability to meet their land management objectives than issues associated with in season fund transfers (USDA Forest Service 2014). In 2014 there has been an active effort within the US Congress and Obama administration to fund a portion of wildfire management expenditures similar to current natural disaster funding under the Federal Emergency Management Agency. By funding wildfire suppression for large wildland fires under such an approach the uncertainty associated with changes in annual fire expenditures due to different levels of fire activity and the downward funding trends on non-fire programs would be simultaneously addressed.

Under certain conditions wildland fire can promote significant ecological benefits and reduce the risk that high intensity fires will spread to areas of high value into the future (Noss et al. 2006). Although there is broad scientific consensus regarding the need for more fire on the landscape, almost all fire management entities within western states require and promote aggressive suppression of all wildfires. Within federal land management agencies knowledge and experience allowing wildfires to burn for beneficial effects has been growing over time and is established within the land management objectives of the different federal agencies. The published 2009 interpretation of US Federal Wildfire Policy (Fire Executive Council 2009) expanded the opportunity of managers to better consider beneficial aspects of wildfire when determining fire management strategies. However, achieving increased beneficial wildfire opportunities in practice has been far more challenging. Existing land and fire management plans have not sufficiently considered the role of existing beneficial use on future fire management opportunities (Doan et al. 2006; Calkin et al. 2011) and managers are subject to a status quo bias that makes them reluctant to select beneficial use strategies when full suppression has been previously established (Wilson et al. 2011). Despite increased scientific recognition of the need to allow wildfire in fire adapted ecosystems to burn for resource benefit and examples of successful wildland fire use programs suppression persists as the dominant fire management strategy.

Landscape hazard & value susceptibility

Concurrent with the increased wildfire activity has been a substantial increase in residential development within the Wildland Urban Interface (WUI) (Theobald and Romme 2007). Despite the vigorous fire suppression near human communities (Gebert et al. 2007; Liang et al. 2008; Gude et al. 2013) the number of homes destroyed by wildfire continues on an upward trend. Wildfire-related insured losses in the United States in the 10 year span of 2002 to 2011 totaled $7.9 billion (US). This represents a $6.2 billion increase over the previous decade (Haldane 2013). Although wildfire risk reduction near the WUI has been targeted by public land management agencies, failure to consider the conditions under which loss occurs may limit the effectiveness of these investments in reducing residential property loss (Calkin et al. 2014). Further, certain landscape conditions and fire regimes are more amenable to fuel treatments and community capacity and institutional factors have a strong influence on social acceptability (Spies et al. 2014) and capacity to act thus requiring context specific risk reduction strategies (Moritz et al. 2014).

A primary tool for land managers to address increasing landscape hazard is hazardous fuels reduction treatments. Within the federal fuel treatment programs a majority of funding is directed towards treatments within the WUI. However, it has been well demonstrated that residential home destruction is primarily determined by the immediate surrounding (30–60 m) of the home, known as the home ignition zone (HIZ) (Cohen 2000, 2010). The HIZ principally determines home ignition potential during extreme wildland fire behavior and includes the home construction characteristics and its immediate surroundings, in most case largely on private land. Focusing solely on wildland vegetation without consideration and mitigation of HIZs furthers the illusion of WUI protection without homeowner engagement. Prioritizing public investments on fuel reduction efforts and wildfire suppression in and around the WUI reduces the true cost of housing location decisions thus incentivizing development in high wildfire hazard areas and need for increased future investment (Cleetus and Mulik 2014). Significant progress has been made in the development of community based wildfire protection plans however, land use zoning restrictions and requirements to reduce structure ignitability remain contentious issues (Jakes et al. 2011).

Beyond the increasing risk to human development, the increasing population within the western US drives increased demand for municipal water. In the western US, half of all water originates on lands administered by the US Forest Service (Brown et al. 2008). Severe wildfires can result in significant and costly impacts to municipal watersheds, ranging from increased sedimentation to increased likelihood of debris flows to damage of water delivery infrastructure and interruption of service, prompting additional interest in risk mitigation options (Warziniack and Thompson 2013; Tillery et al. 2014).

Prescribed fire has been well recognized within the scientific literature (Hann and Bunnell 2001; Graham et al. 2004; Martinson and Omi 2013) as a fuel modification and restoration technique that is highly effective at mitigating wildfire behavior. Hann and Bunnell (2001) proposed that fire and land management planning focus fuel treatment efforts on reducing the current level of departure of forested lands from their historical fire regime to address increasing wildfire concerns within the United States. Although challenges emerged to define and manage towards departure many of the original concepts are present in the current emphasis on restoring the condition of dry frequent fire pine forests in the western US. There is general scientific agreement that fuel loading and stand density have reduced the resilience of these forests by making them more susceptible to wildfire and insect and disease outbreaks (Allen et al. 2002; Hessburg et al. 2005; Noss et al. 2006). In general society has a preference for active management to address wildfire risk and degrading ecological conditions of fire adapted forests (McCaffrey et al. 2012). However, societal factors such as inadequate funding, conflicts among objectives and priority of resource values as well as limited public understanding have created significant delay in the implementation of these programs to restore forest resilience to fire (Franklin et al. 2014; Rideout et al. 2014). Additionally, the current scale of fuel treatment is far less than required to achieve landscape resilience (North et al. 2012). Along with societal barriers, conflicting priorities within land management plans and mandates within environmental regulations such as the Clean Air Act and the Endangered Species Act limit the ability to achieve substantial reduction of wildfire hazard (Calkin et al. 2011).

Future fires and maladapation to risk

Pyne (2011) summarizes the condition of fire management in the US as follows: "What is striking about the American style of fire is how technically robust it is, and how politically dysfunctional and inept in practice so much of it has become." In the face of the increasing wildfire risk and highly damaging events, political responses have typically focused on increasing the suppression response (Busenberg 2004). Wildfire suppression without a commensurate program to address the fuel accumulation resulting from the aggressive suppression policy represents a major policy error in federal fire management (Busenberg 2004). This policy error was propagated with mounting impacts as federal land management agencies acquired new resources and influence to reinforce the established institutional wildfire management approach of aggressive suppression. This self-reinforcing action has caused the impacts of the original policy failure to gradually escalate over time. The self-reinforcing nature of aggressive suppression response is not unique to wildfire management in the US and has been demonstrated using a system dynamics modelling approach to wildfire management in Portugal (Collins et al. 2013). Further the authors identified three primary reasons why transformation of this self-reinforcing and costly behavior is unlikely to occur organically due to 1) managerial incentive structures focusing on short term results, 2) challenges to identifying when mitigation resulted in damages averted and the inability to take credit, and 3) the financial interests of established firefighting organization in maintaining current policy. Spies et al. (2014) argued that challenges such as heterogeneity in wildfire behavior and effects, human behavior and values and weak landscape feedbacks to humans create a highly complex system that may lead to maladaptive behavior and unintended consequences of fire policy.

State agencies charged with wildfire management typically have less flexibility in their suppression policy. With few exceptions most western state agencies are directed to suppress wildfires at the smallest size feasible. Further, as many of the states have experienced highly damaging events the typical response has been to increase the pre-established emphasis on aggressive suppression of all wildfires. For example, the state of Colorado experienced several highly damaging wildfire events including the Fourmile Canyon Fire of 2011 (169 homes destroyed, zero fatalities), High Park Fire (259 homes destroyed, one fatality), Waldo Canyon Fire of 2012 (346 homes destroyed, zero fatalities), and the Black Forest Fire of 2013 (486 homes destroyed, two fatalities). Additionally, the Lower North Fork Fire in 2012, an escaped prescribed fire conducted by the Colorado State Forest Service, burned 23 homes and killed three people. In a response to this event responsibility for wildfire management was moved from the Colorado State Forest Service to the Department of Public Safety. Additionally, the state has implemented new restrictions and additional requirements for the application of prescribed burning and approved a bill to fund a fleet of state owned aerial wildfire suppression resources. The intent of the investment in aerial resources was stated as follows: "Keep all wildfires with values at risk smaller than 100 acres and to suppress all fires in Wildland Urban Interface (WUI) areas at less than ten acres, 98% of the time." (Colorado Department of Fire Prevention and Control 2014). This demonstrates an instance when a wildfire agency defines target conditions of preparedness and capability below the large fire-disaster levels. Highly damaging wildfires such as the Waldo Canyon, Fourmile, and Black Forest Fires occur during the 2 percent of events where suppression is not effective (Calkin et al. 2014).

Breaking the cycle

Policies and actions to reduce the cycle of ever increasing wildfire suppression effort, management costs, and resource losses will be challenging to implement. Inertia of the existing social systems habituated by the current management paradigm is entrenched in social expectations and agencies' cultures. Significantly reducing fuel loading associated with the current condition of the forested landscape through active management would require huge capital investment and conflict with other existing environment regulations. The 2009 Interpretation of Federal Wildfire Policy opened up opportunities to manage wildfires for resource benefit within the wildfire suppression program. However, in recent years managing fires by allowing them to burn under certain conditions to achieve resource benefit has proved challenging (Hubbard 2012).

Concepts of system resilience may provide an interesting lens to examine current challenges in wildfire management. Resilience thinking extends previous efforts for sustainable resource management such as ecosystem analysis and adaptive management by further emphasizing the critical linkages between social and ecological systems (Rist and Moen 2013). Resilience can be defined as "the capacity of a system to absorb disturbance, re-organize, and keep functioning in much the same way as before" (Walker 2013). Managing towards resilience can be characterized as actions that maintain a desirable state or transform existing structure to achieve a more desirable state (Walker et al. 2006). Resilience by itself at one spatial and temporal scale does not imply that a system is in a beneficial state. Many types of ecological or social systems may be highly resilient, but result in reduced ecological condition or social wellbeing. For example cheat grass invaded ecosystems in the Western US are highly resilient to disturbance. Thus, resilience requires looking at systems at different scales – achieving resilience at one level may require transformation at other levels (Walker 2013). The current wildfire management policy in the US has slowly evolved over the last century but has been relatively consistent; aggressively suppress all wildland fires. Altering the trajectory of risk will require system transformation. There are several possible trajectories for the future of wildfire management. As describe earlier, maintaining the status-quo has obvious implications in terms of increasing wildfire risk, increased damage and loss due to wildfire, along with significant consequences to the structure and functioning of public land agencies and their ability to meet their core missions. Alternatively, wildfire management could be driven to transform by the public demanding an alternative wildfire management approach and/or by transformative actions initiated within federal agencies themselves.

Transformation is frequently initiated by a series of highly salient events that create a dramatic shift in public expectations and demands from existing social structures. The scale of event that would lead to US society demanding a transformation of current wildfire policy remains unknown. O'Neill and Handmer (2012) suggest that the Black Saturday events in Victoria and New South Wales Australi, where 172 civilians were killed represent such a transformative event. The significant media and public attention caused by the scale of loss of life resulted in a critical examination of bushfire management in general and more specifically the implication of the Stay and Defend Policy and prescribed fire program. Despite the increased political and media focus associated with the events of Black Saturday, there is some concern that homeowner response to wildfire may not have changed and that existing public behavior has not transformed in light of these events. Specifically, approximately one-third of respondents to a survey targeted at Australian households at risk from bushfire just one year after the Black Saturday fires was to wait and see what happens during the fire, but leave if threatened by the fire (Rhodes 2011). Essentially this represents a strategy of evacuating late which is in complete contrast to the stated policy in Australia. Thus, it is probably too early to determine if in fact the Black Saturday events result in a transformation of Australian fire management policy and implementation.

Systems can adaptively transform in ways that allow the system to better handle changing conditions, stress, hazards, risk or opportunities in the future (see Smit and Wandel 2006). Such an adaptive transformation would require a range of alternative approaches to be pursued including increased levels and improved efficiency of fuel reduction treatments, alternative management structures that encourage less aggressive suppression strategies that incur broader ecological outcomes and reduce future wildfire hazard, improved risk sharing among public land managers and interface communities through more fire resistant structure design, modified fuel conditions adjacent to communities, and zoning restrictions of further development within the most fire prone areas (Cleetus and Mulik 2014).

If wildfire suppression funding was handled under a disaster relief funding mechanism, continued stress on fuel reduction funding and other active management that may reduce wildfire risk could be alleviated. However, unless the scale of fuels funding were dramatically increased along with relief from conflicting environmental regulations it is unlikely that the scale of fuel modification will be accelerated to reduce the wildfire hazard in the short term. Therefore, although it does not appear that fuels treatments activities could replace wildfire in achieving ecological resilience of forested ecosystems in the western US, they will likely play a critical role in positive adaptive management response. Recognizing this reality, Reinhardt et al. (2008) identified the importance of fuel treatment in creating landscapes where fire can occur without devastating consequences; recognizing that we cannot, nor should not expect fuel treatment programs to replace naturally occurring wildfire. Achieving this objective requires focusing treatments such that, in the event of a wildfire, spread and intensity is such that significant loss to highly valued resources is minimized.

Reduced fuel loading and improved forest health is a necessary but not sufficient condition for reducing wildfire risk. The range of ecosystem conditions, fire regimes and land uses suggests a one size fits all approach to wildfire and fuels management is neither desirable nor feasible (Keeley et al. 2009). In many communities fuels reduction activities may be quite challenging due to a range of factors such as rapid regrowth, high management cost, and complex ownership patterns. Further in some areas such as lodgepole pine forests in the interior west, the natural fire regime is infrequent high intensity fire and reasonable fuels reduction treatments are limited. Under these circumstances reducing the susceptibility of highly valued resources to wildfire may be more appropriate. As Dombeck et al. (2004) state, "communities need to shoulder greater responsibility for regulating sprawl and for encouraging proactive efforts by homeowners to reduce the risk of home ignition during wildfire."

Within the Southeastern US a culture of aggressive prescribed fire has developed gradually over the past 50 years with the reduction of fire hazard as the primary management objective across ownerships (Fowler and Konopik 2007). By maintaining fuel conditions through regular burning and the development of a community of practitioners forest lands are generally maintained in a condition of relatively low hazard with management cost far less on a unit basis than equivalent treatments in the Western US. For example, National Forest land in the southeastern US (US Forest Service Region 8) represents only 10 percent of the total Forest Service land base, but between 2007 and 2012 the region accounted for almost 70 percent of the Agency's prescribed burned area while accounting for only 6 percent of large fire suppression costs (USDA Forest Service internal reports).

By allowing naturally ignited wildfires to burn to achieve resource benefit (formerly known as wildland fire use), fuel conditions may be improved thus reducing wildfire risk in adjacent areas. The reduced wildfire risk may further increase the area and weather conditions where wildfires need not be aggressively suppressed. In several wilderness areas throughout the Western US, such as the Selway-Bitteroot and Bob Marshall Wilderness Areas in Montana, and the Gila Wilderness in New Mexico, once reduced fuel loading and barriers to fire spread associated with past fire scars have been achieved, wildfire can play a role in maintaining these landscapes (Teske et al. 2012; Parks et al. 2014). Despite the critical role past fire scars have played in the suppression of large fires (Graham 2003; Cochrane et al. 2012), extensive application of the use of naturally ignited wildfires has not yet been well demonstrated outside large wilderness areas. Recognition that fire exclusion is neither desirable nor possible in many regions of the country is an essential (Moritz et al. 2014) and will be a critical component of any effort to transform wildfire management.

Conclusions

Examination of ecological and human community resilience in fire adapted ecosystems suggests a social ecological system that is under considerable stress. Increasing wildfire hazard due to fuel accumulation and climate related stressors and increasing vulnerability of developed residential communities and natural resource values suggest a future of increasing risk and management cost unless the current management paradigm is transformed. For transformation to succeed it will require increased recognition of the consequences associated with the current paradigm, social acceptance of alternative fire management strategies, and alteration of the culture of public agencies charged with wildfire suppression. Specifically actions include: 1) enhanced risk sharing among affected partners; 2) modification of managerial incentive structures and enhanced training; 3) land management treatments that directly address local risk factors and align with broad risk reduction strategies; 4) reduction of the uncertainty around outcomes from less aggressive suppression response through improved decision support; and 5) enhanced consideration of long term impacts of current decisions (visioning).

Fortunately, just like the current wildfire suppression paradox, many of the necessary changes are themselves self-reinforcing. The self-reinforcing nature of the use of fire to achieve land management objectives has been well demonstrated on selected public lands (Parks et al. 2014) and the Southeastern US has effectively used prescribed burning to lower wildfire risk and mitigation costs.

A one size fits all approach to reducing wildfire risk does not exist. Each high risk landscape will require a suite of actions to modify the trajectory of increasing suppression demand and loss. For society to demand transformation of the current wildfire management paradigm some sort of tipping point event may need to occur. However, adaptive transformation where public agencies working alongside private citizens challenge the supremacy of the existing wildfire suppression model in favor of long term economically efficient risk mitigation strategies is possible.

Competing interests
The authors declare that they have no competing interests.

Authors' contributions
DC developed the conceptual model and drafted the manuscript. MT contributed content on risk assessment and decision science and manuscript editing. MF provided wildfire historical context and review of wildfire behavior along with manuscript editing. All authors read and approved the final manuscript.

Author details
[1]Forestry Sciences Laboratory, US Forest Service Rocky Mountain Research Station, 800 East Beckwith, Missoula, MT 59801, USA. [2]Fire Sciences Laboratory, US Forest Service Rocky Mountain Research Station, 5775 Highway 10 West, Missoula, MT 59808, USA.

References

Allen CD, Savage M, Falk DA, Suckling KF, Swetnam TW, Schulke T, Stacey PB, Morgan P, Hoffman M, Klingel JT (2002) Ecological restoration of southwestern ponderosa pine ecosystems: a broad perspective. Ecol Appl 12:1418–1433

Arno SF, Brown JK (1991) Overcoming the paradox in managing wildland fire in western wildlands. University of Montana, Montana Forest and Conservation Experiment Station, Missoula, MT, pp 40–46

Brockway DG, Gatewood RG, Paris RB (2002) Restoring fire as an ecological process in shortgrass prairie ecosystems: initial effects of prescribed burning during the dormant and growing seasons. J Environ Manage 65(2):135–152

Brown TC, Hobbins MT, Ramirez JA (2008) Spatial distribution of water supply in the Coterminous United States. J Am Water Resour As (JAWRA) 44(6):1474–1487

Busenberg G (2004) Wildfire management in the United States: the evolution of a policy failure. Rev Policy Res 21(2):145–156

Calkin DE, Gebert KM, Jones JG, Neilson RP (2005) Forest Service large fire area burned and suppression expenditure trends. 1970–2002. J For 103(4):179–183

Calkin D, Finney MA, Ager AA, Thompson MP, Gebert KG (2011) Progress towards and barriers to implementation of a risk framework for Federal wildland fire policy and decision making in the United States. For Policy Econ 13(5):378–389

Calkin DE, Venn TJ, Wibbenmeyer MJ, Thompson MP (2013) Estimating US federal wildland fire managers' preferences toward competing strategic suppression objectives. Int J Wildland Fire 22(2):212–222

Calkin DE, Cohen JD, Finney MA, Thompson MP (2014) How risk management can prevent future wildfire disasters. Proc Natl Acad Sci 111(2):746–751

Canton-Thompson J, Gebert KM, Thompson B, Jones G, Calkin D (2008) External human factors in incident management team decisionmaking and their effect on large fire suppression expenditures. J For 106(8):416–424

Cleetus R, Mulik K (2014) Playing with fire: how climate change and development patterns are contributing to the soaring costs of western wildfires., Available at: http://www.ucsusa.org/assets/documents/global_warming/playing-with-fire-report.pdf. Accessed September 9, 2014

Cochrane MA, Moran CJ, Wimberly MC, Baer AD, Finney MA, Beckendorf KL, Eidenshink J, Zhu Z (2012) Estimation of wildfire size and risk changes due to fuels treatments. Int J Wildland Fire 21(4):357–367

Cohen JD (2000) Preventing disaster, home ignitability in the wildland-urban interface. J For 98(3):15–21

Cohen J (2010) The wildland-urban interface fire problem. Fremontia 38(2, 3):16–22

Collins R, de Neufville R, Claro J, Oliveira T, Pacheco A (2013) Forest fire management to avoid unintended consequences: a case study of Portugal using system dynamics. J Environm Manage 130:1–9

Colorado Department of Fire Prevention and Control (2014) Special report: Colorado Firefighting Air Corps., Available at: http://dfs.state.co.us/. Accessed October 30, 2014

Doan D, O'Lauglin J, Morgan P, Miller C (2006) Barriers to wildland fire use: a preliminary problem analysis. Int J Wilderness 12:36–38

Dombeck MP, Williams JE, Wood CA (2004) Wildfire policy and public lands: Integrating scientific understanding with social concerns across landscapes. Conserv Biol 18(4):883–889

Donovan GH, Brown TC (2005) An alternative incentive structure for wildfire management on national forest land. For Sci 51(5):387–395

Donovan GH, Noordijk P (2005) Assessing the accuracy of wildland fire situation analysis (WFSA) fire size and suppression cost estimates. J For 103(1):10–13

Donovan GH, Prestemon JP, Gebert K (2011) The effect of newspaper coverage and political pressure on wildfire suppression costs. Soc Nat Res 24(8):785–798

Eidenshink J, Schwind B, Brewer K, Zhu Z, Quayle B, Howard S (2007) A project for monitoring trends in burn severity. Fire Ecol (Special Issue) 3:3–21

Finney MA (2005) The challenge of quantitative risk assessment for wildland fire. For Ecol Manage 211:97–108

Fire Executive Council (2009) Guidance for implementation of federal wildland fire management policy., http://www.nifc.gov/policies/policies_documents/GIFWFMP.pdf. Last Accessed July 23, 2014

Fowler C, Konopik E (2007) The history of fire in the southern United States. Human Ecol Rev 14(2):165–176

Franklin JF, Hagmann RK, Urgenson LS (2014) Interactions between societal goals and restoration of dry forest landscapes in western North America. Landscape Ecol. DOI 10.1007/s10980-014-0077-0

Gebert KM, Calkin DE, Yoder J (2007) Estimating suppression expenditures for individual large wildland fires. West J Appl For 22:188–196

Graham RT (2003) Hayman Fire Case Study. Gen. Tech. Rep. RMRS-GTR-114. U.S. Department of Agriculture, Forest Service, Rocky Mountain Research Station, Ogden, UT, pp 1–32

Graham RT, McCaffrey S, Jain TB (2004) Science basis for changing forest structure to modify wildfire behavior and severity. USDA For. Serv. General Technical Report RMRS_GTR120

Gude PH, Jones K, Rasker R, Greenwood MC (2013) Evidence for the effect of homes on wildfire suppression costs. Int J Wildland Fire 22(4):537–548

Haldane M (2013) Insurers, government grapple with costs of growth in wildland urban interface. Insurance J. Available at: www.insurancejournal.com/news/national/2013/08/15/301833.htm. Accessed September 9, 2014

Hann WJ, Bunnell DL (2001) Fire and land management planning and implementation across multiple scales. Int J Wildland Fire 10(4):389–403

Hessburg PF, Agee JK, Franklin JF (2005) Dry forests and wildland fires of the inland Northwest USA: contrasting the landscape ecology of the pre-settlement and modern eras. For Ecol Manag 211:117–139

Hubbard J (2012) 2012 Wildfire Guidance. Available at: https://www.documentcloud.org/documents/407523-2012-wildfire-guidance-memo-may-25.html. Accessed April 13, 2015

Hyde KD, Dickinson MB, Bohrer G, Calkin DE, Evers L, Gilbertson-Day J, Nicolet T, Tague C (2013) Research and development supporting fire effects prediction

for fire and fuels management: status and needs. Int J Wildland Fire 22(1):37–50
Interagency Working Group (2001) Review and update of the 1995 federal wildland fire management policy. National Interagency Fire Center, Boise, ID
Jakes PJ, Nelson KC, Enzler SA, Burns S, Cheng AS, Sturtevant V, Williams DR, Bujak A, Brummel RF, Grayzeck-Souter S, Staychock E (2011) Community wildfire protection planning: is the Healthy Forests Restoration Act's vagueness genius? Int J Wildland Fire 20(3):350–363
Keane RE, Ryan KCVTT, Allen CD, Logan J, Hawkes B (2002) Cascading effects of fire exclusion in the Rocky Mountain ecosystems: a literature review. Gen. Tech. Rep. RMRS-GTR-91. U.S. Department of Agriculture, Forest Service, Rocky Mountain Research Station, Fort Collins, CO, p 24
Keeley JE, Aplet GH, Christensen NL, Conard SG, Johnson EA, Omi PN, Peterson DL, Swetnam TW (2009) Ecological Foundations for Fire Management in North American Forest and Shrubland Ecosystems. USDA For. Serv. General Technical Report: PNW-GTR-779
Knapp PA (1996) Cheatgrass (*Bromus tectorum* L.) dominance in the Great Basin Desert: history, persistence, and influences to human activities. Global Environ Chang 6(1):37–52
Liang J, Calkin DE, Gebert KM, Venn TJ, Silverstein RP (2008) Factors influencing large wildland fire suppression expenditures. Int J Wildland Fire 17:650–659
Lydersen JM, North MP, Collins BM (2014) Severity of an uncharacteristically large wildfire, the Rim Fire, in forests with relatively restored frequent fire regimes. Forest Ecol Manag 328:326–334
Maguire LA, Albright EA (2005) Can behavioral decision theory explain risk-averse fire management decisions? For Ecol Manag 211(1):47–58
Martinson EJ, Omi PN (2013) Fuel treatments and fire severity: a meta-analysis. USDA For Serv Research Paper RMRS_RP103www
McCaffrey S, Toman E, Stidham M, Shindler B (2012) Social science research related to wildfire management: an overview of recent findings and future research needs. Int J Wildland Fire 23:567–576
Moritz MA, Batllori E, Bradstock RA, Gill MA, Handmer J, Hessburg PF, Leonard J, McCaffrey S, Odion DC, Schoennagel T, Syphard AD (2014) Learning to coexist with wildfire. Nature 515(7525):58–66
National Interagency Fire Center (2014) Available at: https://www.nifc.gov/fireInfo/fireInfo_documents/SuppCosts.pdf. Accessed February 23, 2015
North MP, Collins BM, Stephens SL (2012) Using fire to increase the scale, benefits, and future maintenance of fuels treatments. J For 110(7):392–401
Noss RF, Beier P, Covington WW, Grumbine RE, Lindenmayer DB, Prather JW, Schmiegelow F, Sisk TD, Vosick DJ (2006) Recommendations for integrating restoration ecology and conservation biology in ponderosa pine forests of the Southwestern United States. Restor Ecol 14(1):4–10
O'Neill SJ, Handmer J (2012) Responding to bushfire risk: the need for transformative adaptation. Environ Res Lett 7:014–018
Parks SA, Miller C, Nelson CR, Holden ZA (2014) Previous fires moderate burn severity of subsequent wildland fires in two large western US wilderness areas. Ecosystems. doi: 10.1007/s10021-013-9704-x
Pechony O, Shindall DT (2010) Driving forces of global wildfires over the past millennium and the forthcoming century. P Natl Acad Sci 107(45):19167–19170
Pyne SJ (1982) Fire in America. A cultural history of wildland and rural fire. Princeton University Press, Princeton, NJ, USA
Pyne S (2011) The wallow fire: a ruinous burn. Wildfire Magazine 2011:14–17
Reinhardt ED, Keane RE, Calkin DE, Cohen JD (2008) Objectives and considerations for wildland fuel Treatment in the Interior Western United States. For Ecol Manag 256:1997–2006
Rhodes A (2011) Opinion: ready or not? Can community education increase householder preparedness for bushfire? Aust J Emerg Manag 26:6–10
Rideout DB, Ziesler PS, Kernohan NJ (2014) Valuing fire planning alternatives in forest restoration: Using derived demand to integrate economics with ecological restoration. J Environ Manag 141:190–200
Rist L, Moen J (2013) Sustainability in forest management and a new role for resilience thinking. For Ecol Manag 310:416–427
Scott JH, Thompson MP, Calkin DE (2013) A wildfire risk assessment framework for land and resource management. Gen. Tech. Rep. RMRS-GTR-315. U.S. Department of Agriculture, Forest Service, Rocky Mountain Research Station, Fort Collins, CO, USA, p 83
Scott JH, Helmbrecht DJ, Thompson MP (2014 [In Press]) Assessing the expected effects of wildfire on vegetation condition on the Bridger-Teton National Forest, Wyoming, USA. Research Note RMRS-RN-XX. U.S. Department of Agriculture, Forest Service, Rocky Mountain Research Station
Short KC (2014) A spatial database of wildfires in the United States, 1992–2011. Earth Syst Sci Data 6:1–27
Show SB, Kotok EI (1924) The role of fire in the California pine forests. Bulletin, 1294U.S. Department of Agriculture, Washington, D.C, p 80
Smit B, Wandel J (2006) Adaptation, adaptive capacity and vulnerability. Global Environ Change 16(3):282–292
Spies TA, White EM, Kline JD, Fischer AP, Ager A, Bailey J, Bolte J, Koch J, Platt E, Olsen CS, Jacobs D, Shindler B, Steen-Adams MM, Hammer R (2014) Examining fire-prone forest landscapes as coupled human and natural systems. Ecol Soc 19 (3):9, http://dx.doi.org/10.5751/ES-06584-190309
Steelman TA, McCaffrey S (2013) Best practices in risk and crisis communication: Implications for natural hazards management. Nat Hazards 65(1):683–705
Stephens SL, Ruth LW (2005) Federal forest-fire policy in the United States. Ecol Appl 15(2):532–542
Stewart OC (2002) Forgotten fires: native Americans and the transient wilderness. Univ Oklahoma Press, Norman
Sudworth GB (1900) The Stanislaus and Tahoe Forest Reserves and adjacent territory. pp 499–561 In: Twenty First Annual report of the United States Geological Survey to the Secretary of the Interior 1899–1900. Charles D. Walcott, Director. Part V- Forest Reserves. Washington DC, USA
Teske CC, Seielstad CA, Queen LP (2012) Characterizing fire-on-fire interactions in three large wilderness areas. Fire Ecol 8(2):82–106
Theobald DM, Romme WH (2007) Expansion of the US wildland urban interface. Landscape Urban Plan 83(4):340–354
Thompson MP (2014) Social, institutional, and psychological factors affecting wildfire incident decision making. Soc Nat Res 27(6):636–644
Thompson MP, Calkin DE (2011) Uncertainty and risk in wildland fire management: a review. J Environ Manag 92:1895–1909
Thompson MP, Calkin DE, Finney MA, Gebert KM, Hand MS (2013) A risk-based approach to wildland fire budgetary planning. For Sci 59(1):63–77
Tillery AC, Haas JR, Miller LW, Scott JH, Thompson MP (2014) Potential postwildfire debris-flow hazards—A prewildfire evaluation for the Sandia and Manzano Mountains and surrounding areas, Central New Mexico: U.S. Geological Survey Scientific Investigations Report 2014–5161, p 24. with appendix, http://dx.doi.org/10.3133/sir20145161
USDA Forest Service (2014) The rising cost of fire operations: effects on the forest service's non-fire work., Available at: http://www.fs.fed.us/sites/default/files/media/2014/34/nr-firecostimpact-082014.pdf. Accessed October 30, 2014
Venn TJ, Calkin DE (2011) Accommodating non-market values in evaluation of wildfire management in the United States: challenges and opportunities. Int J Wildland Fire 20(3):327–339
Walker B (2013) What is Resilience? Project Syndicate., http://www.project-syndicate.org/commentary/what-is-resilience-by-brian-walker
Walker BH, Gunderson LH, Kinzig AP, Folke C, Carpenter SR, Schultz L (2006) A handful of heuristics and some propositions for understanding resilience in social-ecological systems. Ecol Soc 11(1):13
Warziniack T, Thompson M (2013) Wildfire risk and optimal investments in watershed protection. West Econ Forum 12(2):19–28
Wibbenmeyer MJ, Hand MS, Calkin DE, Venn TJ, Thompson MP (2013) Risk preferences in strategic wildfire decision making: a choice experiment with U.S. wildfire managers. Risk Analy 33(6):1021–1037
Williams J (2013) Exploring the onset of high-impact mega-fires through a forest land management prism. Forest Ecol Manag 294:4–10
Wilson RS, Winter PL, Maguire LA, Ascher T (2011) Managing wildfire events: risk-based decision making among a group of federal fire managers. Risk Anal 31:805–818
Wright HA (1982) Fire Ecology: United States and Southern Canada. John Wiley & Sons, New York, NY, USA

12

Long-term landscape changes in a subalpine spruce-fir forest in central Utah, USA

Jesse L. Morris[1*], R. Justin DeRose[2] and Andrea R. Brunelle[3]

Abstract

Background: In Western North America, increasing wildfire and outbreaks of native bark beetles have been mediated by warming climate conditions. Bioclimatic models forecast the loss of key high elevation species throughout the region. This study uses retrospective vegetation and fire history data to reconstruct the drivers of past disturbance and environmental change. Understanding the relationship among climate, antecedent disturbances, and the legacy effects of settlement-era logging can help identify the patterns and processes that create landscapes susceptible to bark beetle epidemics.

Methods: Our analysis uses data from lake sediment cores, stand inventories, and historical records. Sediment cores were dated with radiometric techniques (^{14}C and $^{210}Pb/^{137}Cs$) and subsampled for pollen and charcoal to maximize the temporal resolution during the historical period (1800 CE to present) and to provide environmental baseline data (last 10,500 years). Pollen data for spruce were calibrated to carbon biomass (C t/ha) using standard allometric equations and a transfer function. Charcoal samples were analyzed with statistical models to facilitate peak detection and determine fire recurrence intervals.

Results: The Wasatch Plateau has been dominated by Engelmann spruce forests for the last ~10,500 years, with subalpine fir becoming more prominent since 6000 years ago. This landscape has experienced a dynamic fire regime, where burning events are more frequent and of higher magnitude during the last 3000 years. Two important disturbances have impacted Engelmann spruce in the historical period: 1) high-grade logging during the late 19th century; and (2) a high severity spruce beetle outbreak in the late 20th century that killed >90 % of mature spruce (>10 cm dbh).

Conclusions: Our study shows that spruce-dominated forests in this region are resilient to a range of climate and disturbance regimes. Several lines of evidence suggest that 19th century logging promoted a legacy of simplified stand structure and composition such that, when climate became favorable for accelerated beetle population growth, the result was a landscape-scale spruce beetle outbreak. The lasting impacts of settlement-era landscape history from the Wasatch Plateau, UT may be relevant for other areas of western North America and Europe where sufficient host carrying capacity is important in managing for resistance and resilience to outbreaks.

Keywords: Spruce beetle, Climate change, *Dendroctonus*, Engelmann spruce, Logging, Western North America, Wildfire

* Correspondence: jlmorris@uidaho.edu
[1]Department of Forest, Rangeland, and Fire Sciences, College of Natural Resources, University of Idaho, Moscow, ID 83844, USA
Full list of author information is available at the end of the article

Background

During the 21st century, the impacts of global environmental change are projected to alter the frequency and severity of landscape disturbances, including wildfire and insect outbreaks (Raffa et al. 2008; Flannigan et al. 2009). In Western North America (WNA), fire regimes are likely to intensify in response to lengthening of fire season from diminished snowpack (Westerling et al. 2006). Warming temperatures will also promote outbreaks of native bark beetles (Curculionidae: Scolytinae) and other forest insects and pathogens, leading to additional tree mortality (Bentz et al. 2010). Individual and successive disturbances influence the capability of ecosystems to return to pre-disturbed states and can also facilitate alternative successional pathways, which can exacerbate turnover rates during periods of rapid climate transition (Turner 2010). Conifer-dominated forests in WNA face an uncertain future from concomitant climate warming and disturbance regime intensification (Rehfeldt et al. 2006; Westerling et al. 2006). Some bioclimatic models predict that warming temperatures will drive Engelmann spruce-subalpine fir forests (*Picea engelmannii-Abies lasiocarpa*) to be replaced by grasslands (Poaceae) in the coming century (Notaro et al. 2012). Recent studies suggest that unusually severe disturbances can promote transitions of forested landscapes to grasslands (e.g. Savage and Mast 2005; Odion et al. 2010).

Retrospective ecological records, such as those derived from lake sediments, are essential to understand baseline vegetation conditions and past disturbance regime variability. Lake sediment records span decadal to millennial timescales, and these data help to assess recent and future disturbance impacts, ecosystem services, biodiversity, and changes in vegetation composition relative to baseline conditions (Willis and Birks 2006; Willis et al. 2010; Dawson et al. 2011; Jeffers et al. 2015). The dynamics of wildfire disturbances can be reconstructed reasonably well in paleoecological studies through the analysis of charcoal preserved in sedimentary records (Whitlock et al. 2010), including information about the recurrence, severity, and fuel sources of past fire events (Enache and Cumming 2007; Calder et al. 2015). The reconstruction of bark beetle disturbances centers on the host-specific requirements of the disturbance agent, and abrupt, relative changes in vegetation composition (i.e. host to non-host) that can be detected in pollen assemblages (Anderson et al. 2010; Morris and Brunelle 2012).

In this paper we use pollen and charcoal proxies from two lake sediment cores, forest demographic data, and historical records to understand the dynamics and interactions of natural (beetles, wildfire) and anthropogenic (logging) disturbances on a subalpine spruce-fir forest in central Utah. Our study aims to use retrospective data to understand the interactions of long-term climate change and disturbances by addressing the following three questions: 1) How did the dynamics of a spruce-fir forest respond to climate change and disturbance over the Holocene? 2) How does understanding human modifications to a spruce-fir landscapes help inform our understanding of a recent, high-severity spruce beetle outbreak? and 3) What do reconstructions of host-specific biomass (C t/ha) reveal about the antecedent conditions in a spruce-fir ecosystem that supported a high-severity outbreak?

Methods

Study area

The Wasatch Plateau is a steeply sloped, fault-block landform encompassing an area of 2477 km^2, mostly situated above 3000 m in central Utah, USA. Prevailing climate of this landform is generally continental, with a complex bimodal precipitation regime shaped by heterogeneous topography and multiple moisture delivery pathways (Mock 1996). The Wasatch Plateau was used for timber harvesting and livestock grazing during the 19th century settlement-era (Hall 2001). By the early 20th century, the results of early land use practices resulted in severely denuded landscapes and the Wasatch Plateau was considered among the most detrimentally impacted rangelands in WNA (Hall 2001). While early commercial logging practices tended to favor harvesting of large-diameter class sizes of straight-grained species like ponderosa pine (*Pinus ponderosa*) and Douglas-fir (*Pseudotsuga menziesii*), the steep escarpments of the Wasatch Plateau has little mid-montane habitat to support prized commercial species and therefore Engelmann spruce was often selected (Ellison 1954; Hall 2001; Gill 2007). Also, an important associate in many spruce-fir forests of the Northern and Central Rocky Mountains, lodgepole pine (*Pinus contorta*), is absent from the Utah 'Plateaus' (i.e., Wasatch Plateau) that characterize the western flank of the Colorado Plateau (Critchfield and Little, 1966). Like other merchantable species, spruce of larger size classes were harvested first (i.e., high-grading), resulting in homogenized forest composition dominated by the relatively short-lived subalpine fir and understory spruce (Arno and Hammerly 1984; Hall 2001). By the late 20th century, the subalpine forest zone was composed of dense stands of Engelmann spruce that released (or regenerated) during the settlement-era logging.

In the late-1980s endemic populations of spruce beetle (*Dendroctonus rufipennis*) established in the Nine Mile Creek drainage of the Wasatch Plateau following a slope failure event (Dymerski et al. 2001). Spruce beetle rapidly increased to outbreak levels in response to the warm temperatures of the 1990s (Hebertson and Jenkins 2008), causing >90 % mortality in mature Engelmann spruce (> 10 cm in diameter) across ~75,000 ha of subalpine forest (Dymerski et al. 2001). The epidemic persisted until mature hosts were exhausted, resulting in a collapse of the spruce beetle population and outbreak cessation. To investigate the vegetation history of the Wasatch Plateau we collected sediment cores from two subalpine lakes that experienced the high severity outbreak during the 1990s. Blue Lake (39° 3' 20.33" N, 111° 30' 17.43" W) and Emerald Lake (39°4'26.72"N, 111° 29'50.964"W) (Fig. 1) were selected because they matched four criterion: (1) high-severity damage from spruce beetle (>90 % spruce mortality) in surrounding forest; (2) limited surface water inflow/outflow; (3) absence of constructed dams and/or berms; and (4) no known wildfire events during the historical period.

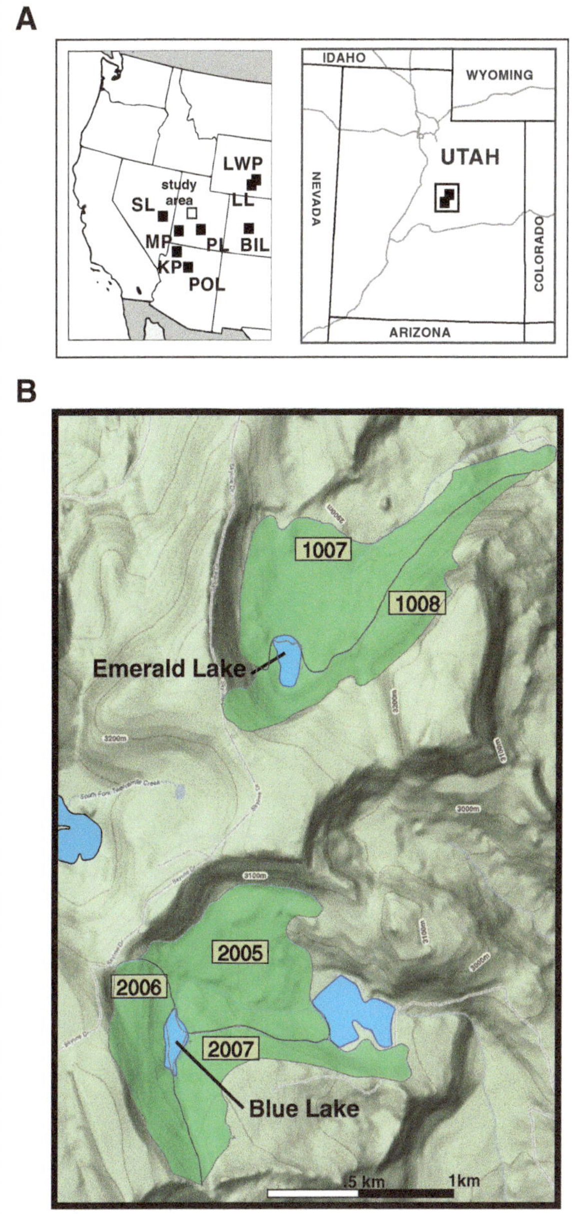

Fig. 1 Map depicting the study area and proximal sites discussed here: Bison Lake (BIL; Anderson 2011), Kaibab Plateau (KP; Weng and Jackson 1999; Little Windy Hill Pond (LWHP; Minckley et al. 2012), Long Lake (LL; Carter et al. 2013); Markagunt Plateau (MP; Anderson et al. 1999; Morris et al. 2013a), Purple Lake (PL; Morris et al. 2013b), Potato Lake (POL; Anderson 1993), and Stella Lake (SL; Reinemann et al. 2009). Location and forest inventory plots within the catchments of Emerald Lake (1007 and 1008) and Blue Lake (2005, 2006, and 2007) are also depicted. See Table 2 for inventory summaries

Age-depth models

Chronology for the upper 24 cm of sediment collected from Emerald Lake and Blue Lake was established through the analysis of the radiometric isotopes of lead (^{210}Pb) and cesium (^{137}Cs) (Fig. 2; Table 1). In preparation for ^{210}Pb/^{137}Cs analysis, 12 samples representing 2-cm increments of the Emerald Lake core were weighed and dried in a muffle furnace at 100 °C to remove water content. For Blue Lake, subsamples were prepared for each contiguous cm for the upper 24 cm. Next, dehydrated samples were submitted for analysis to the USGS Laboratory in Denver, CO. The 1963 CE peak in ^{137}Cs is derived from aboveground nuclear weapons testing and this distinct peak is instructive for interpreting the ^{210}Pb profile (Morris and Brunelle 2012). The uppermost cm (i.e. core top) was assigned to the year of core retrieval and used in age model construction, which corresponds to 2007 CE (common era) for Emerald Lake and 2005 CE for Blue Lake.

Accelerator mass spectrometry (AMS) ^{14}C dates were obtained from plant macrofossils isolated during charcoal analysis (see below) (Fig. 2; Table 1). For Emerald Lake, eight macrofossil samples were analyzed at the Center for Applied Isotope Studies (CAIS) at the University of Georgia USA. For Blue Lake, two macrofossil samples were analyzed at Beta Analytical, Inc. located in Miami, Florida USA (Morris et al. 2010). The INTCAL09 calibration set (Calib 7.1) was used to convert ^{14}C ages to calendar years before present (years hereinafter) (Reimer et al. 2013). Due to a reversal, one AMS date from Emerald Lake was rejected (Table 2). Therefore, seven dates were used to build the Emerald Lake age–depth model using CLAM script in the R software package (Blaauw 2010). Age-depth relations for Blue Lake were calculated using a linear interpolation.

Pollen sampling and analysis

For the high temporal resolution necessary to track the land use and outbreak history, we analyzed every contiguous cm for Emerald Lake and Blue Lake for the uppermost sediments (24 cm). For Emerald Lake sediments below 24 cm, pollen was analyzed at 4 cm (~ 100 year) intervals. At Blue Lake they were analyzed at 8 cm (~250 year) intervals. Pollen analysis followed standard acid digestion procedures (Fægri

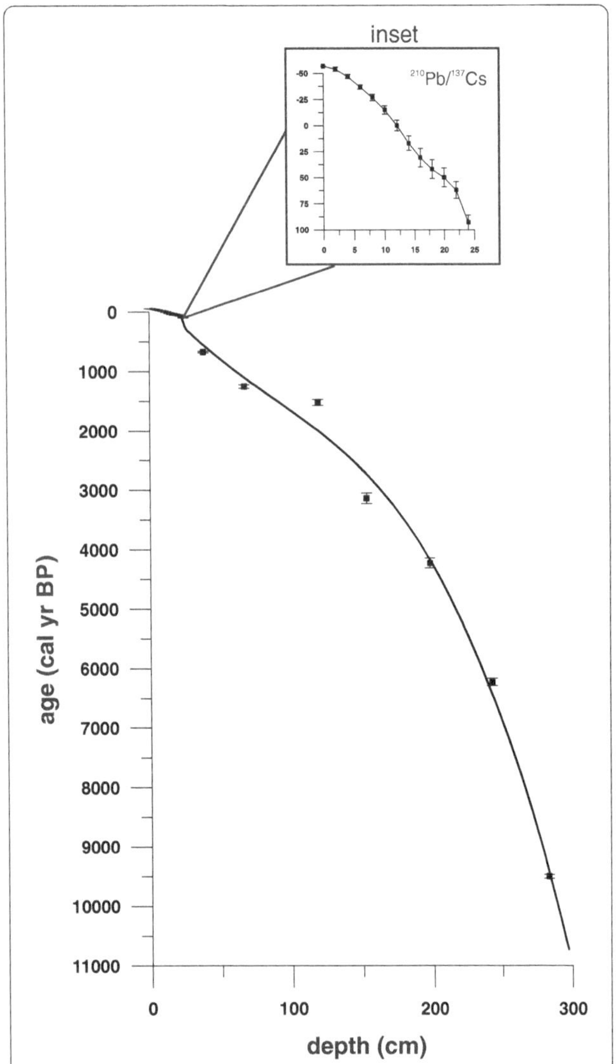

Fig. 2 Summary plot of age-depth relationships for Emerald Lake, Utah based on twelve $^{210}Pb/^{137}PCs$ dates (inset) and seven ^{14}C dates. Age-depth model was created using CLAM software (Blaauw 2010) in R

et al. 1989), and known concentrations of exotic club moss spores (i.e. *Lycopodium*) were introduced to each sample to establish pollen accumulation rates (PAR) (grains/cm^2/year). PAR were calculated by multiplying raw pollen counts by sedimentation rate (cm^2/year) (Giesecke et al. 2010). Pollen residues were classified and tabulated using light microscopy at 500 × until a minimum of 300 terrestrial grains were counted. Pollen identification was assisted by relevant keys and literature (e.g. Kapp et al. 2000). The ratio of spruce to fir pollen was calculated as (a-b)/(a + b), where a = spruce and b = fir. We assume that spruce and fir pollen types, which are generally not diagnostic to species-level assignment, are reflective of Engelmann spruce and subalpine fir in this environmental setting.

Charcoal sampling and analysis

For Emerald Lake, subsamples for charcoal analysis (5 cm^3) were analyzed for each contiguous cm and were prepared by standard conventions (Whitlock and Millspaugh 1996). Sediments were screened through 125-µm and 250-µm nested sieves and tabulated using light microscopy at 40 × magnification. Fraction sizes > 250-µm do not travel far from their source and therefore represent catchment-scale fire history (Clark 1988; Gardner and Whitlock 2001). Charcoal counts were converted to concentration (particles/cm^3) and then influx (particles/cm^2/year). We minimized variations in the record from changes in sediment deposition rates by binning concentrations using the median sediment deposition time (25 years), which were converted to charcoal accumulation rates (CHAR, particles/cm^2/year) and then decomposed into background (BCHAR) and peak components (Higuera et al. 2009). By decomposing CHAR, the low frequency variation is reflected in BCHAR while high frequency deviations (i.e. positive) from BCHAR represent charcoal delivery originating from a significant fire episode. To determine peaks above BCHAR, we first applied a Lowess smoother robust to outliers, within a 500-year window width and then subtracted background values for each time interval from the total CHAR accumulation (for each interval) (Higuera et al. 2010). The peak accumulations in charcoal (i.e., intervals with CHAR values above BCHAR) were tested for significance using a Gaussian distribution (>95th percentile). We applied this procedure to every 500-year overlapping portion of the record, thereby producing a unique threshold for each subsample. Once all peaks were identified using this criterion, we screened our results to eliminate peaks that might have resulted from statistically insignificant variations in CHAR (Gavin et al. 2006). Specifically, if the maximum CHAR values corresponding to a detected peak had a >5 % chance of coming from the same Poisson-distributed population as the minimum charcoal count of the antecedent 25-years period, then we rejected the peak (Higuera et al. 2010).

Forest inventory data

The USFS collected district-specific stand inventories in 1994 CE during the spruce beetle outbreak in the Emerald Lake and Blue Lake catchments (Fig. 1, Table 2). At each plot individual trees were noted for species and status (live or dead). When dead trees were encountered, the cause of death noted (e.g., spruce beetle). For each tree on the plot, height and diameter at breast height (dbh) were recorded and increment cores were collected for a subset of the tallest trees to determine stand age and site potential productivity (*sensu* Alexander 1967). We calculated total aboveground biomass (C/kg) for each tree using

Table 1 Summary of age-depth relationships for Emerald Lake, Utah

Depth (m)[a]	Core ID	^{14}C Age ± 1σ[b]	2σ age range (cal yr BP)[c]	Relative area under distribution	Assigned ^{210}Pb age (Year AD)[d]	Assigned age (cal yr BP)	Material dated	Lab ID Number
0.00	EL07A				2007 ± 0[e]	−57	core top	n/a
0.02	EL07A				2004 ± 2	−54	gyttja	EL07A0102
0.04	EL07A				1997 ± 2	−47	gyttja	EL07A0304
0.06	EL07A				1987 ± 2	−37	gyttja	EL07A0506
0.08	EL07A				1977 ± 3	−27	gyttja	EL07A0708
0.10	EL07A				1965 ± 4	−15	gyttja	EL07A0910
0.12	EL07A				1950 ± 5	0	gyttja	EL07A1112
0.14	EL07A				1933 ± 5	17	gyttja	EL07A1314
0.16	EL07A				1919 ± 9	31	gyttja	EL07A1516
0.18	EL07A				1908 ± 9	43	gyttja	EL07A1718
0.20	EL07A				1900 ± 9	50	gyttja	EL07A1920
0.22	EL07A				1888 ± 8	62	gyttja	EL07A2324
0.24	EL07A				1857 ± 7	93	gyttja	EL07A2123
0.38	EL07A	735 ± 15	666-688	1.000		677	macro	4781
0.67	EL07A	1280 ± 15	1223-1275	0.594		1249	macro	4783
0.81[f]	EL07C	2451 ± 27	2362-2542	0.547		2452	macro	4785
1.20	EL07C	1617 ± 26	1474-1562	0.620		1518	macro	4787
1.54	EL07C	2958 ± 24	3057-3209	0.973		3133	macro	4789
1.99	EL07C	3843 ± 27	4153-4300	0.787		4227	macro	4791
2.44	EL07C	5381 ± 28	6174-6281	0.804		6228	macro	4793
2.84	EL07C	8483 ± 26	9471-9530	1.000		9501	macro	4795

[a]depth below the sediment-water interface
[b]AMS ^{14}C age determinations from Center for Applied Isotope Studies (CAIS) at the University of Georgia
[c]^{14}C ages calibrated to calendar years before present (Cal yr BP) with INTCAL09 in Calib 7.0 (Reimer et al. 2013)
[d]^{210}Pb/^{137}Cs ages from USGS Denver determined by Constant Rate of Supply (Appleby et al., 1979) adjusted for 1963 AD ^{137}Cs peak
[e]Core top sample assigned to year of core retrieval
[f]date rejected due to reversal

standard allometric equations from individual tree height and diameter (Ter-Mikaelian and Korzukhin 1997). Tree-level carbon biomass estimates were summed for total and species-specific estimates at the plot-level (C t/ha) (Table 2). The plots were then averaged for a catchment-level estimate of Engelmann spruce biomass (C t/ha), which was then calibrated to PAR using a transfer function (Seppä et al. 2009). This method provides a long-term reconstruction of spruce biomass (C t/ha), as a proxy for Engelmann spruce stand conditions, in the Emerald Lake and Blue Lake catchments.

Table 2 Summary of stand attributes for Wasatch Plateau, Utah

	Blue Lake	Emerald Lake
Watershed size (ha^{-1})	132	116
Number plots (stocked)	34	26
Age of dominant Picea	117(8)	140(11)
Total basal area m^2 ha^{-1} (se)	43.9(6.8)	24.7(4.9)
Picea basal area m^2 ha^{-1} (se)	36.5(6.4)	20.5(4.5)
Quadratic mean diameter of Picea engelmannii > 25.4 cm (se)	43.5(3.5)	33.4(4.4)
Trees per hectare > 12.7 cm (se)	513(92)	501(119)
Picea engelmannii biomass tonnes/ha (se)[1]	44.5(8.1)	29.1(6.8)

[1]Biomass value used for transfer function in Fig. 5

Results

Baseline environmental conditions

The pollen-based vegetation reconstruction from the Emerald Lake catchment indicated that Engelmann spruce (15–20 %) has been dominant over the last 10,500 years (Fig. 3). Subalpine fir became prevalent (7–10 %) after ~6000 years ago while pine fluctuated (25 to 11 %) from 10,500 years ago to ~6000 years ago, before rebounding to higher values (25 %) after ~3000 years ago. Throughout the record, the pollen of sagebrush (*Artemisia* spp.) and asters (*Asteraceae* spp.) were present, consistently ranging between 15–20 % and 4–8 %, respectively. Meadow-rue (*Thalictrum* spp.), cattail (*Typha* spp.), sedges (Cyperaceae), and willows (*Salix* spp.) occurred in greater abundance during the early record (~9000 years ago) and then decreased towards modern. Grass pollen was complacent throughout the Emerald Lake record, ranging between 4 and 10 %.

From charcoal, the Emerald Lake record suggests that fire became progressively more prevalent and of higher severity from the early record until modern (Fig. 4). The mean fire return interval (MFRI) during the early record (6000 to 9000 years ago) ranged between 300 and 350 years between events. The MFRI increased to 150–200 years beginning at ~3000 years. As MFRI increased, peak magnitudes also increased (pieces/cm^2/year) and accordingly, the BCHAR trajectory rose over the course of the record by an order of magnitude (0.02 to 0.2 pieces/cm^2/year). A conspicuous departure from the generally increasing trend occurred at around 1000 years ago, where BCHAR declines abruptly (< 0.1 pieces/cm^2/year), before resuming an increasing trajectory.

Historical era

Over the last 200 years (since 1800 CE), forest composition at Emerald Lake and Blue Lake were

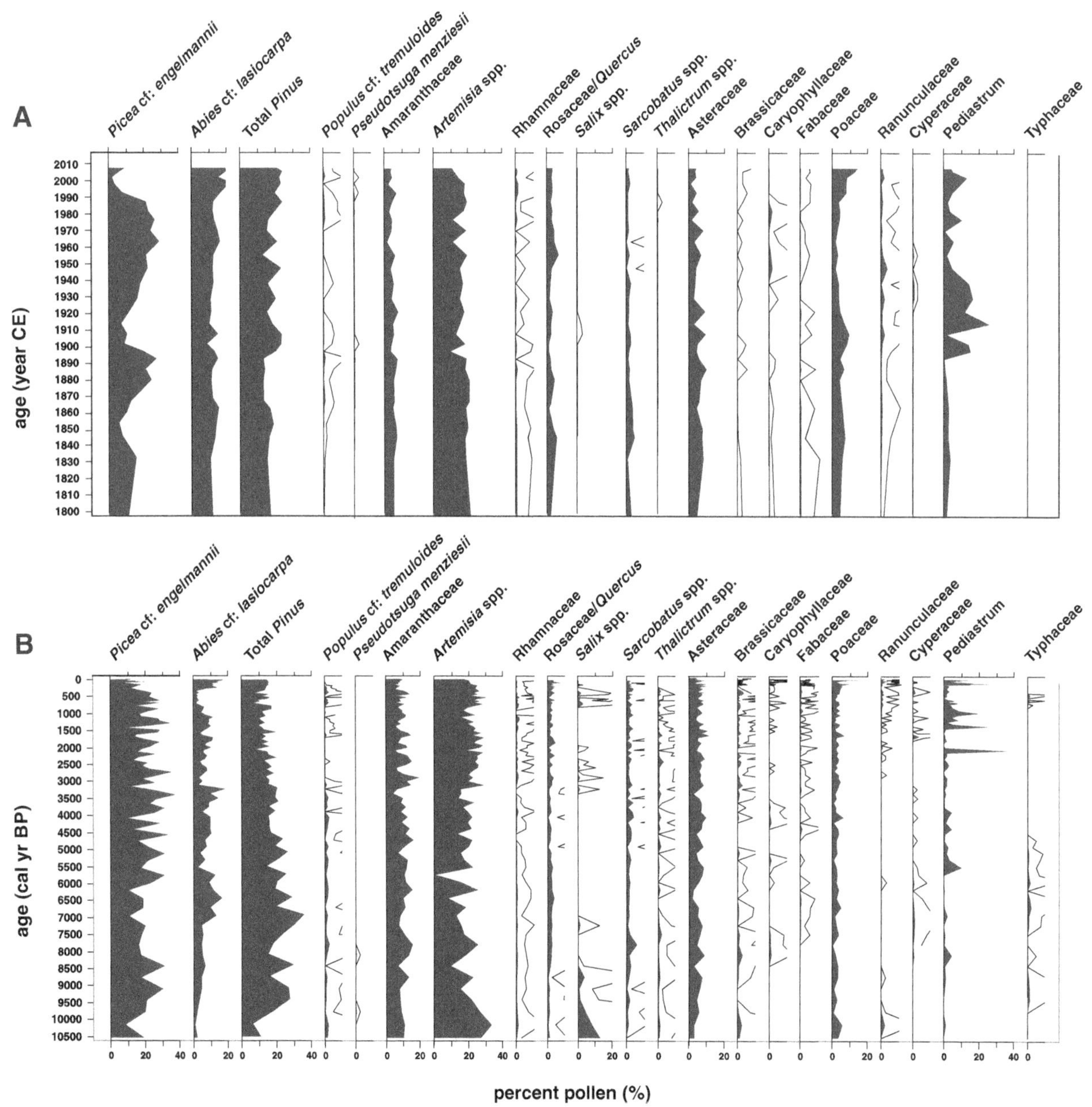

Fig. 3 Pollen diagram portraying vegetation change at Emerald Lake. Panel (**a**) depicts the historical period (1800 CE to present) and panel (**b**) shows the Holocene (10,500 years before present)

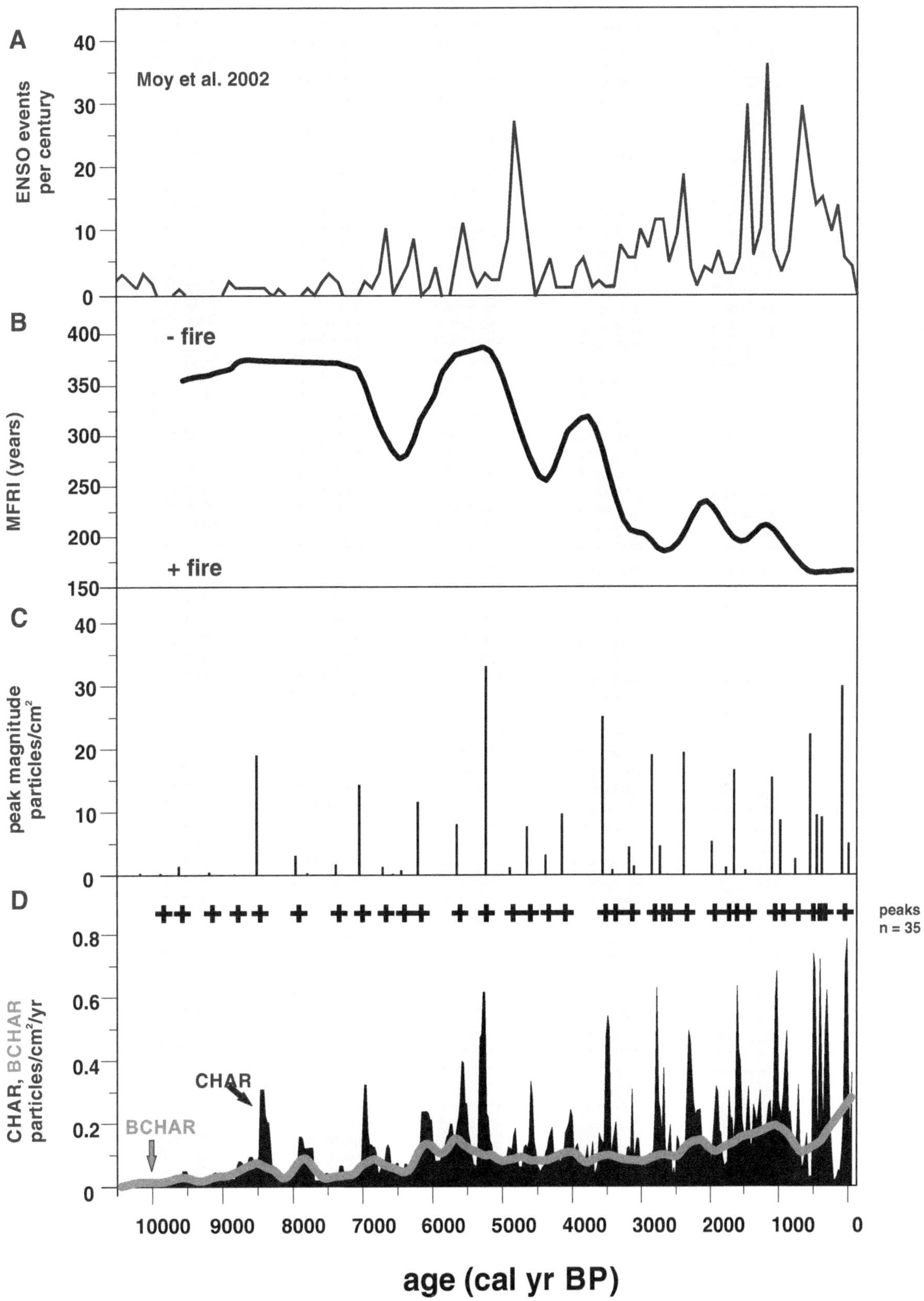

Fig. 4 Line plots showing the charcoal-based fire reconstruction for Emerald Lake. Panel (**a**) presents reconstructed ENSO activity (Moy et al. 2002), panel (**b**) depicts the fire return intervals (time since fire), panel (**c**) provides histograms of peak magnitude, and panel (**d**) shows charcoal accumulation rates (CHAR) in black silhouette plot, background charcoal (BCHAR) in gray line plot, and peak events (+). Charcoal data were analyzed using the CharAnalysis software package (Higuera et al., 2009)

compositionally similar with spruce-fir ratios averaging ~0.25 and spruce biomass ranging between 10 and 30 t/ha for non-outbreak periods (Fig. 5). For this time period spruce biomass was well below the immediate pre-outbreak level (1950–1985 CE). Conspicuous and abrupt declines in spruce biomass at ca. 1850 CE were likely from logging. Similarly, rapid reductions in biomass and spruce-fir ratios characterized the recent spruce beetle outbreak in both catchments. Concurrent reductions in the spruce-fir ratios corroborate a rapid shift from spruce to fir dominance at both Blue Lake and Emerald Lake. Box plots provide evidence that spruce biomass has generally been much lower relative to modern at Emerald Lake over the last 10,500 years, while at Blue Lake was more uniform during the last 5000 years (Fig. 5).

Discussion

Long-term vegetation and disturbance histories, such as those produced from sediment cores, are useful for characterizing recent events in the context of longer term dynamics. We use pollen and charcoal from two lake sediment cores, Emerald Lake and Blue Lake, along with forest demographic data and historical records to understand the dynamics and potential interactions of natural (beetles, wildfire) and anthropogenic (logging) disturbances over the Holocene, but with particular focus on the historical era. The longer-term data helped establish the precedence for spruce-fir forest and fire dynamics in the study area. Incorporating the historical narrative improved our understanding of possible anthropogenic influences on simplified forest composition, which ultimately contributed to the landscape-scale spruce beetle outbreak. The calibrated biomass transfer function represents another line of evidence that aided in teasing apart natural from anthropogenic causes of forest composition shifts in this study, and might be helpful in other regions. Finally, our results help to temper recent model-based claims that the subalpine spruce-fir forests of central Utah are particularly vulnerable to intensifying disturbance regimes and warming temperatures (Rehfeldt et al. 2006; Notaro et al. 2012).

Baseline environmental context

By focusing on the dynamics of spruce-fir systems over the Holocene, our study raises the question, how does long-term environmental data provide context for the resilience of spruce-fir forests to climate change and disturbance? The Emerald Lake record indicates that for the last 6000 years, this catchment has been dominated by forests of Engelmann spruce and subalpine fir, including common understory taxa, such as sagebrush and grass, that are compositionally similar to modern pollen representations of subalpine vegetation communities (Anderson et al. 1999; Morris et al. 2013a). In general, the vegetation community is essentially a classic WNA spruce-fir forest, which feature dense spruce-fir stands

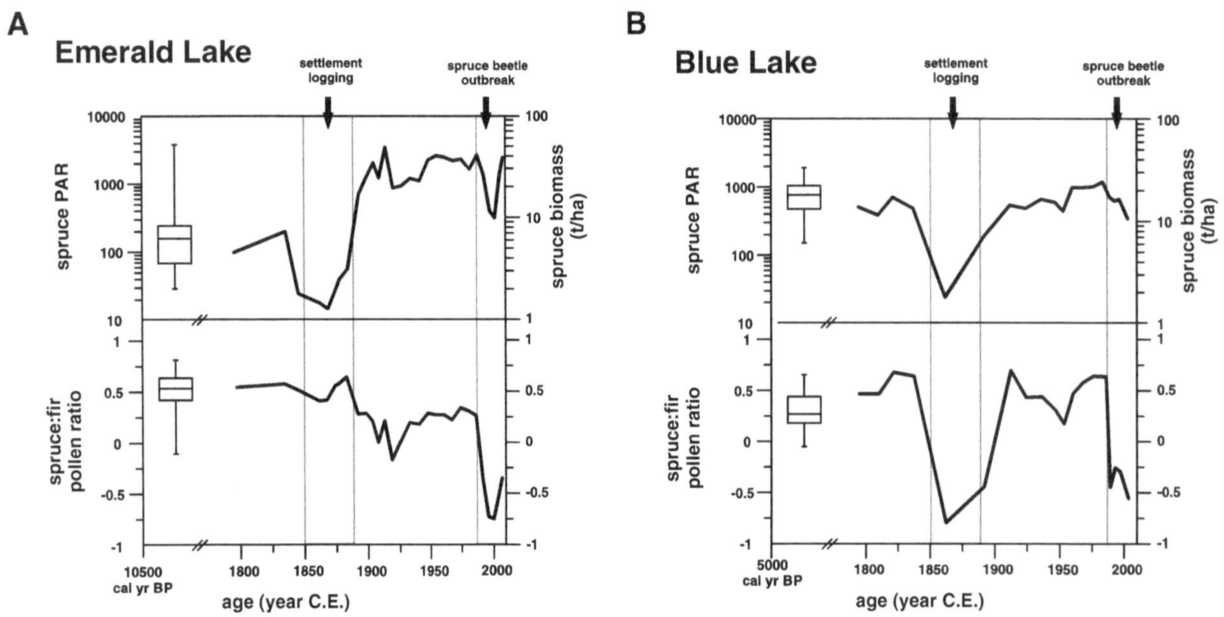

Fig. 5 Line and box plots for reconstructed Engelmann spruce pollen accumulation rates (PAR) calibrated to carbon biomass (t/ha) at Emerald Lake (panel **a**) and Blue Lake (panel **b**). Lower plots shows the ratio of spruce to fir, which is indicative of disturbances that preferentially cause spruce mortality. The shaded boxes highlight periods of documented disturbances that favored spruce mortality, such as settlement-era logging and a spruce beetle outbreak. Box plots indicate the range of variability recorded over the length of the sediment records, 10,500 cal yr BP for Emerald Lake and 5000 cal yr BP for Blue Lake, to facilitate comparison with the historical range of variability (post-1800 CE)

interspersed with meadows of grass, sagebrush, and other subalpine meadow taxa (Peet 1981). Over longer temporal scales, encroachment of extra-local species is improbable in our study region due to the vast expanses of low-elevation sagebrush steppe, effectively creating 'sky islands' of conifers that are common throughout WNA (e.g., O'Connor et al. 2014). Therefore, prior to anthropogenic modification, immigration of extra-local taxa to colonize new and/or vacant niches created by climate change and/or disturbance would be unlikely.

The Emerald Lake fire history indicates that MFRI ranged between 300 and 350 years between 3000 and 10,500 years ago, and after 3000 years ago, the MFRI decreased to 150–200 years (i.e. fire became more frequent towards modern). Intensifying fire regimes over the last 3000 years contrasts other high-resolution fire histories from sites located south of the Wasatch Plateau. Purple Lake (Fig. 1) occurs at a similar elevation to Emerald Lake (3200 m asl), and the ~9500 year record there indicates a decreasing trajectory of fire activity during the last 3000 years (Morris et al. 2013b). Where the Emerald Lake fire history indicates a stepwise increase in charcoal accumulation beginning around 3000 years ago, the Purple Lake fire history shows a transition to essentially no fire from 3000 years ago until 1850 CE when European settlers arrives in the region. Morris et al. (2013b) suggest that declining fire activity at Purple Lake in last 3000 years is due to increasing winter snowpack in response to increasing frequency and intensity of El Niño-Southern Oscillation (ENSO) (Fig. 4a, see also Anderson 2011) that is known to be an important control on wildfire in central Utah (Schoennagel et al. 2005). The contrasting charcoal records from Emerald Lake and Purple Lake could be explained by placing the Wasatch Plateau into the northern (and Purple Lake in the southern) realm(s) of influence of the ENSO dipole where the boundary has historically occurred between 40 and 42 ° N latitude (Wise 2010). The increasing charcoal stratigraphy at Emerald Lake, which occurs at ~41 °N, resembles the fire history of lake records occurring farther north of the Wasatch Plateau, such as Little Windy Hill Pond and Long Lake (Fig. 1) (Minckley et al. 2012; Carter et al. 2013). Interestingly, both Little Windy Hill Pond and Long Lake are located in the Central Rocky Mountains in lodgepole pine forests, which is a subalpine plant community absent from the Wasatch Plateau and elsewhere on the western Colorado Plateau.

The pollen record from Emerald Lake and other high elevation lakes in the region, such as Purple Lake and Morris Pond (Morris et al. 2013a, b), together suggest that Engelmann spruce was resilient to a range of climate and fire regimes during the last 9500 years, which provides context for interpreting forecasts of bioclimatic models (Notaro et al. 2012). Paleoenvironmental reconstructions from mid-elevation forest sites in interior WNA suggest (Fig. 1) that Engelmann spruce became extinct, or locally rare, during warm intervals between 4000 and 6000 years ago (Anderson 1993; Anderson et al. 1999; Weng and Jackson 1999). However, the vegetation records from Emerald Lake and other sites (Morris et al. 2013a, b) indicate that Engelmann spruce was unaffected at high elevations. If current warming trends continue unabated and surpass thermal conditions experienced regionally in the last ~10,000 years, it is conceivable that some spruce habitat could be lost, potentially hastened by severe disturbance (e.g. Savage and Mast 2005; Odion et al. 2010).

Historical era

Our study seeks results improved our understanding of how human modifications to spruce-fir landscapes simplified species composition, an important antecedent condition for high-severity spruce beetle outbreaks. On the Wasatch Plateau, pollen records from Emerald Lake and Blue Lake clearly recorded the impacts of 19th century spruce harvesting (Fig. 5). At both sites spruce biomass plummets coincident with regional settlement during the mid-19th century, decreasing by 95 % at Blue Lake and 88 % at Emerald Lake. At Emerald Lake the spruce:fir pollen ratio suggests that some spruce remained on the landscape, likely smaller class sizes. In general, the settlement logging event simplified stand composition by removing primarily spruce, which was favored due to its larger size, superior wood quality, and resistance to heart rot in comparison to fir (Blanchette 1982), which is supported by the age class structure of the inventory data (Table 2). The mean dbh age of spruce from the 1994 USFS stand survey for both lakes (Blue Lake $\bar{x}$ age = 117 ± 8 or ca. 1877, Emerald Lake $\bar{x}$ age = 140 ± 11 or ca. 1854) indicates that the majority of spruce in both stands regenerated (or more likely released) shortly after the settlement-era (Table 2). For instance, around 1880 the spruce:fir ratio at Emerald Lake increases to its highest values in the historical period, suggesting abundant spruce regeneration and/or release. While spruce can regenerate in open environments and grow rapidly, it also has the unique ability to remain in the seedling bank for many centuries (DeRose and Long 2010) awaiting an opportunity (via canopy removal) to accede to the overstory. Topography and the timing and severity of the harvest likely influenced the ability of spruce to regenerate on the site. The loss of topsoil and alterations to the subalpine microclimate from logging and overgrazing may also have inhibited spruce establishment in the short term at some sites on the Wasatch Plateau (Hall 2001; Arno and Hammerly 1984; Klemmedson and Tiedemann 1998).

While Engelmann spruce can dominate a site after a large-scale disturbance that exposes mineral soil, such as fire, reoccupation occurs over centennial time-scales (Alexander 1967, Aplet et al. 1988), and was likely a relatively rare condition for pre-settlement spruce-fir forests of the interior WNA (Alexander 1967; Long 1994). Alternatively, gap-phase dynamics - small gaps created by windthrow, endemic insect populations, or avalanches can be responsible for the initiation of new age cohorts (Whipple and Dix 1979; Veblen 1986). These gaps perpetuate spruce-fir compositional dominance, rather than large-scale disturbances, which are infrequent and climate-limited (Veblen et al. 1994; Schoennagel et al. 2005). As a result, the demographics of spruce and fir are in equilibrium owing to abundant fir regeneration but short lifespan, juxtaposed against sparse regeneration of significantly longer-lived spruce. Based on this characterization of natural spruce-fir forests, Schmid and Frye (1977) suggested minor spruce mortality would occur, but a major epidemic would not be possible until ~500 years into stand development. Preferential harvesting of spruce would also alter the demographic balance of spruce-fir, initially promoting the short-lived fir. Ultimately, as long-lived spruce, released during settlement harvesting, came to dominate, the stand would become increasingly susceptible to a beetle outbreak.

We determined that reconstructions of host-specific biomass (C t/ha) revealed important information regarding the vulnerability of subalpine ecosystems that supported a high-severity spruce beetle outbreak. Approximately 140 years after settlement, the cumulative effects of high-grade logging and range deterioration likely resulted in simplified forest structure and composition, dominated by mature spruce. For instance, in the decades leading up to the spruce beetle outbreak (i.e. 1950–1985), average spruce biomass at both lakes was greater than at any other period over the last 200 years (Fig. 4). At Blue Lake, between 1800 and 1994, spruce biomass increased from 12 to 45 t/ha^{-1}, while at Emerald Lake spruce biomass increased from 3 to >30 t/ha^{-1}. Therefore, when climate became advantageous for spruce beetle fecundity in the 1990s (Hebertson and Jenkins 2008), the broad availability of susceptible spruce hosts (> 10 cm dbh) provided increasing beetle populations with widespread, suitable host substrate to facilitate a landscape-scale outbreak.

In response to the ~1990s outbreak, living spruce biomass was drastically reduced, decreasing by 84 % and 86 % at Blue Lake and Emerald Lake, respectively. The outbreak is conspicuous in the vegetation reconstructions as precipitous reductions in both the average spruce/fir pollen ratio (which reflects compositional dominance by spruce) and the reconstructed spruce biomass (Fig. 5). For example, at Blue Lake the spruce/fir ratio dropped from 0.5 to −0.5 and from 0.4 to −0.75 at Emerald Lake in less than a decade. Morris et al. (2013c) demonstrated using a generalized linear mixed model, reductions in the spruce to fir pollen ration is a robust response variable ($p < 0.001$) to spruce beetle outbreaks, shifting by a factor of 2:1 between pre-outbreak (1950 and 1985) and outbreak periods (1985 to 2005).

While detecting the occurrence or severity of past bark beetle events in lake sediments may be challenging given the short temporal duration of outbreak events relative to the typical resolution of sedimentary records (Morris and Brunelle 2012), compositionally related metrics, such as host biomass, can help corroborate modern records (e.g. Dymerski et al. 2001), and help interpret longer records (this study). The biomass transfer function we used in our analysis (Seppä et al. 2009) provides a method to detect past bark beetle outbreaks by first screening for the probability of beetle disturbances based on host abundance. The method can also help in the interpretation of other proxy indicators of past outbreaks, such as shifts in host/non-host ratio, non-pollen palynomorphs of beetle obligates (e.g., blue stain fungi), and preserved beetle remains found in sediments (Brunelle et al. 2008). For example, a number of paleoecological studies of bark beetles have focused on detecting insect disturbances in a similar fashion as wildfire events, by identifying a residual product of the disturbance event in the sedimentary record such as charcoal (e.g. Whitlock and Larsen 2001). However, screening sediments for beetle remains has not proven to be a reliable approach for detecting insect outbreaks because even epidemic populations of bark beetles do not generally leave behind diagnostic remains (Morris et al. 2015). Where diagnostic remains are lacking, biomass transfer functions offer another line of evidence to aide in reconstructing past outbreaks with greater certainty.

Conclusions

High elevations of the Wasatch Plateau have been dominated by Engelmann spruce forests for the last ~10,500 years, with subalpine fir becoming more prominent 6000 years ago. This landscape has experienced a dynamic fire regime, where burning events are more frequent and of higher magnitude during the last 3000 years. Despite this shift towards a more intense fire regime, we observed little evidence for species turnover likely due to the improbability of species encroachment from the surrounding expanse of sagebrush steppe. Our evidence, considered alongside other sites in the region, show that high elevation Engelmann spruce ecosystems are resilient to a broad range of climate and disturbance regimes (fire and beetle outbreaks).

The Emerald Lake and Blue Lake records suggest that anthropogenic ecosystem modification from extensive logging during the mid-1800s resulted in historically unprecedented forest structure and composition. The legacy of this logging event promoted landscape-level susceptibility to a severe spruce beetle outbreak. Our data suggest that the simplified stand structure and composition found in the late 20th century on the Wasatch Plateau were relatively high in a longer-term context, resulting in a landscape particularly susceptible to bark beetles. Homogeneous coniferous forests are prevalent elsewhere in WNA because the region was settled relatively synchronously and resource exploitation occurred in a similar manner. We speculate that anthropogenic landscape modification through logging could serve as a forest composition simplifying mechanism that would provide suitable, even-aged hosts in landscapes where other native beetles occurring in WNA, such as *D. pseudotsugae* (Douglas-fir beetle) and *D. ponderosae* (mountain pine beetle). While climate-mediated outbreaks are the ultimate driver of outbreaks, perhaps tree mortality would be less severe in the absence of historic logging events that simplified stand structure and composition.

Competing interests
The authors declare that they have no competing interests.

Authors' contributions
JLM conceived and carried out the study and performed field and laboratory analysis. JLM and RJD performed data analysis and drafted the manuscript. AB coordinated field work and contributed to drafting of the manuscript. All authors read and approved the final version.

Acknowledgements
We are grateful to Diane Cote and Steve Munson for providing stand inventory data from the Manti-LaSal National Forest used in our analysis. We thank Arttu Paarlahti for assistance with data visualization. This paper was prepared in part by an employee of the US Forest Service as part of official duty and is therefore in the public domain.

Author details
[1]Department of Forest, Rangeland, and Fire Sciences, College of Natural Resources, University of Idaho, Moscow, ID 83844, USA. [2]USDA Forest Service, Forest Inventory Analysis, Rocky Mountain Research Station, Ogden, UT 84401, USA. [3]Department of Geography, University of Utah, Salt Lake City, UT 84112, USA.

References
Alexander RR (1967) Site indexes for Engelmann spruce. USDA Forest Service Report RM-32: 7 pp.
Anderson L (2011) Holocene record of precipitation seasonality from lake calcite $\delta^{18}O$ in the central Rocky Mountains, United States. Geology 39:211–4
Anderson RS (1993) A 35000 year vegetation and climate history from Potato Lake, Mogollon Rim, Arizona. Quaternary Res 40:351–9
Anderson RS, Hasbargen J, Koehler PA, Feiler EJ (1999) Late Wisconsin and Holocene subalpine forests of the Markagunt Plateau of Utah, southwestern Colorado Plateau, USA. Arct Antarct Alp Res 31:366–78
Anderson RS, Smith SJ, Lynch AM, Geils BW (2010) The pollen record of a 20th century spruce beetle (*Dendroctonus rufipennis*) outbreak in a Colorado subalpine forest, USA. For Ecol Manag 260:448–55
Aplet GH, Laven RD, Smith FW (1988) Patterns of community dynamics in Colorado Engelmann spruce-subalpine fir forests. Ecology 69:312–9
Appleby PG, Oldfield F, Thompson R, Huttunen P, Tolonen K. (1979) ^{210}Pb dating of annually laminated lake sediments from Finland. Nature 280:53–55
Arno SF, Hammerly RP (1984) Timberline: Mountain and Arctic Forest Frontiers. The Mountaineers, Seattle, WA
Bentz BJ et al (2010) Climate change and bark beetles of the western United States and Canada: direct and indirect effects. Bioscience 60:602–13
Blaauw M (2010) Methods and code for 'classical'age-modelling of radiocarbon sequences. Quat Geochron 5:512–8
Blanchette RA (1982) Decay and canker formation by *Phellinus pini* in white and balsam fir. Can J Forest Res 12:538–44
Brunelle A, Rehfeldt GE, Bentz B, Munson AS (2008) Holocene records of Dendroctonus bark beetles in high elevation pine forests of Idaho and Montana, USA. For Ecol Manag 255:836–46
Calder WJ, Parker D, Stopka CJ, Jiménez-Moreno G, Shuman BN (2015) Medieval warming initiated exceptionally large wildfire outbreaks in the Rocky Mountains. Proc Natl Acad Sci 112:13261–6
Carter VA, Brunelle A, Minckley TA, Dennison PE, Power MJ (2013) Regionalization of fire regimes in the Central Rocky Mountains, USA. Quat Res 80:406–16
Clark JS (1988) Particle motion and the theory of charcoal analysis: source area, transport, deposition, and sampling. Quat Res 30:67–80
Critchfield WB, Little EL (1966) Geographic distribution of the pines of the world. US Department of Agriculture, Forest Service Publication 991.
Dawson TP, Jackson ST, House JI, Prentice IC, Mace GM (2011) Beyond predictions: biodiversity conservation in a changing climate. Science 332:53–8
DeRose RJ, Long JN (2010) Regeneration response and seedling bank dynamics on a *Dendroctonus rufipennis*-killed *Picea engelmannii* landscape. J Veg Sci 21:377–87
Dymerski AD, Anhold JA, Munson AS (2001) Spruce beetle (*Dendroctonus rufipennis*) outbreak in Engelmann spruce (*Picea engelmannii*) in central Utah, 1986–1998. West N Am Nat 61:19–24
Enache MD, Cumming BF (2007) Charcoal morphotypes in lake sediments from British Columbia (Canada): an assessment of their utility for the reconstruction of past fire and precipitation. J Paleolimn 38:347–63
Ellison L (1954) Subalpine vegetation of the Wasatch Plateau, Utah. Ecol Monogr 24:89–184
Fægri K, Kaland PE, Kryzywinski K (1989) Textbook of pollen analysis, 4th edn. John Wiley, Chichester, UK
Flannigan MD, Krawchuk MA, de Groot WJ, Wotton BM, Gowman LM (2009) Implications of changing climate for global wildland fire. Int J Wildland Fire 18:483–507
Gardner JJ, Whitlock C (2001) Charcoal accumulation following a recent fire in the Cascade Range, northwestern USA, and its relevance for fire-history studies. The Holocene 11:541–9
Gavin DG, Hu FS, Lertzman K, Corbett P (2006) Weak climatic control of stand-scale fire history during the late Holocene. Ecology 87:1722–32
Giesecke T, Fontana SL, Knaap WO, Pardoe HS, Pidek IA (2010) From early pollen trapping experiments to the Pollen Monitoring Programme. Veg Hist Archaeobot 19:247–58
Gill RA (2007) Influence of 90 years of protection from grazing on plant and soil processes in the subalpine of the Wasatch Plateau, USA. Rangeland Ecol Manag 60:88–98
Hall M (2001) Repairing mountains: restoration, ecology, and wilderness in twentieth-century Utah. Environ Hist 6:584–610
Higuera PE, Brubaker LB, Anderson PM, Hu FS, Brown TA (2009) Vegetation mediated the impacts of postglacial climate change on fire regimes in the south-central Brooks Range, Alaska. Ecol Monog 79:201–19
Higuera PE, Gavin DG, Bartlein PJ, Hallett DJ (2010) Peak detection in sediment–charcoal records: impacts of alternative data analysis methods on fire-history interpretations. Int J Wildland Fire 19:996–1014
Jeffers ES, Nogué S, Willis KJ (2015) The role of palaeoecological records in assessing ecosystem services. Quaternary Sci Rev 112:17–32
Hebertson EG, Jenkins MJ (2008) Climate factors associated with historic spruce beetle (Coleoptera: Curculionidae) outbreaks in Utah and Colorado. Environ Entomol 37:281–92
Kapp RO, Davis OK, King JE (2000) Pollen and Spores, 2nd edn. American Association of Stratigraphic Palynologists, New York
Klemmedson JO, Tiedemann AR (1998) Soil-vegetation relations of recovering subalpine range of the Wasatch Plateau. West N Am Nat 58:352–62
Long JN (1994) The middle and southern Rocky Mountain Region. Regional Silviculture of the United States. John Wiley & Sons, New York, NY
Minckley TA, Shriver RK, Shuman BN (2012) Resilience and regime change in a southern Rocky Mountain ecosystem during the past 17000 years. Ecol Monog 82:49–68

Mock CJ (1996) Climatic controls and spatial variations of precipitation in the western United States. J Climate 9:1111–25
Morris JL, Brunelle AR, Munson AS (2010) Pollen evidence of historical forest disturbance on the Wasatch Plateau, Utah. West N Am Nat 70:175–88
Morris JL, Brunelle AR (2012) Pollen accumulation in lake sediments during historic spruce beetle disturbances in subalpine forests of southern Utah, USA. The Holocene 22:961–74
Morris JL, Brunelle A, DeRose RJ, Seppä H, Power MJ, Carter V, Bares R (2013) Using fire regimes to delineate zones in a high-resolution lake sediment record from the western United States. Quat Res 79:24–36
Morris JL, Brunelle A, Munson AS, Spencer J, Power MJ (2013b) Holocene Vegetation and Fire Reconstructions from the Aquarius Plateau, Utah, USA. Quat Int 310:111–23
Morris JL, le Roux PC, Macharia AN, Brunelle A, Hebertson EG, Lundeen ZJ (2013c) Organic, elemental, and geochemical contributions to lake sediment deposits during severe spruce beetle (*Dendroctonus rufipennis*) disturbances. For Ecol Manag 289:78–89
Morris JL, Mustaphi CJC, Carter VA, Watt J, Derr K, Pisaric, MFJ, Anderson RS, Brunelle AR (2015). Do bark beetle remains in lake sediments correspond to severe outbreaks? A review of published and ongoing research. Quaternary Int 387:72-86.
Moy CM, Seltzer GO, Rodbell DT, Anderson DM (2002) Variability of El Niño/Southern Oscillation activity at millennial timescales during the Holocene epoch. Nature 420:162–5
Notaro M, Mauss A, Williams JW (2012) Projected vegetation changes for the American Southwest: combined dynamic modeling and bioclimatic-envelope approach. Ecol Appl 22:1365–88
O'Connor CD, Falk DA, Lynch AM, Swetnam TW (2014) Fire severity, size, and climate associations diverge from historical precedent along an ecological gradient in the Pinaleño Mountains, Arizona, USA. For Ecol Manag 329:264–78
Odion DC, Moritz MA, DellaSala DA (2010) Alternative community states maintained by fire in the Klamath Mountains, USA. J Ecol 98:96–105
Peet RK (1981) Forest vegetation of the Colorado front range. Vegetatio 45:3–75
Raffa KF, Aukema BH, Bentz BJ, Carroll AL, Hicke JA, Turner MG, Romme WH (2008) Cross-scale drivers of natural disturbances prone to anthropogenic amplification: the dynamics of bark beetle eruptions. Bioscience 58:501–17
Rehfeldt GE, Crookston NL, Warwell MV, Evans JS (2006) Empirical analyses of plant-climate relationships for the western United States. Int J Plant Sci 167: 1123–50
Reinemann SA, Porinchu DF, Bloom AM, Mark BG, Box JE (2009) A multi-proxy paleolimnological reconstruction of Holocene climate conditions in the Great Basin, United States. Quaternary Res 72:347–358.
Reimer PJ, Bard E, Bayliss A, Beck JW, Blackwell PG, Bronk Ramsey, Buck CE, Cheng H, Edwards L, Friedrich M, Grootes PM, Guilderson TP, Haflidason H, Hajdas I, Hatté C, Heaton TJ, Hoffmann DL, Hogg AG, Hughen KA, Kaiser KF, Kromer B, Manning SW, Niu M, Reimer RW, Richards DA, Scott EM, Southon JR, Staff RA, Turney CSM, van der Plicht J. (2013) IntCal13 and Marine13 radiocarbon age calibration curves 0–50,000 years cal BP. Radiocarbon 55: 1869–1887.
Savage M, Mast JN (2005) How resilient are southwestern ponderosa pine forests after crown fires? Can J For Res 35:967–77
Schmid JM, Frye RH (1977) Spruce beetle in the Rockies. USDA Forest Service General Technical Report RM-49: 38 pp.
Schoennagel T, Veblen TT, Romme WH, Sibold JS, Cook ER (2005) ENSO and PDO variability affect drought-induced fire occurrence in Rocky Mountain subalpine forests. Ecol Appl 15:2000–14
Seppä H, Alenius T, Muukonen P, Miller PA, Ojala AEK (2009) Calibrated pollen accumulation rates as a basis for quantitative tree biomass reconstructions. The Holocene 19:209–20
Ter-Mikaelian MT, Korzukhin MD (1997) Biomass equations for sixty-five North American tree species. For Ecol Manag 97:1–24
Turner MG (2010) Disturbance and landscape dynamics in a changing world. Ecology 91:2833–49
Veblen TT (1986) Age and size structure of subalpine forests in the Colorado Front Range. Bull Torrey Bot Club 113:225–40
Veblen TT, Hadley KS, Nel EM, Kitzberger T, Reid M, Villalba R (1994) Disturbance regime and disturbance interactions in a Rocky Mountain subalpine forest. J Ecol 82:125–35
Weng C, Jackson ST (1999) Late Glacial and Holocene vegetation history and paleoclimate of the Kaibab Plateau, Arizona. Palaeogeog, Palaeoclim, Palaeoecol 153:179–201
Westerling AL, Hidalgo HG, Cayan DR, Swetnam TW (2006) Warming and earlier spring increase western US forest wildfire activity. Science 313:940–3
Whipple SA, Dix RL (1979) Age structure and successional dynamics of a Colorado subalpine forest. Am Midl Nat 101:142–58
Whitlock C, Millspaugh SH (1996) Testing the assumptions of fire-history studies: an examination of modern charcoal accumulation in Yellowstone National Park, USA. The Holocene 6:7–15
Whitlock C, Larsen CPS (2001) Charcoal as a fire proxy. In Tracking environmental change using lake sediments. Springer, The Netherlands, pp 75–97
Whitlock C, Higuera PE, McWethy DB, Briles CE (2010) Paleoecological perspectives on fire ecology: revisiting the fire-regime concept. Open Ecol J 3:6–23
Willis KJ, Birks HJB (2006) What is natural? The need for a long-term perspective in biodiversity conservation. Science 314:1261–5
Willis KJ, Bailey RM, Bhagwat SA, Birks HJB (2010) Biodiversity baselines, thresholds and resilience: testing predictions and assumptions using palaeoecological data. Trends Ecol Evol 25:583–91
Wise EK (2010) Spatiotemporal variability of the precipitation dipole transition zone in the western United States. Geophys Res Lett 37: 10.1029/2009GL042193.

13

Which trees should be removed in thinning treatments?

Timo Pukkala[1*], Erkki Lähde[2] and Olavi Laiho[2]

Abstract

Background: In economically optimal management, trees that are removed in a thinning treatment should be selected on the basis of their value, relative value increment and the effect of removal on the growth of remaining trees. Large valuable trees with decreased value increment should be removed, especially when they overtop smaller trees.

Methods: This study optimized the tree selection rule in the thinning treatments of continuous cover management when the aim is to maximize the profitability of forest management. The weights of three criteria (stem value, relative value increment and effect of removal on the competition of remaining trees) were optimized together with thinning intervals.

Results and conclusions: The results confirmed the hypothesis that optimal thinning involves removing predominantly large trees. Increasing stumpage value, decreasing relative value increment, and increasing competitive influence increased the likelihood that removal is optimal decision. However, if the spatial distribution of trees is irregular, it is optimal to leave large trees in sparse places and remove somewhat smaller trees from dense places. However, the benefit of optimal thinning, as compared to diameter limit cutting is not usually large in pure one-species stands. On the contrary, removing the smallest trees from the stand may lead to significant (30–40 %) reductions in the net present value of harvest incomes.

Keywords: Continuous cover forestry, Tree selection, High thinning, Optimal management, Spatial distribution, Spatial growth model

Background

A tree is financially mature for cutting when its relative value increment falls below the guiding rate of interest (Davis and Johnson 1987; Knoke 2012). However, the value increment may improve in the future, due to e.g. changes in the proportions of timber assortments that can be obtained from the tree. Therefore, Duerr et al. (1956) advice to calculate the relative value increment for several coming time periods and classify the tree as financially mature if the highest projected rate of value increase is smaller than the guiding rate of interest.

In Finland and many other countries the main timber assortments are pulpwood and saw log, of which saw log is more valuable. Figure 1 (bottom) shows that, in Finnish conditions, the consequence of unequal prices of different timber assortments is the existence of peaks in relative value increment when the stem attains sufficient dimensions for pulpwood log, first saw log, second saw log etc. At later ages, when most of the volume is already saw log, the peaks gradually disappear and the relative value increment decreases monotonously with increasing tree age and size. If the tree has passed all the value jumps, it is enough to analyze the current value increment to judge whether the tree is financially mature for cutting. It is noteworthy, however, that different trees of the stand do not reach financial maturity at the same age or diameter (Fig. 1 bottom; Knoke 2012). This is because of genetic variation, spatial variation in site productivity, and differences in the competitive positions of the trees. The last factor can be taken into account in financial analysis if distance-dependent growth models are used to predict increment.

Figure 1 (top) shows that a regular spatial distribution of trees makes it possible to maintain sufficient value increments with larger average tree diameter as compared

* Correspondence: timo.pukkala@uef.fi
[1]University of Eastern Finland, PO Box 111, 80101 Joensuu, Finland
Full list of author information is available at the end of the article

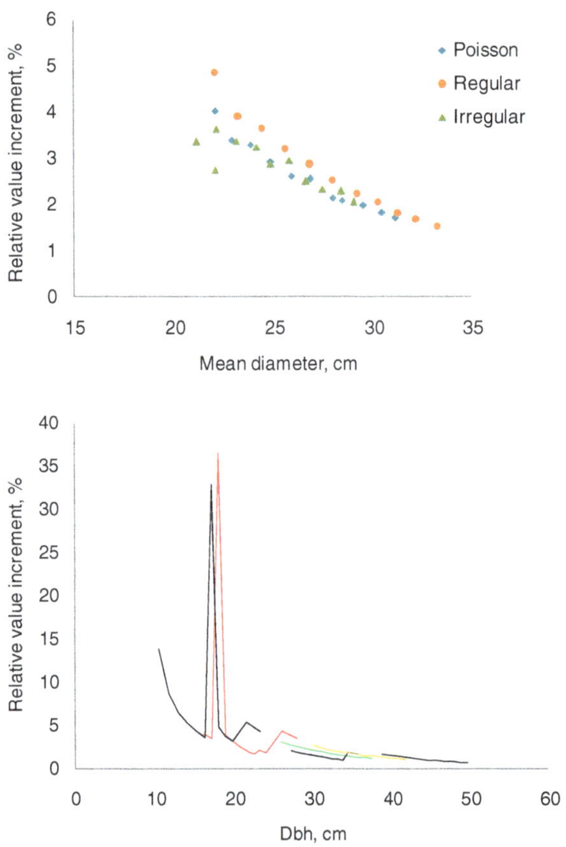

Fig. 1 Top: Relative value increment of uneven-sized spruce stands in 50-year simulations when there are no cuttings and the spatial distribution of trees is Poisson, regular (systematic) or irregular (aggregated). Bottom: development of relative value increments of six trees in the Poisson distributed stand in 50-year simulation

to more aggregated spatial arrangements. This means that, on the average, a tree reaches financial maturity at smaller diameter in irregular spatial distribution. However, some of the trees in an aggregated spatial tree distribution, growing in sparsely populated places or being surrounded by small trees, may grow better than a similar tree would grow in a regular stand, which postpones the financial maturity of these trees.

As Davis and Johnson (1987) pointed out, the effect of tree removal on the growth of remaining trees should also be taken into account when deciding when a tree should be harvested and which trees should be removed. Removing a large tree leads to improved value increment in smaller trees. This calls for cutting large trees earlier than their relative value increment suggests. The removal of a large tree may improve the productivity of several remaining trees and increase the relative value increment of the whole residual stand.

The capital invested in wood production consists of the value of the trees plus the value of bare land. If bare land has a positive value this means that a tree is financially mature at higher relative value increment than indicated by the guiding rate of return. This is because the value increment must be compared to the opportunity cost of the tree and the piece of land occupied by the tree. Bare land value and the effect of tree removal on the growth of remaining trees both increase the rate of value increment at which cutting is optimal decision.

The above analysis suggests that it is optimal to remove the largest trees in a thinning treatment. However, unequal competitive status of trees, as well as unequal effect of tree removal on the growth of surrounding trees, makes the decision more complicated than for instance applying diameter limit cutting. A detailed analysis of optimal tree selection needs a distance dependent growth model, or at least the calculation of value increments and the effects of tree removal at individual tree level.

Although the overall principles that should determine tree selection in thinning have been understood and described already several decades ago (Duerr et al. 1956; Davis and Johnson 1987) few studies have actually optimized the selection of removed trees. An exception is the study of Pukkala and Miina (1998) who optimized a tree selection rule which was based on the effect of tree removal on the competitive positions of remaining trees. It was found that it was optimal to thin from above, i.e. remove large trees.

This study proposed a more straightforward approach to optimizing tree selection: the criteria of the cutting rule were the tree's stumpage value, its value increment, and the effect of removal on the growth of surrounding trees. It was assumed that increasing stumpage value, decreasing relative value increment and increasing competitive effect increase the likelihood of removal. Another hypothesis was that the optimal order of tree removal is less strongly correlated with tree diameter in irregular spatial distributions, as compared to regular tree arrangements. A third hypothesis was that, when net present value is maximized, it is optimal to remove trees at higher relative value increment than the discount rate that is used to calculate net present value.

Methods

Three different sample plots of Norway spruce (50 m by 50 m) with the same diameter distribution but different spatial distribution of trees were generated for the analyses. The first plot had a Poisson distribution of trees (henceforth referred to as Poisson stand). The x and y coordinates of trees were drawn from uniform distribution. The second stand was regular, and the third stand was very irregular (aggregated). Regular spatial distributions can be easily achieved by silvicultural treatments and the irregular distribution might be a result of removing all birches from a naturally emerged mixture of birch and spruce. Spruces may be very irregularly distributed in such stands. The mean diameter of each initial stand was about 22 cm, stand basal area was 17 $m^2 \cdot ha^{-1}$ and the number of trees per hectare was 1200. The stands were assumed to grow on mesic site in Central Finland.

The models of Pukkala et al. (2013) for diameter increment, tree survival and ingrowth were used in simulation. The models are based on about 60,000 diameter increment and survival observations in different stand types. The models can be used in both even-sized and uneven-sized stands. However, the models are not spatial. When the models were used in this study, the predictors which describe competition (stand basal area, *G*, and basal area in larger trees, *BAL*) were calculated from trees that were within 10 meters from the tree for which predictions were calculated. This is justified because most of the modelling data of Pukkala et al. (2013) were measured on plots with approximately 10-m radius. However, to make the growth simulator distance dependent, both *G* and *BAL* were calculated as the average *G* ($m^2 \cdot ha^{-1}$) or *BAL* ($m^2 \cdot ha^{-1}$) within 10 m, within 10/2 m (5 m), and within 10/3 m (3.33 m). As a result, the closest trees had a larger influence on *G* and *BAL*. This is in accordance with several studies, which show that the effect of neighbor trees on the growth of a subject tree decreases with increasing distance (see e.g. Miina and Pukkala 2000).

Mortality was simulated by comparing the tree's survival probability to random number distributed uniformly between 0 and 1. If the random number was larger than the survival probability, the tree was assigned as dead. Ingrowth (number of ingrowth trees) was predicted with the unmodified model of Pukkala et al. (2013) but spatial criterion was used to choose the places for ingrowth trees. Fifty candidate positions were generated for each ingrowth tree and the competition index proposed by Miina and Pukkala (2000; their Equation 6b)was calculated for each location. The tree was placed to that location which had the lowest completion index. The procedure mimics the observed dynamics of spruce stands (e.g. Eerikäinen et al. 2007) in which ingrowth trees appear in openings and places with little competition by larger trees.

The optimization problem consisted of selecting the thinning intervals and the weights of the three criteria of the following tree selection rule:

$$RemovalScore = w_1 Value + w_2 RelativeValueIncrement + w_3 BALeffect \quad (1)$$

where *Value* is the stumpage value of the stem, *RelativeValueIncrement* is the predicted 5-year value increment divided by the stumpage value of the stem, *BALeffect* is the total reduction in the BALs of neighbor trees in case the tree is removed, and w_1, w_2 and w_3 are optimized parameters which determine the effect of the three criteria on the order of tree removal. The BAL effect was calculated in the same way as BALs were calculated in growth prediction, i.e. by giving more weight to close neighborhood. A removed tree affected most within 3.33 m, somewhat less within 3.33–5 m, still less within 5–10 m and not at all beyond 10 meters.

The optimized parameters were thinning intervals and parameters w_1, w_2 and w_3 of the tree selection rule. The problem formulations correspond to continuous cover management since planting was not an option when stand development was simulated. The remaining basal area was calculated with the following model, which is based on 20,583 optimized cuttings of 6,861 stands located in different parts of Finland. All optimizations meet the constraints of Finnish forestry legislation:

$$\begin{aligned} G_{remain} = {} & \exp(8.149 - 4.656D^{0.1} + 0.0221G_{total} \\ & + 0.211G_{pine}/G_{total} + 0.254G_{spruce}/G_{total} \\ & - 0.0327\sqrt{r} - 0.0333MT - 0.0992VT \\ & - 0.164CT) \end{aligned} \quad (2)$$

where G_{remain} is remaining basal area ($m^2 \cdot ha^{-1}$), *D* is basal-area-weighted mean diameter of trees (cm), G_{total} is stand basal area before thinning ($m^2 \cdot ha^{-1}$), G_{pine} is basal area of pine ($m^2 \cdot ha^{-1}$), G_{spruce} is basal area of spruce ($m^2 \cdot ha^{-1}$), *r* is discount rate (%), and *MT*, *VT* and *CT* are indicator variables for mesic, sub-xeric and xeric site, respectively.

When a thinning was simulated, trees were removed according to their removal score (Eq. 1) until the remaining basal area was equal to the value calculated with Eq. 2. In another set of optimizations, also the remaining basal was optimized. These optimizations may not always meet the current legal limits of Finland.

A 10-m wide buffer zone was generated around the plot when stand development was simulated (when computing the predictors of the models), and the buffer was removed after completing a simulation time step. The buffer was generated by assuming that the plot was surrounded by similar plots on all sides. Since the models that were used in growth simulation have five-year time step, stand development was simulated in 5-year steps.

Computation of the removal scores of trees involved the calculation of the stumpage value, 5-year value increment and BAL effect for every tree. To obtain the stumpage value, taper models (Laasasenaho 1982) were used to calculate assortment volumes, which were multiplied by their unit prices. The assortments were saw log (50 €·m^{-3}, minimum top diameter 16 cm, minimum length 4 m), and pulpwood (15 €·m^{-3}, minimum top diameter 9 cm, minimum length 2 m). To calculate value increment, the diameter (dbh) and height of the tree were incremented by five-year growth, and assortment volumes corresponding to the incremented dimensions were calculated with the taper model.

The obtained stumpage values, value increments and BAL effects were used to calculate the removal scores for all trees, and the tree with the highest score was removed. Since a tree removal may affect the competitive influences, value increments and removal scores of remaining trees, the BAL effects and value increment predictions of all remaining trees were updated after every tree removal. This involved removing the buffer, generating the buffer again, and calculating the BAL effects and value increments again. Removing and adding the buffer after every tree removal was based on the assumption that the forest that surrounds the plot is thinned simultaneously with the plot.

Three next thinnings were optimized in the analyses of this study. The *NPV* of the ending growing stock (residual stand after the third thinning) was predicted with the model (see Pukkala 2015b). The model prediction, once discounted to the starting year of simulation, gives the *NPV* of all incomes and costs that are later than the last optimized cutting. It has been shown (Pukkala 2015a, 2015b) that using the model for estimate the *NPV* of distant cuttings has no major effect on the optimization results for the next cuttings, as compared to a higher number of optimized cuttings.

Since the simulation involves stochasticity in mortality and ingrowth, every simulation that was conducted during the optimization run was repeated 10 times and the mean *NPV* of the repeated simulations was used as the objective function (returned to the optimization algorithm). The direct search method of Hooke and Jeeves (1961) was used in optimization. Every optimization was repeated 5 times, each direct search starting from the best of 100 random combinations of optimized variables. The best solution (highest *NPV*) was taken as the optimal solution. *NPV* was calculated with 3 % discount rate.

Results and discussion

Remaining basal area not optimized

Optimizations in which the remaining basal area was not optimized suggested immediate thinning in all three stands, another thinning after 10 years, and a third thinning 10 years later (Table 1, boldface). However, there was some variation between repeated optimizations in the cutting intervals, especially in the irregular stand. Looking at the NPVs of the solutions suggests that those solutions that propose intervals other than 10 years may be sub-optimal, i.e., the algorithm has converged to local optimum.

The ranking of the three stands in terms of *NPV* was logical. The regular stand produced the highest *NPV* whereas the highly irregular stand produced clearly smaller economic benefit than the other stands (Table 1).

Table 1 Net present values and optimal values of decision variables for Poisson, regular and irregular stand in 5 repeated optimizations when the remaining basal area of thinning was calculated with a model (Eq. 2)

NPV (€·ha^{-1})	Years to 1st thinning	Years to 2nd thinning	Years to 3rd thinning	w_1 (stem value)	w_2 (value increment)	w_3 (BAL effect)
	Poisson stand					
9133	**0**	**10**	**10**	**0.37**	**−0.25**	**0.38**
9121	0	10	15	0.24	−0.47	0.29
9127	0	10	10	0.48	−0.50	0.02
9032	0	15	15	0.43	−0.27	0.29
9105	0	10	10	0.29	−0.05	0.66
	Regular stand					
9319	**0**	**10**	**10**	**0.35**	**−0.31**	**0.34**
9237	0	10	20	0.49	−0.27	0.24
9304	0	15	10	0.78	−0.04	0.18
9289	0	15	15	0.46	−0.29	0.26
9313	0	10	15	0.30	−0.13	0.57
	Irregular stand					
8247	0	10	10	0.42	−0.13	0.45
6852	0	25	45	0.46	−0.27	0.26
8266	**0**	**10**	**10**	**0.32**	**−0.43**	**0.25**
7511	0	30	10	0.38	−0.30	0.32
8238	0	10	35	0.50	0.01	0.52

Optimization result with the highest NPV is indicated with boldface

The signs of the parameters of the removal score function were logical: high stem value and high BAL effect (high reduction in the competition of remaining trees) increased the score, and high relative value increment decreased it. Valuable trees with low value increment and strong competitive effect on other trees were the first ones to leave. As a result, thinnings from above were conducted, as can be seen from the maps of Fig. 2.

A closer inspection of the diameters of the removed and remaining trees (Fig. 3) revealed that the thinnings of the regular stand resembled diameter limit cutting. There were only one or two diameter classes which included both remaining and removed trees. In the first thinning of the Poisson forest, there were 4 diameter classes (8 cm diameter range) having both remaining and removed trees. This range was 12 cm in the irregular stand. Large trees were left in sparse places and rather small trees were removed from dense places.

The diagrams of Fig. 4 (bottom right) show that, in the regular stand, the order of removal followed decreasing breast height diameter fairly closely. The selection score correlated closely with both dbh and relative value increment, suggesting that either of these variables alone could be used as the harvesting criterion in a regular stand. In the other stands, the order of removal did not follow decreasing dbh equally closely. In both Poisson and regular stands, most trees removed in the first thinning were larger than 25 cm.

The situation was different on the irregular stand, in which several removed trees were smaller than 25 cm. To have a sufficient remaining basal area, some large trees were left to continue growing (Figs. 2 and 3). In the irregular stand, the removal score did not correlate strongly with dbh or relative value increment. In this stand, the BAL effect of tree removal varied much more than in the other stands (Fig. 4 bottom left) and had a stronger influence on the removal score than in the other stands. This cannot be concluded from the weights of BAL effect in Table 1 since the effect of the criterion depends on both the weight and the range of variation in the criterion variable in a particular stand and thinning. As a conclusion, when thinning an irregular stand, more importance should be given to the reduction of competition due to tree removal.

Table 1 shows that repeated optimizations result in different weights of the criteria of the removal score,

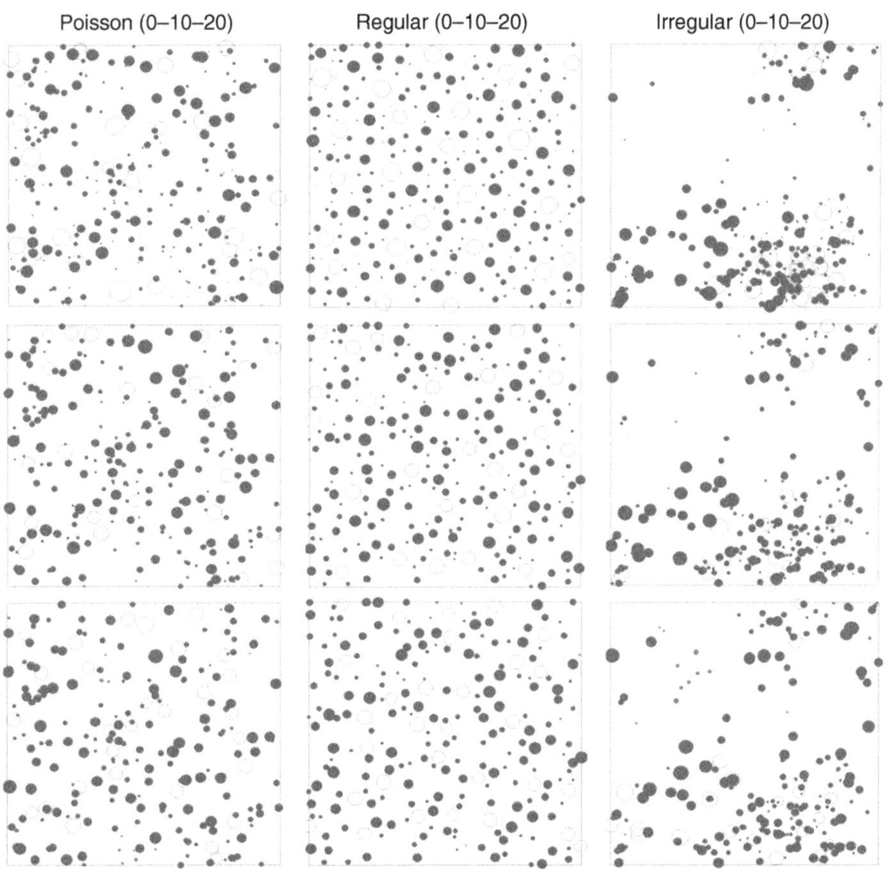

Fig. 2 Remaining (filled circles) and removed (open circles) trees in the first (top), second (middle) and third (bottom) thinning of a Poisson-distributed, regular and irregular stand. Ingrowth trees are shown with lighter tone (red). Thinning years are given within parentheses

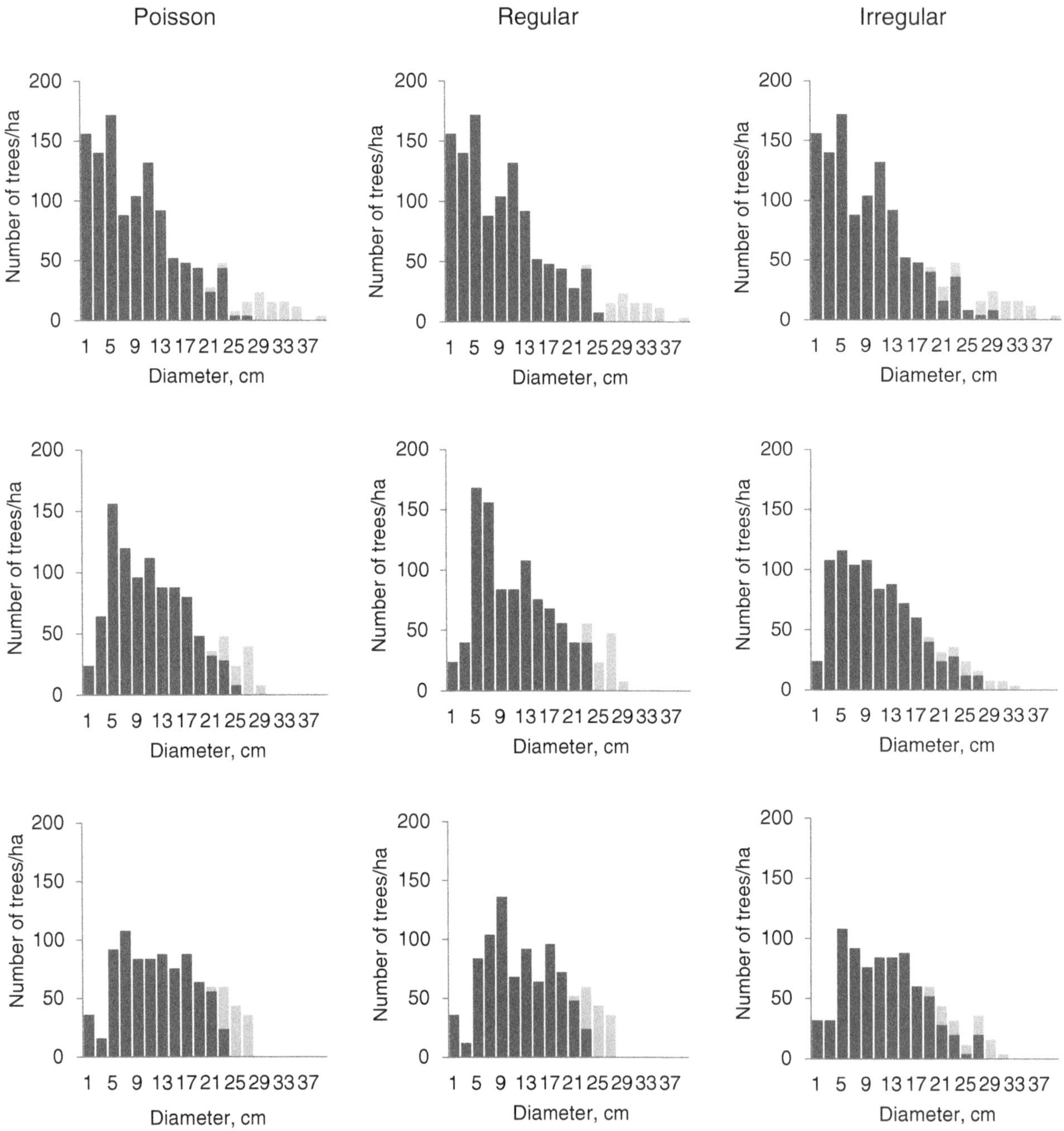

Fig. 3 Number of remaining (dark green) and removed (grey) trees in the first (top), second (middle) and third (bottom) thinning of a Poisson-distributed, regular and irregular stand when NPV is maximized with 3 % discount rate

suggesting that there is much uncertainty in this respect. However, since the three criteria correlate with each other, the removal score may be fairly similar with different weights. This means that, although there is uncertainty about the exact weights of the removal criteria, there is less uncertainty about the removal order of trees. This can be seen from Fig. 5, which shows the locations of trees removed in the first thinning of the Poisson forest when the five different solutions of Table 1 were used to select the trees. Almost the same trees were removed when using different values for weights w_1, w_2 and w_3. Three trees were removed in only 1 solution and another three trees were removed in less than five solutions. All the other trees were removed in all solutions. Therefore, high variation in the obtained weights of the removal criteria bring only little uncertainty in tree selection.

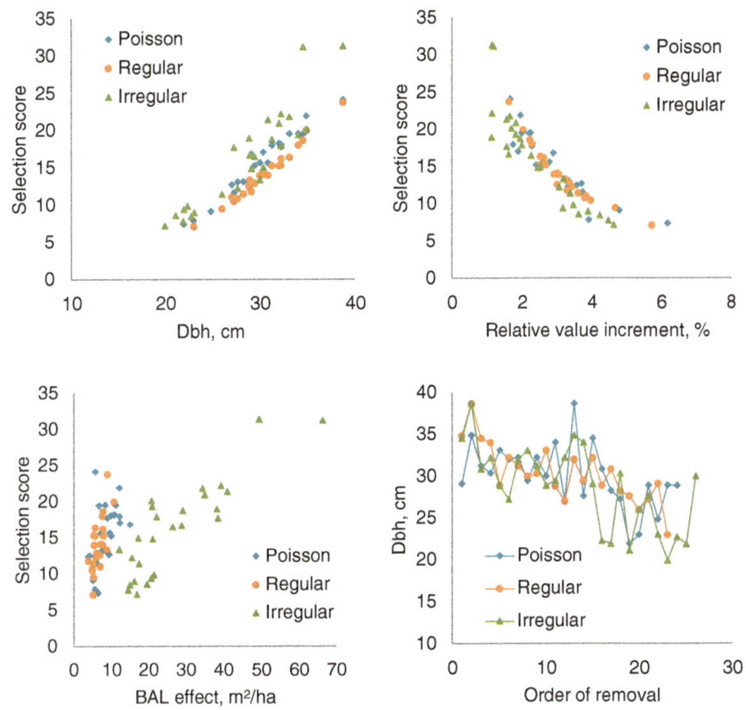

Fig. 4 Correlation of the tree selection score of the first thinning with dbh, relative value increment and BAL effect (reduced competition of other trees if the tree is removed). The bottom-right diagram shows the dbh of the removed trees as a function of the order of removal (trees on the left are removed first and trees on the right are the last trees removed)

Another simulation was done by applying the optimal weights of the thinning rules (boldface rows in Table 1) of the three different stands to the first thinning of the Poisson stand (Fig. 6). The results show again that nearly the same trees were removed from the Poisson forest when applying the optimal tree selection rule of regular, irregular or Poisson stand. This result implies that the same tree selection rule could be used in all spatial distributions, which may sound counterintuitive in the light of the diagrams of Fig. 4. However, even though the

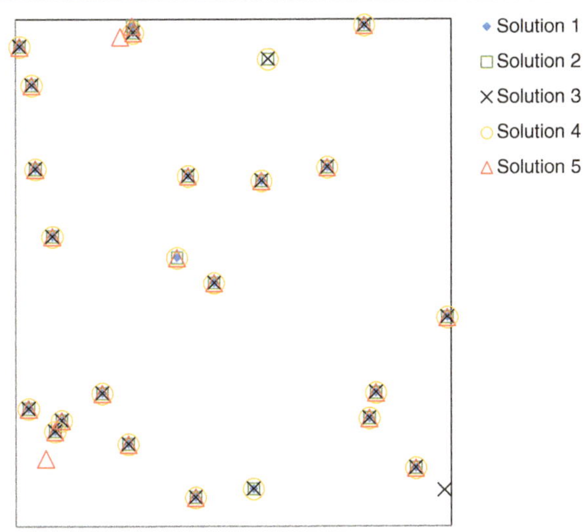

Fig. 5 Location of trees removed in the first thinning of the Poisson stand when tree selection is based on five different solutions of the optimization problem. When the symbols of different solutions coincide, the same tree is removed in all solutions

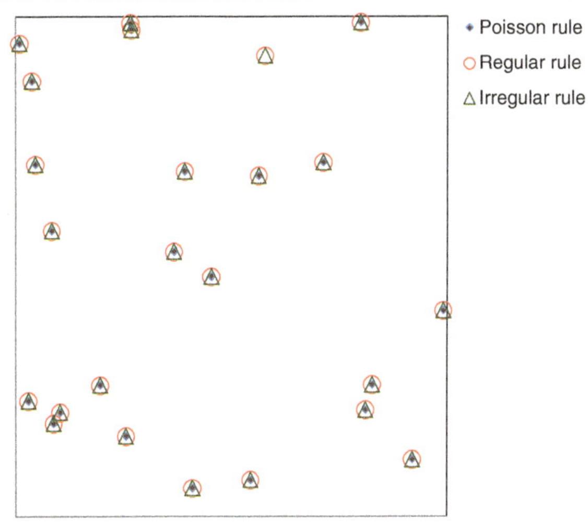

Fig. 6 Location of trees removed in the first thinning of the Poisson stand when tree selection is based on the optimal tree selection rule obtained for Poisson stand, regular stand and irregular stand. When the symbols of different rules coincide, the same tree is removed with all rules

weight of e.g. BAL effect would be the same in all stands, the influence of BAL effect on tree removal would be stronger in the irregular stand, due to greater variation of BAL effect in this stand.

Remaining basal area optimized

When the remaining basal area was optimized together with the thinning intervals and tree selection rule, the *NPV* of the optimal schedule increased by 4–5 % as compared to optimizations in which remaining basal area was calculated with Eq. 2 (Tables 1 and 2). The remaining basal areas were now 6.5–8.9 $m^2 \cdot ha^{-1}$ whereas they were 10–11 $m^2 \cdot ha^{-1}$ when calculated with the model. The thinning intervals became longer in the Poisson and regular stands. It can be concluded that the requirement for a certain minimum residual basal area decreased the profitability of forest management in the three analyzed stands.

As a consequence of stronger thinnings, more ingrowth appeared in the stands, especially in the Poisson and regular stands, as compared to previous optimizations (Fig. 7). The thinning was extended to smaller diameters (Fig. 8), and the diameter range that had both remaining and removed trees, became narrower. This means that when remaining basal area was not restricted, optimal thinning resembled more diameter limit cutting, also in the aggregated spatial distribution.

The optimality of diameter limit cutting was further inspected by removing the trees according to dbh, starting from the largest tree ('High thinning' in Fig. 9). For comparison, simulations were also conducted so that the smallest trees were removed in the thinning treatments ('Low thinning' in Fig. 9). The results show that systematic diameter limit cutting was not much worse than optimal tree selection. Thinning from below would be a clearly inferior management approach.

Effect of discount rate

The previous optimizations used 3 % discount rate. When remaining basal area was optimized in the Poisson stand (together with thinning intervals and tree selection rule), lower discount rate (1 %) led to clearly longer thinning intervals, higher pre-thinning basal areas, larger mean tree diameters, and slightly higher post-thinning basal areas and mean diameters, compared to 3 % discount rate (Fig. 10). Increased discount rate (5 %) resulted in 10-year thinning interval, lower remaining basal area and smaller mean tree size of the residual stand. When the discount rate was 1 %, the largest trees of the stand were 35 cm in dbh when the thinning was conducted. With 5 % discount rate the thinning was conducted when the largest trees reached 25 cm breast height diameter.

The effect of discount rate was the most clear in growing stock value (Fig. 10, bottom left). When the discount rate was 1 %, the stumpage value of the growing stock was 8000 €$\cdot ha^{-1}$ at the second thinning and 13 000 €$\cdot ha^{-1}$ at the third thinning. With 5 % discount rate the pre-thinning growing stock value was only about 2200 €$\cdot ha^{-1}$.

When *NPV* was maximized with 1 % discount rate, it was optimal to thin the stand when its relative value increment was 2–3 %. With 3 % discount rate the stand was thinned at 4.5–6 % value increment, and at with 5 % discount rate at 8 % value increment. When the discount rate was 3 %, in the first thinning of the Poisson stand about half if the removed trees had a relative value increment of about 2 % and the rest had 3–6 % rate of value increment (Fig. 11, top). However, the first thinning, which was immediately, was most probably later than its optimal timing would have been. In later thinnings, 50 % of removed trees were removed at about 4 % rate of value increment and the rest were removed at higher, up to 7 % value increment.

In the second thinning of the Poisson stand, trees were removed at 2–7 % value increments when discount rate was 1 %, at 3–7 % increments when discount rate was 3 % and at 5–14 % value increments when discount rate was 5 % (Fig. 11, bottom). This is in line with the hypothesis of the study, according to which it is optimal to remove a tree at higher relative value increment than the guiding rate of interest. This is because of the opportunity cost of bare land and the fact that tree removal improves the relative value increment of remaining trees.

Conclusions

The results suggest that it is nearly optimal to select the trees that are removed in a thinning treatment on the basis of breast height diameter, starting from the largest tree. However, in irregular spatial distributions, the competition faced by the tree and the effect of removal on

Table 2 Net present values and optimal values of decision variables for Poisson, regular and irregular stand when the remaining basal area (*G*) was optimized together with thinning intervals and tree selection rule

Stand	*NPV* (€$\cdot ha^{-1}$)	1st thinning		2nd thinning		3rd thinning		w_1 (stem value)	w_2 (value increment)	w_3 (BAL effect)
		Years	*G*	Years	*G*	Years	*G*			
Poisson	9477	0	6.9	15	8.9	20	7.4	0.33	−0.45	0.22
Regular	9680	0	6.5	15	7.7	20	7.6	0.68	−0.25	0.07
Irregular	8671	0	6.9	10	8.6	10	6.8	0.83	−0.09	0.08

Fig. 7 Remaining (filled circles) and removed (open circles) trees in the first (top), second (middle) and third (bottom) thinning of a Poisson-distributed, regular and irregular stand when the remaining basal area is optimized. Ingrowth trees are shown with lighter tone (red). Thinning years are given in parentheses

the growth of surrounding trees should also be taken into account. The degree of irregularity of the irregular stand of this study was so high that such stands are rarely encountered in managed forests. Therefore, diameter limit cutting seems to be a sufficient approach in most stands. However, there are other types of irregularity, which are more common. For example, the stand may have sub-areas of predominantly large trees while other sub-areas are occupied by smaller trees. In this case, diameter limit cutting leads to openings and unutilized growing space. Since decreased competition increases the dbh of financial maturity, it would most probably be better to leave some large trees to continue growing. This means that a tree selection rule that incorporates several criteria is more likely to work better (than dbh alone) in a wide range of stand structures. In mixed stands, relative value increment or a more complicated tree selection rule is certainly better than using only dbh to select the removed trees (Knoke 2012). This is because of differences in the inherent growth rates, assortments dimensions, and assortment prices of different species.

Although there are very few previous studies on optimal tree selection, several recent results on pre-and post-cutting diameter distributions in economically optimal uneven-aged management support the conclusion that optimal cutting resembles diameter-limit cutting (Tahvonen et al. 2010; Tahvonen 2011; Pukkala et al. 2014; Pukkala 2015a, 2015b). The American studies conducted during the 1980s lead to similar conclusions (Haight 1985, 1987; Haight and Getz 1987). Also long-term silvicultural trials support the conclusion than diameter limit cutting is more profitable than single-tree selection. For example, in the Vessari experiment located in Central Finland, the net present value (calculated with 3 % discount rate) of diameter limit cutting was 13750 €·ha^{-1} whereas it was only 10250 €·ha^{-1} in single tree selection during a 40-year monitoring period. In the nearby Honkamäki experiment the NPV of diameter limit cutting was 10500 €·ha^{-1} and the NPV of single tree selection was 7800 €/ha (Pukkala et al. 2012).

The optimizations of this study were done for continuous cover management. However, the same principles of analyzing the financial maturity of trees also apply to

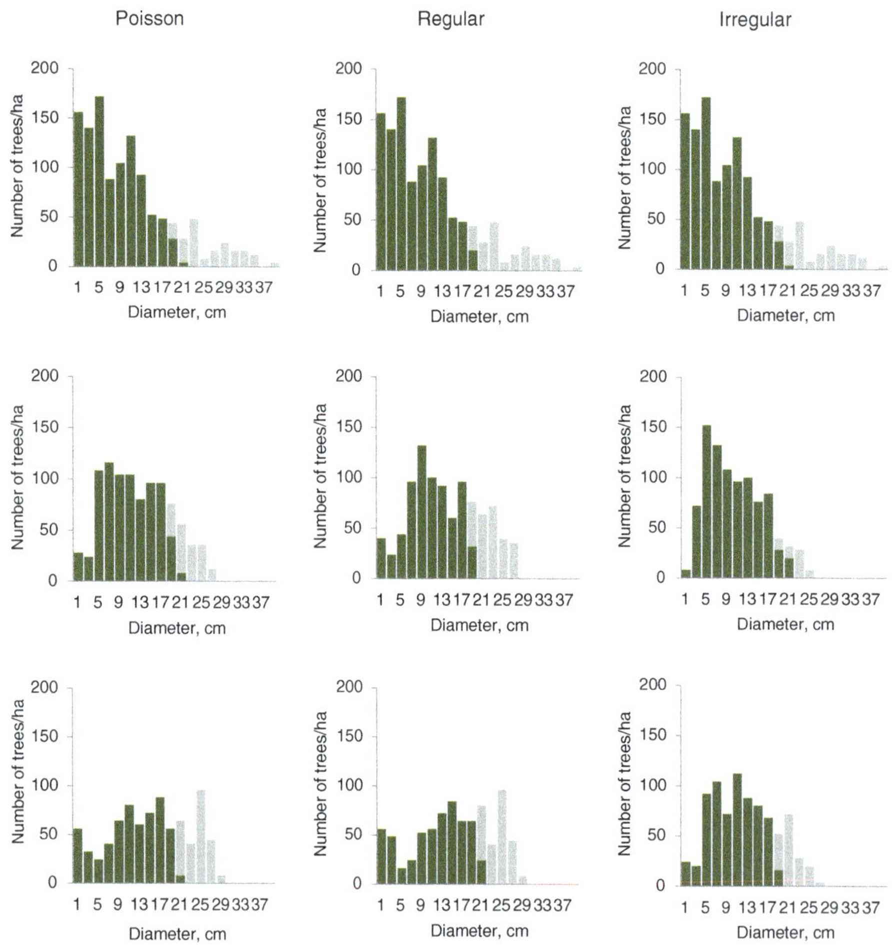

Fig. 8 Number of remaining (dark green) and removed (grey) trees in the first (top), second (middle) and third (bottom) thinning of a Poisson-distributed, regular and irregular stand when NPV with 3 % discount rate is maximized and remaining basal area is one of the optimized variables

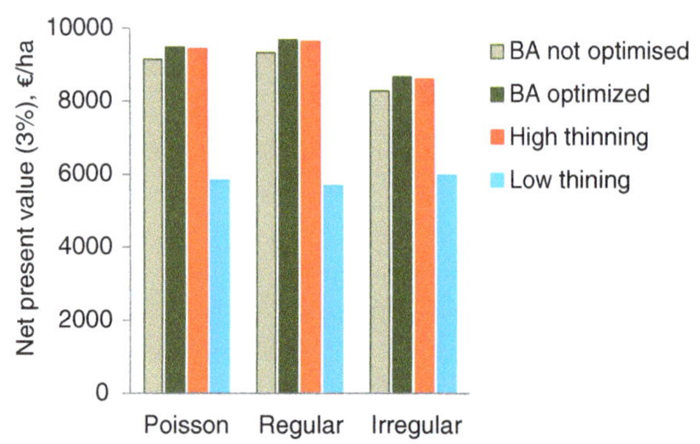

Fig. 9 Net present value when tree selection is based on optimized rule (BA not optimized, BA optimized) or on dbh (High thinning, Low thinning). In high and low thinning, the stand is thinned to the optimal remaining basal area

even-aged management. Several studies have shown that the optimal thinning of a certain stand would be rather similar in even-aged management and continuous cover forestry (e.g. Pukkala 2015b). In the study of Pukkala et al. (2014), which optimized the cuttings of 200 different stands representing different stand structures without the limitation to pursue either even-aged management or continuous cover forestry, 97–99 % of thinnings were high thinnings similar to those that were found optimal in this study. However, some studies (Valsta 1992; Hyytiäinen et al. 2005; Pukkala 2015b) have found that, in even-aged management when a forced clear-felling belongs to the management schedule, it is sometimes optimal to remove trees from both ends of the diameter distribution.

Tree quality, health and vigor are additional characteristics which should affect tree selection. However, these criteria are difficult to include in simulation and optimization studies. As suggested already Möller (1922), trees whose vigor is decreased should be removed

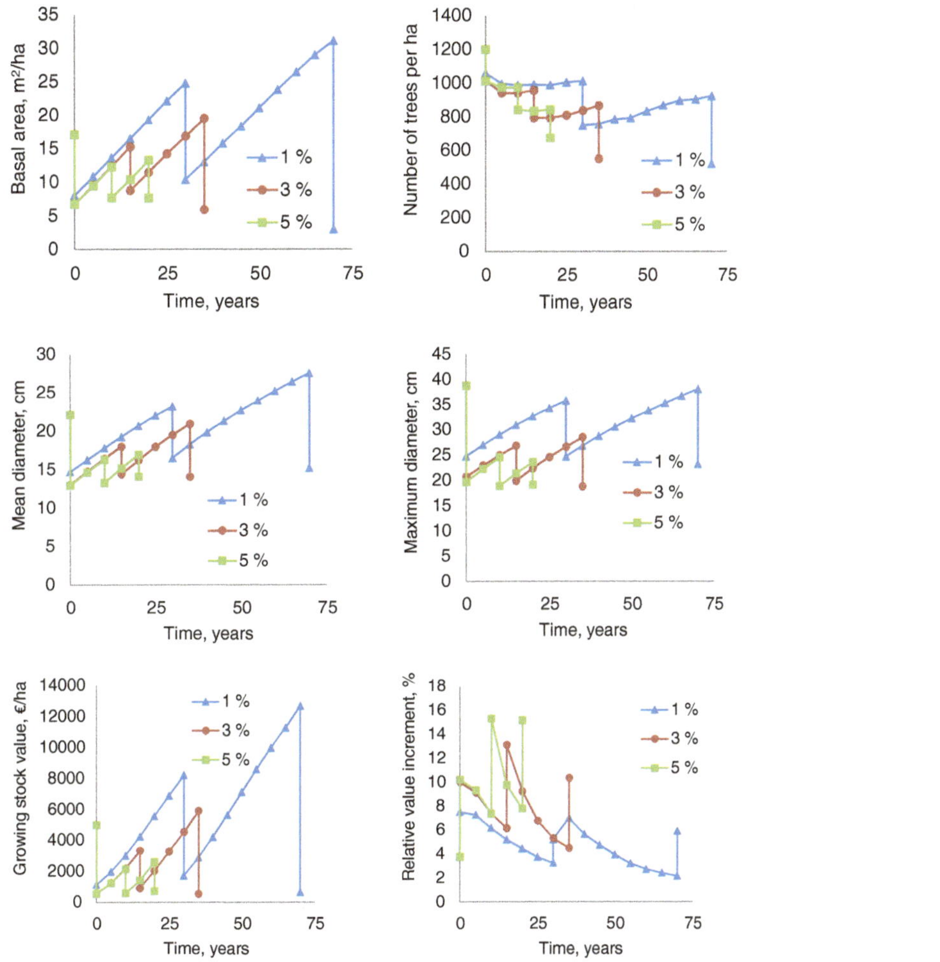

Fig. 10 Development of stand basal area, number of trees per hectare, mean diameter, maximum diameter, growing stock value and relative value increment in the optimal management schedule of the Poisson forest when NPV is maximized with 1, 3 and 5 % discount rate

in thinnings. If low-quality trees overtop smaller and better-quality individuals, they should also be removed, as commonly done in forestry practice. If the quality of all trees is low (or equal), the criteria proposed in this study can be used.

The optimal tree selection rules that were developed in this study, all lead to thinning from above. The comparison of Fig. 9 also shows that thinning from below may not be economically justified. This can also be concluded from earlier literature that discusses financial maturity (Duerr et al. 1956; Davis and Johnson 1987; Knoke 2012). Also several optimization studies show that high thinning is in most cases more profitable than low thinning (e.g., Haight and Monserud 1990; Valsta 1992; Tahvonen et al. 2013). In fact, it is hard to find economic arguments which would justify the use of low thinning.

The study used a non-spatial model in spatial simulation. It was justified by the fact that the area of computing the competition variables (*G* and *BAL*) corresponds to the area of the sample plots in the modelling data of Pukkala et al. (2013). However, there is one difference: in the modelling data, *G* and *BAL* were computed within 300-m^2 plots (around 10 m radius) for all trees of the plot, not only for trees near plot center. These values of *G* and *BAL* may not describe the competition that edge trees face in the best possible way because trees outside the plot also create competition. This may be called as "sampling error" in the calculation of *G* and *BAL* for the edge trees in individual-tree growth modelling. Sampling error results in weaker relationship (underestimated influence) between growth and *G* or *BAL* in the growth model. In the current study, this underestimation was counteracted by the distance-dependent computation of *G* and *BAL*, which increased their variation. As a result, the predictions may in fact be better than when calculating *G* and *BAL* in the same way as they were computed in the data preparation step of growth modelling.

The simulations of this study included stochasticity in mortality and ingrowth. Therefore, every simulation

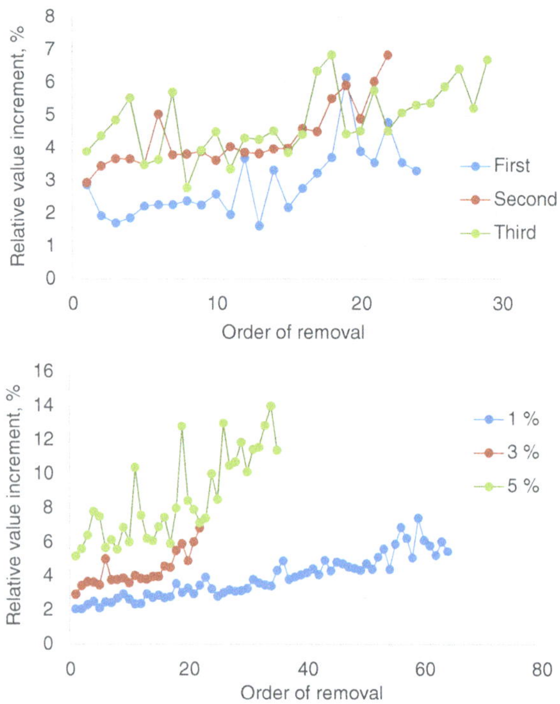

Fig. 11 Predicted value increments of removed trees in the first, second and third thinning of the Poisson stand when discount rate is 3 % (top), and in the second thinning of Poisson stand when discount rate is 1, 3 or 5 %. The relative value increments of remaining trees have been up-dated after every tree removal (for example, the relative value increment of the last tree removed is calculated for a situation in which all trees of the removal, except the last, have been removed)

was repeated 10 times and the mean *NPV* of the 10 simulations was used as the value of the management schedule. Ten simulations is a small number in stochastic simulation. However, few trees die in managed forests, and dead trees are usually small. Ingrowth begins to affect harvest removals and stand value gradually, its effect being minimal during the first decades although ingrowth is critically important for the long-term sustainability of continuous cover forestry. Therefore, it can be concluded that the two sources of stochasticity (mortality and ingrowth) did not have any significant effect on the results of this study.

Competing interests
The authors declare that they have no competing interests.

Authors' contributions
TP conducted the analyses. EL and OL participated in the writing of the manuscript. All authors have read and approved the final manuscript.

Author details
[1]University of Eastern Finland, PO Box 111, 80101 Joensuu, Finland. [2]Joen Forest Program Consulting, Rauhankatu 41, 80100 Joensuu, Finland.

References
Davis LS, Johnson KN (1987) Forest management. Third edition. McGraw-Hill Inc. p 790
Duerr WA, Fedkiw J, Guttenberg S (1956) Financial maturity: A guide to profitable timber growing. US Dep Agric Tech Bull 1146:74
Eerikäinen K, Miina J, Valkonen S (2007) Models for the regeneration establishment and the development of established seedlings in uneven-aged, Norway spruce dominated stands of southern Finland. Forest Ecol Manage 242:444–461
Haight RG (1985) A comparison of dynamic and static economic models of uneven-aged stand management. Forest Sci 31(4):957–974
Haight RG (1987) Evaluating the efficiency of even-aged uneven-aged stand management. Forest Sci 33(1):116–134
Haight RG, Getz WM (1987) Fixed and equilibrium problems in uneven-aged stand management. Forest Sci 33(4):908–931
Haight RG, Monserud RA (1990) Optimizing any-aged management of mixed-species stands: II. Effects of decision criteria. Forest Sci 36(1):125–144
Hooke R, Jeeves TA (1961) "Direct search" solution of numerical and statistical problems. J ACM 8:212–229
Hyytiäinen K, Tahvonen O, Valsta L (2005) Optimum juvenile density, harvesting and stand structure in even-aged Scots pine stands. Forest Sci 51:120–133
Knoke T (2012) The economics of continuous cover forestry. In: Pukkala T, Gadow KV (eds) Continuous Cover Forestry. Springer. ISBN 978-94-007-2201-9. pp 167–193
Laasasenaho J (1982) Taper curve and volume equations for pine spruce and birch. Communicationes Instuti Forestalis Fenniae 108:1–74
Miina J, Pukkala T (2000) Using numerical optimization for specifying individual-tree competition models. Forest Sci 46(2):277–281
Möller A (1922) Der Dauerwaldgedanke: sein Sinn und seine Bedeutung. Springer, Berlin, p 84
Pukkala T (2015a) Optimizing continuous cover management of boreal forest when timber prices and tree growth are stochastic. Forest Ecosyst 2(6):1–13
Pukkala T (2015b) Plenterwald, Dauerwald, or clearcut? Forest Policy Econ 2016(62):125–134, http://dx.doi.org/10.1016/j.forpol.2015.09.002
Pukkala T, Miina J (1998) Tree-selection algorithms for optimizing thinning using a distance-dependent growth model. Can J Forest Res 28:693–702
Pukkala T, Lähde E, Laiho O (2012) Continuous cover forestry in Finland – Recent research results. In: Pukkala T, Von Gadow K (eds) Continuous Cover Forestry. Springer., pp 85–128
Pukkala T, Lähde E, Laiho O (2013) Species interactions in the dynamics of even- and uneven-aged boreal forests. J Sustain Forest 32:1–33
Pukkala T, Lähde E, Laiho O (2014) Optimizing any-aged management of mixed boreal forest under residual basal area constraints. J Forest Res 23(3):727–636
Tahvonen O (2011) Optimal structure and development of uneven-aged Norway spruce forests. Canadian Journal of Forest Res 42:2389–2402
Tahvonen O, Pukkala T, Laiho O, Lähde E, Niinimäki S (2010) Optimal management of uneven-aged Norway spruce stands. Forest Ecol Manage 260:106–115
Tahvonen O, Pihlainen S, Niinimäki S (2013) On the economics of optimal timber production in boreal Scots pine stands. Can J Forest Res 43:719–730
Valsta L (1992) A scenario approach to stochastic anticipatory optimization in stand management. Forest Sci 38(2):430–447

14

Optimal plot design in a multipurpose forest inventory

Helena M. Henttonen[1] and Annika Kangas[2*]

Abstract

Background: We explore the factors affecting the optimal plot design (size and type as well as the subsample tree selection strategies within a plot) and their relative importance in defining the optimal plot design in amultipurpose forest inventory. The factors include time used to lay out the plot and to make the tree measurements within the plot, the between-plot variation of each of the variables of interest in the area, and the measurement and model errors for the different variables.

Methods: We simulate different plot types and sizes and subsample tree selection strategies on measuredtest areas from North Lapland. The plot types used are fixed-radius, concentric and relascope plots. Weselect the optimal type and size first at plot level using a cost-plus-loss approach and then at cluster level byminimizing the weighted standard error with fixed budget.

Results: As relascope plots are very efficient at the plot level for volume and basal area, and fixed-radius plots for stems per ha, the optimal plot type strongly depends on the relative importance of these variables. The concentric plot seems to be a good compromise between these two in many cases. The subsample tree selection strategy was more important in selecting optimal plot than many other factors. In cluster level, the most important factor is the transfer time between plots.

Conclusions: While the optimal radius of plots and other parameters were sensitive to the measurement times and other cost factors, the concentric plot type was optimal in almost all studied cases. Subsample tree measurement strategies need further studies, as they were an important cost factor. However, their importance to the precision was not as clear.

Keywords: Sample, Plot, Forest inventory, Measurement, Cost, Loss

Background

Optimal inventory sampling design is a very important goal in National Forest Inventories (Mandallaz 2007). The inventory design is optimized in a sense that we wish to have the highest accuracy given a fixed budget or we wish to have the lowest cost for a given accuracy. Optimization is possible, if we make assumptions concerning the population. In an analytical setting, we need to be able to anticipate the population variance (Mandallaz & Ye 1999). It is even possible to optimize the measurements of trees in the plots, for instance to determine how many subsample trees (i.e. second-phase sample trees) to measure out of the total number of tally trees (i.e. first-phase sample trees), if we can anticipate the error in the volume estimates of the tally trees.

Defining optimal sample plot size and type analytically would require that we can anticipate the effects of the plot size and type on the population (or between-plot) variance. If the expected between-plot variation can be expressed as a function of plot size (see Freese 1961, Zeide 1980) the optimal plot size can be calculated analytically. However, such a function can only be an approximation of the between-plot variation as the relationship depends on the characteristics of the population such as spatial pattern of the trees, which cannot fully be described with a model.

In addition the expected costs, measured with time consumption as a function of plot size are needed for optimization. In fixed-radius plots the number of trees in a plot is proportional to plot area, but the time needed to check the borderline trees is proportional to the perimeter (Zeide 1980). In relascope plots, time consumption is inversely proportional to the fixed angle defined by the relascope factor. Although Kulow (1966) and

* Correspondence: Annika.kangas@luke.fi
[2]Natural Resources Institute Finland (Luke), Economics and Society unit, Yliopistokatu 6, 80100 Joensuu, Finland
Full list of author information is available at the end of the article

Grosenbaugh & Stover (1957) compared the coefficient of variation using both fixed-radius and relascope plots, they did not compare the overall efficiencies of these two types of plots related to the time spent.

While many factors affecting the accuracy can be accounted for analytically, some aspects like the spatial pattern, are more difficult. The analytical calculations usually assume a random pattern (Mandallaz 2007). Likewise, the number of subsample trees and the selection of measurements taken from each tree (e.g. height and/or upper diameter) can be difficult to account for in detail in an analytical setting. Therefore, the optimal plot size and type has most often been defined by simulating sampling in an accurately measured and mapped forest area. In the earliest studies, simulation was carried out by measuring a grid of small cells and building larger sample plots as their combination (Johnson & Hixon 1952, Mesavage & Grosenbaugh 1956). In later studies, computer simulation based on mapped data has been utilized (e.g. Kulow 1966). In a simulation based on real data, the optimal plot size is heavily dependent on the forest conditions on the area, which makes definite conclusions difficult (Mesavage & Grosenbaugh 1956).

Optimal sampling design and optimal plot design (size and type) depends highly on the purpose of an inventory. It is easy in principle to define an optimal inventory for one variable of interest such as biomass or volume with regard to measurement costs and accuracy. When the number of characteristics of interest increases, the task gets more complicated as the optimal plot number, size and type are likely to be different for each characteristic. For instance, class variables such as land use and its changes could be determined from a very small plot or even point, but volume and biomass require a larger plot. Thus, prioritizing the forest characteristics is needed if an optimal plot is to be determined.

The estimation method is also likely to have an effect: if we assume a design that is based purely on field plots, the optimal plot size and type are likely to be very different from a case where auxiliary information such as remote sensing information is used in stratification (e.g. Tomppo et al. 2014), traditional regression estimation, model-assisted estimation or model-based estimation. In these cases, the variation between plots may not be the decisive factor, but rather the correlation between the forest characteristics and the remote sensing data.

The results may also depend on the specific criterion used for defining the optimum. One option is to minimize some criterion like standard error of the estimate for a given budget constraint such as amount of time (Johnson & Hixon 1952, Mesavage & Grosenbaugh 1956). Using this approach, Johnson & Hixon (1952) concluded that while long and narrow rectangular plots tended to have smaller between-plot variation, the time needed to lay out such plots was larger. Thus, the most efficient plots for a given amount of time were compact plots.

Another way to define the optimal plot size is to use a cost-plus-loss (CPL) approach (Hamilton 1978, Ståhl 1994). It means that the losses due to poor estimates (possibly resulting sub-optimal decisions) are calculated as a function of the uncertainty involved and these losses are added to the measurement costs described as a function of measurement time. This criterion would be ideal, if the losses due to poor estimates could be accurately defined. Often the losses are described as a function of the standard error or some other criterion (Barth & Ståhl 2012), but they could also be calculated for an actual decision problem (Eid et al. 2004). When the inventory is multipurpose, the cost-plus-loss method is more complicated (see Burkhart et al. 1978). If we were able to define the losses due to the poor estimates for each of the variables of interest (i.e. give relative weight to the errors of each variable), it is possible.

Total measurement costs can be calculated as a function of time used for each sample plot. The time depends on: 1) the time required to go to the plot and lay out the plot; 2) the total number of trees to be measured and 3) the measurements carried out for each tree. Laying out the plot means defining the plot center (or center for several sub-plots) and determining which trees belong to the plot(s). For circular or relascope plots that means checking the distance of borderline trees from the plot center with a measuring tape or an (optical) rangefinder (e.g. Loetsch et al. 1973).

The measurements needed for each tree depend on the characteristics of interest (e.g. volume, biomass, stems per ha). Typically not all characteristics needed are measured on all trees within a plot. The diameter at breast height (d1.3) is measured for all tally trees, but height, upper diameters, age, and growth are measured only for subsample trees. Thus, the measurement time also depends on the number of subsample trees within each plot, and the number of measurements carried out on each tree. As biomass and volume require additional subsample tree measurements compared to stems per ha, also the time consumption needs to be defined separately for each of the variables.

The precision of the sample plot measurements in describing the forest stand can be measured using the standard errors of the estimators of given forest characteristics, which depends, in part, on the spatial variation of the characteristics of interest within the forest. In general, the bigger the sample plot area, the larger the proportion of total variation that falls within the plot, and consequently the smaller the standard errors (e.g. Loetsch et al. 1973, Koivuniemi 2003).

Measurement and model errors for the variables used to calculate the characteristics of interest have also an effect on precision (e.g. Päivinen 1987, Ståhl et al. 2014). Their combined effect again depends on the number of subsample tree measurements and the models/methods available to generalize the subsample tree measurements to the tally trees. It may be assumed that the errors in volume / biomass for subsample trees are negligible, but not for the tally trees. It is quite possible that the model which is most efficient when all measurements are assumed error-free is not the most efficient when these errors are included (Eid 2003). Therefore, it would be best to select the models used for generalizing the subsample tree characteristics to tally trees simultaneously with deciding the number of subsample trees and the variables measured for each of them.

The aim of this study is to analyze optimal sample plot type and size with a simulation study and explore the relative effects of different factors on optimal plot measurement strategy in the special conditions of North Lapland. The study region is partially located close to the northern timberline, where clustered spatial patterns of trees challenge the planning of an efficient forest inventory. The studied plot types were fixed-radius plots with varying radii, a combination of two concentric plots with varying radii and varying diameter limits for the larger radius, and relascope plots with varying relascope factor and maximum radii. The forest characteristics concerned were volume, basal area and stems per ha. The class variables such as forest/non-forest were excluded from the study.

Material

Measurements of 50 m x 50 m test areas were carried out in 2002 in Inari, North Finland (Figure 1). The measured areas were sampled from the plots of the 8th National Forest Inventory. In total, 18 test areas were measured, together with the planar coordinates of the

Fig. 1 The locations of the mapped test areas

trees (with d1.3 ≥ 2.5 cm) mapped with tachymeter (SOKKIA SET 4C), as well as d1.3 (in two perpendicular directions), height (h, m), and upper diameter at a height of 6 m (d6, cm) (for trees ≥ 8 m tall). An example of our data is given in Fig. 2 illustrating the spatial distribution of trees and their diameters in a mature Scots pine *(Pinus sylvestris L.)* stand with birch (*Betula pubescens*) undergrowth. Volumes for the trees were calculated using volume functions (Laasasenaho 1982). For trees with heights ≥ 8.1 m, volumes were estimated as a function of d1.3, h, and d6. When the height of a tree was less than 8.1 m, volumes were estimated as a function of d1.3 and h.

Methods

Analysis of the point patterns of trees on the 50 m x 50 m areas was carried out using the R package *spatstat* (Baddeley and Turner 2005). Our main interest was in assessing whether the point patterns could be considered random (Poisson). For this purpose, we carried out a simultaneous (simultaneous for different values of the distance *r*) Monte Carlo test for Ripley's K–function and L-function, which is a variance stabilizing transformation of K. Inhomogeneity was taken into account by modelling trends as a function of coordinates. In areas divided into two different stands, the stand was used as an indicator variable in modelling inhomogeneity.

Analysis of the optimal plot design was carried out at two levels: plot level and cluster level. The cluster is interpreted here as a combination of *m* plots, but any

Fig. 2 An example of a mapped test area with a clustered spatial pattern. The sizes of circles are proportional to the diameters of the trees

specific spatial arrangement for the cluster is not determined. If the spatial arrangement were specified, the cluster could also be interpreted as a plot with *m* subplots.

In the plot level analysis, the optimal plot design was defined by minimizing the cost-plus-loss (*CPL*) defined for the *p* variables of interest for one plot. The general function to be optimized is

$$\begin{aligned} OP &= \min CPL_{\{size,type,strategy\}} \\ &= \min\left(l_1 + \ldots + l_p + c\right) \end{aligned} \tag{1}$$

where the losses are a function of RMSE as

$$l_p = w_p RMSE_p \tag{2}$$

where w_p is the weight given to the RMSE of a given variable *p*. Costs *c* are defined as a function of the time for transfer between the plots (LT used at cluster level, assumed to be 15 min/plot), the time needed to check the borderline trees (BT assumed to be 0.5 min/tree), tally each tree in the plot (TT assumed to be 0.5 min/tree) and measure the subsample tree characteristics from each the subsample tree (ST assumed to be 4.5 min/tree) as

$$c = LT + BTn_1 + TTn_2 + STn_3 \tag{3}$$

where n_2 is the number of tally trees, n_1 is the number of borderline trees and n_3 he number of subsample trees (Päivinen 1987). A tree was defined as borderline tree if its distance from the plot center differed less than 0.5 m from the radius for a tree with a given size.

The plot level analysis was carried out so that with each plot type and size, we simulated $N = 1000$ randomly located plots (simulated plots) within each mapped 50 m x 50 m area. The accuracy of each characteristic within each area was analyzed as

$$RMSE_j = \sqrt{\sum_{i=1}^{N} \frac{\left(\hat{y}_{ji} - y_j\right)^2}{N}} \tag{4}$$

where $\hat{y}_{ji}$ is the observed forest characteristics from the area *j* and simulated plot *i* and y_j is the true value of the characteristics calculated from the whole area *j*. Relative RMSE was calculated by dividing the RMSE with the mean across the test areas (Table 1). The plot-level analysis $RMSE_{\ p}$ (equation 2) was the average of these RMSEs in the 18 test areas. The variables of interest were plot volume (V, m^3/ha), basal area (G, m^2/ha) and stems per ha (N). The simulated plots were located within the test areas so that the center point was at least 11 m from the edge so that edge corrections were not used. The possible bias resulting from this is included in the RMSE.

Table 1 Mean volumes, stem per ha and basal areas of the 50 m x 50 m mapped test areas

Testarea	V,m³/ha	N,/ha	G,m²/ha
1	43.4	668	8.4
2	70.6	728	13.1
3	8.2	316	1.9
4	17.0	224	3.3
5	45.0	904	9.3
6	19.0	480	4.5
7	118.8	424	17.9
8	92.4	260	12.1
9	63.9	356	11.2
10	4.0	324	1.4
11	10.6	424	2.8
12	20.3	980	5.0
13	64.6	1276	12.5
14	25.9	1388	6.5
15	44.5	1412	9.8
16	21.0	960	4.7
17	92.9	984	12.6
18	24.8	692	5.1
Mean	43.7	711.1	7.9

In the cluster level analysis, the average MSE from the plot level analysis was used as the "within-test-area" variation (Var_w). In addition, the total variation included the "between-test-areas" variation among the 18 test areas, defined as

$$Var_b = \sum_{j=1}^{18} \frac{\left(y_j - \bar{y}\right)^2}{18} \tag{5}$$

where $\bar{y}$ is the overall mean of forest characteristic in these 18 areas (Table 1). We calculated the total variation as

$$Var_{total} = Var_b + Var_w \tag{6}$$

where the within-area variation depends on the plot type and size but the between area variation does not.

The total variation can be used to simulate a situation where both the optimal sample plot size and the optimal number of sample plots are selected. It can thus be used to analyse if it is more useful to select a large number of small plots or a small number of large plots. In this study, we analysed the optimal plot type and size for one cluster consisted of m plots. The budget for measuring one cluster was fixed to one day work, approximately 420 min. We calculated the affordable cluster plot number m as

$$m_{case} = \frac{B}{c_{case}} \tag{7}$$

where the c_{case} is the measurement time needed (Equation 3) for one simulated plot with a given type and size and a given subsample tree measurement strategy, and B is the total budget of measurements (in minutes per day). The relative standard error of the mean of cluster for volume (V, m³/ha), basal area (G m²/ha) and stems per ha (N) was calculated using the simple random sampling (SRS) formula with the number of plots m_{case} fitting the budget and the total variation Var_{total} depending on the plot type and size as

$$SE\left(\hat{\bar{y}}_{case}\right) = \sqrt{\frac{Var_{total,y}}{m_{case}}} \tag{8}$$

Relative error was calculated by dividing with the mean across the areas (Table 1) to standardize the units. The optimal plot size, type and number of plots in a cluster was defined as the one with the minimum weighted mean of the SE%s as

$$OP = \min_{\{size,type,strategy,m_{case}\}} \left(w_V SE\%_V + w_N SE\%_N + w_{BA} SE\%_{BA}\right) \tag{9}$$

We examined three different plot types. The first type was a fixed-radius plot with radius varying from 3 to 11 m. The second type was a combination of two concentric plots with the larger radius varying from 5 to 11 m and the smaller from 3 to 7 m. The diameter limit (*DL*) for trees included to the larger plots varied from 5 cm to 15 cm. The third type was a relascope plot where the relascope factor (*RF*) varied from 1 to 3 m²/ha. The relascope plots were restricted to maximum radius (*rmax*) varying from 6 m to 11 m. Thus, the radius of the inclusion zone for trees with diameter larger than $100\, rmax \sqrt{RF} / 50$ was always equal to *rmax*. For the estimation of mean values per hectare using relascope plot with maximum radius see e.g. Tomppo et al. (2011). The specific plot designs tested are presented in Table 2.

We assumed that diameter on tally trees is measured in two directions. Here we tested two subsample tree selection strategies. In the first one (fixed strategy or S1), tally trees with $d1.3 > 25$ cm were measured as subsample trees, along with all tally trees closer than 1 m to the plot center. The assumption here is that large trees are more important subsample trees than small trees, as they contribute more to the plot volume and the variance of their volume estimates is higher than that of small trees (see discussion below). In the second strategy (relascope strategy or S2) the subsample trees were selected using a relascope factor 5 m²/ha, also assuming that large trees are more important than small trees. We

Table 2 The tested combinations of plots

Plot type	Radiusm	Radius 2m	Relascopefactor RF m^2	Diameter limit DLcm
fixed-radius	3,4,5,6,7,8,9,10 and 11	-	-	-
concentric	11	7	-	5, 7.5,10,12.5,and 15
	9	6	-	5, 7.5,10,12.5,and 15
	7	5	-	5, 7.5,10,12.5,and 15
	6	4	-	5, 7.5,10,12.5,and 15
	5	3	-	5, 7.5,10,12.5,and 15
relascope	6,7,8,9,10 and 11	-	1	-
	6,7,8,9,10 and 11	-	1.5	-
	6,7,8,9,10 and 11	-	2	-
	6,7,8,9,10 and 11	-	2.5	-
	6,7,8,9,10 and 11	-	3	-

assumed that the volume of the subsample trees could be measured error free (in fact there is error but it is assumed negligible), while for tally trees we assumed an error.

For every tree in the 18 test areas, the volume was calculated using d1.3, d6 and h as predictor variables (Laasasenaho 1982) and this was assumed to be the measured volume. Using these volumes, a simpler and less precise model

$$v = \beta_0 + \beta_1 d1.3 + \beta_2 d1.3^2 + \varepsilon \tag{10}$$

was fitted using only d1.3 as a predictor variable. The fitted model has $R^2 = 0.9556$ and RMSE = 0.02626 m^3. The model errors were heteroscedastic (Fig. 3), which is typical for a volume model. In the analysis, the volumes of the tally trees were estimated using this simple model (10), while for the subsample trees the above mentioned measured volumes were used, to describe the effect of not measuring the height and upper diameter of each tree in the plot.

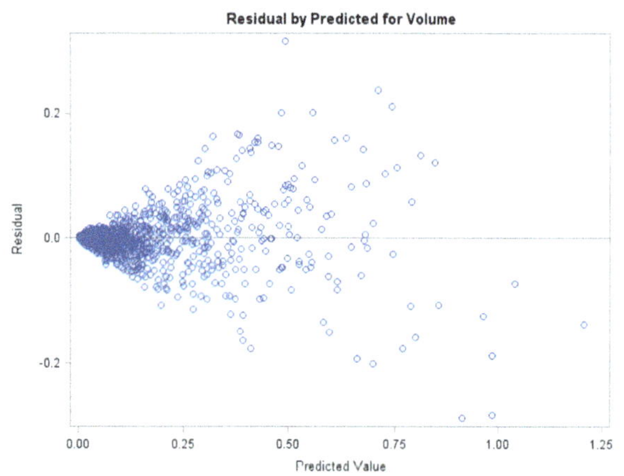

Fig. 3 The residuals of volume model (Equation 10)

Results

The fixed-radius plots included much more measured trees than the other two plot types. In fixed-radius plot with radius 11 m, the maximum number of trees (in the 1000 replications and 18 test areas) to be measured reached 80, while in relascope plots the maximum number was below 40 and in concentric below 60 (Fig. 4). The maximum number of borderline trees was 2.39 on average for fixed-radius plots and only 0.96 for relascope plots. With fixed-radius plots, the proportion of borderline trees varied from 8.9 to 30.4 %, for relascope plots from 14.3 to 25.1 % and for concentric plots from 9.6 to 25.9 %. The variation in the number of measured subsample trees between the plot types and sizes was quite low (see also Fig. 6).

When the average relative RMSE (Equation 4) was plotted as a function of measurement times (min) for different plot types and forest characteristics, it was clear that the relascope plot type was very efficient for volume and basal area, while the fixed-radius plot was best for stems per ha (Fig. 5). The concentric sample plots seemed to be a very useful compromise, which was near optimal for all characteristics.

The strategy to select the subsample trees with an relascope factor 5 (S2) was clearly distinguishable from the strategy to select all large trees and trees closest to plot center (S1) with longer measurement times for the plot (Fig. 5). The reason for this can be seen from the number of subsample trees measured in each case with these two strategies: the relascope strategy on average produced about 0.5 more subsample trees per plot (Fig. 6). With smaller radii the difference could be as much as 1.0 subsample trees, while the differences petered out with larger plot sizes. For plots with the smallest radii

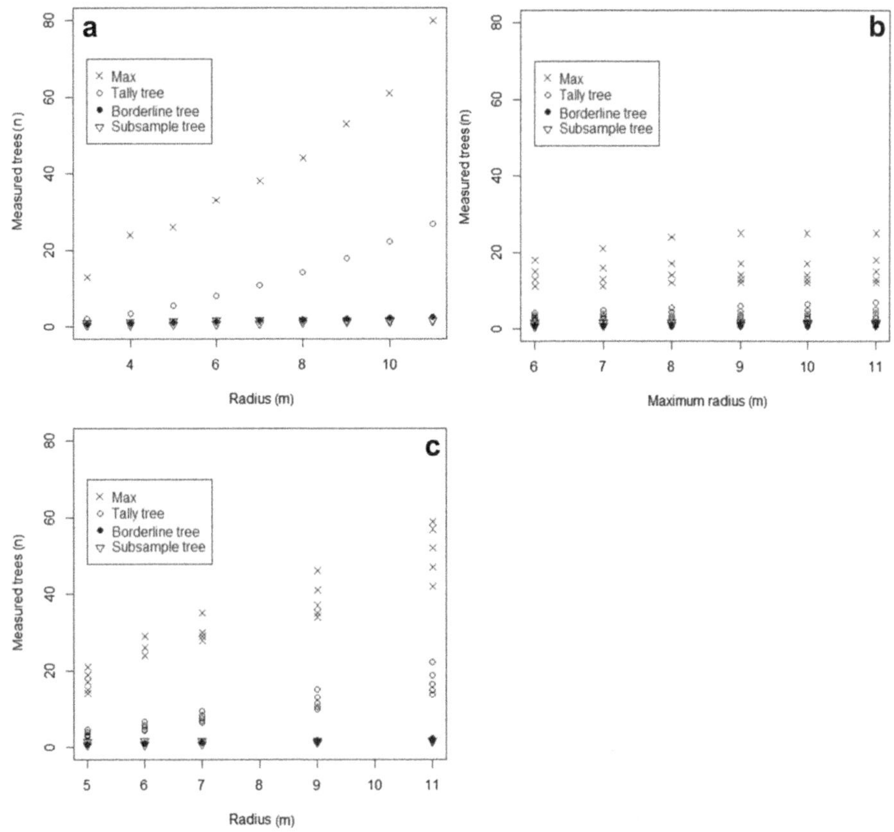

Fig. 4 The maximum number of trees in a plot, and average number of tally trees, borderline trees and subsample trees as a function of plot radius (m) with the two subsample tree selection strategies with fixed-radius (**a**), relascope (**b**) and concentric (**c**) plots. In relascope plots the variation within each relascope factor is due to varying maximum radius and in concentric plots the variation within each radius is due to varying diameter limit

the relascope strategy (S2) seemed a little more efficient with respect to the RMSE of volume while the fixed strategy (S1) was more efficient for larger radii plots (Fig. 7).

To select an optimal plot type and size for a fixed number of plots (i.e. at plot level), a cost-plus-loss analysis was carried out (Equation 1), with weight $w_p = 0.08$ for the RMSEs (Equation 2) of all three characteristics considered. For fixed-radius plots the optimal strategy of 6 m radius was very clear (CPL 20.72). The relascope strategy for selecting subsample trees was clearly less efficient than the fixed strategy (Fig. 8). For relascope plots, the differences between the subsample tree selection strategies were also very clear, as well as the differences between the relascope factors. On the other hand, the CPL did not appear to depend on the maximum radius. The optimal relascope plot had a relascope factor 1 m^2/ha and a maximum radius 7 m (CPL 21.42). With concentric plots, the dependency on radius was similar but less pronounced than with fixed-radius sample plots. The optimal plot radius was a little bit larger (7 m), but trees with d1.3 less than 15 cm were only measured within 5 m plot. This plot type produced the smallest CPL (19.81). The effect of varying the subsample tree measurement strategy was much larger than that of the diameter limit.

If the weight of RMSE for stems per ha was tripled (ceteris paribus), the optimal radius for fixed-radius plots was 7 m. In this case, the smallest CPL was obtained with fixed-radius plots (27.40). Thus, when stems per ha is important enough, the fixed-radius plot is the most efficient. For the concentric plot the optimal diameter limit changed from 15 to 5 cm. If the weight of the volume RMSE was tripled (ceteris paribus), the maximum radius of relascope plots increased to 8 m. The optimal plot was a concentric sample plot with radii 9 / 6 m and diameter limit of 15 cm (CPL = 28.31).

A more marked change occurred when the relative importance of losses compared to costs was reduced to 0.01 for all variables (Equation 2). In that case, the optimal fixed-radius plot radius was 3 m, the optimal relascope factor 3 m^2/ha with a maximum radius 6 m, and

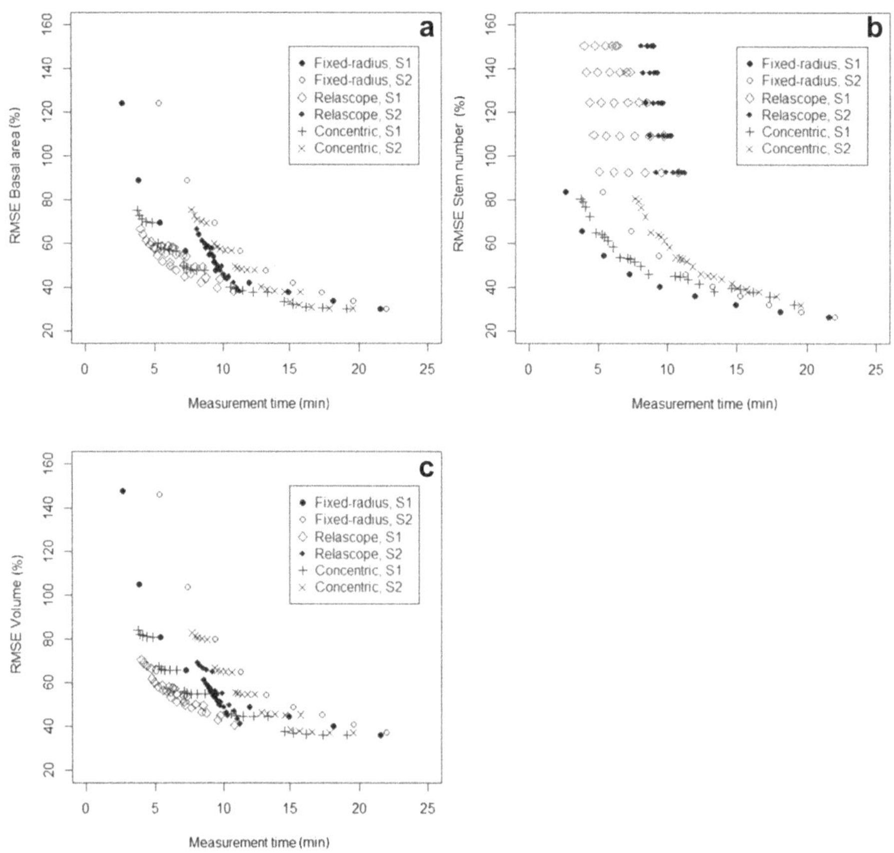

Fig. 5 The average relative RMSE as a function of measurement times (min per plot) for different plot types for basal area (**a**), stems per ha (**b**) and volume (**c**) with the two subsample tree selection strategies (S1 = fixed, S2 = relascope). In relascope plots, the vertical variation is due to varying relascope factor and the horizontal due to varying maximum radius. In fixed and concentric plots the vertical variation is due to the varying radii and the horizontal due to the diameter limit

for concentric plots the optimal radii were 5 /3 m with the diameter limit of 15 cm. In this case, the concentric plot had the smallest CPL (6.14) but the relascope plot was very close (6.18). That means that for all plot types, the optimal plot size was the smallest considered. When the weight of losses was increased compared to costs (weight 0.2 for all variables), the optimal radius for the fixed-radius plot was 8 m, the optimal relascope factor was 1 m^2/ha with a maximum radius of 10 m, and the optimal concentric plot had radii of 11 / 7 m with a diameter limit 15 cm. It also had the smallest CPL (36.69).

When the budget was fixed to one day's worth of measuring minutes (420) and the sample plot number, size and type within a cluster could all be decided at the same time, the optimal combination was to measure 19 concentric sample plots with radii 7/5 m with a diameter limit of 10 cm (Fig. 9). When the time for transfer between the plots (LT) was reduced from 15 to 10 min, the optimal number of plots increased from 19 to 25 and the diameter limit increased to 15 cm (Fig. 10). On the other hand, when LT was increased to 20 min, the optimal number of plots reduced to 14 and the optimal radii of the concentric plots increased to 9/6 m and the diameter limit increased to 12.5 cm. When the transfer time from plot to plot shortens, it is better to measure a larger number of smaller plots and vice versa.

When the measurement time of one tally tree (in Equation 3) was increased to 0.7 min and that of a subsample tree to 7 min, the optimal number of measured sample plots per cluster was reduced to 18, the optimal plot radii to 6/4 m with a diameter limit of 7.5 cm (Fig. 11). Thus, the longer it takes to measure one tree, the smaller the optimal plot size. However, the number of plots is affected less than when the transfer time is changed (Figs. 10 and 12).

The analysis of the point patterns showed that 10 out of 18 point patterns could be considered random (Poisson). However, six of these areas included parts from

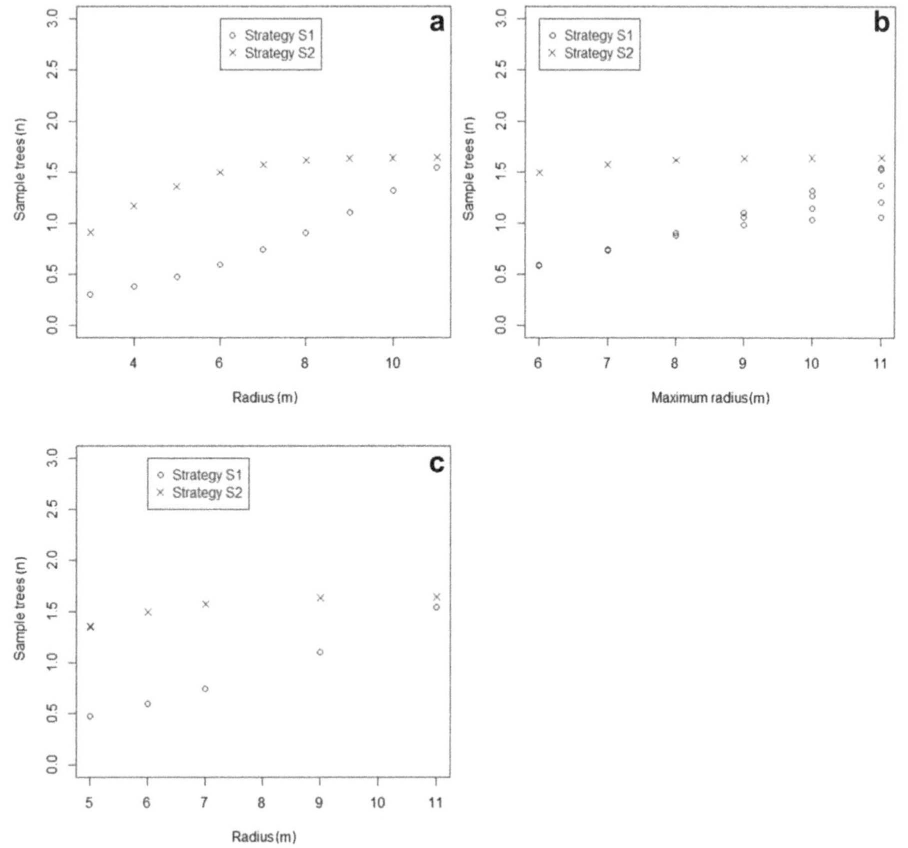

Fig. 6 Number of subsample trees as a function of plot radius (m) with the two subsample tree selection strategies (S1 = fixed, S2 = relascope) for fixed-radius (**a**), relascope (**b**) and concentric (**c**) plots. In relascope plots the variation within each maximum radius is due to varying relascope factor

more than one stand, which may have affected the spatial pattern. Seven of the point patterns were assessed as clustered. However, the clusters seemed to be quite small (< 2 m) and they were probably due mainly to birch (*Betula pubescens*) clones. Only one area showed evidence of a regular pattern. We compared the average relative RMSE's (Equation 4) of the stems per ha (Fig. 13), volume, and basal area when the test areas were classified into the different point patterns with the 6 areas divided between different stands excluded. Sampling was carried out with a fixed-radius plot with plot radius varying from 3 to 11 m. The differences between clustered and Poisson patterns were in average small, but the variation between areas was higher in clustered patterns. The relative RMSE's in the area with the regular point pattern seemed to be less sensitive to the radius of the plot.

Discussion

In this study, we analyzed the effect of plot type (fixed-radius plots, a combination of two concentric plots with a varying diameter limit, and relascope plots with varying maximum radius), different plot size (varying radii or relascope factor) and two different strategies for measuring subsample trees within plots (either all trees with d1.3 > 25 cm and all trees within 1 m from the plot center, or with relascope factor 5 m^2/ha). We examined three different variables, volume, basal area and stems per ha in order to reach a compromise solution that would be suitable for many other variables as well. We did not include class variables such as forest/non-forest classification or forest site or type classification, although these are important variables in forest inventory. In plot-level considerations, a very small plot or even a point would be optimal for many of these variables. Thus, including these variables to the calculations would make more sense if the whole design were optimized rather than just the plot type and size.

Relascope plots were most efficient for volume and basal area, but not as efficient for stems per ha. For stems per ha, fixed-radius plot were optimal. When the weight for stems per ha is increased enough, the fixed-radius plot becomes optimal overall. If we considered an inventory purely for stems per ha or basal

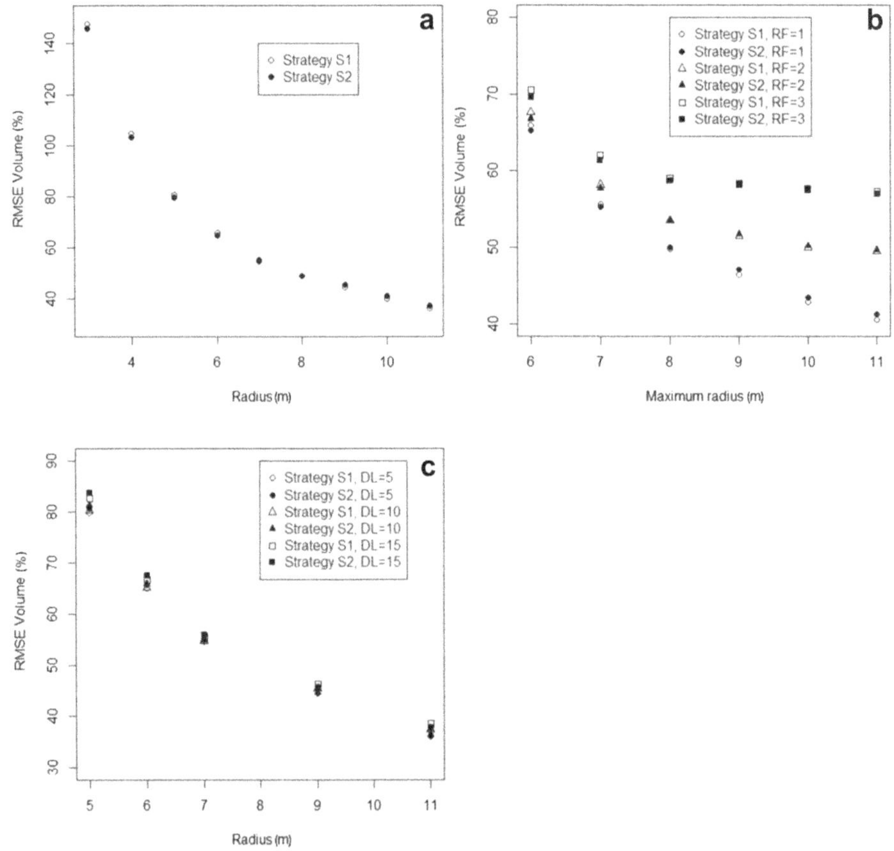

Fig. 7 Relative RMSE of volume as a function of plot radius (m) in two subsample tree selection strategies (S1 = fixed, S2 = relascope) for fixed-radius (**a**), relascope (**b**) and concentric (**c**) plots. RF = relascope factor and DL = diameter limit

area, subsample trees would not be needed at all. In fixed sized plots, measuring the diameters would neither be necessary except for borderline trees. Subsample tree selection strategies would thus be irrelevant for such an inventory. However, if the subsample tree measurement costs were removed, and the measuring cost of each tally tree would be reduced (some time would be needed to record the species and check the borderline trees), the conclusion would still be the same: relascope plot type is the best for basal area and fixed-radius plot is the best for stems per ha. While in principle there are no advantages in using fixed radius over variable radius (Stage & Rennie 1994), in relascope plots the effective plot size for small trees constituting most of the stems per ha is so small that the relascope plot type is very inefficient for stems per ha. Concentric sample plots were a good compromise between efficiency and accuracy. It also turned out that the optimal plot radius in the tested area was somewhat smaller than the one used in current Finnish NFI, 9 m.

We studied two different subsample tree selection strategies. The strategy of measuring all large trees along with trees close to the plot center produced, on average, from 0.5 to 1.5 subsample trees in the different variations of the concentric plots, while the relascope selection strategy produced from 1.4 to 1.6 subsample trees. Although the difference may seem to be small, the strategies differed quite a lot with respect to measurement time. On the other hand, the differences in relative RMSE of volume were not large. Both the strategies acknowledge that the largest trees have largest variation in the volume estimates (Fig. 3), which makes them more attractive as subsample tree candidates. So, the relascope strategy with a larger relascope factor could have been more efficient still. The model was estimated from all the trees measured from the test area, and it therefore produced zero mean error for the whole area, but not necessarily within each diameter class. However, possible bias is implicitly accounted for in the simulations.

The results of the measurement strategy suggest that very few subsample trees would be needed. However, in this study volume was the only characteristics which required subsample tree measurements. Other variables, such as (total) biomass, might require more subsample

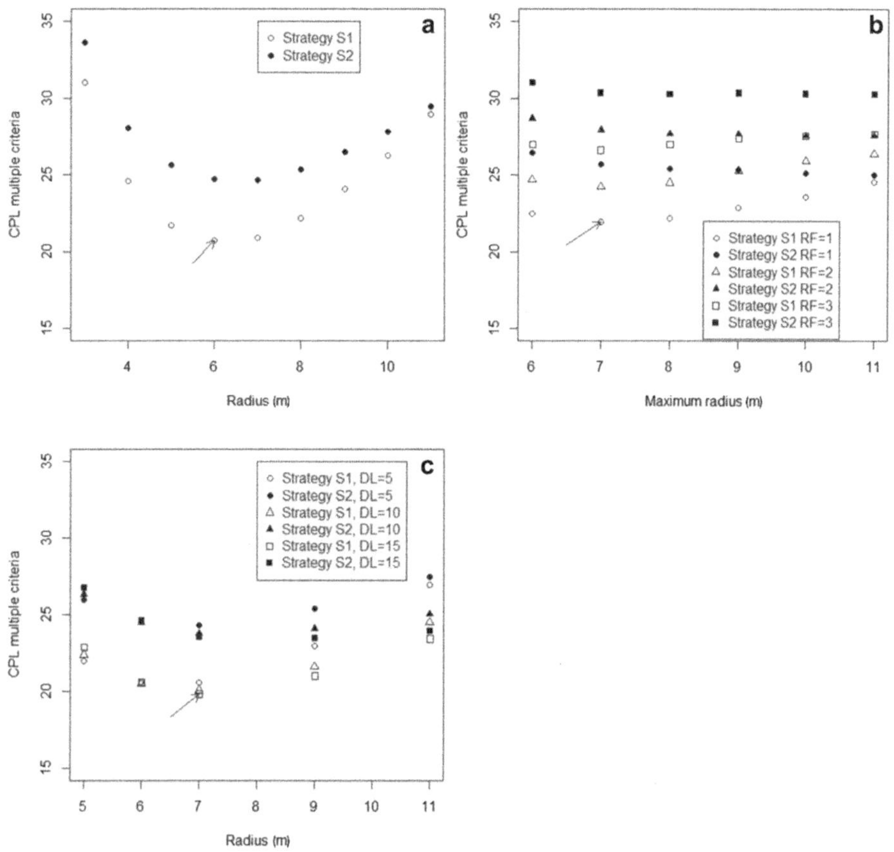

Fig. 8 Cost-plus-loss as a function of plot radius (m) with weight of 0.08 each for RMSE of G, N and V in the loss function in two subsample tree selection strategies (S1 = fixed, S2 = relascope) for fixed-radius (**a**), relascope (**b**) and concentric (**c**) plots. The optimal plot size (minimum CPL assuming fixed plot number) with each plot type is marked with an arrow

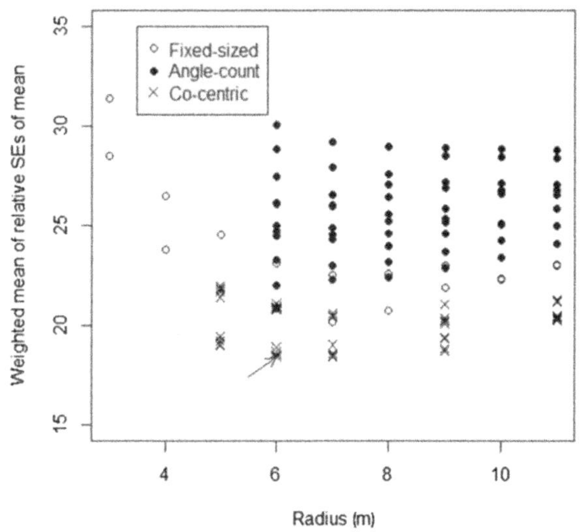

Fig. 9 The weighted mean of relative standard errors for mean volume, basal area and stems per ha as a function of plot radius (maximum radius for relascope plots). The number of plots is based on a budget of 420 min for one cluster. Assumed costs (Equation 3) are LT = 15, BT = 0.5, TT = 0.5 and ST = 4.5 min. The optimal number of plots for a cluster using the optimal plot type and size (marked with an arrow) is 19

trees as biomass models using only d1.3 as predictor are generally less precise than similar volume models. We also considered only temporary plots here. If we had analysed growth of the trees using permanent plots, more subsample tree measurements might prove to be needed, as the estimated growth per tree would be more reliable. These issues remain to be studied in the future.

The optimal plot size and number is quite sensitive to the assumed times to move from plot to plot or of measuring the trees. The concentric plot type was the best plot type for both plot-level and cluster-level calculations, and practically irrespective of the changes in the parameters in the cost function or the weights of different variables. On the other hand, this result can depend on the conditions in Lapland, and in other condition such as southern Finland or tropical areas some other plot type would be optimal.

We did not consider the effect of diameter distribution in this study, but it may also have an effect on the optimal plot type and size. In our northern data, 45 cm was the largest diameter at breast height (the maximum diameter within one test area varied from 19 to

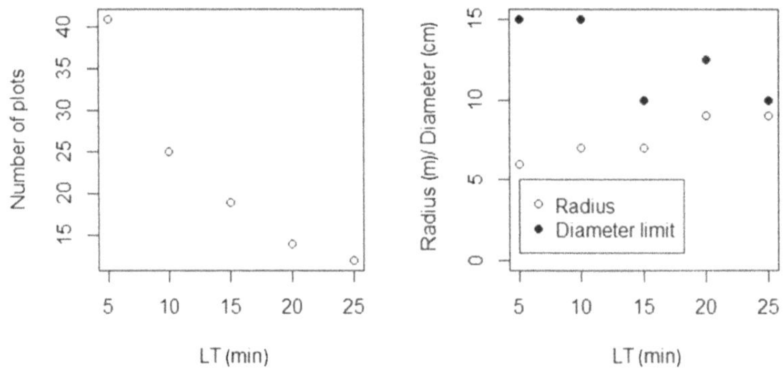

Fig. 10 Sensitivity of the optimal number of plots, diameter limit and plot radii for concentric plot as a function of time for transfer between the plots (LT, min)

45 cm). If the variation had been greater, the variation among the smaller sized plots would most likely have been higher. This also remains to be studied in the future.

We did a preliminary analysis about the effect of point pattern on accuracy. The areas with clustered patterns seemed to have higher between-stands variation in RMSE, although the size of the tree clusters was quite small compared with the tested plot radii. The RMSE on one study area with regular point pattern seemed to be less sensitive to the plot radii than the Poisson and clustered patterns. This might be of importance in the planning of inventories in the future, since the area and volumes of regular planted forests is rapidly increasing in Finland. The effect of point patterns on optimal sampling needs further study and modelling efforts.

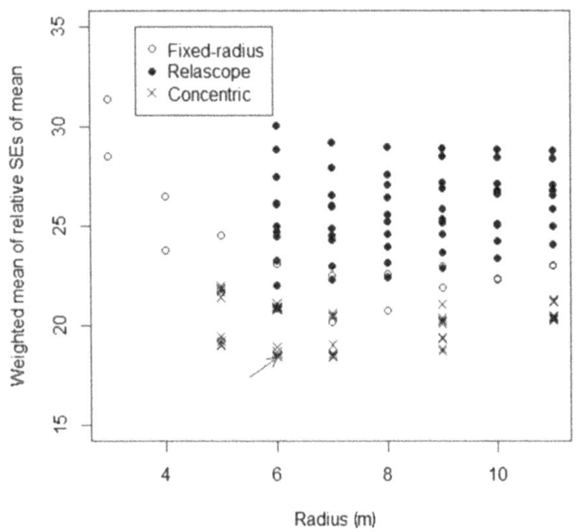

Fig. 11 The weighted mean of relative standard errors for mean volume, basal area and stems per ha as a function of plot radius (maximum radius for relascope plots). The number of plots is based on a budget of 420 min for one cluster. Assumed costs (Equation 3) are LT = 15, BT = 0.7, TT = 0.7 and ST = 7.0 min. The optimal number of plots for a cluster using the optimal plot type and size (marked with an arrow) is 18

If remote sensing material was used as auxiliary data and a model-assisted or model-based framework was employed instead of e.g. simple random sampling, a larger plot size might be optimal (see e.g. Hofstadt et al. 2015). This is because we assume the correlation between the remote sensing data and plot data to be higher with larger plots due to e.g. co-registration errors. Moreover, remote sensing registers crowns rather than stems, and crowns of trees included into the plot will often be partly outside the plot boundaries, and respectively the crowns of trees not included into the plot will be partly inside the plot boundaries. Within larger plots, the effect of crown overlapping should be smaller. This also remains to be studied in the future.

In this study, we searched for an optimal plot type and size at plot level, i.e. for the case when the number of plots is fixed, and for a case where the optimal number of plots in a cluster was defined simultaneously with the plot type and size, i.e. for the case where the number of clusters is fixed. The analysis is valid for a wide range of sampling designs used. However, the resulting optimal plot design could be sub-optimal if also the sampling design and total plot number were simultaneously optimized. For instance, it might be better to measure a large number of small clusters (like half day clusters) rather than a small number of large clusters. Or it might be better to measure less and larger plots if remote sensing material were used as auxiliary data. Unfortunately the data we had available is not large enough for such analysis.

The relative importance of the optimal plot type and size in defining the optimal sampling design has not been defined. Based on our results, we would recommend that

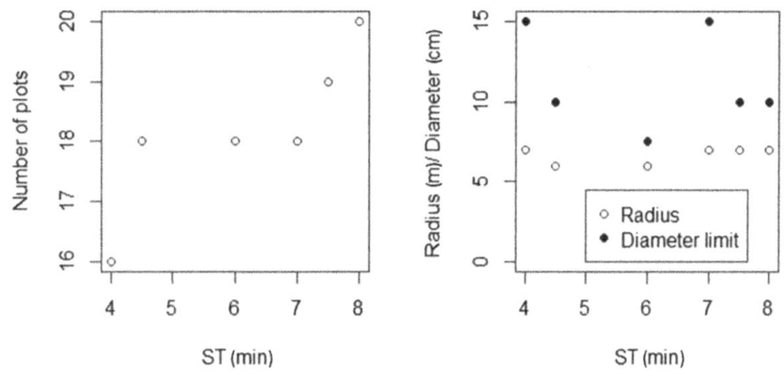

Fig. 12 Sensitivity of the optimal number of plots, diameter limit and plot radii for concentric plot as a function of time needed to measure one subsample tree (ST, min)

the whole chain of decisions from measurements, plot type, plot size, number of plots (total and/or within a cluster), number of clusters, cluster design (spatial arrangement of the plots within a cluster), sampling design and the estimation method should be simultaneously defined. Such analysis would, however, require a very large area that is measured in detail, with very large costs. Nowadays, a simulated forest might be a better option (e.g. Päivinen 1987). The design has often been optimized using a forest map based on a satellite image (e.g. Tomppo et al. 2010, 2011), but while that approach allows for selecting the optimal cluster design and number of plots within a cluster, it does not include enough information for selecting the optimal plot type. A mapped forest area based on individual tree detection from a lidar data (Holopainen et al. 2013) might provide a good starting point for a data where total optimization is possible.

Conclusions

While the optimal radius of a plot and other design parameters were quite sensitive to the measurement time and other cost factors, the concentric plot type was optimal in almost all studied cases. It is important to select a plot size that would be near optimal in many different conditions. Here, for instance, a 6–7 m radius and 10 cm diameter limit was optimal or near optimal option in most calculations. Yet, it needs to be noted that the results were calculated for Northern Finland, and elsewhere a separate optimality analysis would be needed.

The more weight is given to the costs compared to the RMSEs of the variables of interest, the smaller the optimal plots with a fixed plot number. With fixed budget, having more, smaller plots is optimal, if the transfer time between the plots is short. However, the distance between the plots within a cluster and therefore also the transfer time needs to be selected long enough to avoid high autocorrelation between the plots.

Subsample tree selection and measurement strategies need further studies, as subsample trees are a quite important cost factor but their importance to the accuracy of the final results was not as clear. The errors for tally trees had little impact on the accuracy of volume, but when other variables such as volume growth is analyzed, the subsample tree measurements may be of greater importance.

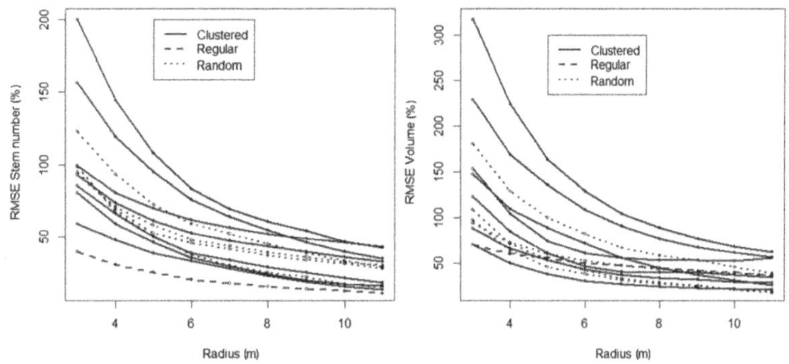

Fig. 13 RMSE (%) of stems per ha and mean volume for test areas with different point patterns as a function of plot radius (m) for fixed-radius plot. Test areas completely within one stand included

Competing interests
The authors declare that they have no competing interests.

Authors' contributions
HH was responsible for designing the experiment and collecting the data, analyzing the point patterns, and main responsible for coding the simulator for sample plots. AK was main responsible for calculating and analyzing the results and writing the article. Both authors read and approved the final manuscript.

Acknowledgements
This study was funded by Natural Resources Institute Luke.

Author details
[1]Natural Resources Institute Finland (Luke), Economics and Society unit, Jokiniemenkuja 1, 01370 Vantaa, Finland. [2]Natural Resources Institute Finland (Luke), Economics and Society unit, Yliopistokatu 6, 80100 Joensuu, Finland.

References
Baddeley A, Turner R (2005) Spatstat: an R package for analyzing spatial point patterns. J Stat Software 12(6):1–42. ISSN: 1548–7660. URL: http://www.jstatsoft.org/article/view/v012i06
Barth A, Ståhl G (2012) Determining sample size in national forest inventories by cost-plus-loss analysis: an exploratory case study. Eur J For Res 131:339–346
Burkhart HE, Stuck RD, Leuschner WA, Reynolds MA (1978) Allocating inventory resources for multiple-use planning. Can J Forest Res 8:100–110
Eid T (2003). Model validation by means of cost-plus-loss analyses. In: Amaro A, Reed D, Soares P (eds) Modelling forest systems. Cambridge USA: CABI Publishing. pp 295–305.
Eid T, Gobakken T, Næsset E (2004) Comparing stand inventories for large areas based on photo-interpretation and laser scanning by means of cost-plus-loss analyses. Scand J For Res 19:512–523
Freese F (1961) Relation of plot size to variability: an approximation. J For 59:679
Grosenbaugh LR, Stover WS (1957) Point-sampling compared with plot-sampling in Southeast Texas. For Sci 3:2–14
Hamilton DA (1978) Specifying precision in natural resource inventories. In: Integrated inventories of renewable resources: proceedings of the workshop. Tuscon, Arizona Usa: USDA Forest Service, General technical report RM-55: 276–281.
Hofstadt EH, Gobakken T, Solberg S, Kangas A, Ene L, Mauya E, Næsset E (2015) Relative efficiency of ALS and InSAR for biomass estimation in Tanzanian rainforest. Remote Sens 7:9865–9885
Holopainen M, Kankare Vi, Vastaranta M, Liang, X, Lin Y, Vaaja M, Yu X, Hyyppä J, Hyyppä H, Kukko A, Tanhuanpää T, Alho P (2013) Tree mapping using airborne, terrestrial and mobile laser scanning – A case study in a heterogeneous urban forest. Urban Forestry & Urban Greening 12:546–553. doi:10.1016/j.ufug.2013.06.002 DOI:10.1016/j.ufug.2013.06.002#doilink
Johnson FA, Hixon HJ (1952) The most efficient size and shape of plot to use for cruising in old-growth douglas-fir timber. J For 1:17–20
Koivuniemi, J (2003) Metsiköihin ja paikannettuihin koealoihin perustuvan kuvioittaisen arvioinnin tarkkuus. Summary: The accuracy of the compartmentwise forest inventory based on stands and located sample plots. Doctoral thesis. Publications of the Department of Forest Resource Management 36. University of Helsinki, Helsinki, p 160
Kulow DL (1966) Comparison of forest sampling designs. Journal of Forestry July 469–474.
Laasasenaho J (1982) Taper curve and volume functions for pine, spruce and birch. Commun. Inst. For. Fenn. 108. p 72
Loetsch F, Zöhrer F, Haller KE (1973) Forest Inventory Volume 2. BLV Verlagsgesellschaft, München, p 469
Mandallaz D, Ye T (1999) Forest inventory with optimal two-phase, two-stage sampling schemes based on the anticipated variance. Can J Forest Res 29:1691–1708
Mandallaz D (2007) Sampling techniques for forest inventories. Chapman & Hall. p 256
Mesavage C, Grosenbaugh LR (1956) Efficiency of several cruising designs on small tracts in North Arkansas. Journal of Forestry September 569–576.
Päivinen R (1987) Metsän inventoinnin suunnittelumalli. [A planning model for forest inventory, In Finnish]. University of Joensuu publications in Sciences, Joensuu, N:o 11, p 179
Stage AR, Rennie JC (1994) Fixed-radius plots vs. variable-radius plots. J For 92:20–24
Ståhl G (1994) Optimizing the utility of forest inventory activities. Ph.D. thesis, Swedish University of Agricultural Sciences, Department of Biometry and Forest Management, Umeå.
Ståhl G, Heikkinen J, Petersson H, Repola J, Holm S (2014) Sample-based estimation of greenhouse gas emissions from forests - a new approach to account for both sampling and model errors. For Sci 60(1):3–13
Tomppo E, Gschwanter T, McRoberts RE, Lawrence M, Editors (2010). National forest inventories – pathways for common reporting. Springer. ISBN 978-90-481-3232-4.
Tomppo E, Heikkinen J, Henttonen HM, Ihalainen A, Katila M, Mäkelä H, Tuomainen T, Vainikainen N (2011). Designing and conducting a forest inventory - case: 9th National Forest Inventory of Finland. Springer, Managing Forest Ecosystems 21, p 270. ISBN 978-94-007-1651-3
Tomppo E, Malimbwi R, Katila M, Mäkisara K, Henttonen HM, Chamuya N, Zahabu E, Otieno J (2014) A sampling design for a large scale forest inventory: case Tanzania. Can J Forest Res 44:931–948
Zeide B (1980) Plot Size Optimization. For Sci 26:251–257

15

Susceptible conditions for debarking by deer in subalpine coniferous forests in central Japan

Hayato Iijima* and Takuo Nagaike

Abstract

Background: Recently, deer have expanded their distribution to higher altitude ranges including subalpine forests. However, culling deer and construction of deer fence in subalpine forests are difficult because of steep slopes and complex topography. Thus it is necessary to clarify the factors which are associated with debarking by deer for the effective protection of subalpine forests. In this study, we examined which factors are associated with debarking by sika deer (*Cervus nippon*) in subalpine coniferous forests.

Methods: We conducted our survey in Minami-Alps National Park, central Japan. We established 24 10 m × 40 m plots and surveyed the occurrence of debarking on saplings >30 cm in height and <3 cm in diameter at breast height (DBH) and on trees >3 cm in DBH, as well as sapling density within each plot. Minimum distances to nearest grassland of plots were calculated (tentatively assuming grassland would attract deer and would cause high debarking pressure in the surrounding subalpine forests).

Results: The mean percentage of debarked live saplings was higher than that of live trees. The mean percentage of debarked saplings which had already died was 81.6 %. Debarking of saplings increased with lower elevation, taller sapling size, and marginally increased near grassland. Sapling density was lower in plots with low basal area of conspecific trees near grassland and differed among species. Sapling density marginally decreased with decreasing elevation and increasing stand tree density. Debarking of trees was positively related to small DBH and low elevation, and marginally increased near grassland and differed among species.

Conclusions: Our results suggest that tall saplings in subalpine forests of low elevation or near subalpine grassland were susceptible to debarking by deer and monitoring of these areas may permit the early detection of the impacts of deer in subalpine coniferous forests.

Keywords: *Abies*, *Cervus nippon*, Debarking, Grassland, *Picea*, Sapling density, Subalpine region

Background

In recent years, the population densities of large ungulates, especially deer species, have increased worldwide (Stewart and Burrows 1989; Fuller and Gill 2001; Rooney 2001; Apollonio et al. 2010; Iijima et al. 2013). An increase of deer density has been shown to result in more prevalent debarking of trees (Akashi and Nakashizuka 1999; Nagaike and Hayashi 2003; Iijima and Nagaike 2015) and browsing of saplings and understory vegetation (Gill and Beardall 2001; Beguin et al. 2009; Suzuki et al. 2013) in forests. To deal with the overabundance of deer and its effect on forest ecosystems, control of nuisance deer and the construction of fences to protect vegetation have been conducted. However, such countermeasures are difficult to perform in subalpine forests because of difficult access (e.g., steep slopes, complex topography, great distances from roads). Deer seasonally migrate to higher elevations, where delayed bud flush caused by low temperature provides fresh leaves even in mid-summer and they can escape from predators and more intense hunting pressure in low-elevation areas (Mysterud et al. 2011). These factors also have caused the number of deer in many higher elevation areas to

* Correspondence: hayato.iijima@gmail.com
Yamanashi Forest Research Institute, Saisyoji 2290-1, Fujikawa, Yamanashi 400-0502, Japan

increase (Takatsuki 2009), and debarking in subalpine forests has increased (Yokoyama et al. 2001; Takeuchi et al. 2011) in recent years. Therefore, factors correlated with debarking by deer in subalpine forests should be clarified to help effectively conserve these forests.

Many studies have examined the factors influencing debarking and browsing in lower elevation forests. Debarking was shown to depend on the size of trees (e.g., Nagaike and Hayashi 2003; Koda and Fujita 2011; Borkowski and Ukalski 2012), their species (e.g., Kay 1993; Akashi and Nakashizuka 1999; Moore et al. 1999; Takeuchi et al. 2011), proportion of coniferous stands (Ligot et al. 2013), the distance from forest road (McLaren et al. 2000), and snow depth (Iijima and Nagaike 2015). In addition to these factors, spatial variation of deer impact was observed in lower elevation forests: the higher the deer density, the higher the proportion of debarked trees (Iijima and Nagaike 2015) and browsed saplings (Akashi et al. 2011) and the lower the sapling density (Beguin et al. 2009). The spatial variation of deer density across low-elevation areas was explained by the presence of attractive habitat (e.g., artificial grassland; Kamei et al. 2010; Iijima et al. 2013). However, the attractive habitat in subalpine forests, which are located on steeper slopes and have more complex topography than lower elevation areas, has not been well studied.

Takeuchi et al. (2011) reported that elevation was an important factor affecting expansion of deer density into subalpine zones. In addition to elevation, the amount of attractive habitat (e.g., subalpine grassland) may affect deer density. Deer generally prefer herbaceous plants to woody plants (Takatsuki 1986; Winnie 2012), and severe browsing of herbaceous plants in subalpine grasslands by deer has been reported (Schütz et al. 2003; Nagaike 2012; Nagaike et al. 2014). Thus, subalpine forests which were surrounded by large subalpine grasslands would be more susceptible for debarking by deer. Thus far, however, direct evaluation of the relationship between these factors and the intensity of debarking in subalpine forest is rare.

The objective of this study was to clarify the conditions which are correlated with debarking by sika deer in subalpine forests in order to improve the effective management and conservation of these subalpine forests. We hypothesized that the occurrence of debarking and sapling density would increase and decrease in low altitude and abundant grassland.

Methods

Study area

The present study was conducted in subalpine forests in Minami-Alps National Park (357.5 km^2) in central Japan (Fig. 1), where sika deer (*Cervus nippon*) density has been increasing in recent years (Izumiyama and Mochizuki 2008; Izumiyama et al. 2009). The park is characterized by steep slopes and a concentration of many mountains within a small area (Fig. 1). Subalpine coniferous forests dominate from 2000 to 2500 m elevation, mainly

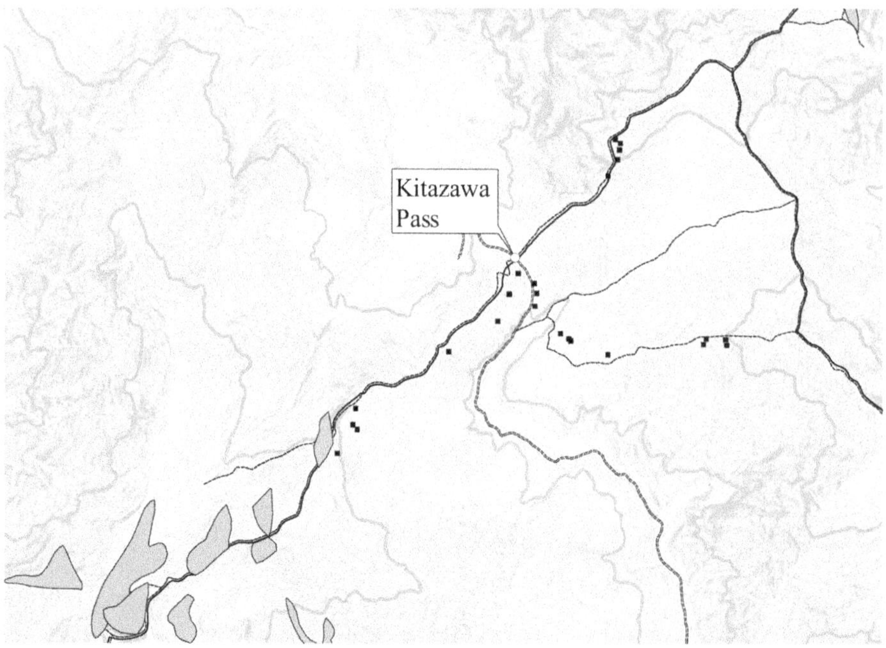

Fig. 1 Map of the study area. Gray polygons indicate the locations of subalpine grasslands (from the Natural Environment Information geographic information system provided by the Biodiversity Center of Japan: http://www.biodic.go.jp/trialSystem/top.html). Solid and dashed lines denote prefectural boundaries and trails, respectively. Solid squares indicate surveyed plots. This map was drawn by QGIS (QGIS Development 2015)

composed of *Abies mariesii, Abies veitchii, Picea jezoensis* var. *hondoensis*, and *Tsuga diversifolia* (Fig. 2). Subalpine grasslands consisting of herbaceous species and dwarf tree community (*Pinus pumila*) are present at elevations higher than 2500 m, and mixed conifer-broadleaf forests are present below 2000 m. Sika deer are widely distributed in Japan. Their shoulder heights range from 90 to 190 cm and they have a highly varied diet (they eat graminoids in northern Japan and eat leaves and fruits in southern Japan; Ohdachi et al. 2009). They can reach bark and leaves up to ca. 2 m in height.

Field surveys

We established 24 rectangular plots (each of 10 m × 40 m) in natural coniferous forests at elevations of 2000 m (low-elevation zone) and 2500 m (high-elevation zone) and surveyed in September 2011 (Fig. 1). For all trees with the diameter at breast height (DBH) > 3 cm in the plots, we noted the species and measured DBH and the presence of debarking. Furthermore, for saplings survey, 8 subplots (each of 1 m × 2 m) were established at 10-m intervals along the two long sides of each plot (8 subplots per a plot). Height and the occurrence of debarking on saplings >30 cm in height and <3 cm in DBH were measured and recorded in each subplot. In this study, we could not determine the timing of the occurrence of debarking and did not distinguish new and old debarking.

The minimum distance from each plot to subalpine grassland

The location of subalpine grassland in this region was obtained from the Natural Environment Information geographic information system (http://www.biodic.go.jp/trialSystem/top.html, accessed 20 April 2015) provided by the Biodiversity Center of Japan (Fig. 1). The minimum distance from each plot to subalpine grassland was calculated by QGIS ver 2.8.1 (Quantum GIS Development 2015).

Statistical analysis

Sapling density and the occurrence of debarking of saplings and trees were analyzed using a generalized linear

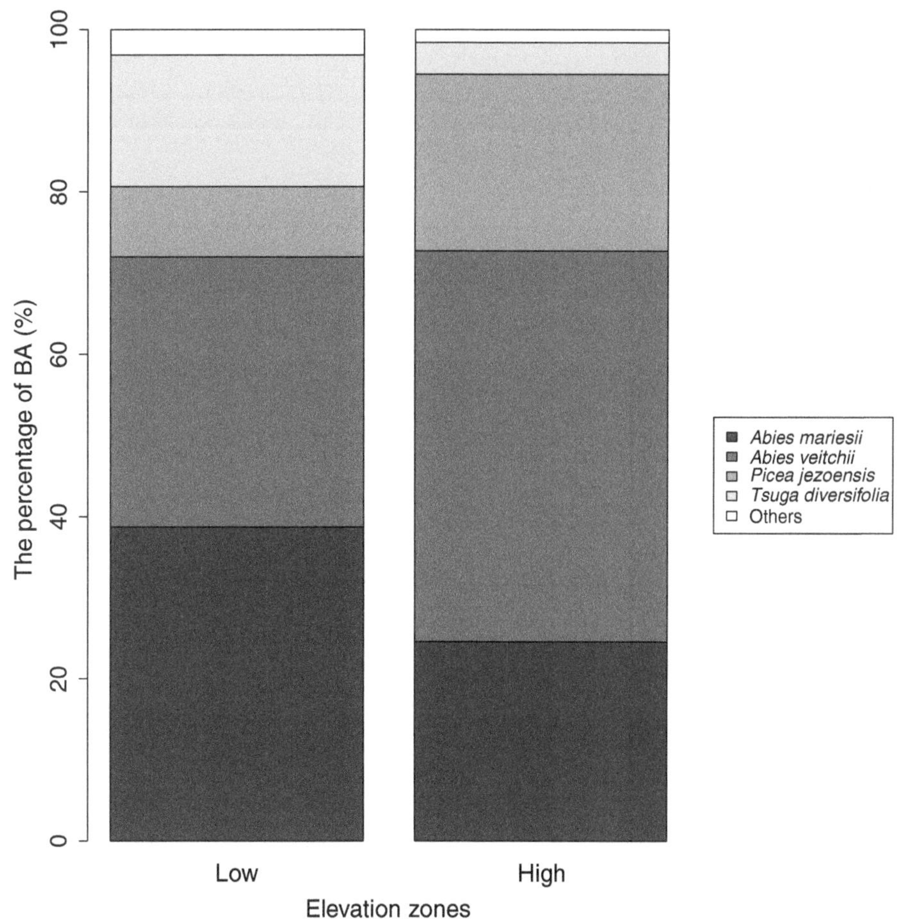

Fig. 2 Species composition of the dominant trees at low- and high-elevation zones. BA is basal area of trees at breast height. "Others" include *Larix kaempferi, Betula corylifolia, Betula ermanii, Prunus maximowiczii*, and *Sorbus commixta*

mixed model (GLMM). For all models, we adopted each plot as random effect.

For analysis of sapling density, minimum distance from each plot to subalpine grassland, elevation zone, tree density in each plot (indicating understory light conditions), and species of each sapling were used as explanatory variables. We used the number of saplings as response variable. The means of the number of saplings and sapling density were equal because surveyed areas of the number of saplings were same for all plots. We also added basal area (BA) of conspecific trees (as an index of seed source) as an explanatory variable because the abundance of the seed source may affect sapling density. A Poisson distribution was adopted as the error distribution with a log-link function. *Picea jezoensis* was omitted from the species data because none of the surveyed individuals had been debarked, which made it impossible to estimate the coefficient.

For analysis of the occurrence of debarking of saplings, minimum distance from each plot to subalpine grassland, elevation zone, species, and sapling height were included as explanatory variables. A binomial distribution was adopted as the error distribution with a logit link function.

For analysis of the occurrence of debarking of trees, minimum distance from each plot to subalpine grassland, elevation zone, DBH, and species were used as explanatory variables. A binomial distribution was adopted as the error distribution with a logit link function.

All analyses were conducted using R ver 3.1.3 (R Core Team 2015) and the glmmML package (Brostrom and Holmberg 2011). Model selection was performed using the Akaike information criterion (AIC, Akaike 1973) in conjunction with a backward elimination procedure. We determined that explanatory variables used in conjunction with all models which had delta AIC compared to the lowest AIC < 2 are significant and explanatory variable used in conjunction with any models which had delta AIC compared to the lowest AIC < 2 are marginally significant. At the same time, we calculated Akaike weight (w) and the relative importance of variable (Burnham and Anderson 2002) for comparing the importance of each variable. Akaike weight is defined as

$$w_i = \frac{\exp\left(-\frac{\Delta_i}{2}\right)}{\sum_{r=1}^{R} \exp\left(-\frac{\Delta_i}{2}\right)}$$

where $\Delta_i = AIC_i - AIC_{min}$ is the difference between an AIC of each model and the minimum AIC among all candidate models (including null model and full model). This value, referred to as the Akaike weight, provides a relative weight of evidence for each model. The relative importance of predictor variable can be calculated as the sum of the Akaike weights over all of the models in

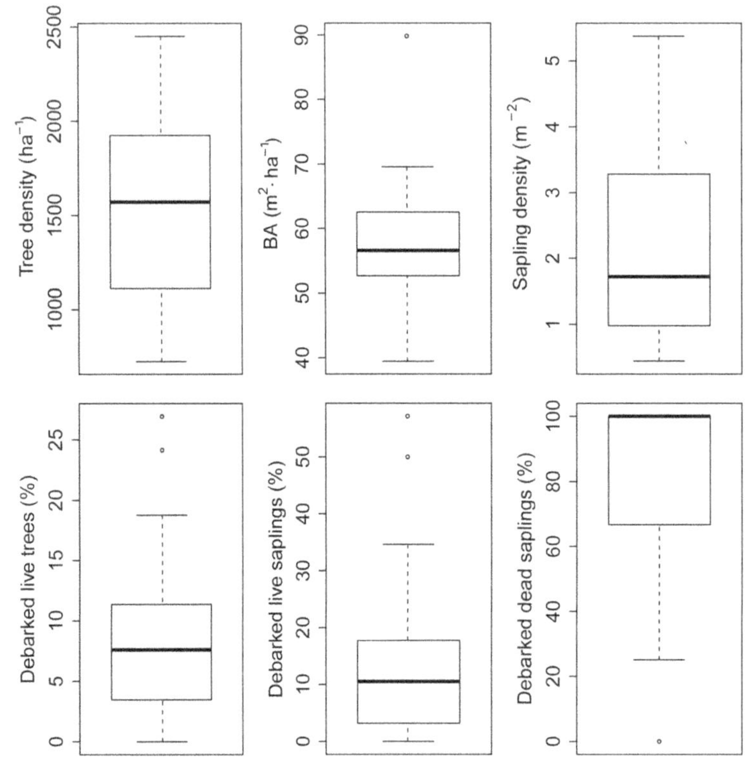

Fig. 3 Stand conditions

Table 1 AIC and coefficients of models which had ΔAIC < 2 compared to the lowest AIC model for explaining the occurrence of debarking on saplings

AIC	ΔAIC	Elevation zone[a] 2500 m	Dist. grassland	Height of saplings	Species[b,c]	
					Abies veitchii	*Tsuga diversifolia*
398.028	0.000	−2.634		0.037		
399.976	1.948	−2.619	-1.269×10^{-4}	0.037		

[a]Coefficient of elevation zone was calculated as a relative value when the coefficient of the low-elevation zone (2000 m) was set to 0. [b]Coefficient of species was calculated as a relative value when the coefficient of *Abies mariesii* was set to 0. [c]*Picea jezoensis* was excluded from the analysis of saplings because no *P. jezoensis* saplings were debarked, so it was impossible to estimate the coefficient

which the parameter of interest appears. The relative importance range from 0 (least important) to 1 (most important).

Results

The percentages of debarked live trees differed among plots (Fig. 3). The differences among sapling types was large, with dead saplings being debarked at very high rates (mean = 81.6 %; Fig. 3).

Debarking of saplings was significantly related to a taller height of saplings and lower elevation zone, marginally related to minimum distance to grassland, and was not affected by species (Table 1). Sapling density significantly increased with increasing BA of conspecific trees and increasing minimum distance to grassland and marginally increased with increasing elevation and decreasing tree density (Table 2). Sapling density of *A. mariesii* was significantly higher than that of other species (Table 2). Debarking of trees was significantly related to small DBH and low elevation zone, and marginally decreased with long distance to grassland. Debarking of trees marginally differed by species (Table 3). Overall, Size of saplings and trees had strong effects on the occurrence of debarking (Table 4). Elevation zone, minimum distance to grassland, and species were marginal factors of the occurrence of debarking of saplings and trees and sapling density (Table 4).

Discussion

We showed the relationship between debarking of trees and saplings in subalpine coniferous forests and various factors. The percentages of debarked live saplings were higher than those of live trees, and the percentages for dead saplings were much higher than those for live saplings (Fig. 3). These results suggest that debarking by deer first occurs on saplings.

Tree and sapling sizes affected the occurrence of debarking but the effects of sizes showed opposite patterns for trees and saplings (Tables 1 and 3). Higher percentages of debarked small trees were also reported in many previous studies (Akashi and Nakashizuka 1999; Yokoyama et al. 2001; Nagaike and Hayashi 2003; Jiang et al. 2005; Kiffner et al. 2008; Takeuchi et al. 2011; Borkowski and Ukalski 2012), although the reasons were not clarified. One possibility for the selective debarking of small trees is the relative ease of debarking smaller trees (Ando et al. 2004). In contrast to trees, the relationship between sapling size and the occurrence of debarking has rarely been examined. Takatsuki and Gorai (1994) compared the size structure of trees between two distinct areas in Japan that differed in deer density. They showed that forests with high deer density lacked saplings > 90 cm in height, whereas there were abundant saplings shorter than 90 cm in forests both under high and low deer densities. In the present study, one reason for the low debarking intensity on small saplings (Table 1) is that they are more difficult to find than large saplings. Consequently, saplings >30 cm in height and < 3 cm in DBH might be more susceptible to debarking by deer, suggesting that these might be a possible optimal size for debarking by sika deer.

P. jezoensis saplings appear to be less palatable for sika deer because no *P. jezoensis* saplings was debarked and we found marginal effect of species on debarking of trees (Table 3). Generally, *Abies* species were preferentially debarked over co-occurring *Picea* species (Heuze et al.

Table 2 AIC and coefficients of models which had ΔAIC < 2 compared to the lowest AIC model for explaining sapling density

AIC	ΔAIC	Elevation zone[a] 2500 m	BA[b] of conspecific trees	Dist. grassland	Species[c]			Tree density
					Abies veitchii	*Picea jezoensis*	*Tsuga diversifolia*	
1382.409	0.000		0.048	9.310×10^{-4}	−0.555	−2.675	−1.237	
1382.624	0.214	0.745	0.048	9.561×10^{-4}	−0.559	−2.676	−1.236	−0.018
1383.631	1.272	0.210	0.048	9.372×10^{-4}	−0.557	−2.680	−1.239	
1384.286	1.877		0.048	9.311×10^{-4}	−0.554	−2.672	−1.239	−0.002

[a]Coefficient of elevation zone was calculated as a relative value when the coefficient of the low-elevation zone (2000 m) was set to 0. [b]BA, basal area. [c]Coefficient of species was calculated as a relative value when the coefficient of *Abies mariesii* was set to 0

Table 3 AIC and coefficients of models which had ΔAIC < 2 compared to the lowest AIC model for explaining the occurrence of debarking on trees

AIC	Δ AIC	Elevation zone[a] 2500 m	DBH[b] of trees	Dist. grassland	Species[c]		
					Abies veitchii	*Picea jezoensis*	*Tsuga diversifolia*
526.047	0.000	−0.684	−0.014	-4.146×10^{-4}			
526.261	0.214	−0.771	−0.014	-4.196×10^{-4}	0.127	−1.379	−0.947
527.907	1.856	−0.592	−0.014				
527.974	1.926	−0.666	−0.014		0.076	−1.453	−0.959

[a]Coefficient of elevation zone was calculated as a relative value when the coefficient of the low-elevation zone (2000 m) was set to 0. [b]DBH, diameter at breast height. [c]Coefficient of species was calculated as a relative value when the coefficient of *Abies mariesii* was set to 0

2005), and the different palatability among tree species suggests that deer debarking may alter the species composition of forests. However, *P. jezoensis* has been heavily debarked in some areas in Japan (Yokoyama et al. 2001; Ando et al. 2003), indicating that palatability of debarking may be affected by surrounding vegetation (Moore et al. 1999; McLaren et al. 2000). Therefore, *P. jezoensis* saplings at the sites examined in the present study may be debarked in the future if the deer density increases.

In addition to size and species as factors for debarking, elevation zones and grassland area also were associated with the damage to subalpine forests by deer. Low sapling density and high risk of debarking of trees and saplings were found in low elevation plots (Table 2). Other studies in Japan also found the intensities of debarking on trees and saplings in subalpine coniferous forests to be greater at lower elevations (Takeuchi et al. 2011). In our study area, sika deer migrate to high-elevation areas in summer and to low-elevation areas from late autumn to late spring (Izumiyama and Mochizuki 2008) although there are some deer in low-elevation area even in summer. Then, opportunities of debarking for saplings and trees were larger in low-elevation zone than in high-elevation because deer stay longer time in low elevation-area than in high elevation-area. In addition to elevation, we found that the large subalpine grassland caused low sapling density (Table 2) and high risk of the occurrence of debarking of saplings and trees (Tables 1 and 3). Deer would likely be attracted to subalpine grassland because it has numerous herbaceous plants that serve as rich food sources (Takatsuki 1986; Winnie 2012). Trees and saplings in subalpine forests near subalpine grassland would be eaten by deer before sprout of grasses in early spring and after defoliation of grasses in late autumn. Some sika deer migrate to high-elevation area before sprout of grasses and after defoliation to escape the competition with other deer in low-elevation areas (Shin-ichiro Hamasaki, personal communication). Minimum distance to subalpine grassland, however, had only a marginal effect on the occurrence of debarking of saplings and trees (Tables 1 and 3). As the result of intensive debarking on saplings (Fig. 3), low sapling density near subalpine grassland occurred (Table 2) and then it might be obscure the relationship between the distance to grassland and the occurrence of debarking of saplings. In the future, the intensity of debarking of tree would increase if deer density remains at this level or increases.

Table 4 Relative importance of variables in models

	Debarking of saplings	Sapling density	Debarking of trees
Elevation zone	1.000	0.506	0.785
BA of conspecific species	Not included	1.000	Not included
DBH of trees	Not included	Not included	1.000
Dist. grassland	0.274	1.000	0.674
Height of saplings	1.000	Not included	Not included
Species	0.137	1.000	0.463
Tree density	Not included	0.457	Not included

Conclusions

Our analysis revealed that large saplings within the range of 30 cm in height and 3 cm in DBH in subalpine forests were most susceptible for debarking by sika deer. In addition to the size, saplings in low elevation zone or near subalpine grasslands were more susceptible to debarking by sika deer. This information may be useful to improve monitoring deer impact, or determining where to construct deer fences in the subalpine region. At the same time, deer population control in the subalpine region should be conducted in forests of low elevation zone or near subalpine grasslands.

Competing interests

This study was funded by the Comprehensive Research Organization for Science and Technology of Yamanashi Prefectural Government and Mitsui & CO., LTD. The authors declare that they have no competing interests about non-financial aspects.

Authors' contributions

HI and TN substantially contributed about setting the experimental design and collecting data. HI and TN deeply discussed and approved the content of manuscript. Both authors read and approved the final manuscript.

Acknowledgements
We thank Chiaki Ootsu for field assistance, as well as the staff of the "Komorebi-sansou" mountain hut.

References

Akaike H (1973) Information theory and an extension of the maximum likelihood principle, in 2nd International Symposium on Information Theory. Tsahkadsor, Armenian SSR, pp 267–281

Akashi N, Nakashizuka T (1999) Effects of bark-stripping by sika deer (*Cervus nippon*) on population dynamics of a mixed forest in Japan. For Ecol Manage 113:75–82

Akashi N, Unno A, Terazawa K (2011) Effects of deer abundance on broad-leaf tree seedling establishment in the understory of *Abies sachalinensis* plantations. J For Res 16:500–508

Ando M, Yokota H, Shibata E (2003) Bark stripping preference of sika deer, *Cervus nippon*, in terms of bark chemical contents. For Ecol Manage 177:323–331

Ando M, Yokota H, Shibata E (2004) Why do sika deer, *Cervus nippon*, debark trees in summer on Mt. Ohdaigahara, central Japan? Mamm Study 29:73–83

Apollonio M, Andersen R, Putman R (2010) European ungulates and their management in the 21st century. Cambridge University Press, NY.

Beguin J, Pothier D, Prévost M (2009) Can the impact of deer browsing on tree regeneration be mitigated by shelterwood cutting and strip clearcutting? For Ecol Manage 257:38–45

Borkowski J, Ukalski K (2012) Bark stripping by red deer in a post-disturbance area: the importance of security cover. For Ecol Manage 263:17–23

Brostrom G, Holmberg H (2011) Generalized linear models with clustered data: fixed and random effects models. Comput Stat Data Anal 55:3123–3134

Burnham KP, Anderson DR (2002) Model selection and multimodel inference: a practical information-theoretic approach, 2nd edn. Springer, New York

Fuller RJ, Gill RMA (2001) Ecological impacts of increasing numbers of deer in British woodland. Forestry 74:193–199

Gill RMA, Beardall V (2001) The impact of deer on woodlands: the effects of browsing and seed dispersal on vegetation structure and composition. Forestry 74:209–218

Heuze P, Schnitzlerm A, Klein F (2005) Consequences of increased deer browsing in winter on silver fir and spruce regeneration in the southern Vosges Mountains: implication for forest management. Ann For Sci 62:175–181

Iijima H, Nagaike T, Honda T (2013) Estimation of deer population dynamics by Bayesian state-space model with multiple abundance indices. J Wildl Manage 77:1038–1047

Iijima H, Nagaike T (2015) Appropriate vegetation indices for measuring the impacts of deer on forest ecosystems. Ecol Ind 48:458–463

Izumiyama S, Mochizuki T (2008) Seasonal range use of sika deer, which inhabits the subalpine zone in the southern Japan Alps. Bull Shinshu Univ Alps Field Center 6:25–32 (in Japanese with English summary)

Izumiyama S, Mochizuki T, Takii A (2009) GPS tracking of Sika deer which inhabits the sub-alpine zone in the Southern Japan Alps. Bull Shinshu Univ Alps Field Center 7:63–71 (in Japanese with English summary)

Jiang Z, Ueda H, Kitahara M, Imaki H (2005) Bark stripping by sika deer on Veitch fir related to stand age, bark nutrition, and season in northern Mount Fuji district, central Japan. J For Res 10:359–365

Kamei T, Takeda K, Koh K, Izumiyama S, Watanabe O, Ohshima K (2010) Seasonal pasture utilization by wild sika deer (*Cervus nippon*) in a sown grassland. Grass Sci 56:65–70

Kay S (1993) Factors affecting severity of deer browsing damage within coppiced woodlands in the south of England. Biol Cons 63:217–222

Kiffner C, Rößiger E, Trisl O, Schulz R, Rühe F (2008) Probability of recent bark stripping damage by red deer (*Cervus elaphus*) on Norway spruce (*Picea abies*) in a low mountain range in Germany: a preliminary analysis. Silva Fennica 42:125–134

Koda R, Fujita N (2011) Is deer herbivory directly proportional to deer population density? Comparison of deer feeding frequencies among six forests with different deer density. For Ecol Manage 262:432–439

Ligot G, Gheysen T, Lehaire F, Hébert J, Licoppe A, Lejeune P, Brostaux Y (2013) Modeling recent bark stripping by red deer (*Cervus elaphus*) in South Belgium coniferous stands. Ann For Sci 70:309–318

McLaren BE, Mahoney SP, Porter TS, Oosenbrug SM (2000) Spatial and temporal patterns of use by moose of pre-commercially thinned, naturally-regeneration stands of balsam fir in central Newfoundland. For Ecol Manage 133:179–196

Moore NP, Hart JD, Langton SD (1999) Factors influencing browsing by fallow deer *Dama dama* in young broad-leaved plantations. Biol Cons 87:255–260

Mysterud A, Loe LE, Zimmermann B, Bischof R, Veiberg V, Meisingset E (2011) Partial migration in expanding red deer populations at northern latitudes: A role for density dependence? Oikos 120:1817–1825

Nagaike T, Hayashi A (2003) Bark-stripping by sika deer (*Cervus nippon*) in *Larix kaempferi* plantations in central Japan. For Ecol Manage 175:563–572

Nagaike T (2012) Effects of browsing by sika deer (*Cervus nippon*) on subalpine vegetation at Mt. Kita, central Japan. Ecol Res 27:467–473

Nagaike T, Ohkubo E, Hirose K (2014) Vegetation recovery in response to the exclusion of grazing by sika deer (*Cervus nippon*) in seminatural grassland on Mt. Kushigata, Japan. ISRN Biodiversity: Article ID 493495

Ohdachi SD, Ishibashi Y, Iwasa MA, Saitoh T (2009) The Wild Mammals of Japan. Shoukadoh Book Sellers, Kyoto

QGIS Development Team (2015) Quantum GIS Geographic Information System. Open Source Geospatial Foundation Project. http://qgis.osgeo.org

R Core Team (2015) R: A Language and Environment for Statistical Computing. R Foundation for Statistical Computing, Vienna, Austria, http://www.R-project.org/

Rooney TP (2001) Deer impacts on forest ecosystems: a North American perspective. Forestry 74:201–208

Schütz M, Risch AC, Leuzinger E, Krüsi BO, Achermann G (2003) Impact of herbivory by red deer (*Cervus elaphus* L.) on patterns and processes in subalpine grasslands in the Swiss National Park. For Ecol Manage 181:177–188

Stewart GH, Burrows LE (1989) The impact of white-tailed deer *Odocoileus virginianus* on regeneration in the coastal forests of Stewart Island, New Zealand. Biol Cons 49:275–293

Suzuki M, Miyashita T, Kabaya H, Ochiai K, Asada M, Kikvidze Z (2013) Deer herbivory as an important driver of divergence of ground vegetation communities in temperate forests. Oikos 122:104–110

Takatsuki S (1986) Food habits of sika deer on Mt. Goyo, northern Honshu. Ecol Res 1:119–128

Takatsuki S, Gorai T (1994) Effects of sika deer on the regeneration of a *Fagus crenata* forest on Kinkazan Island, northern Japan. Ecol Res 9:115–120

Takatsuki S (2009) Effects of sika deer on vegetation in Japan: a review. Biol Cons 142:1922–1929

Takeuchi T, Kobayashi T, Nashimoto M (2011) Altitudinal differences in bark stripping by sika deer in the subalpine coniferous forest of Mt. Fuji. For Ecol Manage 261:2089–2095

Winnie JAJ (2012) Predation risk, elk, and aspen: tests of a behaviorally mediated trophic cascade in the Greater Yellowstone Ecosystems. Ecology 93:2600–2614

Yokoyama S, Maeji I, Ueda T, Ando M, Shibata E (2001) Impact of bark stripping by sika deer, *Cervus nippon*, on subalpine coniferous forests in central Japan. For Ecol Manage 140:93–99

16

Managing project risks and uncertainties

Mike Mentis

Abstract

This article considers threats to a project slipping on budget, schedule and fit-for-purpose. Threat is used here as the collective for risks (quantifiable bad things that can happen) and uncertainties (poorly or not quantifiable bad possible events). Based on experience with projects in developing countries this review considers that (a) project slippage is due to uncertainties rather than risks, (b) while eventuation of some bad things is beyond control, managed execution and oversight are still the primary means to keeping within budget, on time and fit-for-purpose, (c) improving project delivery is less about bigger and more complex and more about coordinated focus, effectiveness and developing thought-out heuristics, and (d) projects take longer and cost more partly because threat identification is inaccurate, the scope of identified threats is too narrow, and the threat assessment product is not integrated into overall project decision-making and execution. Almost by definition, what is poorly known is likely to cause problems. Yet it is not just the unquantifiability and intangibility of uncertainties causing project slippage, but that they are insufficiently taken into account in project planning and execution that cause budget and time overruns. Improving project performance requires purpose-driven and managed deployment of scarce seasoned professionals. This can be aided with independent oversight by deeply experienced panelists who contribute technical insights and can potentially show that diligence is seen to be done.

Keywords: Budget; Fit-for-purpose; Management; Project; Risks; Schedule; Threats; Uncertainties

Background

This article considers the risks and uncertainties to completing projects within budget, on time and fit-for-purpose. What are the risks and uncertainties? When, where and how do they commonly impact on projects? How might controls be applied to reduce slippage?

Review

Over the past decade 'planning fallacy' (underestimating the time and cost to complete a task) has been argued as the soul of project slippage (Flyvbjerg et al. 2003; Ansar et al. 2013; Flyvbjerg 2014) and Kahneman (2012). Though attributed to Kahneman and Tversky (1979), a related idea of planning shortfalls is stated by Parkinson's Law that 'a task will take all the time you give it', and a bit more (Parkinson 1958). 'Fit-for-purpose' (the project meets the needs for which it was undertaken to satisfy) is less explicitly addressed than time and cost, but implicitly shortcomings in the quality of the product translate into more time and cost (or less benefit). The Flyvbjerg-Kahneman remedy is to use reference class forecasting (the performance of previous similar projects as the basis for predicting), and realistic and transparent scheduling and estimation of costs and benefits. This article addresses the shortcomings of project planning and execution, focussing on risk and uncertainty controls particularly on the environmental and social aspects of infrastructure, natural resource and socio-economic projects in developing countries.

The framework for this article is the constraints analysis in Figure 1.

The project: risk and uncertainty

Various contemporary perceptions of the nature of risk are described by Blennow et al. (2013). In this article *risk* is a bad thing that might happen and for which the probability and the consequence can be quantified. Bad things that potentially can happen in a project and that are poorly to not quantifiable are *uncertainties. Threat* is here the collective term for *risk* and *uncertainty.* Here good things are not *threats*, but that they might not come about constitutes a *risk* or *uncertainty* of obstruction to good things eventuating. This is consistent with

Correspondence: mmentis@saol.com
Business & the environment, Postnet Suite 10102, Private Bag X7005, 3650 Hillcrest, South Africa

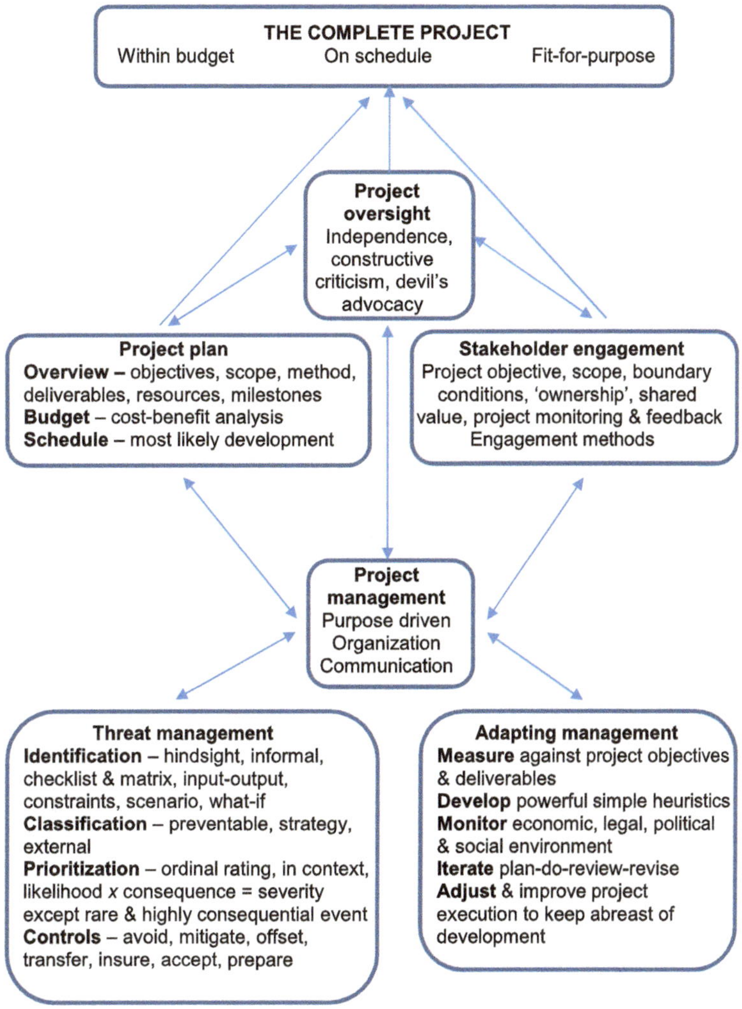

Figure 1 Framework for the generic project.

what management tries to do – remove obstacles or ease constraints as explained in *The Goal* (Goldratt 1986). While *risk* is often used loosely on projects, the stricter usage here does not invalidate conventions (e.g. likelihood *x* consequence = severity), as considered below.

Are the *threats* that most materially affect projects *risks* or *uncertainties*? Consider four South African infrastructure construction projects (Table 1). For the three water supply projects the degree of achievement is generously stated by the principal. The environmental and social impact assessment (ESIA) for the Olifants Water Resource Development Project (ORWRDP) started in 2004. Obtaining environmental authorization took more than two years. This delayed dam and water delivery construction. On the Mooi-Mgeni Transfer Scheme (MMTS), delays in environmental authorization, started in 1999, set back achievement more than stated in Table 1, constructing the transfer pipeline started only in late 2013, and the land dwellers

Table 1 Examples of project setbacks

Project	2012-2013 target	2012-2013 achievement	Setback causes
ORWRDP2C – water supply pipeline*	30%	8%	Land acquisition resettlement metal pipe supply
Mokolo-Crocodile water augmentation project*	45%	36%	Weather, industrial action
Mooi-Mgeni water transfer scheme*	70%	50%	Unforeseen conditions
National Road 3 – De Beers Pass section	Submit ESIA	ESIA not submitted	Failure to submit ESIA

*From South African Department of Water & Sanitation annual report 2012–2013.
'Targets' are intended percentage project completion, and 'achievement' actual percentage completion.

displaced by inundation of Spring Grove Dam have not yet moved into their replacement housing. The Mokolo-Crocodile Water Augmentation Project (MCWAP) has been delayed further by a big flood in March 2014, and later a dam failure. The national road project was designed, and its servitude purchased, more than 40 years ago. The road was never built. The ESIA has been in preparation for 4 years, it is now released for public comment, it has yet to be submitted to the authorities, it might still be contested in court, and construction could be subject to industrial action. In the meanwhile costs are mounting – no return on already incurred costs, and the accidents and lives lost on the present dangerous road.

Students of project management might interpret Table 1 by the 'break-fix' model (Flyvbjerg 2014). The project stalls (the break) for want of anticipating threats, so a new version of the project is generated (the fix) with revised budget and timelines. The previous cost and completion date are erased from memory. Iterated 'break-fix' can make any project look good.

Time-consuming ESIA preparation and authorization, stakeholder intervention, court injunction, industrial action, rare events and defective construction are the types of *threat* that are chief project show-stoppers. They are poorly quantifiable. They are not *risks*, but *uncertainties*.

Project oversight

Project proponents typically employ professionals to manage and execute. Why then would oversight be required?

Projects can slip for many reasons. Flyvbjerg et al. (2003) consider project politics, planning and execution, and attribute slippage to ambition, egos, vested interests, changes in currency exchange rates, altering or developing project scope and design, underestimated cost, overestimated benefits, optimism that everything goes according to plan (EGAP), and disregard of safety requirements and the full environmental and social externalities. van Marrewijk et al. (2008) present a more benign view with project participants in an alliance collaborating to realise a shared vision. The cultural setting of the project might indeed determine degrees of autocracy, accountability, groupthink, project loyalty, etc., and in many projects many participants perform professionally to complete the project satisfactorily. In hydropower projects the amount of slippage did not correlate with the degree of autocracy/democracy, but slippage occurred possibly for different reasons under the differing cultures (Ansar et al. 2013). Both autocratic and democratic project leaders have vested interests. They want to know whether the project is running to plan, what shortcomings exist, how achievement might be improved, and neither will want to advertise cost and time overruns. Even if the most able are appointed as leaders, oversight institutions and mechanisms are warranted if shortcomings are to be avoided. Shortcomings range from limitations in expertise, experience and capacity, through vested interests, to corruption and dishonesty. In practice these are not simply told apart. No one is omniscient. All people are dishonest. Unless projects are transparent and publicly reported upon then society's default assumption – understandable and justified by the behavioural economists' research (Ariely 2012) – is that there is something to hide.

The World Bank prescribes that construction projects which it funds be reviewed periodically by panels of independent engineering, environmental and social experts. Financial institutions adopting the Equator Principles require peer review for ESIA, construction and operations phases of funded projects. Kaplan and Mikes (2012) and the Flyvbjerg-Kahneman school advocate the 'outside view'. To some project proponents these panels are unwarranted and unwanted expense and interference. However, a perceived utility of the panels is illustrated by them continuing after World Bank loans close, and by them being adopted on projects without World Bank funding. The panellists act as sounding boards and devils' advocates, examining and criticizing designs, explaining why things might fail, and suggesting improvements, with savings running to US$ millions. Also, the mere fact of surveillance moderates excesses of ambition and optimism, and is recommended by Ariely (2012) to curtail dishonesty that everyone is prone to if not watched.

The project plan

Key components of the project plan are objectives and scope, what will be delivered and how, resources including budget, and schedule. Objectives and scope are dealt with under stakeholder engagement below, and the other components here. The boundary conditions of our generic project are related. Time is money, and so is fit-for-purpose. If project completion is delayed it likely costs more, because the contractor incurs more cost, and because the benefits, revenue streams and loan payback are postponed. If the product is not fit-for-purpose, further time and expense are incurred to effect remedy which, if not possible, decreases benefit.

From Table 1, project delays, and hence increased costs, are common. In four megaprojects construction costs were 125% to 180% of the initial estimate (Flyvbjerg et al. 2003). While some projects come in close to budget the Flyvbjerg figures understate the overrun in the typical public and public private partnership projects (PPPs), as illustrated for hydropower projects (Ansar et al. 2014) and Table 2. Ansar et al. (2013) found cost overruns increase with project size and duration. This is not evident from Table 2, but the Flyvbjerg-Kahneman school would say the projects in Table 2 are of different types.

In some business projects there is an impression of thorough cost-benefit analysis (CBA) or equivalent. It is

Table 2 Cost overruns in construction projects

Project	Cost ($billions)		Cost overrun (%)
	Initial estimate	Possible final	
Gautrain	3.38	3.5	104
New Multi Products Pipeline	2.25	2.4	107
Medupi Power Station	12.24	22.2	181
Ingula Pumped Storage Scheme	1.4	3	214
Gauteng Freeway Improvement Project	3.1	10.4	335
Nklandla - President Zuma's residence	0.00431	0.0274	636

Values are in 2014 US dollars. Estimates from information in the public domain.

not the same on PPPs. Why is that? In some public projects the benefit might be very large, and the project is managed not so much to hone the cost-benefit ratio but to expedite the project while containing the cost. The benefit of a 555 km long fuel pipeline connecting economic hubs and costing US$2.4 billion might be infinitely better than a daily 65 km long convoy of 30 000 litre tankers on an already busy freeway (Table 2, second item). Yet in many cases it can be vital to determine the benefits accurately in order to know what cost can be afforded yet the project remain viable. Suppose a pumped storage scheme. The prices of buying power (to pump water when there is surplus power on the grid) and selling power (when stored water is released to generate hydropower), and the amount of power to be bought and sold, must be known to set an upper limit to the construction, mitigation and operation costs. It is not enough for the ESIA to conclude qualitatively that 'there are no fatal flaws' – all impacts can in theory be mitigated, but at what cost? How many households must be relocated? What area of arable land will be lost? How many livelihoods will be destroyed? What will mitigation – probably spanning decades – cost? The cost might be unaffordable, in which case the facile 'no fatal flaws' is mistaken.

Done thoroughly at feasibility, with 'reference class forecasting' (Ansar et al. 2013), CBA provides a necessary though incomplete guide on project go-ahead. Conventionally it is numerical. Inexactness of number inputs into the CBA can be entered as statistical distributions and Monte Carlo techniques used to produce expected values with confidence limits. The most likely development (MLD) with variation about it should be adopted rather than 'planning fallacy' (Kahneman 2012) or optimistic EGAP (Flyvbjerg et al. 2003) that is unlikely ever to arise. Contemporary sensitivity analysis software helps detect where good data are needed, and which are the most critical cost and benefit aspects of the model. If the confidence limits from the Monte Carlo modelling are wide it means that even the numbers are warning that the project is risky. If the confidence limits are narrow there are still the uncertainties that the numbers have not captured, in particular about the future. Circumstances will change but precisely how is unknowable. To illustrate, return to our example of pumped storage schemes. These are huge investments with long operating lives over which payback is required. Yet even now a pilot pumped-heat electricity storage is being developed, anticipated to store energy at 75% of the cost of conventional pumped storage. Might today's pumped storage schemes become obsolete before capital outlay is repaid? Ansar et al. (2014) urge agile energy alternatives with shorter time horizons than conventional hydropower projects. To limit planning fallacy not only is experience of previous similar ventures relevant, but the proponents should not assume their new project is manned by smarter operators and thereby immune to past perils (Kahneman 2012).

Having done the CBA using reference class forecasting, and though there will remain greater or lesser uncertainties, the project proponents and managers must work within budget and schedule. Providing cost and time reserves or contingencies will, by Parkinson's Law, guarantee project overrun. I recall one build-operate-transfer project in which the concessionaire had to find his own funding. Is it coincidence or selective memory that there was no project slippage? Follow-up CBAs, conducted through the construction and operation phases, can help keep the project viable (e.g. whether to cut costs, where, by how much, what tariff increases to seek, if and when to abort the project).

Stakeholder engagement

In the previous section project objective and scope were mentioned as key project components. What is the purpose of the project? Project aims can be too vague, too numerous, misdirected or unachievable. A project can solve the wrong problem. What stakeholders *want* is not enough. What the project proponents *think* the stakeholders need, can be misperception. Projects must meet real needs, yet often they do not, constituting waste of donor, investor or taxpayer funds. Involuntary stakeholders must be identified and properly informed, which may extend to teaching them how to assess the impacts of a project on their lives, properties and livelihoods. A few examples follow.

Polak and Warwick (2013) report a project to help poor families in Haiti making charcoal from sugarcane waste, providing fuel, jobs and income. A method to make charcoal was indeed developed. What did not materialize though was an efficient technique to produce a competitive and saleable fuel to benefit buyers and enrich producers. What was needed and not obtained was insight into the circumstances. When are sugarcane waste and labour available? At what cost can the charcoaling equipment be bought to produce a saleable product at what price and quality to pay back the investment and make a profit?

In another example, road users in the Gauteng Freeway Improvement Project (GFIP) were not engaged on the scope, nature of improvement, method of funding, and options for payback. All this was decided by the authorities, because they know best (Table 2, fifth item). Now 80% of road users refuse to register and pay electronic road tolls. In a year the road agency has racked up US $200 million in unpaid toll fees. The users have no ownership over GFIP. There is no social contract.

Polak and Warwick (2013) urge project proponents to clarify the central problem, talk to the people with the problem, and listen to and understand them, to craft the essential intent (McKeown 2014), including scope and boundary conditions. Infrastructure projects are engineer-driven and top-down. Stakeholder engagement in southern Africa is typically authoritarian and only informs and invites comment. No wonder problems arise, as with GFIP. The World Bank's experience with land reform shows that achieving success goes beyond technical land management and depends on the institutional environment and factors that determine investment and profitability (Binswanger 2007). Insight into the lives of the stakeholders requires surveys to seek unbiased perceptions, one-on-one dialogue and forum discussion, and stakeholder buy-in is helped by facilitating meetings where stakeholders decide their priorities, and by involving them in implementation (Decker et al. 2012). For a project to be sustainable, it must be profitable and ideally all stakeholders must benefit (Polak and Warwick 2013; Simanis and Duke 2014). Porter and Kramer (2006) reshape corporate social responsibility from philanthropy to corporate social integration through designing business activities in which society shares value. One of their examples is Nestlé developing a dairy business at Moga, India. Nestlé built refrigerated dairies and sent trucks to collect milk, at the same time dispensing veterinary and agronomic services, paying more for milk, increasing local farmers from 180 to 75 000, prompting competitive dairies and milk factories and spawning an industry cluster.

If stakeholders are not engaged early in dialogue, and not involved in pilot trials and the like, the project becomes 'yours' not 'ours', potential benefits are liable to be overlooked, and *uncertainties* abound that the project will not address real needs, fall short in delivering sustainable benefits, and might be obstructed by civil disobedience, industrial action or court injunction. Participative stakeholder engagement is a necessary measure that, done well, can swing obstructers and objectors into neutral observers, supporters and even 'make-it-happen' allies.

Threat management

Conventionally risk (and uncertainty) management involves two components: (a) assessment comprising threat identification, classification, prioritizing, and devising controls, and (b) applying the controls.

ESIA is the most visible form of threat management for projects. A decade ago Flyvbjerg et al. (2003) observed that ESIA was learning only slowly, and shortcomings concerned (a) lack of accuracy in impact prediction, (b) the narrow scope of impacts and their time horizons, and (c) inadequate organization, scheduling and institutional integration of the ESIA process in overall project decision-making. What has happened in the interim, as reflected by project performance in a developing region?

In South Africa ESIAs are done to obtain government authorization, not to control threats (Mentis 2010). Often the ESIA passes over serious threats. For example for big infrastructure projects the loss of households, cultivated fields and jobs of the poor may be overlooked. Houses are big assets to most people, and poor rural people depend on cultivated fields for their livelihood. Houses can be replaced but in southern Africa arable land is scarce, and restoring the lost livelihood of a poor land-dependent household is difficult. Unskilled labour is not mobile and hard to retrain and retool. Without identifying these issues they are not quantified and included in the CBA, and not considered in control measure affordability and project feasibility. With overlooked threats, the environmental and social management plan (ESMP) is defective, the proposed controls are superficial and might not comply with the law that, for example, in Lesotho requires 'full description', and possibly insufficient to enable a tendering contractor to prepare a threat management budget. Often there is poorly defined responsibility for the fuzzily and incorrectly identified threats so that during the project impacts are not avoided and not mitigated, and the project bequeaths detrimental externalities.

In developing regions there is a skills shortage – a scarcity, at all levels, of professionals experienced in planning, constructing and operating projects. Principals too often appoint impact assessors without experience on that reference class of project – a virtual guarantee for overlooked threats, poorly controlled detrimental externalities, and overruns. Blennow et al. (2013) make the point that forest owners seeing and believing the effects of climate change improve prospects for adopting adaptive measures. Awareness, perceptions, belief, commitment and implementation might be aided by forums (e.g. lessons learnt) that rarely take place even within single organizations undertaking several big projects.

Threat identification

In Table 3 examples of methods to identify threats are presented. These are indeed just examples because there

Table 3 Methods to identify threats

Method	Procedure	Advantages	Disadvantages
Hindsight	Review successes and failures. Error elimination.	Learn by experience.	Not anticipate events. Poor for big irreversible impacts.
Informal	Expert guesswork, intuition, asking stakeholders. System 1 thinking.	Quick, easy, cheap.	Overlook and misunderstand threats. Over-influenced by occasional big sensational events (e.g. airliner crashes), under-influenced by common events (e.g. road accidents).
Checklist and matrix	Adopts threats encountered on previous similar projects.	Formal. Simple. Uses experience.	Limited information. Every project more or less unique so some listed threats possibly irrelevant & other threats overlooked.
Input–output analysis (IOA)	Deconstructs project into components each of which is analysed in terms of inputs, outputs, gains or losses of energy, matter, rights & opportunities. System 2 thinking.	Formal. Addresses specifics of a project. Diligent, defensible and repeatable.	Costly – requires experience, time, money and effort.
Constraints analysis	Identifies & links main obstacles (i.e. constraints, threats) to desired outcome. System 2 thinking.	Concise means of representing cause-effect relations among already identified threats & desired outcome.	Does not of itself identify threats, and best done in conjunction with or after IOA.
Scenario analysis	Assess possible future conditions by considering outcomes of 2–3 contrasting event sequences.	Considers development paths & consequences of possible futures. Appropriate for strategy & external threats.	Does not predict one exact picture of the future. Big picture. Qualitative.
What-if analysis	More focused than scenario analysis, usually quantitative.	Useful for establishing, optimizing & refining costs & benefits.	Usable if options are quantifiable.

is no single right way to assess threats. Rather, good assessment is a mindset and the expert assessor adapts her method to the specific circumstances of a project. Table 3 therefore is intended to illustrate, not prescribe.

Examples of contemporary methods can be categorized into hindsight, informal, checklist and matrix, and input–output, constraints, scenario and what-if analyses. A chief reason why many ESIAs are weak is that threat identification is informal. It involves cursory thinking (System 1 or automated thinking of Kahneman (2012)). For example, the local rural poor people are asked and they say that their cultivated fields are unproductive, not because they do not depend on the fields, but because they want employment on the upcoming construction project. The assessor accepts this at face value. He ignores hindsight, and does not use checklists and matrices – what he or others learnt from previous similar projects. He does not entertain the possibility that this new project might have novel aspects and effects. He does not deconstruct the project into components. He does not scrutinize each component for gains and losses of energy, materials, rights and opportunities. He never tries to assemble his threat identification into a coherent picture of cause and effect, as in a constraints analysis. He never gets into deep and effortful thinking (System 2 of Kahneman (2012)).

Threat classification

The identified threats form the project threat register – a list of what might go wrong. Threat management ultimately needs to recognize which threats are material and warrant attention, and which trivial. Conventionally the threats are categorized according to the classical organization functional units such as engineering, environment, finance, health and safety. Done this way project integration can face problems of 'apples with pears' comparisons. Rather classify threats by criteria that relate to the manner of control. Group threats into preventable, strategy and external types (Kaplan and Mikes 2012).

Preventable threats are internal. They arise within the functioning of the organization. Examples are biodiversity loss, dust, fuel and oil spills, noise, water contamination and waste. They are controllable, and the aim with them is to avoid or eliminate impacts cost-effectively. Checklists and matrix methods, and input–output analysis (IOA) are effective means to identify preventable threats. A rules-based system is developed (e.g. keep site clear of litter and waste at all times). Most ESIAs, environmental authorizations and ESMPs focus on the preventable. In some projects, such as MMTS (Table 1, item 2) in which good rapport was struck between consulting engineer and construction contractor, control of preventable threats was done to a high standard. This fits with the alliance model of van Marrewijk et al. (2008). However, in other projects where the preventable threats were poorly controlled, relations between principal, consulting engineer and contractor were more adversarial and confrontational – of course cause and effect are unclear.

Strategy threats relate to how work is done – commitments, methods, standards, values, etc. The aim of managing strategy threats is to reduce likelihood of occurrence and mitigate consequence if the threat materializes. Unlike with the preventable, strategy threat is not entirely under management control. An example is a strategy to adopt a just-in-time (JIT) inventory. For instance in a construction project the aggregate, cement and sand are not stockpiled months in advance, but procured and delivered as needed. The benefits are reduced cost, small lay-down area and therefore reduced project footprint. The downside is possible disruption in the supply chain – the portion of the threat outside managerial control if an external supplier is used. In constructing a roller-compacted-concrete dam wall, disruption might lead not only to project delays and budget overruns, but interruptions in concrete laying could impair the product. Prevention – not adopting JIT – would seem prudent for roller-compacted-concrete dam walls if outside suppliers are used.

Another strategy example concerns project-affected people. A road agency might disregard loss of jobs that arise because a realigned road causes, say, closure of a filling station, on a rationale that employment destroyed here is recreated there. However, unskilled labour is not mobile, and cannot move readily if 'here' and 'there' are far apart. Hence the agency strategy does not accord with the economist's precept of at least one person better off and none worse off, nor with the World Bank's standard of maintaining or improving livelihoods. The road agency might be sued for lost livelihoods. In contrast the strategy in Lesotho Highlands Water Project (LHWP) is fair and prompt compensation for losses suffered, and no unfortunate precedents. But even with the best plans, the unanticipated can happen – the head of the household loses his identity document or dies, or ownership of the lost asset is disputed, or the losses are unfairly calculated or underestimated so the claimants go to court and the cost balloons, or the loss is unfortunately overrated and this sets a precedent for further claims against the project.

Uncontrollable events fall in the class of external threats. Examples are flood, drought, earthquake, disease epidemic, communications blackout, power failure, price escalation, currency fluctuation, industrial action and political turmoil. Threat management here is to reduce the consequence of eventuation, to develop organizational resilience to Black Swans (unpredictable events of big consequence such as the 2008 global financial meltdown, and the recent seismic events and tsunamis in Indonesia and Japan (Taleb 2008)). The control measures centre about developing resilience and emergency response and preparedness, and design of organizational and operational structure to retain function in the face of adversity and catastrophe. The logic is to design the organization so it is robust and safe-in-failure (Mentis 2014; Linkov et al. 2014).

The approach to strategy and external threats is to stage internal workshops, use embedded expertise in scenario analysis, 'stress test' the plans not only in the business-as-usual context but also for when things are unusual, and supplement with independent review. Strategy and external threats are poorly quantifiable, so they are uncertainties rather than risks. On most projects the ESIAs, environmental authorizations, ESMPs and day-to-day management poorly cover strategy and external threats. Managing the project affected people on LHWP is an exception. Hundreds of households were relocated, water and sanitation programs instituted, living standards maintained, and advice given on investing compensation payments, not without shortcomings and failings, but with policy formulation, planning, budgetary provision, diligent implementation and learning from experience. On other projects, building a fish barrier is done better than relocating displaced land dwellers, culvert design for a new road takes precedence over possible destroyed livelihoods that the road could cause, and a behind-schedule project does not have time to take flood protection precautions and so slips further when flood occurs.

Threat prioritization

Possibly the only limit to what can go wrong with a project is imagination. If the project threat register has been prepared diligently, the list of potential problems will be long. On Sasol's Mozambique-Secunda pipeline 708 threats were identified in an operations assessment. The top 10% were prioritized. Even that was ambitious. Finally only 2% of 708 were addressed. There are many reasons to prioritize. Simplest is that no one has the resources of time, budget and manpower to do everything, and the more resources are distributed across many controls the less effective can be control on each threat. This is argued about life generally by McKeown (2014). Notwithstanding this, environmental authorities want comprehensive ESIAs and ESMPs, and if some omitted item, even trivial, is picked up then environmental authorization is withheld. Yet a thought experiment on the economics of threat control shows that prioritization is imperative. Suppose the following. First, resources of budget, staff and time are finite. Second, control measures have variable effectiveness. There is always a bigger or smaller residual impact – remaining effect after the avoidance and mitigation are done. Figure 2 shows the principle of this. As deployment of resources increases so the residual impact reduces, in the form of a decay curve. This is the best case and assumes that resources are deployed first where they have biggest impact reduction. Of course the cost rises as control measures are increased, as

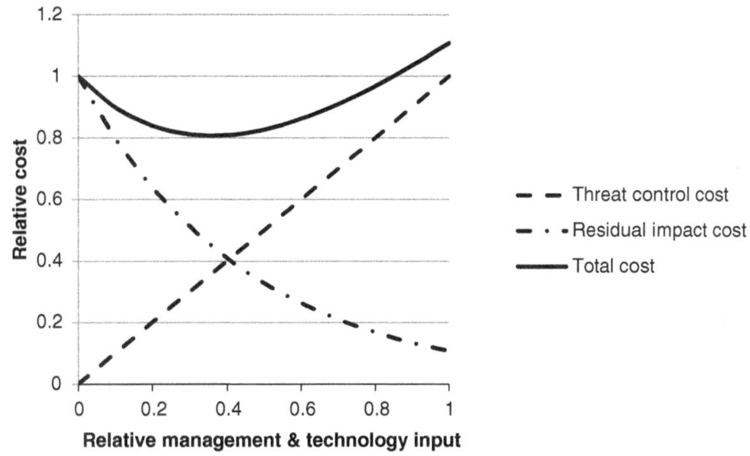

Figure 2 Economics of threat control.

reflected by the line rising from left to right in Figure 2. The total cost (add the decay curve and the straight line) gives a saddle-shaped curve which implies that beyond some moderate level of control costs outweigh benefits. Hence, even if funding were infinite, controlling for all threats is not a proposition.

Examples of methods to prioritize threats are given in Table 4. As with threat identification, there is no single right method, and the expert adapts her technique to the specifics of her project. The 'impact assessor's rating' in Table 4 is in government guidelines, and widely and often uncritically applied. Its rating of consequence is complicated. It implicitly assumes that threats are absolute whereas they are relative to context. For example, in the rich world the loss of a house is easily managed by paying cash compensation, but in the poor world the lost house must be replaced by the project in a complex and costly exercise. On a South African mine transported personnel are required by mine health and safety law to be buckled up, whereas labour on a farm or timber plantation is transported on the back of an open lorry. The 'impact assessor's rating' does not contextualize threat severity, and often it invokes circular logic: a threat is significant because it needs mitigation, and mitigation is necessary so the threat is significant.

Simple qualitative rating might be weak also because it assumes threats are absolute, and because it might be too simple and insensitive. At the other extreme, the probability-cost technique can be expensive to undertake for screening many threats, especially when few threats are ever going to be controlled. Probability-cost also assumes absolutism or otherwise that the expected value

Table 4 Prioritizing threats

Method	Procedure	Advantages	Disadvantages
Impact assessors rating	Rates impacts on ordinal scale for consequence (nature, extent, duration, intensity, irreversibility) & likelihood.	Widely used.	Complex consequence rating. Assumes threats absolute not relative to context. Uses circular logic: threat is significant because it needs mitigation; threat needs mitigation so it is significant. Overlooks Black Swans.
Qualitative rating	Rates threats for likelihood & consequence as low, medium or high, and portrays threat significance as combination of likelihood and consequence.	Simple.	Assumes threats absolute not relative to context. Might not discriminate sufficiently. Overlooks Black Swans.
Ordinal rating related to context	Rates threats for likelihood and consequence on ordinal scale, scores threat severity as rated likelihood x rated consequence, relates severity to society's laws and norms.	Simple. Cost-effective for screening many threats. Enables different threats to be compared in common currency. Relates to context.	Does not value expected loss. Overlooks Black Swans.
Probability-cost	Uses the convention that value of threatened loss is product of probability (p lies between 0.0 and 1.0) and the loss in monetary terms.	Repeatable. Monetary loss directly concerns project cost.	Can be costly to determine. Unsuitable for screening many threats. Assumes threat severity relates to monetary value whereas operative criteria might be legality and stakeholders values and norms. Overlooks Black Swans.

yielded by the product of probability and cost is the supreme measure of severity (Brown et al. 2001; Evans 2012). Dollar threat levels might group together events that differ in legal obligation (e.g. buckling up, as above), society's norms or expectations (e.g. in Gauteng, South Africa, you do not pay your electronic road toll but in other places you do), or commitment in corporate policy (e.g. road agency disregards loss of livelihoods, but LHWP commits to compensate promptly and fairly). By and large, society objects to a project's detrimental externalities irrespective of the actual dollar amount, so dollars might be the wrong criterion in threat screening.

Ordinal rating has several merits. Not all relevant variables can be expressed in ratio or interval data, but it is always possible to convert ratio and interval data to ordinal. Ordinal rating can provide a common currency enabling comparison between different types of threat, and different levels of knowledge on individual threats. An example of an ordinal rating is given in Table 5. It easily accommodates qualitative issues such as whether control of threat is legally required, expected by the stakeholders, and prescribed by corporate policy. The scoring can be calibrated to account for context. For example, once all threats in the project threat register have been scored, they can be summarized in a matrix where cells in the matrix are the number of threats corresponding to the relevant rated likelihood and rated consequence (Figure 3). The locations of a few well known threats in the matrix are then used as a basis to draw a threshold between material threats that warrant control by such criteria as legal requirement, societal expectation or corporate policy, and the other immaterial threats.

Having discriminated between the material and non-material, the question arises: Can the project afford to control the threats rated as material? If not then the project should not proceed. Now the control costs, not the unavoided impact costs, need to be determined. It should be the primary function of ESIAs, CBAs and indeed the whole of threat management to put the project decision-makers in an informed position so the affordability of controlling threats can be determined, as explained above about pumped storage, displaced

Table 5 Rating the likelihood and consequence of identified threats

Likelihood	Score	Consequence
Very unlikely	1	Minor (not detectable)
Fairly unlikely	2	Low (just detectable)
50/50 chance	3	Moderate (affects stress or safety or economics)
Fairly likely	4	High (affects health or well-being or sustainability)
Very likely	5	Very high (life threatening or non-sustainable or legally controlled or socially unacceptable)

		Consequence				
		1	2	3	4	5
Likelihood	5	2	1	0	4	7
	4	0	0	0	0	5
	3	0	0	0	1	4
	2	0	0	0	0	1
	1	0	0	1	0	1

Figure 3 Threat chart. Thick line is threshold between material threats (upper right) and other threats.

households and lost arable land. If the necessary costing of environmental and social controls is not estimated under the aegis of the ESIA, who is going to do it? Few ESIAs get near this, and the claim of Flyvbjerg et al. (2003) that impact assessment does not integrate into overall project decision-making applies.

The above prioritization procedure does not work for Black Swans. Stock market crashes – with likelihood calculated from normal distributions fitted to daily stock market fluctuations – are so improbable as to be implausible (unlikely since the Big Bang), yet they happen (Taleb 2008). Interestingly there are some significant non-random variables that affect our lives and projects. Examples in addition to stock market price variations are weather and stream flow. A feature of Black Swans is not a single cause but contingency of events and circumstances. Airliner crashes often do not have one cause that usually the vigilant crew would easily remedy. Rather there is a combination of mechanical failure, lapse in surveillance and preparedness, bad weather conditions, and so on. Similarly with dams. Exceptional heavy rain is accompanied by, say, an uncommonly strong wind that blows water spilling over the dam into the galleries, pumps have not been maintained so the galleries flood, and the low level release valves cannot be operated. Now add reservoir induced seismicity (the dam was never so full) and the implausible event of dam failure suddenly looms large. The poor weather conditions might also disrupt communications so downstream people cannot be forewarned of impending disaster.

Work through the project threat register and flag threats with low likelihood but possible high consequence. These are potential Black Swans that though not above the threshold in Figure 3 warrant control.

There are several ways of controlling threats: avoid, mitigate, offset, transfer, insure, accept and prepare (Table 6). The troublesome ones are preparedness –

Table 6 Threat control methods

Method	Procedure	Advantages	Disadvantages
Avoid	Reduces or eliminates event occurrence (likelihood). Removes cause.	Prevention is better than cure.	Some events unavoidable or beyond control.
Mitigate	Reduces consequence.	Focus on specific effects. Treat effects of unprevented & unpreventable events.	Treats symptoms not causes. Effects can be difficult, costly or even impossible to reduce. Usually is residual impact.
Offset	Loss of one set of resources is treated by investing in another set preferably similar in nature & extent.	Can be appropriate when avoidance & mitigation fail. Can rescue resources that otherwise would be lost without the offset intervention.	Needs economies of scale & costly even non-feasible if replacement resources in many isolated small patches. Original resource is still lost.
Transfer	Responsibility for threat control formally passed from one role player to another.	Enables the role player most exposed to threat, or most liable to cause impact, to undertake control. Clarifies responsibility & helps avoid control 'falling between the cracks'.	Requires thorough threat identification & prioritization.
Insure	Spreads cost consequence of low likelihood threats to other agencies.	Can be used to fund mitigation, especially big losses to third parties.	Does not avoid events, important in possible irreversible loss of indispensable natural capital. Might reduce incentive to implement precautionary measures.
Accept	Apply no control, usually to low threats.	Cost-effective. Avoids promising to apply unaffordable control.	Danger of irresponsible, negligent & uncaring reputation. Requires carefully worked out project appetite for threat.
Prepare	Structure organization to be resilient. Readiness for emergencies & capacity to control eventuating accepted threats, & unpreventable & unanticipated events.	Stay functional, protect people & assets in face of adversity.	Extra effort and expense in emergency drills, what-if & scenario analyses.

discussed above in relation to external threats – and offset and transfer.

Offsetting involves compensating for one set of lost resources by investing in another, preferably of equivalent type, extent and value. It might work, as in case of a grant to an organization undertaking environmental protection or remediation (e.g. protected area management, reclamation of contaminated or eroded land, restoration of wetland, afforestation to sequester carbon). Offsetting greenhouse gas emissions by growing trees is widely publicized. At Eskom's Ingula Pumped Storage Scheme in South Africa (Table 2, item 4) a condition of environmental authorization was to offset wetland and other resource loss by acquiring, rehabilitating and managing surrounding land, including a large wetland, as protected area. The prospects for effective offset are good because one large area of land is involved and the manager (Eskom) is on site. In a contrasting case of the Mooi-Mgeni Transfer Scheme (Table 1, item 2), offset of wetlands lost by inundation by the Spring Grove Dam is required by the environmental authorities. A dozen or so wetland areas on nearby private land are being viewed. Physical interventions – such as closing drains, erosion control structures and withdrawal of farming – are contemplated. This kind of offset is poorly feasible and has high opportunity cost. Landowners will have to be compensated. The many isolated pieces of land will require management and policing, otherwise land use and condition are liable to return to the pre-offset situation.

The notion of transferring threats might seem to the uninitiated like cheaply disposing of responsibilities. In many recent projects in South Africa the project documents have specified that the project proponent or principal is ultimately responsible for conforming to the conditions of project authorization. As watertight as this might seem, it is a loophole for contractors to evade expending resources on controlling their impacts. The low effectiveness of threat control (<50%) on many projects will continue until the quality of ESIAs improves as a basis for risk transfer. Tenderers for contracts must be informed of the scope of threat control that the principal will transfer explicitly in the contract. If the tenderer is not informed he does not estimate (generous estimates make bids uncompetitive), then when the contract is awarded impacts arise that were not budgeted for. A belated compensation event or variation order is then sought, or the impact is not addressed at all. Either way the project expense increases via more control or detrimental externality costs. With the threat transfer in place, the principal must have the contractor and subcontractors indemnify the principal against contractor and subcontractor negligence and misconduct (Brown et al. 2001). The fact that most projects today have several to many role players means that threat transfer is central to overall project threat control.

Related to transferring threats are organizational design and responsibilities. The typical organogram for southern African projects is shown in Figure 4. The design is fine but delegation of responsibility not. Often the government

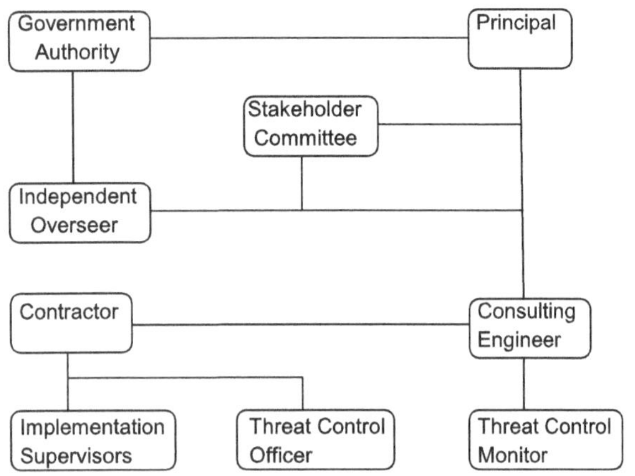

Figure 4 Typical project organogram for threat control.

authorization requires the independent overseer to ensure the project complies with conditions of authorization and laws of the land. Technically the overseer can do no such thing because he has no executive authority and if he did he would lose his independence. Practically this is another loophole for the contractor: compliance is not his responsibility. Project resources are then wasted arguing about responsibilities while consequences of unavoided impacts aggravate. An analogous situation arises with the contractor and his threat control officer. Project documents specify that the threat control officer is responsible for threat control. This is based on the add-on design for threat control where there is an environmental protection department whose responsibility it is to clean up, fix and rehabilitate. This does not work because the rest of the organization then does not give a hoot for the mess left, and it can be unfixable. Rather, like health and safety, threat control must be designed into work procedures. This is the build-in model where the contractor manager is responsible for threat control, not his threat control officer who observes and advises his manager who in turn instructs his supervisors. One client requires each work group to do a mini threat assessment at the start of every working day. The activities, main threats and controls are summarily written down, and everyone signs. If something goes wrong the incident investigation has a head-start. Was the threat identified? Was the control appropriate? Was it applied? Should procedures be revised? Is disciplinary action warranted?

Adapting management

In the mindset of the authorities, project proponents and threat assessors, ESIAs are once-off project approval exercises, rather than tools to protect people, the environment and the project over its lifetime (Mentis 2010). Even if CBAs, ESIAs, ESMPs and related documents were perfect at the time of compilation, the circumstances change: the project moves on, knowledge and experience grow, prices and currencies alter, laws and norms get stricter, the competitive environment shifts, and technology advances. There are several measures that can be adopted to adjust and improve project execution and stay abreast of change.

First, there needs to be measurement against the project objective and deliverables. Because resources are finite and some things are hard to measure, not everything can be measured. A small set of key performance indicators is needed. To contain time and cost here it is imperative to have powerful simple metrics. Yet in project after project armies of scientists are deployed to 'measure everything'. For example, fish are among the parameters in determining the condition of regulated rivers. Common monitoring practice involves deploying the whole fish sampling arsenal on every occasion – boats, electro-shockers, dip nets, fyko nets, seine nets, and hook-and-line. Is it possible to develop a reduced fish monitoring method with skeletal equipment that can be carried in a backpack so even the remote streams in mountainous Lesotho might be sampled to produce data with confidence limits?

Gigerenzer (2007, 2014) urges use of heuristics – rules of thumb – rather than complex and costly models. He questions 'big data', contending that more data can worsen rather than improve decision-making. Gigerenzer's heuristics have potential to save time and cost. To be sure, use of pure gut feelings (System 1 of automated thinking and unthinking action (Kahneman 2012)) is unsuitable for projects funded by investors or taxpayers. Though Gigerenzer finds business executives frequently make decisions by gut feel, and some of these might be brilliant, no statistics are quoted on the success rate of gut feel decisions versus the more carefully rationalized decisions. Are gut feel decisions better than flip of the coin? In cases such as fire-fighting in houses or forests, fast intuitive decisions might be necessary for survival, but in other situations can they offer a cost-effective substitute to slow and complex methods? Is it possible to develop, using Kahneman's System 2 of effortful mental activity, the 'fast and frugal' methods sought by Gigerenzer?

Kahneman and Klein (2009) consider that evaluating the likely quality of an intuitive judgment (heuristic) requires an assessment of the predictability of the environment in which the judgment is made and of the individual's opportunity to learn the regularities of that environment. They distinguish between high-validity systems, where heuristics might develop, and zero-validity situations, such as financial markets and politics, where predictability is low and establishment of expertise constrained.

There is limited incentive to be 'fast and frugal'. Big engineering and construction companies bid for big projects. Contractors get paid a percentage of any subcontracts. There is every reason to go big rather than small. Vested interests (Flyvbjerg et al. 2003) are alive and well.

Iterated plan-do-review-revise is a form of error elimination and a prime way of learning and improving the project. This should take place at least annually, and whenever there is major change in the project, e.g. shift from construction to operation phase. This type of updating rarely happens. The ideal is for the review to include the effectiveness of threat controls, monitoring data, audit and any emergency, accident, incident and near-miss reports, and for the revision to include fresh threat assessment. In many projects the impacts of the project on people and the environment parallel environmental and people effects on the project. For example, if sewer pipelines and water treatment works are not maintained then there are dangers of disease and eutrophication of the environment. At the same time, leaking sewers lower water quality and require subsequent extra purification treatment of the recycled water. If service delivery is poor, for example because of inadequate maintenance and then mechanical breakdown, people can vent their frustration by vandalizing project infrastructure. Understandably there is a reluctance to revisit contemporary ESIAs and ESMPs because of the labour in their preparation, and because their perceived utility is low. However, that should not detract from the benefits of iterated threat management applied throughout the project from pre-feasibility to decommissioning.

In the short time frame, project managers use threats and issues log (Barker and Cole 2007). Issues are threats which have eventuated. The log is a concise list of the current threats and issues, and names and briefly describes each threat or issue, and identifies action, responsibility and time frame. The log is reviewed and revised weekly.

As part of, or in addition to, iterated plan-do-review-revise, attention must be given to emergencies, accidents, incidents, near misses and areas of weak performance against the objectives and deliverables. Possibly the most visible example is the airline industry. The fact of hundreds of tons of metal, plastic, liquid and people hurtling thousands of km through thin air at nearly the speed of sound, mostly without accident or even incident, is truly astonishing. The airline industry boasts the most economical and safest form of transport invented. How did it do it? By relentless pursuit of efficiency and safety. Safe procedures are followed on every flight. Every emergency, accident, incident and even near miss are investigated. No effort is spared to deliver the better safer flight. This is how we got from Kittyhawk to Boeing. In another context, former Anglo American CEO, Cynthia Carroll, could not accept 'mining is inherently dangerous' as the reason for more than 40 deaths a year on Anglo mines (Carroll 2012). In four years she reduced the fatality rate by 60%. In comparison, big infrastructure (dams, pipelines, power stations, railways, roads, transmission lines) is widely regarded as inherently messy, with big environmental and social impacts. But can the practices of the airline industry and Cynthia Carroll be applied? Can project participants review what they are doing – focus on the failings, shortcomings and weaknesses, and fix them? Can projects be transformed from Kittyhawks into Boeings? Reference class forecasting recommended by the Flyvbjerg-Kahneman school helps but is not sufficient to improve procedures at the work face.

Project management

Project performance might be improved further by attending to a few generics: purpose, organization and communication.

Much project activity is not focused on the objective and how the deliverables are achieved (see project plan above), and thereby squanders time and money. Many project plans and reports I review every year are prefaced with pages of background, and burdened generally with marginally relevant material. Often it is hard to find the purpose of the document, and the purpose of the task being considered. How much project resource is consumed compiling these documents? How much is consumed by users searching for the document purpose and trying to use the document? Are the documents indeed read and used? If purpose does not drive and circumscribe thought, word and action then it is easy to stray in Kahneman's System 1 mode. Focus is System 2 thinking. Everything on a project should be purpose driven – what must we deliver, how, by when and with what budget? The adage applies: 'Begin [proceed and finish] with the end in mind' (Covey 1992).

Document organization, or rather the lack of it, is frequently another cost and time sink for projects. The typical project document has, for example, policy, principle, procedure and detail all on the same page. This can be challenging to understand and use, it makes review and revision difficult, and puts error elimination, and improvement from experience, out of easy reach. Policy concerns commitment. Its principles are the laws, customs, expectations and norms on which it is based. Procedure is about how the commitments are met. Detail includes the data. Each aspect has different properties and fulfils different functions. Reviewing and revising or updating any of the individual aspects separately is quite feasible, but mixed up on each of most pages of a document, the distinction between the aspects blurs, and the task of review and revision is so daunting that no one attempts it. Again, if this is how things are written, it probably reflects how project participants think and act – carried along in automatic System 1 mode, and not resorting to deep System 2 thinking that, though demanding to do, saves time and money in the end. An aid here is the precedent of the one-page project manager (Campbell 2007). The principle is to capture project status by the project manager displaying the key issues of objectives, tasks, timeline, cost and owners on one page. The project manager appoints owners of individual tasks each of which would have his own one page to capture the key issues of his responsibility. The task owner might have several assistants or subcontractors, each given ownership of subtasks for each of which there is another one page summary. Project information is thereby hierarchically arranged, and it is possible to zoom in or out depending on the level of information of interest, or the aspect of concern.

The one-page project manager has the dual benefit of helping the project manager and his staff to see the wood for the trees, and of communicating project status and performance to the principal and stakeholders. It is perhaps optimistic to dream that CBAs, ESIAs, ESMPs, audits and the like get condensed into one-page summaries. Currently these documents are poor at communicating. Often they do not have an executive summary, and the purpose of document and purpose of task being addressed are obfuscated by tiresome detail. If the reader is a senior manager she simply does not have time to distil the important issues. In consequence, she proceeds in ignorance of the insight that her staff have expensively gathered. Composing an executive summary has the benefit of clarifying in the mind of the report author what the key issues are to his manager and the project. He should be the best person to do this. He is also challenged to ensure the right message is communicated. That I am asked to review draft reports without executive summaries is indicative of the inefficiency of project management in the developing world.

Discussion

Issues arising from this review are that (a) project slippage is due to *uncertainties* rather than *risks*, (b) while eventuation of some bad things is beyond managerial control, efficient management and execution are the prime means of staying 'within budget, on time and fit-for-purpose', (c) improving project delivery is less about 'bigger and more complex' and more about coordinated effectiveness, focus and developing thought-out heuristics, and (d) projects take longer and cost more partly because threat identification is inaccurate, the identified threats are too narrow, and the threat assessment product is not integrated into overall project decision-making and execution.

Table 1 suffices to dispel the view that project performance is mostly about *risk* management and that *uncertainties* are an aside. The main reasons for slippage are non-numerical:

- the CBA is not done
- the CBA is necessary but insufficient because of the uncertainties that lie outside it
- reference class forecasting is not used
- there is limited transparency
- the stakeholders are not engaged and involved from project inception
- threat identification in the ESIA is defective
- threats are not classified according to type of control and not evaluated relative to the pervading socio-cultural context in a common currency enabling like with like comparisons
- responsibility for threat control or transfer of threats is not explicit in contract documents
- incentive is for project participants to go big rather than be fast and frugal
- error elimination – the learning process that made Toyota and other organizations world leaders – is not used
- project activity is not purpose driven
- project documents are disorganized
- communication is weak.

These observations conflict with project management currently but mistakenly fixated on *risk*. Guides devote pages to numerical economic evaluation with only mention that *uncertainties* exist, then mostly in relation to inexactness of the number inputs to the models (e.g. Sanral 2013). In the words of one project evaluation economist 'we go only by the facts' (i.e. the quantifiable costs and benefits). Publicity given to Black Swans (Taleb 2008) and resilience management (Linkov et al. 2014) might help to redress overreliance on certainty.

It is appropriate to question the management paradigm that if the project slips then management is deficient (Sage et al. 2014). The true effect of management can be difficult to distinguish from good or bad fortune (Taleb 2007). Nevertheless this does not dispose of the need to manage. As a consultant one walks in and out the front door of many an organization, and my experience is summed up by the sign at the entrance gate of the now defunct Duvha Colliery near Witbank, South Africa: 'Through these gates passes the world's finest team of coal miners'. Team members believed in themselves. Their performance was legendary. Why was that? They were ordinary people. The single most important cause was mine leadership. As one employee said to me years later 'We were effective because of the Mine Manager. He did walkabouts. If he found anything broken or short it was fixed, by tomorrow.' To be sure, strategy and external threats may eventuate irrespective of management, and the political and social environment matters on every project. These are givens. If management is viewed as a fixed suite of interventions then these givens could make or break a project. However, the contemporary manager is expected to be adaptable to circumstances (Fernández-Aráoz 2014) in which case it is up to management to succeed or fail with whatever the givens.

There is ongoing belief that good science and technology involve big data, of the ratio or interval sort, used in complex models to guide management infallibly, and anything less is stamp collecting. One is reminded of Laplace's demon – given the means to collect and interpret the data this super-scientist could describe the present in such detail as to be able to reconstruct the past exactly, and predict the future perfectly. If this dream ever materializes it is so far in the future as to be irrelevant to current project management. As things are now, impact assessment and project authorization are overly complicated and getting more so. The cost of an ESIA has soared 20-fold in the past 15 years, with doubtful improvement in project benefits, and in conflict with the experience curve (Henderson's Law) stating that cost falls by a constant percentage with every doubling of product volume. A present challenge is to deliver better value for money which bigger data and greater complexity are not yielding, but which a drive for heuristics might.

The main defects of threat assessment flagged by Flyvbjerg et al. (2003) a decade ago persist. Threat identification is so inaccurate that big threats can be missed altogether. The scope and time horizons of perceived threats may indeed be narrow. Control is often preoccupied with preventable threats and neglectful of strategy and external threats that are chief reasons why EGAP fails and projects slip. Inaccurate narrow ESIAs and ESMPs can hardly contribute to overall project decision-making and execution. The consequently belatedly or weakly controlled, or otherwise unavoided and unmitigated, environmental and social impacts are at financial cost to the project, and probably to future projects because they give threat control a bad name so no one wants to invest resources there.

Conclusion

In conclusion, the main obstacles to projects being within budget, on schedule and fit-for-purpose are not the quantifiable issues generally called *risks*, but *uncertainties*. Almost by definition it is what is poorly known that is problematic. Yet it is not just the unquantifiability and intangibility of uncertainties that cause project slippage, but that they are barely recognized and taken into account in project planning and execution. Improving project performance requires purpose-driven and managed deployment of scarce seasoned professionals. This can be aided with independent oversight of deeply experienced panellists who contribute technical insights and can potentially show that diligence is seen to be done.

Competing interests

The author declares that he has no competing interests.

References

Ansar A, Flyvbjerg B, Budzier A, Lunn D (2014) Should we build more large dams? The actual costs of hydropower megaproject development. Energy Policy (2014), http://dx.doi.org/10.1016/j.enpol.2013.10.069

Ariely D (2012) The (Honest) Truth About Dishonesty. Harper, London, p 314

Barker S, Cole R (2007) Brilliant Project Management. Pearson, Harlow, p 161

Binswanger HP (2007) Empowering rural people for their own development. Agr Econ 37:13–27

Blennow K, Persson J, Wallin A, Vareman N, Persson E (2013) Understanding risk in forest ecosystem services: implications for effective risk management, communication and planning. Forestry 2013: 0, 1–10, doi:10.1093/forestry/cpt032

Brown AR, Lane MR, Martin JH (2001) Triple Bottom Line Risk Management. John Wiley & Sons, New York, p 314

Campbell CC (2007) The One-page Project Manager. John Wiley & Sons, Hoboken, p 140

Carroll C (2012) Getting serious about safety. Harv Bus Rev June 2012: 43-46

Covey SR (1992) The Seven Habits of Highly Effective People. Simon and Schuster, London, p 358

Decker DJ, Riley SJ, Siemer WF (2012) Human Dimensions of Wildlife Management, 2nd edition. Johns Hopkins University Press, Baltimore, p 286

Evans D (2012) Risk Intelligence: How to Live with Uncertainty. Atlantic Books, London, p 276

Fernández-Aráoz C (2014) 21st century talent spotting. Har Bus Rev June 2014:60–73

Flyvbjerg B (2014) What you should know about megaprojects and why: an overview. Project Manage J 45(2):6–19

Flyvbjerg B, Bruzelius N, Rothengatter W (2003) Megaprojects and Risk. Cambridge University Press, Cambridge, p 207

Gigerenzer G (2007) Gut Feelings: Short Cuts to Better Decision Making. Penguin, London, p 280

Gigerenzer G (2014) Risk Savvy: How to Make Good Decisions. Penguin, New York, p 322

Goldratt EM (1986) The Goal: A process of Ongoing Improvement. North River Press, New York, p 273

Kahneman D (2012) Thinking, Fast and Slow. Penguin, London, p 499

Kahneman D, Klein G (2009) Conditions for intuitive expertise. Am Psychol 64 (6):515–526

Kahneman D, Tversky A (1979) Intuitive prediction: biases and corrective procedures. TIMS Studies Manage Sci 12:313–327

Kaplan RS, Mikes A (2012) Managing risks: a new framework. Harvard Business Rev June 2012:48–60

Linkov I, Fox-Lent C, Keisler J, Sala SD, Sieweke J (2014) Perspective: risk and resilience lessons from Venice. Environ Syst Decisions 34:378–382. doi:10.1077/s10669-014-9511-8

McKeown G (2014) Essentialism: The Disciplined Pursuit of Less. Crown Publishers, New York, p 260

Mentis M (2010) Environmental Risk Management in South Africa. Mentis, Hillcrest, p 172

Mentis M (2014) Science writing in the real world. Forest Ecosystems 1:2

Parkinson CN (1958) Parkinson's Law: The Pursuit of Progress. John Murray, London, p 128

Polak P, Warwick M (2013) The Business Solution to Poverty. Berrett-Koehler Publishers, San Francisco, p 245

Porter ME, Kramer MR (2006) Strategy and society: The link between competitive advantage and corporate social responsibility. Harvard Business Rev 2006:78–92

Sage D, Dainty A, Brookes N (2014) A critical argument in favor of theoretical pluralism: Project failure and the many and varied limitations of project management. Int J Project Manage 32(4):544–555

Sanral (2013) Drainage Manual. South African National Roads Agency Ltd, Pretoria, p 464 (www.sanral.co.za)

Simanis E, Duke D (2014) Profits at the bottom of the pyramid. Harv Bus Rev 2014:94–105

Taleb NN (2007) Fooled by Randomness: The Hidden Role of Chance in Life and the Markets. Penguin, London, p 316

Taleb NN (2008) The Black Swan: The Impact of the Highly Improbable. Penguin, London, p 366

van Marrewijk A, Clegg SR, Pitsis TS, Veenswijk M (2008) Managing public-private megaprojects: Paradoxes, complexity, and project design. Int J Project Manage 26:591–600

17

Methods of modelling relative growth rate

Arne Pommerening* and Anders Muszta

Abstract

Background: Analysing and modelling plant growth is an important interdisciplinary field of plant science. The use of relative growth rates, involving the analysis of plant growth relative to plant size, has more or less independently emerged in different research groups and at different times and has provided powerful tools for assessing the growth performance and growth efficiency of plants and plant populations. In this paper, we explore how these isolated methods can be combined to form a consistent methodology for modelling relative growth rates.

Methods: We review and combine existing methods of analysing and modelling relative growth rates and apply a combination of methods to Sitka spruce (*Picea sitchensis* (Bong.) Carr.) stem-analysis data from North Wales (UK) and British Douglas fir (*Pseudotsuga menziesii* (Mirb.) Franco) yield table data.

Results: The results indicate that, by combining the approaches of different plant-growth analysis laboratories and using them simultaneously, we can advance and standardise the concept of relative plant growth. Particularly the growth multiplier plays an important role in modelling relative growth rates. Another useful technique has been the recent introduction of size-standardised relative growth rates.

Conclusions: Modelling relative growth rates mainly serves two purposes, 1) an improved analysis of growth performance and efficiency and 2) the prediction of future or past growth rates. This makes the concept of relative growth ideally suited to growth reconstruction as required in dendrochronology, climate change and forest decline research and for interdisciplinary research projects beyond the realm of plant science.

Keywords: Growth efficiency; Growth coefficient/multiplier; Chapman-Richards growth model; Standardisation; Simultaneous estimations

Background

Growth is a universal and fundamental process of life on earth. The analysis and modelling of plant growth has therefore been a particular concern in plant science as well as in production biology including forestry, agriculture and fishery to name but a few. This research has the important objective to identify growth patterns in response to environmental factors or treatments.

In this context, the concept of relative plant growth, involving the analysis and modelling of plant growth relative to plant size, has proved to be a powerful tool in comparative studies of the growth performance of plants and has a long tradition in plant science (Evans 1972, p. 190ff.; Pommerening and Muszta: Concepts of relative growth – a review, submitted). It first developed at the beginning of the 20th century in what eventually became the British school of plant growth analysis, mainly at Sheffield University (Hunt 1982, p. 1, 16).

Independently of the British school another quantitative plant science group developed at Tharandt/Dresden Technical University in Germany. The Tharandt school characterised the growth of trees by using a variant of the concept of relative plant growth and on this basis eventually developed a population model and a size class model for predicting the growth of trees (Wenk et al. 1990, Wenk 1994). There is also evidence of empirical Russian work in this area (Antanaitis and Zagreev 1969) and particularly remarkable is the detailed Finnish work by Kangas (1968).

Relative growth rate is a standardised measure of growth with the benefit of avoiding, as far as possible, the inherent differences in scale between contrasting organisms so that their performances can be compared on an equitable basis (Hunt 1990, p. 6). Applications of relative growth rates include the study of dry weight, biomass, leaf area, stem volume, basal area and stem

* Correspondence: arne.pommerening@slu.se
Department of Forest Resource Management, Faculty of Forest Sciences, Swedish University of Agricultural Sciences SLU, Skogsmarksgränd, SE-901 83 Umeå, Sweden

diameter. Interestingly, the concept is closely related to plant mortality (Gillner et al. 2013), i.e. low relative growth rates for extended periods of time are good indicators of imminent death. Relative growth rates are also pre-requisites for quantifying and modelling allometric relationships in plants (Gayon 2000).

Assuming that function $y(t)$ models the state of a plant characteristic at time t, for example the weight, area, volume or biomass of a plant, *relative growth velocity* or *instantaneous relative growth rate* (RGR; in forestry termed *relative increment*) can be expressed as

$$p(t) = \frac{\mathrm{d}}{\mathrm{d}t} \log\ y(t) = \frac{\mathrm{d}y}{\mathrm{d}t} \times \frac{1}{y(t)} = \frac{y'(t)}{y(t)} \quad (1)$$

As instantaneous growth rates cannot be measured in practice, the difference between growth characteristics of interest is usually studied at discrete points in time, t_1, $t_2, \ldots, t_n$, which for example are scheduled survey years. In this context, the period between two discrete points in time can be denoted by $\Delta t = t_k - t_{k-1}$ with $k = 2, \ldots, n$. For ease of notation in the remainder of this section we set $y(t_k) = y_k$ and $p(t_k) = p_k$ etc. and assume equidistant time periods. However, the notation can be modified to accommodate unequal time periods (Pommerening and Muszta: Concepts of relative growth – a review, submitted).

According to Blackman (1919), Whitehead and Myerscough (1962) and Hunt (1982, 1990), *periodic relative increment* or *mean relative growth rate*, p_k, over a time period Δt is the difference of the logarithms of y_k and y_{k-1} divided by Δt, see also Causton (1977, p. 213).

$$p_k = \frac{\log\ y_k - \log\ y_{k-1}}{t_k - t_{k-1}} = \frac{\log\ y_k - \log\ y_{k-1}}{\Delta t} = \frac{\log\ (y_k / y_{k-1})}{\Delta t} \quad (2)$$

Considering a short time period, mean relative growth rate is approximately equal to the instantaneous relative growth rate $p(t)$. Blackman (1919) originally referred to equation (2) as "efficiency index" and "specific growth rate", see also Causton and Venus (1981, p. 37). From the last term we can see that equation (2) can be interpreted as the logarithm of the ratio of successive size measurements divided by the corresponding time interval (Pommerening and Muszta: Concepts of relative growth – a review, submitted).

According to Evans (1972, p. 197) and Hunt (1982, p. 17), the current value of a plant characteristic can be calculated from a value in the past based on equation (2) as

$$y_k = y_{k-1} \cdot e^{p_k \cdot \Delta t} \quad (3)$$

Equation (3) is also referred to as Blackman's efficiency index which is supposed "to represent the efficiency of the plant as a producer of new material, and to give a measure of the plant's economy in working" (Blackman 1919).

The exponential term in equation (3) has fascinated plant growth scientists and inspired them to devise special names. Kangas (1968, p. 50f.) coined the name *growth coefficient*, whereas Wenk (1972) suggested the name *growth multiplier*, M_k (equation 4).

$$M_k = e^{p_k \cdot \Delta t} = \frac{y_k}{y_{k-1}} \quad (4)$$

The growth coefficient or multiplier is obviously a function of relative growth rate and can also be defined as the ratio of a particular plant size characteristic at different times. Part of the fascination with M_k stems from the fact that the growth multiplier plays a crucial role in predicting future growth based on relative growth rates (Kangas 1968, p. 19; Wenk et al. 1990, p. 95f.; Murphy and Pommerening 2010).

The allometric coefficient, m_k, mediates relative changes of plant size characteristics, e.g. x and y (where y has the same meaning as in the equations before). m_k is an important part of the concept of relative growth (Gayon 2000). Considering short time periods it is often assumed that the allometric coefficient is constant. Wenk (1978) could show that in such a case the mean relative growth rates p_x and p_y of size characteristics x and y are related as

$$p_{x,k} = 1 - \left(1 - p_{y,k}\right)^{\frac{1}{m_k}}. \quad (5)$$

The objective of this paper is to explore how relative growth rates of individual plants as well as of plant populations can be efficiently analysed and modelled using a system of simultaneous functions of relative growth and allometric relationships developed in different research schools.

Functions of relative growth

Hunt (1982), Wenk et al. (1990, p. 79) and Zeide (1993) give a number of plant growth functions and provide detailed discussions. They are often combinations of power functions and exponential functions (Zeide 1989). Zeide (1993) and Pommerening and Muszta (Concepts of relative growth – a review, submitted) show how they relate to each other. These authors also compiled a number of functions of relative growth, which are reproduced in Table 1.

The functions in Table 1 are based on the original growth functions and on the corresponding functions of absolute growth rate, i.e. the first derivatives of the

Table 1 Frequently used functions of absolute and relative plant growth rate. *a, b, c* are model parameters and the symbol *t* denotes time or age

Function name/source	Growth function	Absolute growth rate	Relative growth rate
Chapman-Richards (Richards 1959)	$a(1-e^{-bt})^c$	$abce^{-bt}(1-e^{-bt})^{c-1}$	$bc\ (1-e^{-bt})^{-1}$
Gompertz (1825)	$ae^{-be^{-ct}}$	$abce^{-ct}e^{-be^{-ct}}$	bce^{-ct}
Kangas (1968)	-	-	$ae^{bt^{-c}}$
Korf (Zeide 1993)	$ae^{-bt^{-c}}$	$abct^{-c-1}e^{-bt^{-c}}$	$bct^{-(1+c)}$
Logistic (Verhulst 1838)	$a(1+ce^{-bt})^{-1}$	$abce^{-bt}(1+ce^{-bt})^{-2}$	$bc\ (c+e^{bt})^{-1}$
Monomolecular (Weber 1891)	$a(1-ce^{-bt})$	$abce^{-bt}$	$bc\ (e^{bt}-c)^{-1}$
Weibull (Zeide 1993)	$a\left(1-e^{-bt^c}\right)$	$abct^{c-1}e^{-bt^c}$	$bct^{(c-1)}\ \left(e^{bt^c}-1\right)^{-1}$

growth functions with respect to time. Most functions of relative growth rate have the advantage that they have fewer model parameters than the corresponding functions of absolute growth rate.

Kangas (1968, p. 69) independently suggested the function listed in Table 1 next to his name for modelling the growth multiplier of equation (4) and referred to it as the *growth coefficient function*.

Modelling individual plant growth

The strategy of modelling individual plant growth is straightforward: **1)** A suitable function of relative growth rate is selected from Table 1 or from other publications. **2)** A primary plant size characteristic is identified, e.g. tree volume. **3)** Secondary plant size characteristics, e.g. tree height and tree diameter, are linked to the function of relative growth rate of the primary plant size characteristic through allometric relations. **4)** The 2–3 model parameters of the function of relative growth rate and the two allometric coefficients are estimated simultaneously through nonlinear regression. Jones et al. (2009, p. 219f.) describe how such more complex types of nonlinear regression can be calculated in R using the function optim.

Wenk et al. (1990, p. 174ff.) selected primary and secondary plant size characteristics in such a way that error propagation was effectively reduced: They identified tree volume as a three-dimensional size characteristic to be the primary characteristic and one-dimensional total tree height and stem diameter as secondary characteristics. However, since this is a generic approach, there is no need to strictly follow this recommendation. Tree volume can for example also be replaced by weight or biomass.

To illustrate this combined methodology we have used stem-analyses data of four Sitka spruce (*Picea sitchensis* (Bong.) Carr.) trees taken from the same forest stand in Clocaenog Forest (North Wales, UK). Stem-analysis data include annual tree size characteristics such as stem volume, total tree height and stem diameter at 1.3 m above ground level. As function of relative growth we selected the well-known and frequently used Chapman-Richards function, but any of the other functions in Table 1 would perform reasonably similar (Pommerening and Muszta: Concepts of relative growth – a review, submitted). As an example, Figures 1 and 2 give a visual impression of

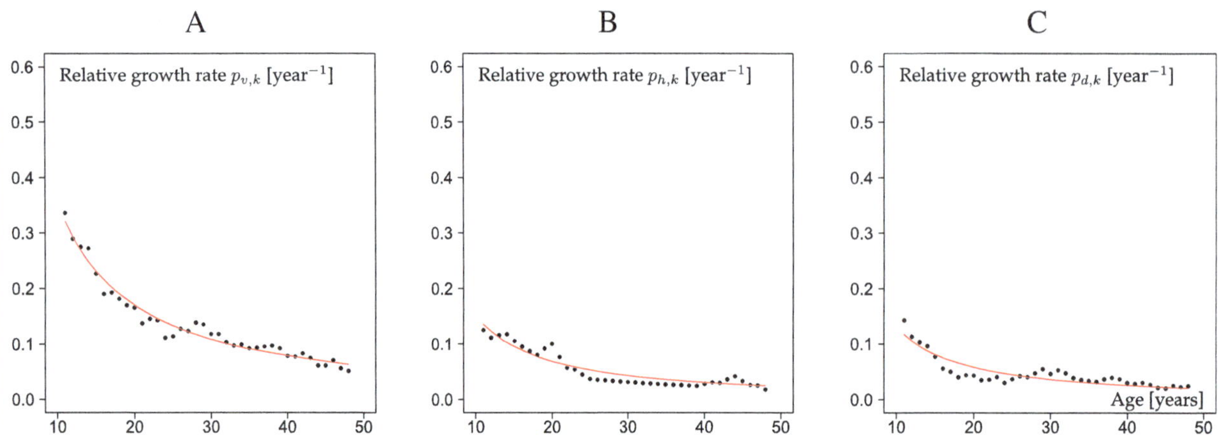

Figure 1 The relative volume (A), height (B), and diameter (C) growth rates over time of tree # 5000 in Clocaenog Forest (North Wales, UK) at Cefn Du (plot 1).

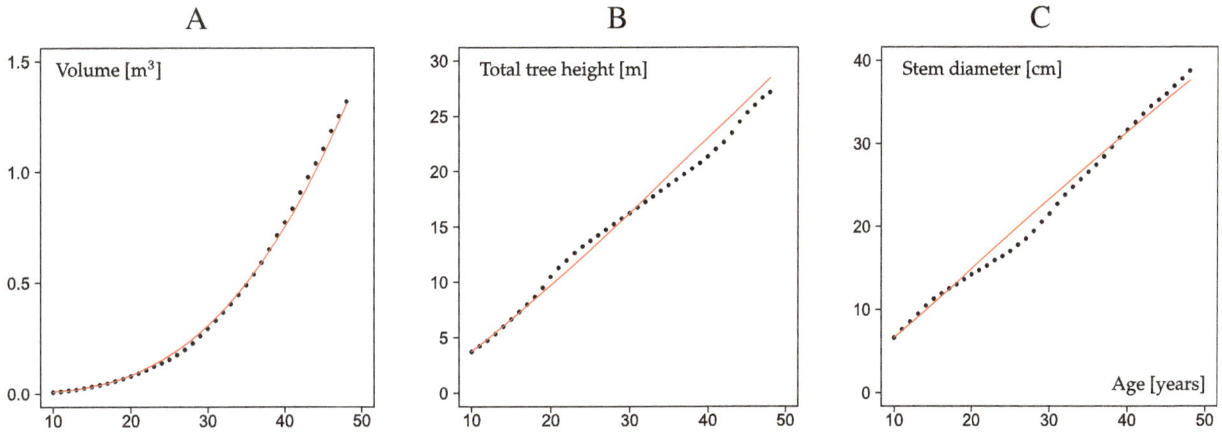

Figure 2 **Volume (A), height (B), and diameter (C) development over time of tree # 5000 in Clocaenog Forest (North Wales, UK) at Cefn Du (plot 1).**

the regression results of tree # 5000. The corresponding model parameters and other summary statistics can be found in Table 2.

The relative growth curve of volume, $p_{v,k}$, in Figure 1 (A) is based on the Chapman-Richards model (see Table 1) with model parameters b and c. The relative height (Figure 1, B) and diameter (Figure 1, C) growth rates, $p_{h,k}$ and $p_{d,k}$, however, have not been modelled independently but linked to $p_{v,k}$ through the allometric coefficients m_1 and m_2 and equation (5).

Figure 2 demonstrates how the relative growth rates of tree # 5000 "translate" to modelling the observed growth characteristics using the growth multiplier of equation (4). Again, any other growth characteristics can be used instead of volume, total height and stem diameter, the growth analyst is completely free to choose.

Figure 3 (A) summarises the curves of relative volume growth rate curves of all four trees.

As expected the three curves decline throughout growth and particularly trees # 2000, 3000 and 4000 appear to have quite similar growth patterns, as is often the case with trees of the same population growing in similar environmental conditions. Their curves form almost parallel lines in the order 4000 > 3000 > 2000, i.e. tree # 4000 has the highest relative growth rate throughout its lifetime and tree # 2000 the lowest. This trend also seems to be reflected by model parameter b, which is responsible for scaling the growth rate (Pienaar and Turnbull 1973), although this requires a more detailed study.

Tree # 5000 exhibits a growth pattern different from the other three trees: Its growth starts with a comparatively low rate, but finishes with a rate markedly higher than those of the other three trees. The curve of Tree # 5000 intersects those of the others approximately halfway through its lifetime.

The data of all four trees start at the same age (10 years) and end approximately at the same age (48–50 years). Trees # 2000 and 5000 are larger than the other two trees at the beginning, whilst trees # 5000 and 4000 are the largest at the end of their lifetime.

Many authors point out that RGR is size dependent, i.e. individuals with a smaller initial size have a larger relative growth rate (Turnbull et al. 2008; Rose et al. 2009; Rees et al. 2010). This can mask important relationships and it may be difficult to tell whether a tree grows slowly because it is large or because it is pursuing a slow growth strategy. To check up on this argument

Table 2 Characteristics and model statistics of four Sitka spruce (*Picea sitchensis* (Bong.) Carr.) trees at Cefn Du (plot 1), Clocaenog Forest (North Wales, UK)

Tree #	*dbh* (cm)		*b*	*c*	Volume (m^3)		Total height (m)		Stem diameter (*dbh*)	
	Min.	Max.			*Bias*	*RMSE*	m_1	*RMSE*	m_2	*RMSE*
2000	6.4	32.2	−0.05251	6.12877	0.00303	0.02478	2.82205	0.01309	1.07431	0.02211
3000	4.2	28.3	−0.05057	6.44977	0.00307	0.01646	2.63452	0.00633	1.04791	0.01835
4000	4.9	32.6	−0.04667	6.40575	0.00249	0.02401	2.78971	0.01296	1.13672	0.01294
5000	6.6	38.7	−0.00822	3.68512	0.00024	0.01248	2.66006	0.01105	1.17268	0.01218

dbh – stem diameter at breast height at 1.3 m above ground level; *b*, *c* – model parameter of the Chapman-Richards model of relative growth rate; m_1, m_2 – allometric coefficients for estimating the relative growth rates of total height and stem diameter, respectively. *Bias* and *RMSE* relate to estimates of relative growth rates.

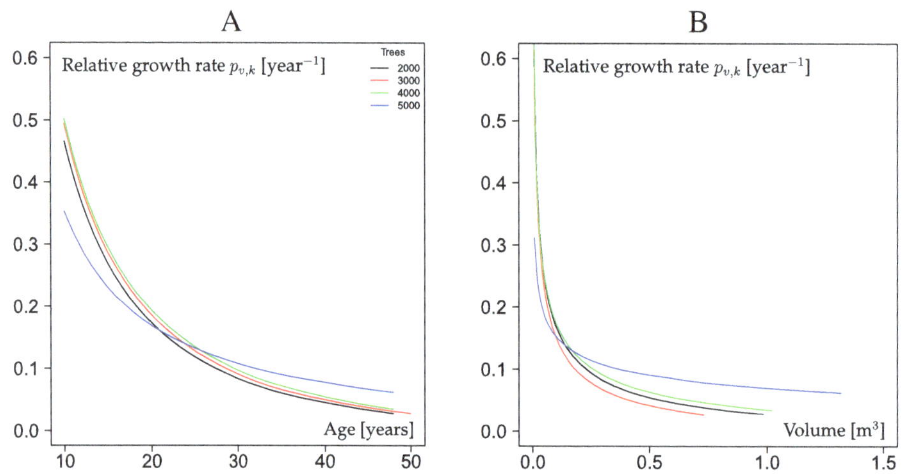

Figure 3 Relative volume growth rates of four trees at Cefn Du (plot 1), Clocaenog Forest (North Wales, UK). A: Relative volume growth rates over time modelled using the Chapman-Richards growth function. **B**: The standardised relative volume growth rates over volume of the same trees based on the same growth model as before.

we can calculate size-standardised RGR. To this end, first the Chapman-Richards growth function (see first row in Table 1) is fitted to the observed size data of the four trees.

Inverting the Chapman-Richards growth curve provides a possibility to convert volume (or any other size characteristic) to time/age, see equation (6) (which obviously is different for other growth functions in Table 1).

$$t = -\frac{ln\left(1-e^{\frac{ln\left(\frac{y}{a}\right)}{c}}\right)}{b} = -\frac{1}{b}\cdot ln\left(1-\left(\frac{y}{a}\right)^{\frac{1}{c}}\right) \quad (6)$$

This conversion of time to a growth characteristic on the abscissa allows depicting size-standardised RGR in Figure 3 (B). We get a better understanding of the differences in relative volume growth rate of the four trees and see more clearly that tree # 5000 due to the lower rate of decline has a superior relative growth rate throughout most of its volume development despite the fact that its initial RGR is much lower than those of the other trees. Also the ordering of trees # 2000 and 3000 is interestingly reversed compared to non-standardised RGR (Figure 3, A). We can also see that tree # 3000 is the tree with the smallest final volume despite having a longer lifespan than all other trees.

As RGRs and growth multipliers are measures of relative growth it seems natural to relate them to relative size variables. The stem diameter growth multiplier of individual trees of the next time step, $M_{d,k+1}$, is for example correlated with their current stem diameters, dbh_k, relative to the current quadratic population diameter, dg_k, i.e. $d_{rel,k} = dbh_k / dg_k$. Figure 4 illustrates this relationship for the conifer trees of a mixed conifer-broadleaved woodland at Coed y Brenin (plot 7, North Wales, UK). As a model function to describe this relationship the Michaelis-Menten saturation curve (Michaelis and Menten 1913; Bolker 2008) was used, as it appears to describe the data well.

Apparently the diameter multiplier decreases with increasing tree size relative to the population mean and levels off with very large ratios when trees are much larger than the population mean. Using the relationship between a relative tree diameter and the growth multiplier

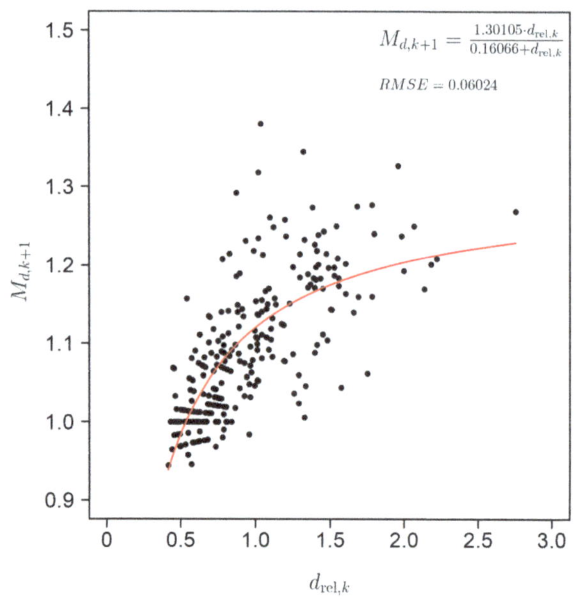

Figure 4 5-years diameter growth multiplier of conifer trees dependent on the ratio of stem diameter and quadratic mean diameter at the experimental plot 7, Coed y Brenin (North Wales, UK).

also has the advantage of predicting future growth from current size and from size dominance at the same time. Coming back to the relationship between relative growth rate and mortality (Gillner et al. 2013), trees in the lower range of $M_{d,k+1}$ carry a high probability of dying.

Modelling of plant populations

In the same way as it is possible to model individual plants one can also model whole plant populations. In this case, the growth rates are calculated from population summary characteristics. Details of this approach are also documented in Wenk and Nicke (1985), Nicke (1988) and in Wenk (1994).

Drawing an analogy to the example used in the previous section we can for example model volume per hectare, top height and quadratic mean diameter of a forest stand. Let us denote the corresponding relative growth rates as $p_{V,k}$, $p_{H,k}$, and $p_{D,k}$ using capital letters for population volume (V), height (H) and diameters (D) in the subscripts. The methodology is almost exactly the same with the exception that there is a need to account for trees leaving the forest stand. These losses can be due to forest management or to natural mortality or can be a combination of both. For simplicity we introduce just one additional function that we simply refer to as loss function here, $l_{V,k}$, to collectively take care of death events as a result of disturbances. This function was suggested by Wenk et al. (1990, p. 159).

$$l_{V,k} = 1 - \frac{Q}{M_{V,k}}, \tag{7}$$

where Q is the ratio $V_{k+1}^{\text{res}}/V_k^{\text{res}}$, i.e. the ratio of successive residual stand volumes (superscript "res" denotes "residual"). The ratio Q defines the loss and must not exceed the volume growth multiplier, as negative values are not defined. With $Q = 1$, $p_{V,k} = l_{V,k}$ and the forest stand does not grow. The loss function is therefore a ratio of two multipliers subtracted from one and can also be estimated with the Chapman-Richards relative growth rate function.

Forest stand volume is then calculated and projected in the following way (considering that the superscripts "prior" define the forest state before disturbance, "lost" the part of the tree population lost and "res" the state of the residual forest after disturbance):

$$\begin{aligned} V_k^{\text{lost}} &= V_k^{\text{prior}} \cdot l_{v,k} \\ V_k^{\text{res}} &= V_k^{\text{prior}} - V_k^{\text{lost}} \\ V_{k+1}^{\text{prior}} &= V_k^{\text{res}} \cdot M_{V,k+1} \end{aligned} \tag{8}$$

In a first step, the volume per hectare of dead trees, i.e. the absolute loss in terms of volume, is calculated as the product of the volume of the forest stand before the disturbance, V_k^{prior}, and the volume loss rate, $l_{V,k}$. Then residual stand volume, V_k^{res}, constitutes the difference between stand volume before disturbance, V_k^{prior}, and the absolute volume loss, V_k^{lost}. Finally the volume of the forest stand before disturbance for the next time step, V_{k+1}^{prior}, is calculated as the product of the residual stand volume of the current time step, V_k^{res}, and the volume growth multiplier of the next time step, $M_{V,k+1}$.

In analogy to volume projection, the development of a density measure such as trees per hectare can be modelled. Stand height is also modelled in a similar way as outlined in the last line of equation (9) and like individual-tree heights.

Finally, the quadratic mean diameter, dg_k^{prior}, of the forest stand before disturbance is calculated from stand basal area before disturbance, G_k^{prior}, and the number of trees per hectare before disturbance, N_k^{prior}, using the following two equations.

$$\begin{aligned} G_k^{prior} &= \frac{V_k^{\text{prior}}}{f_H} \\ dg_k^{\text{prior}} &= 100 \cdot \sqrt{\frac{4 \cdot G_k^{\text{prior}}}{\pi \cdot N_k^{\text{prior}}}} \end{aligned} \tag{9}$$

f_H is a static form height function. Such form height relationships are available for many species and countries.

The equations of this section can be included in a regression routine, for example in the optim function of R mentioned before (Jones et al. 2009, p. 219f.). As a function of relative growth rate we again select the Chapman-Richards function (see Table 1), but any of the other functions is suitable, too.

To illustrate this method we have used Douglas fir (*Pseudotsuga menziesii* (Mirb.) Franco) data from British yield tables (Edwards and Christie 1981), specifically those relating to yield class 24, initial spacing 1.7 m × 1.7 m and crown thinning. Yield tables can be interpreted as tabular summary characteristics of forest stands, whereby the yield classes represent different environmental conditions resulting in a larger or smaller carrying capacity. Naturally any other aggregated plant population data can be selected for this purpose.

The yield table data are provided for five-year intervals and as a consequence relative growth rates and growth multipliers also relate to 5-years periods.

In addition to the yield table data, we have used the UK form height function for Douglas fir suggested by Matthews and Mackie (2006, p. 325):

$$f_H = -0.509255 + 0.426679 \cdot H, \tag{10}$$

where H is stand top height in metres.

Figure 5 (A) shows the relative volume growth rate at forest stand level, $p_{V,k}$, and the relative volume loss rate,

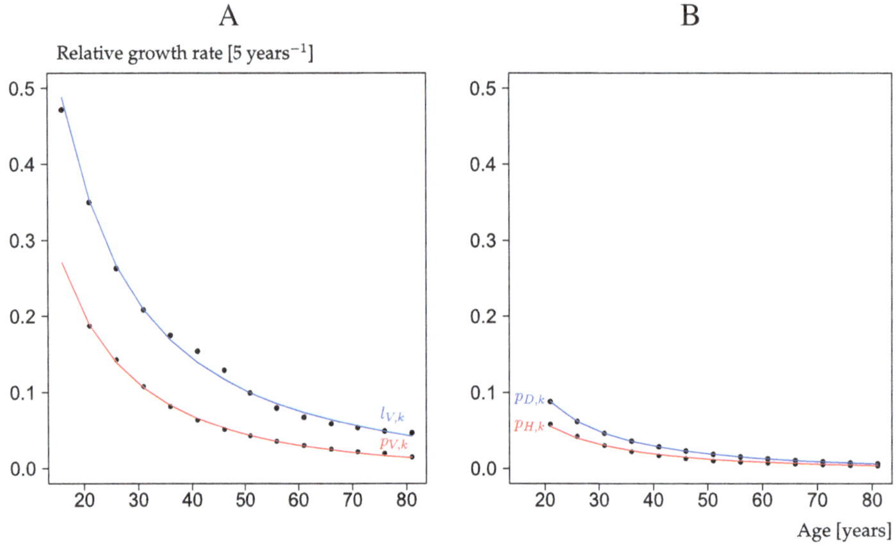

Figure 5 Relative growth rates of the British Douglas fir (*Pseudotsuga menziesii* (Mirb.) Franco) yield class model for yield class 24, initial spacing 1.7 m × 1.7 m and crown thinning (Edwards and Christie 1981). **A**: The relative volume growth rate (red) and relative loss rate (blue) **B**: The corresponding relative growth rates of the quadratic mean stem diameter height (blue) and of top height (red).

$l_{V,k}$. Note that $p_{V,k} < l_{V,k}$ at all times to fulfil the aforementioned requirement that Q must not exceed the volume growth multiplier.

Figure 5 (B) gives the quadratic mean stem diameter and top height relative growth rates over time. They start at a much lower rate than the relative volume growth rate and are related to it by the allometric coefficient m. Table 3 gives data and model summary characteristics.

Figure 6 presents the temporal development of the actual population size characteristics volume per hectare, top height and quadratic mean diameter. Again they have been calculated from relative growth rates as explained in equation systems (8) and (9).

We can now take a look at the relative volume growth rates of the four Douglas fir yield classes 8, 12, 18 and 24 (Figure 7, A). Interestingly the relative growth rates almost do not differ at all, only the curve representing yield class 8 is slightly higher than those of the other three. This result is somewhat counter-intuitive assuming that a yield class of 8 represents the least favourable of all four environmental conditions. The outcome suggests that a forest growing on a worse site is relatively more efficiently growing than a forest benefitting from better site factors.

Using again the Chapman-Richards growth function and equation (6) we can calculate standardised relative growth rates resulting in Figure 7 (B).

We can now clearly see that larger yield class numbers, i.e. more favourable environmental conditions, correspond to higher standardised relative growth rates. The standardisation has again helped us to better discriminate between the differences in relative growth rate of the four populations.

Final discussion and conclusions

Our examples in this paper have demonstrated that by bringing together the approaches in general plant growth analysis and in forest growth and yield science it is possible to advance the concept of relative growth. Modelling of relative growth rates is straightforward and

Table 3 Characteristics and model statistics of four British Douglas fir (*Pseudotsuga menziesii* (Mirb.) Franco) yield tables (Edwards and Christie 1981) relating to initial spacing 1.7 m × 1.7 m and crown thinning

Yield class	*dg* (cm)		*b*	*c*	Volume (m^3·ha^{-1})		Top height (m)		Mean diameter (*dg*)
	Min.	Max.			*Bias*	*RMSE*	*m*	*RMSE*	*RMSE*
24	12.8	76.7	−0.03240	5.68749	0.00011	0.00185	3.64234	0.00153	0.00658
18	12.2	60.9	−0.03147	5.65227	0.00014	0.00181	3.46283	0.00151	0.00560
12	12.4	43.5	−0.02988	5.43637	0.00004	0.00075	3.20122	0.00124	0.00793
8	11.7	32.5	−0.03095	6.16601	−0.00003	0.00072	3.06830	0.00092	0.00757

dg – quadratic mean stem; *b*, *c* – model parameter of the Chapman-Richards model of relative growth rate; *m* – allometric coefficient for estimating the relative growth rate of top height. *Bias* and *RMSE* relate to relative growth rates.

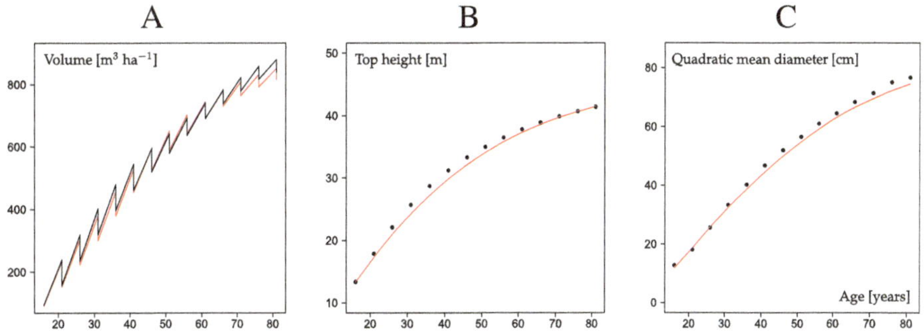

Figure 6 Stand volume (A), top height (B), and quadratic mean stem diameter (C) development over time of the British Douglas fir (*Pseudotsuga menziesii* (Mirb.) Franco) yield class model for yield class 24, initial spacing 1.7 m × 1.7 m and crown thinning (Edwards and Christie 1981).

often the corresponding growth functions have fewer model parameters than the respective versions of absolute growth rates. The recent introduction of size-standardised relative growth rates (Turnbull et al. 2008; Rose et al. 2009; Rees et al. 2010) has proved to be very effective for improving our understanding of growth efficiency and this approach can definitely be recommended.

Modelling relative growth rates serves two purposes, **1)** an improved analysis of growth performance and efficiency and **2)** the prediction of future or past growth rates. The methodology described in this paper also allows "backcasting" by simply dividing plant growth characteristics by the growth multiplier instead of multiplying (Murphy and Pommerening 2010).

The obvious similarity between individual-tree and population modelling of relative growth rates suggests a link between these two modelling levels (see for example Cao 2014). In short, the idea of this approach is to combine the advantages of different modelling resolutions, particularly to employ information of population models (that are mathematically more tractable and statistically more stable) for improving individual-tree models. This strategy is often referred to as *disaggregation*. Zhang et al. (1993) describe a method of disaggregation using relative tree sizes similar to $d_{\mathrm{rel},k}$ shown in this paper. Since this relative size characteristic has also proved useful for modelling growth multipliers, a disaggregation modelling approach for linking the two modelling levels described in this paper is very likely to be successful.

The properties of the concept of relative growth make it a superb choice for growth reconstruction as required in dendrochronology, climate change and forest decline research and we hope that the methodology outlined in this paper will inspire much interdisciplinary research,

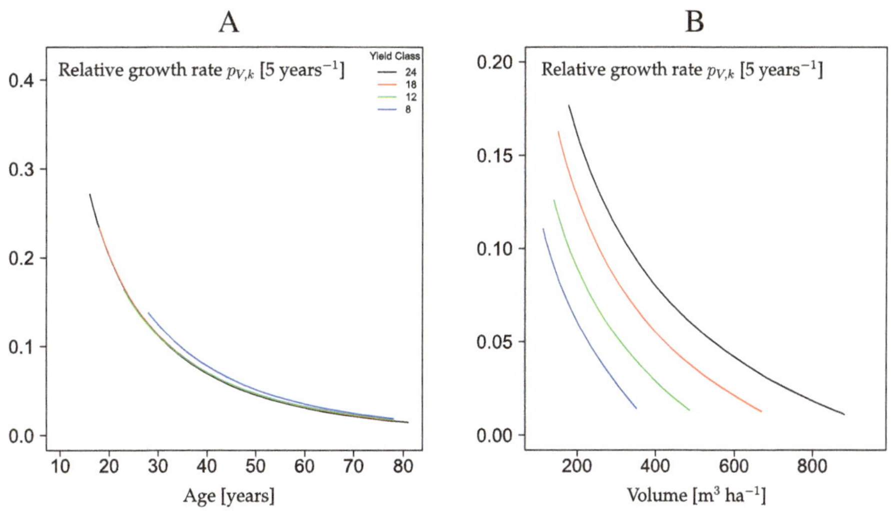

Figure 7 Relative growth rates of four British Douglas fir (*Pseudotsuga menziesii* (Mirb.) Franco) yield tables (Edwards and Christie 1981) relating to initial spacing 1.7 m × 1.7 m and crown thinning. A: Relative volume growth. The growth rates were modelled using the Chapman-Richards growth function. **B**: The standardised relative volume growth rates over volume of the same yield tables based on the same growth model as before.

as the applicability is universal to any phenomenon involving growth processes.

Competing interests

The authors declare that they have no competing interests.

Authors' contributions

AP devised the concept of the article, carried out the data analysis/modelling and wrote the majority of the text. AM was responsible for the mathematical correctness of the equations and also ensured consistent notations. Both authors read and approved the final manuscript.

Acknowledgements

Stephen T. Murphy, Jens Haufe, Owen Davies and Gareth Johnson of the Tyfiant Coed project have carefully analysed and kindly provided the time-series and stem-analysis data used as examples in this review. We are also indebted to the late Günter Wenk who inspired the first author with his enthusiasm of quantifying and modelling relative growth processes.

References

Antanaitis V, Zagreev V (1969) Прирост леса. [Forest increment.]. Издательство лесная промышленность, Moscow
Blackman VH (1919) The compound interest law and plant growth. Ann Bot 33:353–360
Bolker BM (2008) Ecological Models and Data In R. Princeton University Press, Princeton
Cao QV (2014) Linking individual-tree and whole-stand models for forest growth and yield prediction. Forest Ecosyst 1:1–8
Causton DR (1977) A Biologist's Mathematics. Edward Arnold, London
Causton DR, Venus JC (1981) The Biometry of Plant Growth. Edward Arnold, London
Edwards PN, Christie JM (1981) Yield Models for Forest Management. Forestry Commission Booklet 48. Forestry Commission, Edinburgh
Evans GC (1972) The Quantitative Analysis of Plant Growth. Blackwell Scientific Publications, Oxford
Gayon J (2000) History of the concept of allometry. Am Zool 40:748–758
Gillner S, Rüger N, Roloff A, Berger U (2013) Low relative growth rates predict future mortality of common beech (*Fagus sylvatica* L.). For Ecol Manage 302:372–378
Gompertz B (1825) On the nature of the function expressive of the law of human mortality, and on a new mode of determining the value of life contingencies. Phil Trans Royal Soc 115:513–585
Hunt R (1982) Plant Growth Curves. The Functional Approach to Plant Growth Analysis. Cambridge University Press, Cambridge
Hunt R (1990) Basic Growth Analysis: Plant Growth Analysis for Beginners. Unwin Hyman, London
Jones O, Maillardet R, Robinson A (2009) Introduction to Scientific Programming And Simulation Using R. Chapman & Hall/CRC, Boca Raton
Kangas Y (1968) Beschreibung des Wachstums der Bäume als Funktion ihres Alters. [Describing the growth of trees as a function of age.] Acta Forestalia Fennica, 90, Helsinki 1-136.
Matthews RW, Mackie ED (2006) Forest mensuration. A handbook for practioners. Forestry Commission, Edinburgh
Michaelis M, Menten ML (1913) Die Kinetik der Invertinwirkung. [The kinetics of invertase action.]. Biochem Z 49:333–369
Murphy ST, Pommerening A (2010) Modelling the growth of Sitka spruce (*Picea sitchensis* (Bong.) Carr.) in Wales using Wenk's model approach. Allgemeine Forst- und Jagdzeitung 181:35–43
Nicke A (1988) Aufbau eines Bestandesentwicklungsmodells. [Construction of a stand development model.]. Beiträge für die Forstwirtschaft 22:71–74
Pienaar LV, Turnbull KJ (1973) The Chapman-Richards generalization of von Bertalanffy's growth model for basal area growth and yield in even-aged stands. Forest Sci 19:2–22
Rees M, Osborne CP, Woodward FI, Hulme SP, Turnbull LA, Taylor SH (2010) Partitioning the components of relative growth rate: How important is plant size variation. Am Nat 176:E152–E161
Richards FJ (1959) A flexible growth function for empirical use. J Exp Bot 10:290–300
Rose KE, Atkinson RL, Turnbull LA, Rees M (2009) The costs and benefits of fast living. Ecol Lett 12:1379–1384
Turnbull LA, Paul-Victor C, Schmid B, Purves DW (2008) Growth rates, seed size, and physiology: Do small-seeded species really grow faster? Ecology 89:1352–1363
Verhulst P-F (1838) Notice sur la loi que la population pursuit dans son accroissement. [A note on population growth.]. Correspondence Mathematiques et Physiques 10:113–121
Weber R (1891) Lehrbuch der Forsteinrichtung. [Textbook of forest planning.]. Springer-Verlag, Berlin
Wenk G (1972) Zuwachsprognosen, Vorratsfortschreibung und Aufstellung bestandesindividueller Ertragstafeln mit Hilfe von Wachstumsmultiplikatoren. [Increment prognoses, projection of stand volume and stand-specific yield tables on the basis of growth multipliers.]. Wissenschaftliche Zeitschrift der Technischen Universität Dresden 21:1247–1249
Wenk G (1978) Mathematische Formulierung von Wachstumsprozessen in der Forstwirtschaft. [Mathematical formulation of growth processes in forestry]. Beiträge für die Forstwirtschaft 1:25–30
Wenk G (1994) A yield prediction model for pure and mixed stands. For Ecol Manage 69:259–268
Wenk G, Nicke A (1985) Zur Prognose der Bestandesentwicklung. [Projecting stand development.]. Beiträge für die Forstwirtschaft 19:173–176
Wenk G, Antanaitis V, Smelko S (1990) Waldertragslehre. [Forest growth and yield science.]. Deutscher Landwirtschaftsverlag, Berlin
Whitehead FH, Myerscough PJ (1962) Growth analysis of plants. The ratio of mean relative growth rate to mean relative rate of leaf area increase. New Phytologist 61:314–321
Zeide B (1989) Accuracy of equations describing diameter growth. Can J Forest Res 19:1283–1286
Zeide B (1993) Analysis of growth equations. Forest Sci 39:594–616
Zhang L, Moore JA, Newberry JD (1993) Disaggregating stand volume growth to individual trees. Forest Sci 39:295–308

18

Silvicultural alternatives to conventional even-aged forest management - what limits global adoption?

Klaus J Puettmann[1*], Scott McG Wilson[2], Susan C Baker[3], Pablo J Donoso[4], Lars Drössler[5], Girma Amente[6], Brian D Harvey[7], Thomas Knoke[8], Yuanchang Lu[9], Susanna Nocentini[10], Francis E Putz[11], Toshiya Yoshida[12] and Jürgen Bauhus[13]

Abstract

Background: The development of forestry as a scientific and management discipline over the last two centuries has mainly emphasized intensive management operations focused on increased commodity production, mostly wood. This "conventional" forest management approach has typically favored production of even-aged, single-species stands. While alternative management regimes have generally received less attention, this has been changing over the last three decades, especially in countries with developed economies. Reasons for this change include a combination of new information and concerns about the ecological consequences of intensive forestry practices and a willingness on the part of many forest owners and society to embrace a wider set of management objectives. Alternative silvicultural approaches are characterized by a set of fundamental principles, including avoidance of clearcutting, an emphasis on structural diversity and small-scale variability, deployment of mixed species with natural regeneration, and avoidance of intensive site-preparation methods.

Methods: Our compilation of the authors' experiences and perspectives from various parts of the world aims to initiate a larger discussion concerning the constraints to and the potential of adopting alternative silvicultural practices.

Results: The results suggest that a wider adoption of alternative silvicultural practices is currently hindered by a suite of ecological, economic, logistical, informational, cultural, and historical constraints. Individual contexts display their own unique combinations and relative significance of these constraints, and accordingly, targeted efforts, such as regulations and incentives, may help to overcome specific challenges.

Conclusions: In a broader context, we propose that less emphases on strict applications of principles and on stand structures might provide additional flexibility and facilitate the adoption of alternative silvicultural regimes in a broader set of circumstances. At the same time, the acceptance of alternative silvicultural systems as the "preferred or default mode of management" will necessitate and benefit from the continued development of the scientific basis and valuation of a variety of ecosystem goods and services. This publication is aimed to further the discussion in this context.

Review

Much of the history of forest science and management in the last two centuries has focused on optimizing the efficiency of commodity production, mostly of wood for timber, pulp, and fuel. Based on the notion that homogenous products are cheaper to produce and manipulate, management practices have typically led to even-aged, mono-specific or species-poor stands (Puettmann and Ek 1999, Paquette and Messier 2009, FAO 2010, West 2014). Such practices can include planting or natural regeneration, e.g., through coppice or natural or sown seed, combined with weed control and timber stand improvement practices to control species composition, tree size and quality; hereafter called "conventional" forest management. Currently, stands under conventional forest management for commodity production cover about 30% of

* Correspondence: Klaus.Puettmann@oregonstate.edu
[1]Department of Forest Ecosystems and Society, Oregon State University, Corvallis, OR 97330, USA
Full list of author information is available at the end of the article

the global forest land base (FAO 2010). At the same time, silvicultural approaches that can be viewed as alternatives to conventional management have been developed in various parts of the world, but have received less attention at a global scale (e.g., Gayer 1886, Kimmins 1992, Gustafsson et al. 2012, Lindenmayer et al. 2012). However, interest in these alternative approaches has greatly increased in many regions over the last three decades (Figures 1 and 2), particularly in countries with developed economies (Brukas and Weber 2009). Motivations for this recent development have varied across regions, but generally involve increased public scrutiny (Brunson and Steel 1994, Bliss 2000), impetus from forest certification bodies (Baker 2011), and better understanding of the implications of conventional forest management practices on biodiversity and habitat quality (Harlow et al. 1997, Manolis et al. 2000, Bauhus and Schmerbeck 2010), soils, water and nutrient cycling (Swank et al. 2001; Little et al. 2009), landscape visual quality (Ribe 1989; Paquet and Bélanger 1997), and other dimensions of natural capital and ecosystem services provision. Especially in countries with developed economies, public landowners, and, increasingly, many small private landowners do not necessarily rely on forests as their primary income source. Instead, other management goals, including recreational opportunities and landscape amenities, have gained importance (Butler and Leatherberry 2004; Hugosson and Ingemarson 2004; Urquhart and Courtney 2011).

The current variety of alternative silvicultural approaches (Figures 1 and 2) appears to have originated in two distinct contexts: in Europe and North America. Aesthetic and ecological critiques of conventional forest management during the period of 1850–1950 led to an emphasis on irregular size-class distribution and single-tree selection versus clearcuts, thus promoting the plenter forest (Plenterwald) model (Gurnaud 1882; Biolley 1901; Schütz 2002) and other "continuous cover" (Dauerwald) concepts (Möller 1923).

Other early European critiques highlighted concerns about the environmental effects of monocultures, and favored mixed-species stands (Gayer 1886). Similar concerns influenced management in other parts of the world, such as Japan (Marten 2005). Selection and shelterwood silvicultural systems suitable for the management of tropical moist forests were developed during the British and French colonial periods and were deployed in forests in India, Burma, Malaya, and West Africa until the 1960s. Technical development of these approaches has continued

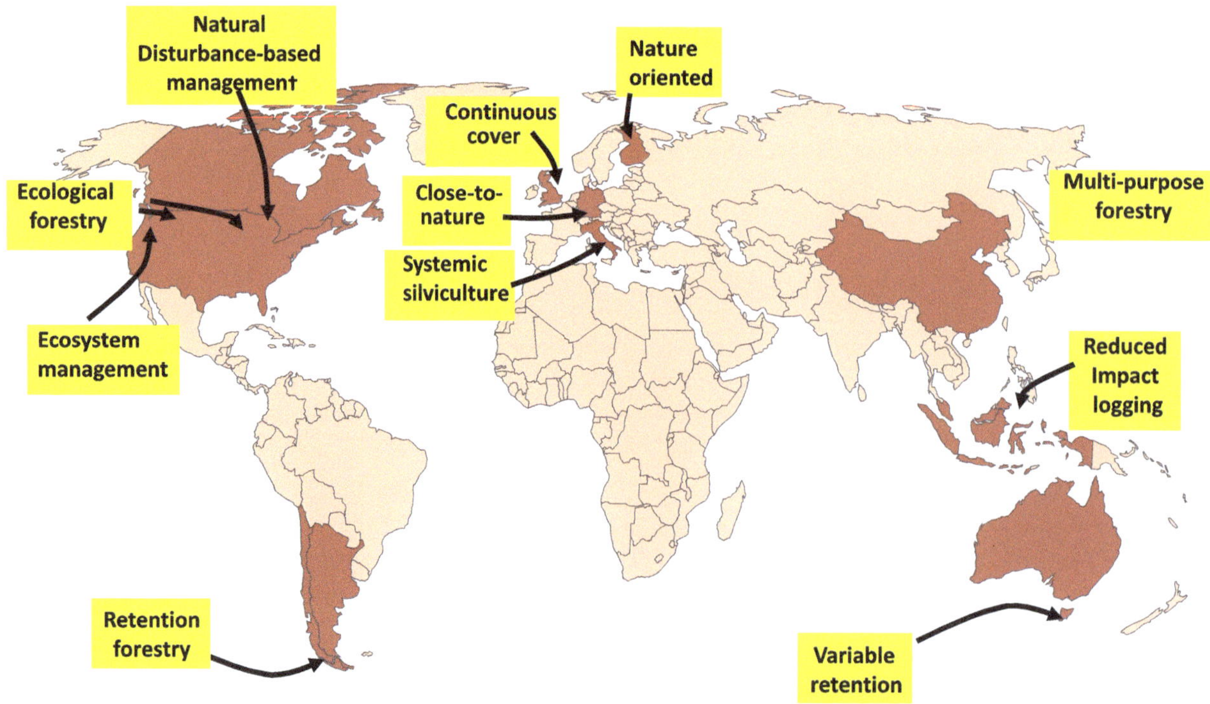

Figure 1 Examples of efforts to apply alternative silvicultural regimes. Countries with notable examples of alternative silviculture practices are in red. Associated text boxes display the common terminology used for alternative silvicultural practices in that region. *For more information about close-to-nature forestry in central Europe see Jacobsen (2001), continuous cover forestry see Mason et al. (2003), ecological forestry in the USA, see Franklin et al. (2007), ecosystem management in the USA see Kohm and Franklin (1997), multipurpose forestry in China see Lu et al. (2013), natural disturbance-based management in Canada see Bergeron et al. (2002), nature oriented forestry in Finland see Kuuluvainen (2009), reduced impact logging in southeast Asia see Putz et al. (2008), retention forestry in Chile see Donoso and Promis (2013), systemic silviculture in Italy see Ciancio and Nocentini (2011), and variable retention in Tasmania see Baker and Read (2011).

Figure 2 Examples of alternative silvicultural practices in different locations. **(a)** Ethiopia, natural regeneration of *Juniperus escelsa* in gaps, **(b)** Chile, recent restoration thinning in second growth *Nothofagus dombeyi* forests in Llancahue, **(c)** Japan, single tree cutting in natural *Quercus crispula* forest with *Abies sachalinensis* regeneration, **(d)** Germany, single-tree selection in mixed-species (*Abies, Picea, Fagus*) mountain forests, **(e)** Australia, aggregated retention in *Eucalyptus* forests in Tasmania, **(f)** Canada, removal of overstory *Populus tremuloides* to release and favor establishment of *Abies* and *Picea* species **(g)** England, silvicultural development in species-enriched *Pinus* and *Larix* plantations, **(h)** USA, variable density thinning with gaps and leave islands in Douglas-fir stands in Oregon, and **(i)** and **(j)** southern China, *Eucalyptus* overstory with *Pinus* underplanted between *Eucalyptus* sprouts and dominant *Pinus* overstory with *Quercus griffithii* and *Castanopsis hystrix* understory, respectively. Photo credit for **(f)** Philippe Duval, all other photos by the authors.

over recent decades, especially in Indonesia and Malaysia. The recent prominence in some European countries of Continuous Cover Forestry (e.g., UK and Ireland; Wilson 2013) and Close-to-Nature forestry (e.g., Germany, Switzerland, Slovenia, Italy) also reflects these historical influences (ProSilva 2014).

A second school of silvicultural approaches originated in the Pacific Northwest of North America under the banner of "New Forestry" (Franklin and Forman 1987) as a reaction to large-scale clearcutting in natural forest ecosystems and its impact on wildlife habitats, visual quality, and other ecosystem functions. Based on an increased understanding of how fire and other natural disturbances invariably leave patches of intact or partially affected forests in affected landscapes, this approach emphasizes the importance of residual structures and organisms (legacies), such as undisturbed vegetation patches, trees, and dead wood left behind after harvesting. These legacies are considered crucial for the speedy recovery of forest ecosystems after human-caused disturbances and thus for a continuation of ecosystem functioning and biodiversity and habitat values. Because this approach focuses on what is actually left behind after harvesting, it has spread to many regions of the world under the broad rubric of "retention forestry" (Lindenmayer et al. 2012).

A distinction between these different schools of alternative silviculture is the relative emphasis on management versus natural disturbances. For example, silviculturists in Central and Southern Europe tend to see a greater role for silvicultural interventions, reflecting the longer history of managed forests in their part of the world. By contrast, the lack of a long management history and an associated focus on natural disturbances in parts of North America and Australia are reflected in an emphasis on natural disturbance patterns as a template for alternative silviculture (Attiwill 1994; Drever et al. 2006; Long 2009). This is based on the notion that the historical range of variability in forest conditions can provide suitable guidance for management decisions (Higgs et al. 2014).

Despite their wide geographical distribution, in most contexts alternative silvicultural practices are not yet considered mainstream. They are often viewed and represented as less accepted alternatives to the conventional forest management paradigm in teaching and field practice (Puettmann et al. 2009). After long-standing arguments about alternative silvicultural systems, especially in the European literature of the last century, it appears that the debate is gaining momentum, as reflected by a number of recent summaries and overviews that discuss basic conceptual frameworks for these approaches, including complexity science (Puettmann 2011; Messier et al. 2013; Filotas et al. 2014), natural disturbance-based systems (Bergeron et al. 2002; Franklin et al. 2007), and legacies and retention (Gustafsson et al. 2012; Lindenmayer et al. 2012).

In many regions, these silvicultural alternatives to conventional forestry are enacted under a variety of social, economic, and ecological conditions and a wide array of ownership objectives, as well as legal and regulatory frameworks. While less applicable to highly intensive, short-rotation plantations (3% of the total forest area are covered by "productive plantations"; FAO 2010) and areas with the primary objective of biodiversity conservation (12%), silvicultural alternatives are most likely to be of interest on land that is designated primarily for either "commodity production" (30%), "multiple use" (24%), "soil and water protection" (8%), or "social services" (4%) (FAO 2010). These last four objectives may not be best achieved by adherence to any single management approach. Instead, a variety of local solutions have developed, including planted forests and forests that are naturally regenerated by seed or vegetative reproduction, reflecting the wide variety of ecological, economic, and social conditions and constraints of different jurisdictions. For an overview of the different efforts and discussions about labels, see Pommerening and Murphy (2004) and Evans (2006); also see Figure 1, including the notation with a list of references.

A basic distinction between conventional and alternative silvicultural approaches is the relative balance of selected values and objectives. Conventional approaches typically emphasize commodity production and view other objectives as constraints, e.g., intrinsic ecosystem values, accounting for natural processes, and maintaining species and structural diversity. In contrast, alternative silvicultural approaches place a unique set of emphases on each value. They regard all values, including non-commodity values, as a basic foundation necessary to achieve high levels and sustainable provision of ecosystem services, including product extraction (*sensu lato*) (Evans 2006).

This maturing view of forest management alternatives expresses itself in a set of five silvicultural principles (Jacobsen 2001; Schütz 2002; Mason et al. 2003; Bauhus et al. 2013):

- *Partial harvest*—use of partial harvesting and avoidance of large clearcuts
- *Natural regeneration*—where possible, preferential use of natural regeneration and native tree species
- *Structural diversity and small-scale variability*—varying management approaches across a range of spatial scales, with a special emphasis on diversity of stand structures at small scales, including single-tree and neighborhood conditions
- *Mixed species*—promotion of mixed-species stands
- *Avoidance of intensive operations*—minimization of intensive site preparation, fertilization, and weed control practices, and reliance on natural process such as self-thinning of seedlings and small saplings.

These principles correspond closely with the "low" and "medium" silvicultural intensity categories described in Duncker et al. (2012), in which management approaches are aligned along a gradient of intervention intensity.

When applying silvicultural alternatives, the focus on variability in stand structures across spatial scales typically results in an emphasis on the development and harvest of individual trees or small patches of trees. This principle may appear on the surface to be similar to selective exploitation harvesting approaches, where activities are limited to harvesting single high-value trees. However, alternative silviculture approaches, as discussed in this paper, are philosophically and practically quite distinct from such practices. In contrast to early selective exploitation, harvesting decisions in alternative silvicultural approaches are driven by an appreciation of the economic and ecological value of retained trees in terms of ensuring the continuity of ecosystem processes and functions. Thus, the application of all five principles aims to maintain or increase the growth potential of the site, conforming to traditional sustained-yield paradigms.

The potential advantages of alternative silvicultural approaches have been hailed in various dimensions, including in terms of biodiversity (Steventon et al. 1998; Fedrowitz et al. 2014), maintenance of "undeniably better resilience" (Schütz et al. 2012), general resilience (O'Hara and Ramage 2013) or, specifically, resilience to climate change (Schütz 1999a). However, limitations to specific adaptation strategies may exist; for example, the introduction of selected or improved genetic material may not be allowed or difficult to achieve in forests managed under these principles (Brang et al. 2014). Alternative approaches have also been claimed to provide other benefits compared to even-aged monocultures, albeit under specific sets of conditions. These conditions include more uniform cash-flow of harvest revenues (Knoke et al. 2001), higher income resulting from the production of high-quality wood (Hanewinkel 2002), more diverse wildlife habitats, especially for late-seral species (Hyvaerinen et al. 2006; Rosenvald and Lohmus 2008), increased wind stability of individual trees (Dvorák et al. 2001, but see Cameron 2002), and improved recreation value and public acceptance (Tönnes et al. 2004; Ribe 2005). This said, there is still much more work to be done in order to investigate and quantify the impacts under specific regional conditions (Gustafsson et al. 2010). Given these reported advantages, however, as well as the fact that many of the public and private landowners who have begun applying these principles are very satisfied with the results, "it is surprising that [alternative silvicultural approaches] are so seldom applied" (Schütz et al. 2012). In the following sections, we hope to provide a basis for a discussion about broader application of alternative silvicultural approaches and ask the following questions:

- What are the major factors that influence the decision to apply alternative silvicultural approaches?
- What is currently limiting the wider application of alternative silvicultural approaches at a global scale?
- What can be done to remove such limitations?

Methods

The authors have extensive experience dealing with research and teaching of alternative silviculture in various parts of the world. An initial collection of arguments by the authors was based on their own experience and familiarity with the relevant literature. This collection highlighted that the answers to questions about factors limiting wider application of alternative silvicultural practices vary between and within regions and between and within ownerships. To improve our insight into such highly complex decision-making environments and to better understand the decision factors, we found it useful to organize the assessment into a framework structured around the five silvicultural principles listed above (partial harvest, natural regeneration, small-scale variability in stand structures, mixed species, avoidance of intensive operations). Within these principles, we determined that the challenges to the implementation of these principles could be grouped into five categories and structured our results and discussion accordingly: ecological; economic; logistical and administrative; educational and informational; and cultural and historical. All authors repeatedly provided insights and feedback regarding the number and distinctions between the silvicultural principles and categories. At the same time, we went through repeated iterations in which the authors interacted to expand and organize the list of challenges. Thus, even though this does not comprise a systematic review of all existing literature (which would go beyond the scope of a single publication), we refined the list of challenges until we felt comfortable that all influential topics were covered in the discussion and provided a solid basis to initiate further discussions.

Results and discussion

Ecological challenges

The emphasis on continuous canopy structures (avoidance of clearcuts) typifies forests growing under natural disturbance regimes characterized by small scale disturbances, e.g., pathogen-induced mortality or mechanical damage (Lutz and Halpern 2006). In much of the world, the spatial scale of natural disturbance regimes can be represented by a J-shaped size frequency distribution, i.e., a combination of frequent small-scale and fewer large-scale disturbances (Veblen et al. 1981; Seymour and Hunter 1999; Shorohova et al. 2009; Yasuda et al. 2013). Examples of disturbances that generate these patterns include wildfires, e.g., in the western United States and Australia, or windstorms, e.g., in storm-prone regions of Western Europe (Schelhaas et al. 2003) or in the temperate forests of southern Chile (Veblen 1985). In such regions, a focus on any single silvicultural approach will only reflect a subset of patterns and dynamics found under natural conditions (Seymour and Hunter 1999). For example, region-wide applications of a management regime that focused only on continuous overstory cover would not promote conditions and patterns associated with severe, larger-scale disturbances, e.g., certain early successional habitat features. In such cases, the emphasis on smaller-scale variability in stand structure may reflect natural conditions found in individual stands, but it may not reflect the overall landscape-scale dynamics. Such concerns are currently discussed regarding the lack of early-seral habitat on federal land in the Pacific Northwest of the USA (Swanson et al. 2011; Franklin and Johnson 2012). Since natural disturbances such as fire and windstorms cannot be totally prevented or controlled in most forest landscapes, concern about the lack of selected disturbance effects may be of little practical relevance in the long run. Exceptions are situations where management artificially reduces the probability of occurrence of large-scale or stand-replacing disturbances. Examples of such situations include the application of short rotations that prevent trees from reaching heights common in older forests, and thus may reduce windthrow. Alternatively, density management aims to reduce tree stress to the remaining trees by ensuring higher light, water, and nutrient levels, and thus may limit competition related mortality (either directly due to drought or indirectly, e.g., insect related mortality due to low vigor, often found in high-density unmanaged stands (e.g., Neumann and Minko 1981). Also, fuel management may reduce the extent of large-scale fire disturbances. Hence, practices that change "natural" disturbance regimes must be carefully assessed in terms of their implications on landscape dynamics, and restoration may be required in some cases. For example, controlled burning as a restoration technique can be beneficial for threatened species in boreal forests where there has been extensive long-term fire suppression (Kouki et al. 2012).

Lack of large openings and early successional conditions in close-to-nature silvicultural systems operating with small canopy gaps can be detrimental to maintenance of light-demanding tree species in ecosystems where these species are mixed with more shade-tolerant species (e.g., Neyland et al. 2012; Bauhus et al. 2013). On the other hand, under selected conditions, such as low-density stands or species with open tree canopies, intensive management with repeated entries can facilitate survival and growth of light-demanding species, e.g., small-group uneven-aged management in Mediterranean pine forests (Ciancio et al. 1986; Barbeito et al. 2008; Calama et al. 2008). However, more often the higher mortality of light-demanding species in shady conditions and the slower responses of suppressed trees after silvicultural release can lead to an increasing dominance of shade-tolerant species (Lüpke and Hauskeller-Bullerjahn 1999; Kneeshaw et al. 2002; Drössler et al. 2013). These trends lead to concerns, as light-demanding species are often desirable in managed forests because of their fast growth rates, desirable timber properties, and dependent biodiversity components. They are also of broader interest under climate change conditions as they can be more drought tolerant than shade-tolerant species (Ninemets and Valladares 2006) and their fast initial growth rates and associated shorter rotations have the effect of reducing long-term risks.

The focus on small-scale variability in stand structure means that trees of different species and sizes grow in close proximity to each other. Such stands may have a lower probability of extensive crown fires, as they do not have continuous high density canopies. However, multilayered stand structures can produce ideal fire ladders (Schoennagel et al. 2004), a great concern in fire-prone areas. The spread of certain pests may also be facilitated in such intimate structures, e.g., large infected trees have the potential to infect nearby regeneration through higher exposure to inocula (Reich et al. 1991; Shaw et al. 2005). Tree-species diversity tends to restrict this threat, however, by reducing host-density for any species-specific pest or pathogen. In addition, smaller trees of one species may be less vulnerable to specific hosts when hidden by taller trees of another species (Castagneyrol et al. 2014). Similar arguments for species diversity at small spatial scales can be made in regards to other perturbations, as species differ in their sensitivity to wind, herbivory, and climatic stresses, such as drought or frost periods (for example, see the mixed species discussion in the *Economic Challenges* section below).

Natural regeneration with the consequent avoidance of the need for financial investments in artificial tree regeneration works well where competitive understory vegetation is sparse. Although dense stand conditions prior to regeneration cutting can reduce the amount of understory vegetation, numerous forest types in the world have understory vegetation that is highly shade tolerant and understory species that can respond very aggressively and competitively to canopy opening, even after partial harvests (e.g. Noguchi and Yoshida 2004; Donoso and Nyland 2005; Nilsson and Wardle 2005; Bose et al. 2014; Dodson et al. 2014). In these forests, regeneration of desirable tree species (especially, but not limited to shade-intolerant species) may be poor or delayed for years or even decades (Hibbs and Bower 2001; Axelsson et al. 2014).

In summary, ecological constraints underline the imperative of allowing more flexibility in adherence to the five principles in order to provide more opportunities for broader applications of silvicultural alternatives. Based on local ecological conditions, it may be necessary to deviate or "stretch" the principles. For example, larger canopy openings or more intense weed-control practices than preferred under alternative silvicultural approaches may be necessary in selected settings to ensure regeneration of desired tree species. Finally, acknowledging the full diversity of natural disturbance and development patterns may put the role of alternative silvicultural practices in a larger perspective.

Economic challenges

Economic considerations are generally strong drivers of land-use decisions, as already made clear by Thünen's (von Thünen 1842) theory on the spatial distribution of forestry and agriculture. A strong focus on economics may not always apply to forestry, where ecological considerations, tradition, social acceptability, culture, lifestyle, and similar factors may dominate management decisions (Kimmins 1992) and long-term economics of decisions are not always clear. However, it is unlikely that alternative silvicultural strategies will be adopted on large scales if they are not financially compatible with economic sustainability.

Economic considerations include the expected profit, uncertainty, and flexibility of management alternatives. The outcome of such economic considerations strongly depends on the specific settings of the land ownership and the services included into the decision matrix (Amsalu et al. 2014). Any application of alternative silvicultural approaches has to be viewed in the context of other management options to consider the opportunity costs for the final silvicultural decisions appropriately. In many places, especially on larger private holdings in the tropics, subtropics, and temperate regions of the Southern Hemisphere, the context includes the economic expectations from short-rotation plantation forestry, generally with exotic tree species, and often aided by political and economic policies (e.g., Niklitschek 2007; Miller Klubock 2014). Even under optimistic expectations, alternative silvicultural approaches are unlikely to reach the short-term financial profitability of intensively managed short-rotation plantations in tropical regions or some temperate regions (Cubbage et al. 2007; Knoke and Huth 2011). In some settings, however, native tree species may compete with exotic tree species when managed in a short-rotation plantation system (Cubbage et al. 2007; Griess and Knoke 2011).

The economic picture for alternative silvicultural methods is more favorable when viewed in a longer-term context, such as typically used in temperate and boreal biomes. For example, Tahvonen et al. (2010) showed the importance of relying on natural regeneration to reduce regeneration costs as a critical criterion that allowed continuous-cover forestry to emerge as the option with the highest net present value in Finnish spruce forests, under nearly all combinations of assumptions used in the study (Note that the study assumed higher regeneration costs due to artificial regeneration in even-aged management regimes). Many other studies support the findings published by Tahvonen et al. (2010) in principle, for example Chang (1981); Haight (1987); Haight and Monserud (1990); Pukkala et al. (2010) and Hyytiäinen and Haight (2010). Another argument often presented in the discussion of economic outcomes of alternative silvicultural practices is that individual trees can be harvested at the time when they peak in terms of their economic values. This can increase the total net present value substantially in stands with an existing wide range of tree values (as determined by species, size, and wood quality) (Knoke 2012). The advantage of harvesting trees at their peak economic value - versus harvesting all trees in a stand at the same time - may be reduced or lost in stands where trees are very homogenous (also, see the discussion about harvesting costs and staffing needs). Thus, it is important to note that the economic attractiveness of alternative silviculture is conditional on the initial state in the short run (Tahvonen et al. 2010). This makes generalizations about comparisons between conventional and alternative silvicultural approaches not very useful for landowners who consider such options for a specific property. The influence of the initial state can be minimized by using a very long (at best an unlimited) time horizon in comparative economic analyses; a requirement that certainly poses additional challenges.

The economic comparison may also change towards alternative silviculture when these approaches result in an increased resilience to natural hazards, e.g., in mixed-species even-aged (Griess et al. 2012; Neuner et al. 2015) or uneven-aged stands (Hanewinkel et al. 2014). Alternative silvicultural approaches may also be more suitable for forest owners with concerns about economic risks, as the risk is lower in stands with higher resilience to disturbances and when timber harvesting is accomplished with multiple operations over time (Roessiger et al. 2013).

Finally, increased stand-level ecological stability in mixed and uneven-aged forests also enhances economic performance of forest operations through improved flexibility because harvest levels can be more easily adjusted to accommodate changes in market conditions (Knoke and Wurm 2006). Management flexibility may be of special importance in a more unpredictable future, e.g., with climate change (Jacobsen and Thorsen 2003; Wagner et al. 2014) or novel pest incidences. Consequently, major issues in economic assessments of silvicultural approaches are the relative emphases on economic risks and management and marketing flexibility.

One argument often used to promote application of single-tree selection and also for alternative silvicultural approaches is the benefit of the production of high value, i.e., large diameter and high-quality trees (e.g., "Dutch cuttings", see Schütz et al. 2012) (often in conjunction with the decreased need for non-commercial thinning operations). As markets have shifted, and in regions where much of the wood-processing industry has re-tooled to efficiently process smaller logs, large logs now may have no premium (per unit of volume), and in some cases may even have price penalties (Fitzgerald et al. 2013). This issue will vary by species and may be alleviated by sawmilling innovation in response to a higher availability of predicted assortments from alternative silviculture. In the meantime, in many regions, the theoretically higher mill recovery from larger diameter logs is often offset by the faster speed when processing homogenous (in terms of size and quality) logs, especially in high-volume sawmills. Concerns also exist about lower log quality for trees that grow in highly variable conditions throughout their lives. This variability can exhibit itself in somewhat adverse branching patterns, including large branch sizes and higher variation in juvenile wood, wood density, and growth rings (Schütz 1999b; Macdonald et al. 2010; Piispanen et al. 2014).

Harvesting costs per unit volume for similar sized trees are typically higher in partial harvesting operations, especially on steep slopes and in mixed-species stands when different species have to be sorted and marketed separately (Pausch 2005). These higher costs are partially due to reduced efficiency of logging when only small amounts are harvested in a single operation. Early adopters also may incur higher costs and time delays when hiring and training specialized logging contractors. The necessary investments into infrastructure, including the development and maintenance of road systems, may also be a financial barrier to managing alternative silviculture, especially for landowners with smaller properties. Similarly, compared to conventional approaches, the proper implementation of alternative silvicultural approaches may result in higher personnel costs due to the higher skill level required, as well as more frequent site visits to assess progress, mark trees, and develop and supervise the implementation of often more complicated management prescriptions. Another economic concern with partial harvesting may be the risk of physical damage to the retained growing stock. This issue is of special concern in difficult terrain and when large-crowned species are felled or species with little decay resistance are left behind (Nill et al. 2011).

In summary, the ability to rely on natural regeneration, the limited need for site preparation and weed-control operations, in conjunction with a higher proportion of larger, high quality trees at harvest time, are viewed as major economic advantages of alternative silvicultural approaches. Thus, ecological conditions that influence natural regeneration, such as seed production, herbivory levels, and competitive conditions in the understory are crucial factors that can determine the financial viability of alternative silvicultural approaches. Risk and flexibility considerations may also constitute strong arguments in favor of alternative silvicultural approaches. However, the economies of scale of industrialized short-rotation management may more than balance these advantages of alternative silviculture in many parts of the world in the short-term, where the production of large amounts of homogenous timber is the main aim of forestry.

Finally, ecosystem services that may be provided more efficiently in forests managed under alternative silvicultural approaches, e.g., visual quality, hydrological services, or wildlife habitat provision, are not necessarily adequately monetized or evaluated (e.g., Ribe 1999; Daniel 2001). Whether or not such values compensate for a potential loss in timber production after conversion to alternative silviculture regimes will vary with the specific situation and assumptions, e.g., soil quality and interest rates, respectively (Amsalu et al. 2014). In many situations, the application of alternative silviculture would only be economically viable through measures such as conservation payments (Cubbage et al. 2007), monetary incentives as part of forest certification efforts (Nasi et al. 2011), or direct payments for ecosystem services (Wunder 2006), such as water supply and regulation (Nahuelhual et al. 2007).

Logistical and administrative challenges

Efficient partial-harvesting operations may require highly trained loggers and specialized machinery. Both can be in scarce supply, particularly in regions where these practices are only emerging. In some areas, forest managers may be able to benefit from loggers with experience in selective harvesting in other settings, such as exploitation cuttings that selectively removed only the highest value trees. Using inexperienced loggers and unsuitable machinery that is not designed for partial harvest operations increases the probability of inadequate prescription application and harvesting damage to residual trees (Jones and Thomas 2004; Putz et al. 2008), both outcomes that can reinforce a perception that alternative silvicultural approaches are inoperable. The damage to retained trees can be very high also in regions with a longer tradition of uneven-aged forest management (Nill et al. 2011). Moreover, advanced regeneration is often cut to improve the visibility for machine operators during harvest procedures or damaged during skidding or forwarding operations (Newton and Cole 2006). In contrast, these concerns can be less prevalent in parts of the world where logging operations are not highly mechanized. For example, manual logging operations with extraction by horses or oxen can be done with relatively little damage to residual trees and regeneration (Bacardit 2014). However, there are notable exceptions, for example where logs are thrown downhill by forest workers through forests on steep slopes, as has been practiced in some parts of China (J. Bauhus, personal observations). Even in stands with little harvesting damage, sudden exposure after partial harvesting can lead to higher evaporative demands (Bladon et al. 2006) and unexpected mortality of residual tress (Yoshida et al. 2006; Aubry et al. 2009).

Health and safety considerations are always important during and after partial harvesting operations due to concerns about the stability of branches and retained trees (Schaetzl et al. 1989). The presence of dead branches or forked trunks leads to safety concerns, but can be taken into account when selecting residual trees. Past scarring, possibly during earlier harvest operations, can also provide indications of possible trunk rot and thus the likelihood of tree fall (Matlack et al. 1993). Training can also reduce workplace safety risks. For example, the injury rate dropped - rather than increased - when variable retention harvesting replaced clearcutting in coastal British Columbia, because of strong organizational focus on worker safety. In contrast, at a larger scale, forests with continuous cover are preferred because of their safety benefits for avalanche protection (Bebi et al. 2009) and protection from landslides (Dhakal and Sidle 2003; Noguchi et al. 2011).

Changing management practices to alternative silvicultural approaches requires the development of different expertise by existing staff and additional educational and training efforts (see below) or, possibly, restaffing. Applying repeated partial-harvesting operations with decisions made at the individual-tree level may initially require a larger workforce until new training programs take effect to reduce or offset the workload. Planning and monitoring processes and inventory systems may have to be modified to more efficiently consider and accurately track the development of tree and stand attributes for anticipated future treatments (Rice et al. 2014).

Alternative silvicultural benefits from continuity of ownership and supervisory management personnel who can oversee the consistent application of silvicultural prescriptions. For example, these approaches may only be feasible if concessions are agreed on beyond the scope of a single harvest to include repeated harvesting operations, and if concessionaires are held responsible for reforestation and other silvicultural operations. Especially on larger private ownerships and in areas with limited infrastructure, the supervision of logging activities is necessary to avoid unauthorized or accidental removal of higher-valued trees intended for retention. Furthermore, the higher diversity of forest conditions (e.g., tree-species combinations – literally dozens in tropical forests - and various stand structural conditions), and the associated lack of simple, standardized instructions and management protocols make it more difficult to cope with rapid turnover of staff. Even where scientific data and understanding already exist from long-term silvicultural trials, foresters often appear to heavily rely on their observations, intuition, and experience when applying alternative silvicultural approaches. Developing and fostering these skills in new personnel is challenging for forest administrators and land owners.

In summary, staffing challenges will require increased training efforts, e.g., in reduced impact logging (RIL, Tropical Forest Foundation 2008). New machinery and modification of harvesting practices may help overcome challenges when implementing alternative silvicultural practices. New technologies for planning and monitoring (e.g., GPS, GIS, remote sensing tools) will reduce the need for staff, but will require a more highly qualified workforce. Finally, the modification of contracting practices, such as involving concessions, may be necessary in many places to facilitate a transition to alternative silvicultural systems.

Informational and educational challenges

Over the last century, many educational programs in forestry schools and continuing professional education programs have focused primarily on conventional forestry practices and have often provided limited exposure to alternative management approaches such as selection systems (Schütz et al. 2012). One reason for this has been the decline in the availability of experienced silvicultural professionals to impart this knowledge. Practical examples of alternative silviculture that can be used as demonstration projects for students and practitioners are still lacking in many regions. This is true especially in countries with developing economies, following the decline of previous colonial and overseas development forestry practices. Related to the lack of practical field examples is the limited capture of traditional ecological knowledge of some indigenous communities (Berkes et al. 2000), although this is changing in some regions. For example in the Maya region of Mexico, new silvicultural systems for the regeneration of *Swietenia macrophylla* were developed as a result of collaborations among forest scientists and local communities (Menzies 2007).

The amount and type of scientific information available to support alternative silviculture reflect the bottom-up development of these approaches. In contrast to conventional forestry operations, which are based on the agricultural model or efficiency paradigm (Puettmann et al. 2009), alternative silvicultural approaches have not received the same level of attention by scientists. Studies that document the long-term response to alternative silvicultural treatments in terms of the full suite of ecosystem goods and services are lacking. Instead, in many instances, foresters adopted these approaches after observing trends in their own forests or visiting forests on which alternative silviculture had already been successfully applied. Thus, the spread of silvicultural alternatives is related to the availability of examples as well as to scientific progress. The limited awareness and utilization of scientific studies covering relevant aspects, such as the physiological basis for seedling and sapling responses to overstory removals or natural regeneration of species mixtures (Puettmann and Ammer 2006), remain a constraint in many regions, particularly in the tropics. New field installations, simulation exercises (Yasuda et al. 2013), and other research programs are currently beginning to provide valuable results (e.g., Thorpe et al. 2010; Neyland et al. 2012). At this time, organizations and forest managers may be (understandably) reluctant to take risks and accept uncertainties associated with trying out unfamiliar management approaches without being able to rely on extensive scientific literature and empirical results from field experiences. This is likely one of the major limitations for a more rapid spread of alternative silviculture.

Just as for forest managers, training opportunities are often lacking (or too expensive) for loggers to obtain information and experience with these new management systems. Successful logging operations require a workforce with a good understanding of the reasons for and basic principles behind different silvicultural treatments. Harvesting operators may also have limited access to innovative machinery and techniques suitable for partial harvesting operations that address small-scale variability in stand structure, are able to deal with a wider variety of species, tree sizes, and qualities, and are designed to protect retained trees (including natural regeneration and site and soil conditions).

Cultural and historical challenges

Schütz et al. (2012) wrote that "willingness" (defined as "readiness to do something" by Merriam-Webster 2014) is one of the main factors limiting the wider application of alternative management regimes. In addition to the reasons for the lack of "readiness" that have been discussed already, the role of strong intellectual, political, and administrative leadership is especially important in shaping public and professional opinions (Brukas and Weber 2009). The degree to which an organizational leadership champions alternative approaches influences the degree to which employees and contractors embrace and apply the practices. Strong support from upper management encourages employees and contractors to overcome operational challenges, as well as prejudices, enabling successful transition to alternative silviculture (Baker 2011). Other reasons may have a deeper, historical background. For example, the expansion of "scientific forestry" (*sensu* Oosthoek 2007; see also Lowood 1990), reflected in controlled management of homogenous, even-aged, mono-cultural plantations from central Europe at the end of the 19th century, was accepted in other countries because it fitted prominent economic ideas about control and efficiency (Lang and Pye 2001; Oosthoek 2007; Cock 2008; Puettmann et al. 2009; Sears and Pinedo-Vasquez 2011). This approach also helped to elevate forestry as an autonomous discipline at universities and research institutions, which facilitated its professional recognition by foresters and society in general. This historical development has led to several phenomena that are still influential today. The resulting expectation of simple, practical operational guidelines cannot easily be met by alternative silvicultural approaches, especially when they emphasize greater silvicultural flexibility (and outcomes) in a wide diversity of forest structures and species compositions. Moreover, the educational focus on conventional forestry has led to "imprinting" or "knowledge lock-in" in many practicing foresters who may consider only the predominant management model as acceptable. An example of this can be found in the management history of beech and black pine forests in the mountains of southern Italy, where selection felling based on traditional local knowledge and applied by private landowners during the last few centuries has been and often still is considered "irrational" by forest administrations and technicians (Ciancio et al. 2006; Nocentini 2009). Thus, the professional and public perceptions of alternative silvicultural approaches may have a major influence on the development of policies and implementation strategies (Brukas and Weber 2009).

Forest policies and practices can still be biased against continuous-forest-cover systems. For example, in Finland and Sweden, forest authorities and professionals have actively encouraged clearcut operations because selective cutting methods applied in the past led to political concerns that private forest owners were devastating their forests (Siiskonen 2007; Brukas and Weber 2009). In 1948, a group of influential Finnish researchers published a declaration against the destructive impacts of selective logging on Finland's forests. Similarly, influential German forestry professors were successful in discrediting the "Dauerwald" (continuous-cover forests) efforts, mainly by stating concerns about the economic performance and management complexity (Pommerening and Murphy 2004). These arguments may exacerbate an existing lack of incentives or desire to switch to alternative silvicultural approaches, given that they are still brought up in current discussions of new laws and regulations.

The "search for simplicity" in silviculture is often also encouraged by staff shortages and overly complex, computerized desk-work environments, where foresters have to deal with multifarious and new, distracting aspects. The reluctance to try more complicated silvicultural prescriptions is also partly the result of highly visible failures of alternative silviculture experiences in some regions. For example, early partial-harvest trials in old-growth *Pseudotsuga menziesii* stands in the Pacific Northwest were pronounced a failure in the 1950s. As noted later by Curtis (1998), "partial cutting trials came to an abrupt end", leading to "the consequent lack of research into alternatives to clearcutting", which "severely handicaps current efforts to meet changing objectives and public concerns". Similarly, "European" silvicultural systems and practices were initially implemented in Australia (and other places around the globe) without an adequate understanding of the ecological functioning of the local ecosystems (Florence 1996). For a long time, silvicultural alternatives were only applied on selected properties, which were viewed as "unique" and in some cases even belittled, making it harder for a broader set of professionals to associate themselves with these practices.

Decisions about management activities are typically the domain of landowners acting within their local ecological, social, and economic contexts. Thus, it is not enough to educate foresters and contractors; it is also necessary to convince private landowners and politicians who have decision power over public land use and policies that impact private landowners. The attitudes of these decision makers regarding alternative silvicultural regimes may be influenced by a variety of factors, including their knowledge and prioritization of desired ecosystem goods and services, as well as the anticipated revenue constraints due to size of the estate. For example, large public or investor-owned properties can afford to spatially separate the production of various forest products or ecosystem services over large areas. This allows them to homogenize portions of their forest holdings for the efficient production of selected forest products or services. In contrast, landholders who manage small and/or fragmented properties may need to obtain a variety of products from their property. In Finland, landowners with small forest land bases often obtain logs of different sizes for firewood or for construction wood and at the same time may harvest berries, mushrooms, and game (Siiskonen 2007). Similarly, in Latin America, campesinos (rural smallholders) may work under the logic of a diversified production to mitigate potential risks. For example, they collect firewood, produce charcoal, sell animals, and produce apple cider—activities that also generate income that may be complemented with off-farm income (Moorman et al. 2013). Alternative silvicultural regimes are typically more suitable in such settings where integrated land use is important because their emphasis on smaller spatial scales facilitates simultaneous production of various ecosystem goods and services (Hein et al. 2006).

The regional or national context also influences decisions about silvicultural approaches. For example, high human population densities may increase the importance of providing high-amenity recreational settings as an ecosystem service, leading to alternative silvicultural approaches that are more visually appealing (Rydberg and Falck 2000). Similarly, mountainous topographies may lead to an emphasis on protection from landslides, e.g., in Japan (Noguchi et al. 2011) and Italy (see discussion about safety above). On the other hand, a well-established and economic and politically powerful forest industry sector can influence the availability of wood processing facility and thus markets for forest products (Miller Klubock 2014). This may limit or bias market accessibility towards products (tree species, sizes, qualities) derived from industrial forest lands through conventional management, effectively reducing the value of a variety of products that could come from forests managed under alternative regimes (for example, see the above discussion about the impact of log sizes on wood value). A strong forest industry sector may also directly or indirectly influence the focus of educational and research themes according to their management interests (Benner and Sandström 2000).

A long history of forest management also means that private and public ownerships are organized and staffed to accommodate conventional approaches, making any transition more difficult. Furthermore, some of the challenges mentioned above may actually be legally embedded in forest policy and regulations. For example, well-intentioned reforestation regulations may specify acceptable tree species and required seedling densities, effectively limiting the choice of species and discouraging natural regeneration (for example, see the Oregon Forest Practices Act, Oregon Department of Forestry 2014). It should be acknowledged that not only environmental conditions, silvicultural options, and markets determine the implementation of alternative management methods, but traditions and stakeholders' views also affect ongoing political discourses (Arts et al. 2010).

Conclusions

In summary, the diversity of settings in which alternative silvicultural approaches are of interest reflects the diversity of challenges that limit their wider implementation. The range of arguments presented here suggests that ecological, economic, logistical and administrative, educational and informational, and cultural and historical challenges are all important. The relative importance of any of these challenges varies among and within geographical regions and ownerships. We suggest that these challenges can be addressed at two levels. At a lower level, each situation can be analyzed for the specific major challenges or limitations. Any such assessment of challenges benefits from viewing alternative regimes as a collection of silvicultural principles. Knowing which principle specifically is related to the major challenge in the situation at hand provides detailed insights that will facilitate discussion of whether that specific principle is necessary or can be modified or ignored, while still achieving overall management objectives. For example, foresters often will forgo natural regeneration if they foresee regeneration failures, and they will plant seedlings—even in close-to-nature forest operations. Having such detailed assessments will allow interested groups, for example landowners or politicians, to initiate efforts to implement policies, procedures, and/or incentives to help overcome these challenges. Examples could include new laws or regulations, subsidies, and scientific and educational efforts that are targeted to specific situations. Although this may be successful in individual settings, it may be an inefficient process for tackling overall problem areas.

At a broader level, our overview provides some general insights that suggest opportunities for more coordinated efforts. It appears obvious that in this context, efforts that allow foresters to take advantage of the widest possible array of silvicultural tools to accommodate a wide variety of local ecological, economic, and social conditions are especially useful. This would be facilitated by a more formal shift of alternative silvicultural approaches away from the focus on desired or acceptable stand structures, often defined as specific diameter distributions or other visual targets, such as photos or sketches of desired stand structures (Larsen and Nielsen 2007) (Figure 2). Instead, a more targeted focus on the suite of desired ecosystem goods and services will allow foresters to integrate a broader set of disturbances, stand dynamics, and successional trends into their management practices. Such a focus will allow more flexibility in the future and encourage practices to be "customized" to optimally fit specific ecological, economic, and social settings without the confining restrictions of labels.

For further progress, especially to overcome information deficits and educational, economic, cultural, and historical challenges, the spread of alternative management approaches would benefit greatly from the continued development of a scientific basis. This basis includes investigations of specific detailed ecological, social, economic, and ethical issues. In addition, developing a larger-scale scientific framework that provides more explicit linkages between basic scientific theories and principles and practical applications of alternative silvicultural approaches will also facilitate progress by allowing better coordination—including internationally—of research and educational activities. This will facilitate an understanding of the distinctions between idiosyncrasies and regularities. It also will help provide answers to questions such as whether experiences are unique to specific situations or how far experiences in one forest can be extrapolated into other settings—questions that were already at the heart of the Dauerwald discussions almost 100 years ago. Furthermore, an expanded scientific basis will be very helpful in developing a coherent and efficient educational and—maybe most important—outreach and policy agendas. Finally, great economic challenges remain for the creation of economic modeling approaches that are more biologically realistic and thus capture the ecological complexity of alternative silvicultural management approaches.

While not a formal review of all pertinent literature, the expertise represented by the authors covering a variety of regions and ecological and social settings allowed us to build a comprehensive framework to provide hypotheses and suggest opportunities to advance the implementation of alternative silvicultural regimes. However, we acknowledge that experiences in specific settings may differ or complement the ones provided by us. We hope that our arguments and such differences will provide the basis for further fruitful discussion.

Competing interests

The authors declare that they have no competing interests.

Authors' contributions

KP initiated the work and coordinated the contributions from the co-authors. SW, SB, PD, LD, GA, BH, TK, YL, SN, FP, TY, and JB provided input into the analytical structure and contributed to writing and revising the manuscript. All authors read and approved the final manuscript.

Acknowledgements

This material is based upon work supported by the National Institute of Food and Agriculture, United States Department of Agriculture, McIntire-Stennis Funds under ID number OREZ-FES-850-P to the lead author and by the Edmund Hayes Professorship. PJ Donoso acknowledges the support from FONDECYT Grant No 1110744. Jürgen Bauhus received funding for this work from the European Union Seventh Framework Programme (FP7/2007-2013) under grant agreement no. 265171. The authors would like to acknowledge many colleagues, practitioners, and students for their contributions.

Author details

[1]Department of Forest Ecosystems and Society, Oregon State University, Corvallis, OR 97330, USA. [2]Consultant Forester and Forest Ecologist, 3 Thorngrove Crescent, Aberdeen, Scotland AB15 7FH, UK. [3]School of Biological Sciences, University of Tasmania, Private Bag 55, Hobart, TAS 7001, Australia. [4]Department of Forests and Society, Faculty of Forest Sciences and Natural Resources, Universidad Austral de Chile, Isla Teja s/n, Valdivia, Chile. [5]Southern Swedish Forest Research Centre, Swedish Agricultural University, PO Box 49, Alnarp SE-23053, Sweden. [6]Haramaya University, P. Box 138, Diredawa, Ethiopia. [7]Institut de recherche sur les forêts, Université du Québec en Abitibi-Témiscamingue, 445, Boulevard de l'Université, Rouyn--Noranda, Québec J9X 5E4, Canada. [8]Institute of Forest Management, TUM School of Life Sciences Weihenstephan, Technische Universität München, Hans-Carl-von-Carlowitz-Platz 2, Freising 85354, Germany. [9]Department of Forest Management and Statistics IFRIT, Chinese Academy of Forestry, Dongxiaofu 2, Xiangshan Road, Beijing, Haidian 100091, China. [10]Department of Agricultural, Food and Forestry Systems, University of Florence, v. S. Bonaventura 13, Florence I-50145, Italy. [11]Department of Biology, University of Florida, P.O. Box 118526, Gainesville, FL 32611-8526, USA. [12]Uryu Experimental Forest, Field Science Center for Northern Biosphere, Hokkaido University, Moshiri, Horokanai 074-0741, Japan. [13]Chair of Silviculture, Faculty of Environment and Natural Resources, University of Freiburg, Tennenbacherstr. 4, Freiburg 79085, Germany.

References

Amsalu DW, Jacobsen JB, Lundhede TH (2014) Economic assessment of use values of near-natural forestry compared with rotational forestry in Denmark. Eur J For Res 133(4):611–622

Arts B, Appelstrand M, Kleinschmit D, Pülzl H, Vissen-Hamakers I (2010) Discourses, actors and instruments in international forest governance. IUFRO World Ser 28:57–73

Attiwill PM (1994) The disturbance of forest ecosystems - the ecological basis for conservative management. Forest Ecol Manag 63:247–300

Aubry KA, Halpern CB, Peterson CE (2009) Variable-retention harvests in the Pacific Northwest: a review of short-term findings from the DEMO study. Forest Ecol Manag 258:398–408

Axelsson EP, Lundmark T, Högberg P, Nordin A (2014) Belowground competition directs spatial patterns of seedling growth in boreal pine forests in Fennoscandia. Forests 5(9):2106–2121

Bacardit P (2014) Evaluación del daño provocado por cortas de selección sobre los árboles residuales en un bosque del Tipo Forestal Siempreverde en Llancahue, provincia de Valdivia. Undergraduate Thesis, Faculty of Forest Sciences and Natural Resources, Universidad Austral de Chile, Valdivia, Chile

Baker SC (2011) Seeking a Balance Between Forestry and Biodiversity – The Role of Variable Retention Silviculture. Insights from Western USA and Canada. FWPA Project Report PG D167-0910. FWPA, Melbourne, p 60

Baker SC, Read SM (2011) Variable retention silviculture in Tasmania's wet forests: ecological rationale, adaptive management and synthesis of biodiversity benefits. Austral For 74:218–232

Barbeito I, Pardos M, Calama R, Cañellas I (2008) Effect of stand structure on Stone pine (*Pinus pinea* L.) regeneration dynamics. Forestry 81:617–629

Bauhus J, Schmerbeck J (2010) Silvicultural Options to Enhance and Use Forest Plantation Biodiversity. In: Bauhus J, van der Meer P, Kanninen M (eds) Ecosystem Goods and Services from Plantation Forests. Earthscan, London, pp 96–139

Bauhus J, Puettmann KJ, Kühne C (2013) Close-to-nature forest management in Europe: does it support complexity and adaptability of forest ecosystems? In: Messier C, Puettmann KJ, Coates KD (eds) Managing forests as complex adaptive systems: building resilience to the challenge of global change. Routledge, New York, pp 187–213

Bebi P, Kulakowski D, Rixen C (2009) Snow avalanche disturbances in forest ecosystems—state of research and implications for management. Forest Ecol Manag 257(9):1883–1892

Benner M, Sandström U (2000) Institutionalizing the triple helix: research funding and norms in the academic system. Res Policy 29:291–301

Bergeron Y, Leduc A, Harvey B, Gauthier S (2002) Natural fire regime: a guide for sustainable forest management of the Canadian boreal forest. Silva Fennica 36:81–95

Berkes F, Colding J, Folke C (2000) Rediscovery of traditional ecological knowledge as adaptive management. Ecol Appl 10:1251–1262

Biolley H (1901) Le jardinage cultural. J Forestier Suisse 52:67–104

Bladon KD, Silins U, Landhäusser SM, Lieffers VJ (2006) Differential transpiration by three boreal tree species in response to increased evaporative demand after variable retention harvesting. Ag For Meteorol 138:104–119

Bliss JC (2000) Public perceptions of clearcutting. J Forestry 98:4–9

Bose AK, Harvey BD, Brais S, Beaudet M, Leduc A (2014) Constraints to partial cutting in the boreal forest of Canada in the context of natural disturbance-based management: a review. Forestry 87(1):11–28

Brang P, Spathelf P, Larsen JB, Bauhus J, Boncina A, Chauvin C, Drössler L, Garcia-Guemes C, Heiri C, Kerr G, Lexer MJ, Mason B, Mohren F, Mühlethaler U, Nocentini S, Svoboda M (2014) Suitability of close-to-nature silviculture for adapting temperate European forests to climate change. Forestry 87(4):492–503, doi:10.1093/forestry/cpu018

Brukas V, Weber N (2009) Forest management after the economic transition—at the crossroads between German and Scandinavian traditions. For Pol Econ 11:586–592

Brunson MW, Steel BS (1994) National public attitudes toward federal rangeland management. Rangelands 16:77–81

Butler BJ, Leatherberry EC (2004) America's family forest owners. J Forestry 102:4–14

Calama R, Barbeito I, Pardos M, del Rio M, Montero G (2008) Adapting a model for even-aged *Pinus pinea* L. stands to complex multi-aged structures. Forest Ecol Manag 256:1390–1399

Cameron A (2002) Importance of early selective thinning in the development of long-term stand stability and improved log quality: a review. Forestry 75(1):25–35

Castagneyrol B, Jactel H, Vacher C, Brockerhoff EG, Koricheva J (2014) Effects of plant phylogenetic diversity on herbivory depend on herbivore specialization. J Appl Ecol 51(1):134–141

Chang SJ (1981) Determination of the optimal growing stock and cutting cycle for an uneven-aged stand. Forest Sci 27:739–744

Ciancio O, Nocentini S (2011) Biodiversity conservation and systemic silviculture: concepts and applications. Plant Biosyst 145:411–418

Ciancio O, Cutini A, Mercurio R, Veracini A (1986) Sulla struttura della pineta di pino domestico di Alberese. Annali dell'Istituto Sperimentale della Selvicoltura, Arezzo, p 17

Ciancio O, Iovino F, Menguzzato G, Nicolaci A, Nocentini S (2006) Structure and growth of a small group selection forest of Calabrian pine in Southern Italy: a hypothesis for continuous cover forestry based on traditional silviculture. Forest Ecol Manag 224:229–234

Cock AR (2008) Tropical forests in the global states system. Int Aff 84(2):315–333

Cubbage F, MacDonagh P, Sawinski J Jr, Rubilar R, Donoso PJ, Ferreira A, Hoeflich V, Morales V, Ferreira G, Balmelli G, Jacek Siry M, Báez N, Alvarez J (2007) Timber investment returns for selected plantations and native forests in South America and the Southern United States. New For 33(3):237–255

Curtis RO (1998) "Selective cutting" in Douglas-fir - History revisited. J Forestry 96:40–46

Daniel TC (2001) Whither scenic beauty? Visual landscape quality assessment in the 21st century. Landsc Urban Planning 54:267–281

Dhakal AS, Sidle RC (2003) Long-term modelling of landslides for different forest management practices. Earth Surf Proc Land 28:853–868

Dodson EK, Burton JI, Puettmann KJ (2014) Multi-scale controls on natural regeneration dynamics after partial overstory removal in Douglas-fir forests in western Oregon, USA. Forest Sci 60(5):953–961

Donoso PJ, Nyland RD (2005) Seeding density according to structure, dominance and understory cover in old-growth forest stands of the evergreen forest type in the coastal range of Chile. 2005. Rev Chil Hist Nat 78(1):51–63

Donoso PJ, Promis A (2013) Silvicultura en Bosques Nativos. Avances en la investigación en Chile, Argentina y Nueva Zelandia. Estudios en Silvicultura de Bosques Nativos, Vol. 1. Ed. Marisa Cuneo, Valdivia, Chile. p 253. https://sites.google.com/site/alvaropromis/Home/libro-silvicultura-bosques-nativos.

Drever CR, Peterson G, Messier C, Bergeron Y, Flannigan M (2006) Can forest management based on natural disturbances maintain ecological resilience? Can J For Res 36:2285–2299

Drössler L, Nilsson U, Lundqvist L (2013) Simulated transformation of even-aged Norway spruce stands to multi-layered forests: an experiment to explore the potential of tree size differentiation. Forestry 87(2):239–248, doi:10.1093/forestry/cpt037

Duncker PS, Barreiro SM, Hengeveld GM, Lind T, Mason WL, Ambrozy S, Spiecker H (2012) Classification of forest management approaches: a new conceptual framework and its applicability to European forestry. Ecol Soc 17(4):51, http://dx.doi.org/10.5751/ES-05262-170451

Dvorák L, Bachmann P, Mandallaz D (2001) Sturmschäden in ungleichförmigen Beständen. Schweiz Zeitschr Forstw 152:445–452

Evans Z (2006) What is ecological forestry?. http://www.forestguild.org/ecological_forestry/Ecological_Forestry_evans_06.pdf. Accessed 10 November 2014

FAO (Food and Agriculture Organization of the United Nations) (2010) Global forest resources assessment 2010. Food and Agriculture Organization of the United Nations, Rome

Fedrowitz K, Koricheva J, Baker DC, Lindenmayer DB, Palik B, Rosenvald R, Beese W, Franklin JF, Kouki J, Macdonald E (2014) Can retention forestry help conserve biodiversity? A meta-analysis. J Appl Ecol 51(6):1669–1679, doi:10.1111/1365-2664.12289

Filotas E, Parrott L, Burton PJ, Chazdon RL, Coates KD, Coll L, Haeussler S, Martin K, Nocentini S, Puettmann KJ, Putz FE, Simard SW, Messier C (2014) Viewing forests through the lens of complex systems science. Ecosphere 5:1, http://dx.doi.org/10.1890/ES13-00182.1

Fitzgerald SA, Oester PT, Parker R (2013) Individual tree selection (ITS) in a northeast Oregon mixed conifer forest. Extension Service, Oregon State University, Corvallis, OR, USA

Florence RG (1996) Ecology and silviculture of eucalypt forests. CSIRO, Melbourne, p 400

Franklin JF, Forman TT (1987) Creating landscape patterns by forest cutting: ecological consequences and principles. Landscape Ecol 1:5–18

Franklin JF, Johnson KN (2012) A restoration framework for federal forests in the Pacific Northwest. J Forestry 110:429–439

Franklin JF, Mitchell RJ, Palik BJ (2007) Natural disturbance and stand development principles for ecological forestry, General Technical Report NRS-19. United States Department of Agriculture, Forest Service, Northern Research Station, Newtown Square, PA

Gayer K (1886) Der gemischte Wald, seine Begrűndung und Pflege, insbesondere durch Horst-und Gruppenwirtschaft. Parey Verlag, Berlin

Griess VC, Knoke T (2011) Can native tree species plantations in Panama compete with Teak plantations? An economic estimation. New For 41:13–39

Griess VC, Acevedo R, Härtl F, Staupendahl K, Knoke T (2012) Does mixing tree species enhance stand resistance against natural hazards? A case study for spruce. Forest Ecol Manag 267:284–296

Gurnaud A (1882) Le controle et le regime forestier. Rev Eaux For 21:1–23

Gustafsson L, Kouki J, Sverdrup-Thygeson A (2010) Tree retention as a conservation measure in clear-cut forests of northern Europe: a review of ecological consequences. Scand J For Res 25:295–308

Gustafsson L, Baker SC, Bauhus J, Beese WJ, Brodie A, Kouki J, Lindenmayer DB, Lõhmus A, Martínez Pastur G, Messier C, Neyland M, Palik B, Sverdrup-Thygeson A, Volney WJA, Wayne A, Franklin JF (2012) Retention forestry to maintain multifunctional forests: a world perspective. Bioscience 62:633–645

Haight R (1987) Evaluating the efficiency of even-aged and uneven-aged stand management. Forest Sci 33:116–134

Haight RG, Monserud RA (1990) Optimizing any-aged management of mixed-species stands: II. Effects of decision criteria. Forest Sci 36:125–144

Hanewinkel M (2002) Comparative economic investigations of uneven-aged and uneven-aged silvicultural systems: a critical analysis of different methods. Forestry 75:473–481

Hanewinkel M, Kuhn T, Bugmann H, Lanz A, Brang P (2014) Vulnerability of uneven-aged forests to storm damage. Forestry 87:525–534, doi:10.1093/forestry/cpu008

Harlow RF, Downing R, Van Lear D (1997) Responses of wildlife to clearcutting and associated treatments in the Eastern United States, Technical Paper no. 19. Department of Forest Resources Clemson University, Clemson, SC

Hein L, van Koppen K, De Groot RS, van Ierland EC (2006) Spatial scales, stakeholders and the valuation of ecosystem services. Ecol Econ 57(2):209–228

Hibbs DE, Bower AL (2001) Riparian forests in the Oregon Coast Range. Forest Ecol Manag 154:201–213

Higgs E, Falk DA, Guerrini A, Hall M, Harris J, Hobbs RJ, Jackson ST, Rhemtulla JM, Throop W (2014) The changing role of history in restoration ecology. Front Ecol Environ 12:499–506

Hugosson M, Ingemarson F (2004) Objectives and motivations of small-scale forest owners; theoretical modelling and qualitative assessment. Silva Fennica 38:217–231

Hyvaerinen E, Kouki J, Martikainen P (2006) Fire and green-tree retention in conservation of red-listed and rare deadwood-dependent beetles in Finnish boreal forests. Conserv Biol 20:1710–1719

Hyytiäinen K, Haight RG (2010) Evaluation of forest management systems under risk of wildfire. European J Forest Res 129:909–919

Jacobsen MK (2001) History and principles of close to nature forest management: a central European perspective. Naconex 3:56–58

Jacobsen JB, Thorsen BJ (2003) A Danish example of optimal thinning strategies in mixed-species forest under changing growth conditions caused by climate change. Forest Ecol Manag 180(1):375–388

Jones TA, Thomas SC (2004) The time course of diameter increment responses to selection harvests in *Acer saccharum*. Can J For Res 34:1525–1533

Kimmins H (1992) Balancing act: environmental issues in forestry. UBC Press, Vancouver, BC

Kneeshaw D, Williams H, Nikinmaa E, Messier C (2002) Patterns of above- and below-ground responses of understory conifer release 6 years after partial cutting. Can J For Res 32:255–265

Knoke T (2012) The economics of continuous cover forestry. In: Pukkala T, Gadow Kv (eds) Continuous cover forestry. Managing forest ecosystems Vol. 23, pp 167–193. doi:10.1007/978-94-007-2202-6_5

Knoke T, Huth A (2011) Modelling forest growth and finance: Often disregarded tools in tropical land management. In: Günter S, Weber M, Stimm B, Mosandl R (eds) Silviculture in the tropics Tropical Forestry 8, part 3, pp 129–142. doi:10.1007/978-3-642-19986-8_11. Springer Verlag, Heidelberg.

Knoke T, Wurm J (2006) Mixed forests and a flexible harvest policy: a problem for conventional risk analysis? Eur J For Res 125(3):303–315

Knoke T, Moog M, Plusczyk N (2001) On the effect of volatile stumpage prices on the economic attractiveness of a silvicultural transformation strategy. Forest Pol Econ 2(3):229–240

Kohm KA, Franklin JF (1997) Creating a forestry for the 21st century: the science of ecosystem management. Island Press, Washington, DC

Kouki J, Hyvarinen E, Lappalainen H, Martikainen P, Simila M (2012) Landscape context affects the success of habitat restoration: large-scale colonization patterns of saproxylic and fire-associated species in boreal forests. Divers Distrib 18:348–355

Kuuluvainen T (2009) Forest management and biodiversity conservation based on natural ecosystem dynamics in northern Europe: the complexity challenge. AMBIO 38:309–315

Lang C, Pye O (2001) Blinded by science: the invention of scientific forestry and its influence in the Mekong Region. Watershed 6:25–34

Larsen JB, Nielsen AB (2007) Nature-based forest management—where are we going?: laborating forest development types in and with practice. Forest Ecol Manag 238:107–117

Lindenmayer DB, Franklin JF, Lõhmus A, Baker SC, Bauhus J, Beese W, Brodie A, Kiehl B, Kouki J, Pastur GM, Messier C, Neyland M, Palik B, Sverdrup-Thygeson A, Volney J, Wayne A, Gustafsson L (2012) A major shift to the retention approach for forestry can help resolve some global forest sustainability issues. Conserv Lett 5:421–431

Little C, Lara A, McPhee J, Urrutia R (2009) Revealing the impact of forest exotic plantations on water yield in large scale watersheds in South-Central Chile. J Hydrol 374:162–170

Long JN (2009) Emulating natural disturbance regimes as a basis for forest management: a North American view. Forest Ecol Manag 257:1868–1873

Lowood HE (1990) The calculating forester: quantification, cameral science, and the emergence of scientific forestry management in Germany. In: Frängsmyr T, Heilbron JL, Rider RE (eds) The quantifying spirit in the eighteenth century. University of California Press, Berkeley, http://ark.cdlib.org/ark:/13030/ft6d5nb455/

Lu Y, Liu X, Lei X, Wang H, Hong L, Guo H (2013) Development of silvicultural models for multi-functional forest management: the systematic consideration and application in experiment center of tropical forestry, Southwestern China. In: Fehrmann L, Kleinn C (eds) Forests in climate change research and police: the role of forest management and conservation in a complex international setting. Proceedings of the 3rd International DAAD workshop. Cuvillier Verlag Goettingen, Germany.

Lüpke B, Hauskeller-Bullerjahn M (1999) Kahlschlagfreier Waldbau: wird die Eiche an den Rand gedrängt? Forst und Holz 54:363–368

Lutz JA, Halpern CB (2006) Tree mortality during early forest development: a long-term study of rates, causes, and consequences. Ecol Monogr 76:257–275

Macdonald E, Gardiner B, Mason W (2010) The effects of transformation of even-aged stands to continuous cover forestry on conifer log quality and wood properties in the UK. Forestry 83:1–16

Manolis JC, Andersen DE, Cuthbert FJ (2000) Patterns in clearcut edge and fragmentation effect studies in northern hardwood-conifer landscapes: retrospective power analysis and Minnesota results. Wildlife Soc B 28:1088–1101

Marten GG (2005) Environmental tipping points: a new paradigm for restoring ecological security. J Policy Studies 20:75–87

Mason B, Kerr G, Pommerening A, Edwards C, Hale S, Ireland D, Moore R (2003) Continuous cover forestry in British conifer forests. Forest Res Annu Rep Acc 2004:38–53

Matlack GR, Gleeson SK, Good RE (1993) Treefall in a mixed oak-pine coastal plain forest: immediate and historical causation. Ecology 82:1559–1566

Menzies NK (2007) Our forests, your ecosystem, their timber. Communities, conservation and the state in community-based forest management. Columbia University Press, New York

Merriam-Webster (2014) Merriam Webster Dictionary. http://www.merriam-webster.com/. Accessed 1 May 2014

Messier C, Puettmann KJ, Coates D (2013) Managing forests as complex adaptive systems: building resilience to the challenge of global change. Routledge, New York

Miller Klubock T (2014) La Frontera: forests and ecological conflicts in Chile's frontier territory. Duke University Press, Durham, NC

Möller A (1923) Der Dauerwaldgedanke. Erich Degreif Verlag, Oberteuringen

Moorman M, Nelson S, Moore S, Donoso P (2013) Stakeholder perspectives on adaptive co-management as a Chilean conservation management strategy. Soc Natur Resour 26:1022–1036

Nahuelhual L, Donoso PJ, Lara A, Núñez D, Oyarzún C, Neira E (2007) Valuing ecosystem services of Chilean temperate rainforests. Environ Dev Sustainabil 9:481–499

Nasi R, Putz FE, Pacheco P, Wunder S, Anta S (2011) Sustainable forest management and carbon in tropical Latin America: the case for REDD+. Forests 2:200–217

Neumann FG, Minko G (1981) The sirex wood wasp in Australian radiata pine plantations. Aust For 44:46–63

Neuner S, Albrecht A, Cullmann D, Engels F, Griess VC, Hahn A, Hanewinkel M, Härtl F, Kölling C, Staupendahl K, Knoke T (2015) Survival of Norway spruce remains higher in mixed stands under a dryer and warmer climate. Global Change Biol 21:935–946

Newton M, Cole EC (2006) Harvesting impacts on understory regeneration in two-storied Douglas-fir stands. West J Appl For 21:14–18

Neyland M, Hickey J, Read SM (2012) A synthesis of outcomes from the Warra Silvicultural Systems Trial, Tasmania, Australia: safety, timber production, economics, biodiversity, silviculture and social acceptability. Aust For 75:147–162

Niklitschek ME (2007) Trade liberalization and land use changes: explaining the expansion of afforested land in Chile. Forest Sci 53(3):385–394

Nill M, Kohnle U, Sauter U (2011) Rindenschäden mit mutmaßlichem Bezug zur Holzernte im Spiegel der Betriebsinventuren in Baden-Württemberg. Forstarchiv 9(6):216–224

Nilsson M-C, Wardle DA (2005) Understory vegetation as a forest ecosystem driver: evidence from the northern Swedish boreal forest. Front Ecol Environ 3:421–428

Ninemets Ü, Valladares F (2006) Tolerance to shade, drought, and waterlogging of temperate Northern Hemisphere trees and shrubs. Ecol Monogr 76:521–547

Nocentini S (2009) Structure and management of beech (*Fagus sylvatica* L.) forests in Italy. iForest 2:105–113, http://www.sisef.it/iforest/show.php?id=499

Noguchi M, Yoshida T (2004) Tree regeneration in partially cut conifer-hardwood mixed forests in northern Japan: roles of establishment substrate and dwarf bamboo. Forest Ecol Manag 190:335–344

Noguchi M, Okuda S, Miyamoto K, Itou T, Inagaki Y (2011) Composition, size structure and local variation of naturally regenerated broadleaved tree species in Hinoki cypress plantations: a case study in Shikoku, south-western Japan. Forestry 84:493–504

O'Hara KL, Ramage BS (2013) Silviculture in an uncertain world: utilizing multi-aged management systems to integrate disturbance. Forestry 86:401–410

Oosthoek J (2007) The colonial origins of scientific forestry in Britain. Essay in: Environmental History Resources. http://www.eh-resources.org/colonial_forestry.html. Accessed 22 February 2015.

Oregon Department of Forestry (2014) Oregon forest practices act. http://www.oregon.gov/ODF/privateforests/pages/fpareforestation.aspx. Accessed 1 May 2014

Paquet J, Bélanger L (1997) Public acceptability thresholds of clearcutting to maintain visual quality of boreal balsam fir landscapes. Forest Sci 43:46–55

Paquette A, Messier C (2009) The role of plantations in managing the world's forests in the Anthropocene. Front Ecol Environ 8:27–34

Pausch R (2005) Ein System-Ansatz zur Darstellung des Zusammenhangs zwischen Waldstruktur, Arbeitsvolumen und Kosten in naturnahen Wäldern Bayerns. Forstliche Forschungsberichte München 199

Piispanen R, Heinonen J, Valkonen S, Mäkinen H, Lundqvist S-O, Saranpää P (2014) Wood density of Norway spruce in uneven-aged stands. Can J For Res 44(2):136–144

Pommerening A, Murphy S (2004) A review of the history, definitions and methods of continuous cover forestry with special attention to afforestation and restocking. Forestry 77:27–44

ProSilva Europe (2014) ProSilva Europe: integrated forest management for resilience and sustainability across 25 countries. http://prosilvaeurope.wordpress.com. Accessed 13 August 2014

Puettmann KJ (2011) Silvicultural challenges and options in the context of global change: simple fixes and opportunities for new management approaches. J Forestry 109:321–331

Puettmann KJ, Ammer C (2006) Trends in North American and European regeneration research under the ecosystem management paradigm. Eur J For Res 126:1–9

Puettmann KJ, Ek AR (1999) Status and trends of silvicultural practices in Minnesota. North J Appl For 16:203–210

Puettmann KJ, Coates KD, Messier C (2009) A critique of silviculture: managing for complexity. Island Press, Washington, DC

Pukkala T, Lähde E, Laiho S (2010) Optimizing the structure and management of une-ven-sized stands of Finland. Forestry 83:129–142

Putz F, Sist P, Fredericksen T, Dykstra D (2008) Reduced-impact logging: challenges and opportunities. Forest Ecol Manag 256:1427–1433

Reich RM, Mielke PW Jr, Hawksworth FG (1991) Spatial analysis of ponderosa pine trees infected with dwarf mistletoe. Can J For Res 21:1808–1815

Ribe RG (1989) The aesthetics of forestry: what has empirical preference research taught us? Environ Manage 13:55–74

Ribe RG (1999) Regeneration harvests versus clearcuts: public views of the acceptability and aesthetics of Northwest Forest Plan harvests. Northw Sci 73:102–117

Ribe RG (2005) Aesthetic perceptions of green-tree retention harvests in vista views: the interaction of cut level, retention pattern and harvest shape. Land-scape Urban Plan 73:277–293

Rice B, Weiskittel A, Wagner R (2014) Efficiency of alternative forest inventory methods in partially harvested stands. Eur J For Res 133:261–272

Roessiger J, Griess VC, Härtl F, Clasen C, Knoke T (2013) How economic performance of a stand increases due to decreased failure risk associated with the admixing of species. Ecol Model 255:58–69

Rosenvald R, Lohmus A (2008) For what, when, and where is green-tree retention better than clear-cutting? A review of the biodiversity aspects. Forest Ecol Manag 255:1–15

Rydberg D, Falck J (2000) Urban forestry in Sweden from a silvicultural perspective: a review. Landscape Urban Plan 47:1–18

Schaetzl RJ, Johnson DL, Burns SF, Small TW (1989) Tree uprooting: review of terminology, process, and environmental implications. Can J For Res 19:1–11

Schelhaas M, Nabuurs GJ, Schuck A (2003) Natural disturbances in the European forests in the 19th and 20th centuries. Glob Change Biol 9:1620–1633

Schoennagel T, Veblen TT, Romme WH (2004) The interaction of fire, fuels, and climate across Rocky Mountain forests. BioScience 54:661–676

Schütz J-P (1999a) Naturnaher Waldbau: gestern, heute, morgen. Schw Z Forstwes 150:1–12

Schütz J-P (1999b) Principles of functioning of mixtures in forests stands; experience of temperate central European forest conditions. In: Olsthoorn AFM, Bartelink HH, Gardiner JJ, Pretzsch H, Hekhuis HJ, Franc A (eds) Management of mixed-species forest; silviculture and economics, IBN Scientific Contribution 15. Inst. For Forestry and Nature Research, Wageningen, pp 219–234

Schütz J-P (2002) Silvicultural tools to develop irregular and diverse forest structures. Forestry 75:329–337

Schütz J-P, Pukkala T, Donoso PJ, von Gadow K (2012) Historical emergence and current application of CCF. In: Pukkala T, von Gadow K (eds) Continuous cover forestry. Springer Verlag, Dordrecht, Heidelberg, p 6

Sears RR, Pinedo-Vasquez M (2011) Forest policy reform and the organization of logging in Peruvian Amazonia. Dev Change 42:609–631

Seymour RS, Hunter ML (1999) Principles of ecological forestry. In: Hunter ML (ed) Maintaining biodiversity in forest ecosystems. Cambridge University Press, Cambridge, pp 499–524

Shaw DC, Chen J, Freeman EA, Braun DM (2005) Spatial and population characteristics of dwarf mistletoe infected trees in an old-growth Douglas-fir – western hemlock forest. Can J For Res 35:990–1001

Shorohova E, Kuuluvainen T, Kangur A, Jogiste K (2009) Natural stand structures, disturbance regimes and successional dynamics in the Eurasian boreal forests: a review with special reference to Russian studies. Ann For Sci 66:201

Siiskonen H (2007) The conflict between traditional and scientific forest management in 20th century Finland. Forest Ecol Manag 249:125–133, doi:10.1016/j.foreco.2007.03.018

Steventon JD, MacKenzie KL, Mahon TE (1998) Response of small mammals and birds to partial cutting and clearcutting in northwest British Columbia. Forest Chron 74(5):703–713

Swank W, Vose J, Elliott K (2001) Long-term hydrologic and water quality responses following commercial clearcutting of mixed hardwoods on a southern Appalachian catchment. Forest Ecol Manag 143:163–178

Swanson ME, Franklin JF, Beschta RL, Crisafulli CM, DellaSala DA, Hutto RL, Lindenmayer DB, Swanson FJ (2011) The forgotten stage of forest succession: early-successional ecosystems on forest sites. Front Ecol Environ 9:117–125

Tahvonen O, Pukkala T, Laiho O, Lähde E, Niinimäki S (2010) Optimal management of uneven-aged Norway spruce stands. Forest Ecol Manag 260:106–115

Thorpe HC, Astrup R, Trowbridge A, Coates KD (2010) Competition and tree crowns: a neighborhood analysis of three boreal tree species. Forest Ecol Manag 259:1586–1596

Tönnes S, Karjalainen E, Löfström I, Neuvonen M (2004) Scenic impacts of retention trees in clear-cutting areas. Scand J For Res 19:348–357

Tropical Forest Foundation (2008) Sustaining tropical forests with reduced impact logging. http://www.tff-indonesia.org/index.php/en/r-i-l/ril-criteria-and-indicators. Accessed 14 April 2015

Urquhart J, Courtney P (2011) Seeing the owner behind the trees: a typology of small-scale private woodland owners in England. Forest Policy Econ 13:535–544

Veblen TT (1985) Forest development in tree-fall gaps in the temperate rain forests of Chile. Natl Geogr Res 1:162–183

Veblen TT, Donoso ZC, Schlegel FM, Escobar RB (1981) Forest dynamics in south-central Chile. J Biogeogr 8:211–247

von Thünen JH (1842) Der isolierte Staat in Beziehung auf Landwirtschaft und Nationalökonomie, 2nd edn. Leopold, Rostock, Germany

Wagner S, Nocentini S, Huth F, Hoogstra-Klein M (2014) Forest management approaches for coping with the uncertainty of climate change: trade-offs in service provisioning and adaptability. Ecol Soc 19(1):32

West PW (2014) Growing plantation forests. Springer, Dordrecht, Heidelberg, London, New York

Wilson SMG (2013) Adoption of alternative silvicultural systems in Great Britain: a review. Q J Forest 100:279–293

Wunder S (2006) Are direct payments for environmental services spelling doom for sustainable forest management in the tropics. Ecol Soc 11:23

Yasuda A, Yoshida T, Miya H, Harvey BD (2013) An alternative management regime of selection cutting for sustaining stand structure of mixed forests of northern Japan: a simulation study. J For Res 18:398–406

Yoshida T, Noguchi M, Akibayashi Y, Noda M, Kadomatsu M, Sasa K (2006) Twenty years of community dynamics in a mixed conifer broad-leaved forest under a selection system in northern Japan. Can J For Res 36:1363–1375

19

Holocene variations of wildfire occurrence as a guide for sustainable management of the northeastern Canadian boreal forest

Ahmed El-Guellab[1], Hugo Asselin[1*], Sylvie Gauthier[2], Yves Bergeron[1] and Adam A. Ali[3]

Abstract

Background: Cumulative impacts of wildfires and forest harvesting can cause shifts from closed-crown forest to open woodland in boreal ecosystems. To lower the probability of occurrence of such catastrophic regime shifts, forest logging must decrease when fire frequency increases, so that the combined disturbance rate does not exceed the Holocene maximum. Knowing how climate warming will affect fire regimes is thus crucial to sustainably manage the forest. This study aimed to provide a guide to determine sustainable forest harvesting levels, by reconstructing the Holocene fire history at the northern limit of commercial forestry in Quebec using charcoal particles preserved in lake sediments.

Methods: Sediment cores were sampled from four lakes located close to the northern limit of commercial forestry in Quebec. The cores were sliced into consecutive 0.5 cm thick subsamples from which 1 cm^3 was extracted to count and measure charcoal particles larger than 150 microns. Age-depth models were obtained for each core based on accelerator mass spectroscopy (AMS) radiocarbon dates. Holocene fire histories were reconstructed by combining charcoal counts and age-depth models to obtain charcoal accumulation rates and, after statistical treatment, long-term trends in fire occurrence (expressed as number of fires per 1000 years).

Results: Fire occurrence varied between the four studied sites, but fires generally occurred more often during warm and dry periods of the Holocene, especially during the Holocene Thermal Maximum (7000–3500 cal. BP), when fire occurrence was twice as high as at present.

Conclusions: The current fire regime in the study area is still within the natural range of variability observed over the Holocene. However, climatic conditions comparable to the Holocene Thermal Maximum could be reached within the next few decades, thus substantially reducing the amount of wood available to the forest industry.

Keywords: Fire occurrence; Holocene; Boreal forest; Northern limit; Forest management

Background

The northern part of the boreal biome contains some of the largest remaining tracts of intact forest in the world (Bradshaw et al. 2009). However, forestry operations have expanded in a south-north fashion over the last few decades (Bouchard and Pothier 2011; Kivinen et al. 2012), and the cumulative impacts of natural disturbances and forest harvesting may affect the ecological integrity of boreal ecosystems. For example, closed-crown spruce-moss forests have shifted to open woodlands in eastern Canada under the combined action of wildfire and clearcut logging (Dussart and Payette 2002; Girard et al. 2008). To lower the probability of occurrence of such catastrophic ecosystem shifts, harvesting rate must decrease when fire frequency increases (Bergeron et al. 2006; Raulier et al. 2013), so that the total disturbance rate (harvesting + wildfire) does not exceed the maximum disturbance rate recorded during the Holocene.

Climate models predict a major rise in mean annual temperature across boreal regions, but only a slight – if any – precipitation increase (IPCC 2014). This could lead to dryer conditions, as higher evapotranspiration

* Correspondence: Hugo.Asselin@uqat.ca
[1]NSERC/UQAT/UQÀM Industrial Chair in Sustainable Forest Management, Université du Québec en Abitibi-Témiscamingue, 445, boulevard de l'Université, Rouyn-Noranda, Québec J9X 5E4, Canada
Full list of author information is available at the end of the article

would offset the precipitation increase (Hély et al. 2010). Hence, forest fires will likely be more frequent and cover larger areas by the end of the 21st century (Goldammer 2013), meaning that the margin of manoeuvre for forestry operations will decrease.

In Quebec (eastern Canada), government authorities have established a northern limit to commercial forestry operations in the boreal forest (Jobidon et al. 2015). North of this limit, forest productivity is deemed too low, and fire frequency too high, to support a profitable forest industry. In a comparative study of Holocene fire regimes along a transect crossing the limit of commercial forestry, Oris et al. (2014b) noted that all regions were characterized by considerably higher fire occurrences during the Holocene Thermal Maximum (HTM; ca. 7000–3000 cal. years BP), followed by a sharp decrease in fire occurrence at the beginning of the Neoglacial period (ca. 3000 cal. years BP) in forests located close to and north of the limit, compared to a more gradual decrease south of the limit. They thus concluded that fire occurrence was more sensitive to climate change close to and north of the limit than further south. These regional patterns, however, hide important between-site variability (see Figure S2 in Oris et al. (2014b)), and additional studies are needed to verify if the same regional fire history can be reconstructed from other sites.

We used high-resolution charcoal analysis to reconstruct the Holocene variability of fire occurrence from 4 small lakes located near the northern limit of commercial forestry in Quebec, ca. 100–200 km east of the lakes studied by Oris et al. (2014b). As wildfire dynamics are mainly controlled by regional climatic patterns in boreal forests (Flannigan and Wotton 2001; Ali et al. 2009; Boulanger et al. 2013), we expected to find a regional fire history similar to that reported by Oris et al. (2014b), despite inter-site variability. Having a clear understanding of the Holocene regional variability in fire occurrence is mandatory to adjust forest harvesting levels so that the total disturbance rate does not exceed the Holocene maximum. We show that fire occurrence at the northern limit of commercial forestry in Quebec was higher during warm and dry periods of the Holocene. According to predictions, climate warming over the next few decades will increase fire occurrence close to the maximum values recorded during the Holocene, thus substantially reducing the margin of manoeuvre of the forest industry.

Methods

Study area

The study area is located within the western black spruce-moss bioclimatic subdomain (Saucier et al. 1998), in northern Quebec, 50–100 km north of Chibougamau (Fig. 1). Mean annual temperature is −0.4 °C and mean annual precipitation is 961.4 mm (Environment Canada 2011). The regional climate history displayed a general progression from a cool early Holocene (<7000 cal. BP) to a warm and dry middle Holocene (Holocene Thermal Maximum: 7000–3500 cal. BP) followed by a cool and wet Neoglacial period (>3500 cal. BP), briefly interrupted (around 1000 cal. BP) by a Medieval Warm Period (Viau and Gajewski 2009). Forests of this region have been dominated by black spruce (*Picea mariana* (Mill.) BSP) throughout the Holocene (Garralla and Gajewski 1992; Payette 1993).

Sampling

Four lakes were selected close to the northern limit of commercial forestry (Fig. 1). The lakes are within 100 km of each other and have similar characteristics: size (<4 ha), absence of inlet or outlet, elevation (376–432 m a.s.l.), water depth (500–571 cm), and local vegetation (dominated by black spruce). Lake sediment cores were sampled in March 2009 at the deepest point of each lake. The water-sediment interface was sampled using a Kajak-Brinkhurst (KB) gravity corer (Glew 1991), and a Livingstone corer was used to sample deeper sediments (Wright et al. 1984). The cores were sliced into consecutive 0.5 cm thick subsamples to maximize temporal resolution. Subsamples were stored in plastic bags in the refrigerator until analysis.

Charcoal quantification

A volume of 1 cm^3 was taken from each subsample and soaked in a 3 % sodium hexametaphosphate – $(NaPO_3)_6$ – dispersing solution for 48 h before sieving through a 150 microns mesh. The material remaining on the sieve was washed with a 10 % sodium hypochlorite (NaOCl) solution to whiten organic matter particles and facilitate distinction of charcoal particles. Charcoal particles larger than 150 microns were counted and measured (area) under a dissecting microscope (20 ×) equipped with a digital camera coupled to an image analysis software (Winseedle, Regent Instruments Inc., Canada). Charcoal particles larger than 150 microns are rarely carried more than 1 km from the fire and are usually assumed to represent local fires (Wein et al. 1987), although long-distance transport (>30 km) of such particles has been reported (Oris et al. 2014a). We nevertheless elected to work with charcoal particles larger than 150 microns to ease comparisons with Oris et al. (2014b).

Age-depth models

Radiocarbon dates were obtained by accelerator mass spectroscopy (AMS) performed on samples from different levels of the sedimentary profiles (Table 1). Macroremains were rare (except for Aurélie lake), and most dates were obtained from bulk sediments (gyttja). All

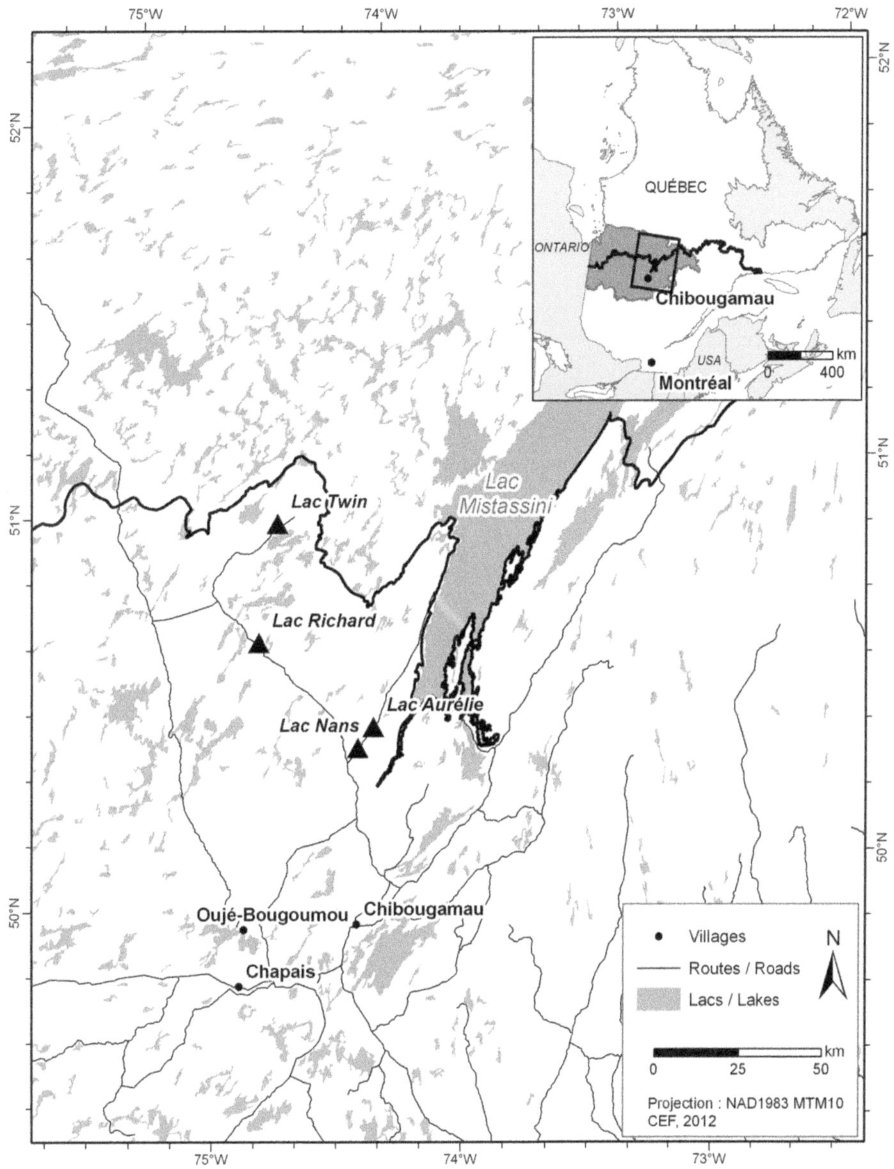

Fig. 1 Location of the 4 sampled lakes (triangles) at the northern limit of commercial forestry (thick black line). The dark gray area in the inset represents the western spruce-moss bioclimatic subdomain

dates were calibrated using the CALIB 6.0.1 program (http://calib.qub.ac.uk/calib/). Age-depth models were produced using MCAgeDepth software version 0.1 (Higuera et al. 2009; available for free at http://code.google.com/p/mcagedepth/) (Fig. 2).

Fire history reconstruction

Charcoal accumulation rates (CHAR) expressed in $mm^2{\cdot}cm^{-2}{\cdot}yr^{-1}$ were calculated by multiplying charcoal concentration ($mm^2{\cdot}cm^{-3}$) by sedimentation rate ($cm{\cdot}yr^{-1}$) for each subsample. To avoid possible influence of sedimentation variations within and between sites, a constant sedimentation rate (20 years, the average of the 4 lakes) was used to compare lakes and to compute the regional fire history (all lakes combined). Version 1.1. of the CharAnalysis software (Higuera et al. 2009; available for free at http://sites.google.com/site/charanalysis/) was used to remove background noise from the CHAR series with a LOWESS function. The remaining peak component was composed of two Gaussian subpopulations: (1) noise inherent to statistical analysis and part of the background noise that would not have been removed previously, and (2) fire peaks, each one representing the occurrence of one or more local fire events (Gavin et al. 2006). The noise subpopulation was removed from the fire peak subpopulation by applying a

Table 1 Radiocarbon dates at different depths in the 4 sediment profiles

Site and depth (cm)	Conventional radiocarbon age ^{14}C BP ± 2 σ	Median probability (range of calibration) cal. years BP	Dated material	Laboratory number
Aurélie Lake				
43–44	2870 ± 30	3007 (2879–3136)	Macroremains	Poz-35983
111–112	3990 ± 35	4443 (4319–4568)	Macroremains	Poz-35984
163–164	4750 ± 35	5457 (5329–5586)	Macroremains	Poz-36014
220–221	6140 ± 40	7047 6931–7163)	Macroremains	Poz-36016
236–237	6490 ± 40	7396 (7317–7476)	Macroremains	Poz-36017
326–327	7460 ± 50	8279 (8185–8373)	Macroremains	Poz-36018
Nans Lake				
20–25	1200 ± 40	1075 (1180–970)	Gyttja	Beta-275126
30–35	1820 ± 30	1675 (1740–1610)	Gyttja	Beta-298239
50–51	3290 ± 40	3480 (3570–3390)	Gyttja	Beta-267031
100–101	4040 ± 40	4555 (4560–4550)	Gyttja	Beta-267032
130–131	4630 ± 40	5420 (5460–5380)	Gyttja	Beta-298237
150–151	4040 ± 40	4490 (4570–4410)	Gyttja	Beta-267033
150–151	4720 ± 40	5555 (5570–5540)	Gyttja	Beta-298238
170–171	5230 ± 40	5950 (6000–5900)	Gyttja	Beta-267034
212–213	7800 ± 40	8505 (8600–8410)	Gyttja	Beta-267035
Richard Lake				
0–5	560 ± 30	510 (530–490)	Gyttja	Beta-293916
20–25	1220 ± 30	1075 (1170–980)	Gyttja	Beta-293917
80–81	3800 ± 30	2325 (4230–420)	Gyttja	Beta-293911
97–98	4870 ± 40	5535 (5600–5470)	Gyttja	Beta-293912
131–132	6770 ± 40	7520 (7580–7460)	Gyttja	Beta-293913
149–150	6820 ± 40	7585 (7620–7550)	Gyttja	Beta-293914
159–160	7560 ± 40	8265 (8350–8180)	Gyttja	Beta-293915
Twin Lake				
24–25	2415 ± 30	2436 (2357–2679)	Gyttja	SacA16557
54–55	3615 ± 30	3923 (3848–4056)	Gyttja	SacA16556
74–75	4105 ± 30	4625 (4480–4806)	Gyttja	SacA16555
94–95	4340 ± 30	4906 (4851–5019)	Gyttja	SaA16554
114–115	4435 ± 30	5028 (4892–5264)	Gyttja	SacA16553
134–135	5245 ± 30	5988 (5931–6169)	Gyttja	SacA16552
164–165	6285 ± 30	7215 (7166–7267)	Gyttja	SacA16551
180–181	6910 ± 50	7742 (7654–7866)	Macroremains	Poz-32158
183–184	7625 ± 45	8419 (8368–8531)	Gyttja	SacA16550

Beta = Beta Analytic Inc; Poz = Poznan Radiocarbon Laboratory; SacA = Laboratoire de mesure du carbone 14

threshold corresponding to the 99th percentile of the noise distribution. General trends in fire occurrence were expressed as number of fires per 1000 years, smoothed with a 100 years moving window using K1D software version 1.2 (Gavin 2010). The composite record illustrating regional fire occurrence was obtained by averaging data from the 4 sites, with a 95 % bootstrap confidence interval.

Results

As expected, Holocene trends in fire occurrence (number of fires per 1000 years) varied between lakes (Fig. 3). At Nans Lake, fire occurrence was higher between 5000 and 3800 cal. BP and around 1000 cal. BP. At Richard Lake, fire occurrence was higher between 4600 and 3600 cal. BP and between 1400 and 800 cal. BP. Most fire events at Aurélie Lake occurred before 3400 cal. BP,

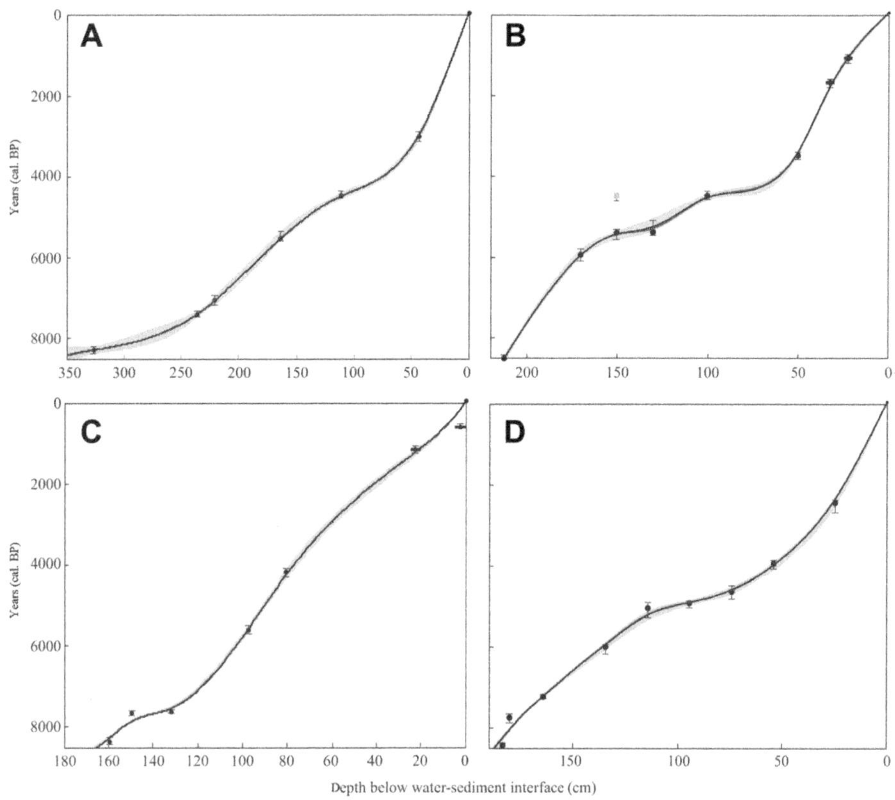

Fig. 2 Age-depth models (cubic spline interpolation) for lakes Aurélie (**A**), Nans (**B**), Richard (**C**), and Twin (**D**). Radiocarbon dates are shown (black dots) with 95 % confidence intervals (whiskers). One of the dates from the lake Nans profile was discarded (grey dot) as it was too young and was likely contaminated by more recent material during sampling

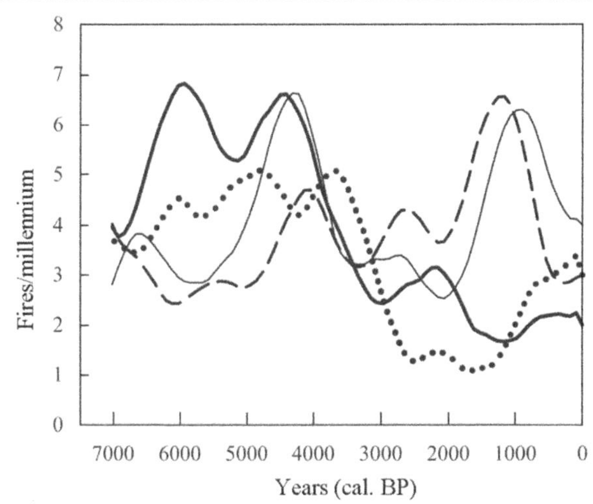

Fig. 3 Fire occurrence (number of fire events per 1000 years) for lakes Aurélie (thick full line), Nans (thin full line), Richard (dashed line), and Twin (dotted line)

and the highest occurrence was recorded between 6800 and 4000 cal. BP. At Twin Lake, fire occurrence was higher before 3200 cal. BP.

The composite curve of regional fire occurrence shows that fire occurrences were generally higher during the middle Holocene (ca. 7000–3500 cal. BP) than during the late Holocene (3500–0 cal. BP), where a brief increase of fire occurrence was nevertheless recorded around 1000 cal. BP (Fig. 4).

Discussion

The long-term fire history at the northern limit of commercial forestry in Quebec fluctuated during the last 7000 years under the influence of regional climate. Fire occurrence was higher between 7000 and 3500 cal. BP, especially between 5500 and 3500 ca. years B.P. This period corresponds to the Holocene Thermal Maximum, previously shown to have been warmer and drier in northern Quebec (Garralla and Gajewski 1992; Viau and Gajewski 2009) and in various locations across North America (Bartlein et al. 1998; Viau et al. 2006). A sharp decrease in fire occurrence occurred in the study area at the beginning of the Neoglacial period (3500–0 cal. BP), especially at lakes Aurélie and Twin, likely in response

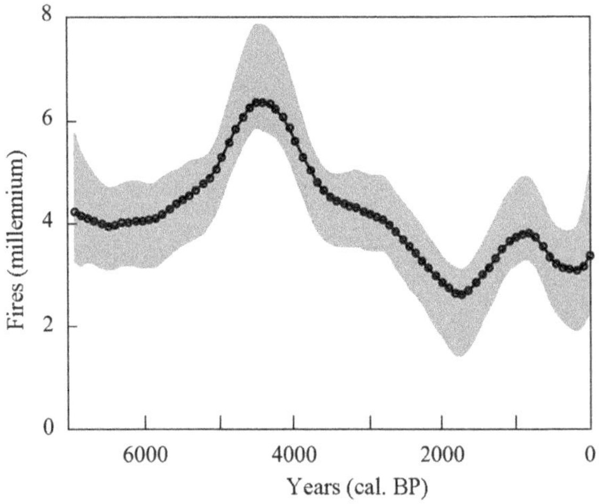

Fig. 4 Regional average of fire occurrence (number of fire events per 1000 years) combining the 4 sampled lakes. The 95 % confidence interval is shown in grey

to a cooler and wetter climate (Garralla and Gajewski 1992; Viau and Gajewski 2009). This general trend is similar to that found further west by Oris et al. (2014b), also at the northern limit of commercial forestry. A brief period of increased fire occurrence was noted in the study area around 1000 cal. BP, coinciding with the Medieval Warm Period (Gajewski 1988; Hunt 2006). The imprint of the Medieval Warm Period was not recorded at all sites, a specificity also noted further south and west in the spruce-moss forest (Ali et al. 2009; Oris et al. 2014b). This could be due to the short duration of this climatic period (a few centuries) making it less likely to affect fire regimes. Alternatively, local factors, such as topography, landscape connectivity, vegetation flammability, or weather could have been more important than regional climate to explain fire occurrence during that period (Gavin et al. 2006; Hu et al. 2006; Long et al. 2007; Ali et al. 2009).

Fire occurrence at the northern limit of commercial forestry in Quebec was variable between sites, but was generally higher during warm and dry periods of the Holocene, especially during the Holocene Thermal Maximum (7000–3500 cal. years BP), when fire occurrence was twice as high as at present. Current fire occurrence is still within the range of natural variability, probably because the amount of precipitation is still high enough to balance the effects of global warming. Fire return intervals (FRI) calculated from the composite chronology of fire occurrences presented here (ca. 167–333 years, based on 3–6 fires per millennium) are slightly higher than – but overlap with – those obtained from spruce-moss forests further south (Cyr et al. 2009: 111–267 years; Ali et al. 2009: 90–230 years) and west (Oris et al. 2014b; 196–312 years). The present-time FRI in our study area (ca. 333 years) is twice as long as that presented by Le Goff et al. (2007) for a particularly dry area located further west (132–153 years). It is however well within the range presented by Mansuy et al. (2010) for a wider study area encompassing both mesic and xeric sites (90–715 years). This is in line with reconstructions from northern Ontario (Canada) that showed significant differences between FRI values from xeric and mesic sites (Senici et al. 2013).

Predicted high temperatures for the middle of the 21st century could be close to those recorded during the Holocene Thermal Maximum (Plummer et al. 2006). If increased temperature is not accompanied by a marked increase in precipitation, the mid-Holocene scenario could be repeated and fire occurrence could increase at the northern limit of commercial forestry in Quebec. The extent to which forest managers will be able to replace wildfire by harvesting activities will be more limited (Bergeron et al. 2006, 2010; Raulier et al. 2013). Indeed, important fire suppression efforts have been deployed in the study area at times when fire activity was at its lowest levels of the past 7000 years.

Conclusions

We showed that fire occurrence was higher during warm and dry periods of the Holocene at the northern limit of commercial forestry in Quebec. Climate predictions for the middle of the 21st century point towards conditions similar to those that prevailed during the Holocene Thermal Maximum, when maximum fire occurrence was recorded. Because sustainable forest management implies that total disturbance (harvesting + wildfire) should not exceed the Holocene maximum, a substantial reduction of the amount of wood available to the forestry industry is to be expected over the next few decades.

Competing interests

The authors declare that they have no competing interests.

Authors' contributions

AE-G carried out the field and laboratory work, performed the analyses and drafted the manuscript. HA, SG, YB and AAA conceived of the study, participated in its design and coordination and helped to draft the manuscript. All authors read and approved the final manuscript.

Acknowledgements

We thank Loic Bircker, Laurent Bremond, Aurélie Genries and Raynald Julien for their assistance during fieldwork. Financial support was provided by the Natural Sciences and Engineering Research Council of Canada and by the CNRS (Paleo2, INSU).

Author details

[1]NSERC/UQAT/UQÀM Industrial Chair in Sustainable Forest Management, Université du Québec en Abitibi-Témiscamingue, 445, boulevard de l'Université, Rouyn-Noranda, Québec J9X 5E4, Canada. [2]Natural Resources Canada, Canadian Forest Service – Laurentian Forestry Centre, 1055 du P.E.P.S., P.O. Box 3800, Sainte-Foy, Québec G1V 4C7, Canada. [3]Centre for Bio-Archeology and Ecology (UMR5059 CNRS), Université Montpellier 2, 163 rue Auguste Broussonet, Montpellier F-34090, France.

References

Ali AA, Carcaillet C, Bergeron Y (2009) Long-term fire frequency variability in the eastern Canadian boreal forest: the influences of climate vs. local factors. Global Change Biol 15:1230–1241

Bartlein PJ, Anderson KH, Anderson PM, Edwards ME, Mock CJ, Thompson RS, Webb RS, Webb T, Whitlock C (1998) Paleoclimate simulations for North America over the past 21,000 years: features of the simulated climate and comparisons with paleoenvironmental data. Quaternary Sci Rev 17:549–585

Bergeron Y, Cyr D, Drever CR, Flannigan M, Gauthier S, Kneeshaw D, Lauzon È, Leduc A, Le Goff H, Lesieur D, Logan K (2006) Past, current, and future fire frequencies in Quebec's commercial forests: implications for the cumulative effects of harvesting and fire on age-class structure and natural disturbance-based management. Can J Forest Res 36:2737–2744

Bergeron Y, Cyr D, Girardin MP, Carcaillet C (2010) Will climate change drive 21st century burn rates in Canadian boreal forest outside of its natural variability: collating global climate model experiments with sedimentary charcoal data. Int J Wildland Fire 19:1127–1139

Bouchard M, Pothier D (2011) Long-term influence of fire and harvesting on boreal forest age structure and forest composition in Eastern Québec. Forest Ecol Manag 261:811–820

Boulanger Y, Gauthier S, Gray DR, Le Goff H, Lefort P, Morissette J (2013) Fire regime zonation under current and future climate over Eastern Canada. Ecol Appl 23:904–923

Bradshaw CJA, Warkentin IG, Sodhi NS (2009) Urgent preservation of boreal carbon stocks and biodiversity. Trends Ecol Evol 24:541–548

Cyr D, Gauthier S, Bergeron Y, Carcaillet C (2009) Forest management is driving the eastern North American boreal forest outside its natural range of variability. Front Ecol Environ 7:519–524

Dussart E, Payette S (2002) Ecological impact of clearcutting on black spruce-moss forests in Southern Québec. Ecoscience 9:533–543

Environment Canada (2011) Normales climatiques au Canada 1971-2000., National Climate Archive. http://www.climate.weatheroffice.ec.gc.ca/climate_normals/

Flannigan MD, Wotton BM (2001) Climate, weather, and area burned. In: Johnson EA, Miyanishi K (eds) Forest fires: behavior and ecological effects. Academic Press, New York

Gajewski K (1988) Late Holocene climate changes in eastern North America estimated from pollen data. Quaternary Res 29:255–262

Garralla S, Gajewski K (1992) Holocene vegetation history of the boreal forest near Chibougamau, Central Quebec. Can J Bot 70:1364–1368

Gavin DG (2010) K1D: multivariate Ripley's K-function for one-dimensional data. Version 1.2. User guide. Department of Geography, University of Oregon, Eugene, OR, USA

Gavin DG, Sheng HF, Lertzman K, Corbett P (2006) Weak climatic control of stand-scale fire history during the late Holocene. Ecology 87:1722–1732

Girard F, Payette S, Gagnon R (2008) Rapid expansion of lichen woodlands within the closed-crown boreal forest zone over the last 50 years caused by stand disturbances in Eastern Canada. J Biogeogr 35:529–537

Glew JR (1991) Miniature gravity corer for recovering short sediment cores. J Paleolimnol 5:285–287

Goldammer J (2013) Vegetation fires and global change: Challenges for concerted international action. A white paper directed to the United Nations and International Organizations. A publication of the Global Fire Monitoring Center (GFMC). Kessel Publishing House, Remagen-Oberwinter, Germany

Hély C, Girardin MP, Ali AA, Carcaillet C, Brewer S, Bergeron Y (2010) Eastern boreal North American wildfire risk of the past 7000 years: a model-data comparison. Geophys Res Lett 37:L14709

Higuera PE, Brubaker LB, Anderson PM, Hu FS, Brown TA (2009) Vegetation mediated the impacts of postglacial climatic change on fire regimes in the south-central Brooks Range, Alaska. Ecol Monogr 79:201–219

Hu FS, Brubaker LB, Gavin DG, Higuera PE, Lynch JA, Rupp TS, Tinner W (2006) How climate and vegetation influence the fire regime of the Alaskan Boreal biome: the Holocene perspective. Mitig Adapt Strateg Glob Change 11:829–846

Hunt BG (2006) The medieval warm period, the little ice age and simulated climatic variability. Clim Dynam 27:677–694

IPCC (2014) Climate change 2014: Synthesis report. Contribution of Working Groups I, II and III to the Fifth Assessment Report of the Intergovernmental Panel on Climate Change. IPCC, Geneva, Switzerland

Jobidon R, Bergeron Y, Robitaille A, Raulier F, Gauthier S, Imbeau L, Saucier J-P, Boudreault C (2015) A biophysical approach to delineate a northern limit to commercial forestry: the case of Quebec's boreal forest. Can J Forest Res 45:515–528

Kivinen S, Berg A, Moen J, Östlund L, Olofsson J (2012) Forest fragmentation and landscape transformation in a reindeer husbandry area in Sweden. Environ Manag 49:295–304

Le Goff H, Flannigan M, Bergeron Y, Girardin MP (2007) Historical fire regime shifts related to climate teleconnections in the Waswanipi area, central Quebec, Canada. Int J Wildland Fire 16:607–618

Long C, Whitlock C, Bartlein PJ (2007) Holocene-scale variations in vegetation and fire history of the Oregon Coast Range, USA. The Holocene 17:917–926

Mansuy N, Gauthier S, Robitaille A, Bergeron Y (2010) The effects of surficial deposit–drainage combinations on spatial variations of fire cycles in the boreal forest of eastern Canada. Int J Wildland Fire 19:1083–1098

Oris F, Ali AA, Asselin H, Paradis L, Bergeron Y, Finsinger W (2014a) Charcoal dispersion and deposition in boreal lakes from 3 years of monitoring: Differences between local and regional fires. Geophys Res Lett 41:6743–6752

Oris F, Asselin H, Finsinger W, Hély C, Blarquez O, Ferland M-E, Bergeron Y, Ali AA (2014b) Long-term fire history in northern Quebec: implications for the northern limit of commercial forests. J Appl Ecol 51:675–683

Payette S (1993) The range limit of boreal tree species in Québec-Labrador: an ecological and palaeoecological interpretation. Rev Palaeobot Palyno 79:7–30

Plummer DA, Caya D, Frigon A, Côté H, Giguère M, Paquin D, Biner S, Harvey R, De Elia R (2006) Climate and climate change over North America as simulated by the Canadian RCM. J Climate 19:3112–3132

Raulier F, Le Goff H, Gauthier S, Rapanoela R, Bergeron Y (2013) Introducing two indicators for fire risk consideration in the management of boreal forests. Ecol Indic 24:451–461

Saucier J-P, Bergeron J-F, Grondin P, Robitaille A (1998) Les régions écologiques du Québec méridional (troisième version). Ministère des Ressources naturelles du Québec. Ministère des Ressources naturelles du Québec, Québec, Canada

Senici D, Lucas A, Chen HYH, Bergeron Y, Larouche A, Brossier B, Blarquez O, Ali AA (2013) Multi-millennial fire frequency and tree abundance differ between xeric and mesic boreal forests in central Canada. J Ecol 101:356–367

Viau AE, Gajewski K (2009) Reconstructing millennial-scale, regional paleoclimates of boreal Canada during the Holocene. J Climate 22:316–330

Viau AE, Gajewski K, Sawada MC, Fines P (2006) Millennial-scale temperature variations in North America during the Holocene. J Geophys Res 111:D09102

Wein RW, Burzynski MP, Sreenivasa BA, Tolonen K (1987) Bog profile evidence of fire and vegetation dynamics since 3000 years BP in the Acadian forest. Can J Bot 65:1180–1186

Wright HE Jr, Mann DH, Glaser PH (1984) Piston corers for peat and lake sediments. Ecology 65:657–659

20

Adaptive economic and ecological forest management under risk

Joseph Buongiorno[1*] and Mo Zhou[2]

Abstract

Background: Forest managers must deal with inherently stochastic ecological and economic processes. The future growth of trees is uncertain, and so is their value. The randomness of low-impact, high frequency or rare catastrophic shocks in forest growth has significant implications in shaping the mix of tree species and the forest landscape. In addition, the fluctuations of wood prices influence greatly forest revenues.

Methods: Markov decision process models (MDPs) offer a rigorous and practical way of developing optimum management strategies, given these multiple sources of risk.

Results: Examples illustrate how such management guidelines are obtained with MDPs for combined ecological and economic objectives, including diversity of tree species and size, landscape diversity, old growth preservation, and carbon sequestration.

Conclusions: The findings illustrate the power of the MDP approach to deal with risk in forest resource management. They recognize that the future is best viewed in terms of probabilities. Given these probabilities, MDPs tie optimum adaptive actions strictly to the state of the forest and timber prices at decision time. The methods are theoretically rigorous, numerically efficient, and practical for field implementation.

Keywords: Risk; Management; Economics; Ecology; Markov; Optimization

Background

Forest planning and decision making models are often deterministic. They assume that the future is known, or they reduce uncertain variables to their expected values, with the hope that the results will work, at least "on average". However, the future is highly uncertain[a], and it can only be described in terms of probabilities, be they "objective probabilities" derived from factual data, or "subjective probabilities" reflecting a personal belief (de Finetti 1937).

There is a large literature concerning the role of risk in forest ecosystems (see e.g. Kant and Alavalapati 2014, p. 307–369). But there are few applications of Markov decision process models (MDPs) in forestry, a paucity also observed in the general operations research literature (White 1993). Nevertheless, MDPs are powerful and general methods that deserve more attention in forest management. This was recognized early on in Lembersky and Johnson (1975), and Lembersky (1976) who dealt with applications to the management of forest plantations. Their pioneering work was followed by several applications to the uneven-aged management of various forest types for financial objectives (Kaya and Buongiorno 1987), ecological objectives (Zhou and Buongiorno 2006), and combined financial and ecological objectives (Lin and Buongiorno 1998; Rollin et al. 2005; Zhou et al. 2008a, b).

The purpose of this paper is to describe how stochastic simulation and MDPs can be used to deal with decision making under risk in forestry. The data and the specific applications described below deal with mixed loblolly pine (*Pinus taeda* L.)-hardwood stands in the southern United States. However, the methods are general and can be applied to a variety of forest ecosystems, ranging from even-aged monospecific plantations to mixed-species uneven-aged forests, and from single stands to large forest areas. The objective throughout is to develop simple adaptive management guidelines that can be readily applied in the field.

* Correspondence: jbuongio@wisc.edu
[1]University of Wisconsin-Madison, Madison, USA
Full list of author information is available at the end of the article

The paper is organized as follows. The next section presents the methods: stochastic forest growth models and their reduction to Markov chains which are then used for prediction and optimization. This is followed by the results which show first some of the consequences of (wrongly) predicting the evolution of forest ecosystems without taking into account the effects of random ecological, climatic, or economic shocks. Next are shown predictions of the long-term evolution of forest ecosystems with and without management and the implications for economic and ecological criteria. And, MDP results optimizing decisions adaptively for discounted or undiscounted economic and ecological objectives, and combinations thereof. The discussion and conclusion deal with limitations of the methods and potential improvements.

Methods

The guiding principle of the methods is to represent the evolution of a forest ecosystem subject to various impact variables, such as prices and catastrophic disturbances, with Markov chains: probability matrices describing the frequencies of transition of each variable between discrete states. After the transition probabilities between the system states have been determined, the Markov model is used to predict future system states subject to specific management policies. Furthermore, decision process models (MDPs), optimization techniques based on Markov chains, are used to determine the best management policy for a particular objective. A policy is a set of rules that prescribe a decision for each observed system state at decision time. Various adaptive strategies are obtained in this way depending on the objective function which may deal with discounted or undiscounted criteria and constraints that may limit the decision domain.

Markov forest model

The different possible states of a forest stand are described in a practical manner by the level of basal area in trees of different sizes and species. Basal area per hectare, the sum of the cross sectional area of the trees, typically measured at breast height, is one of the simplest measures of forest stand density and composition. To be practical, the number of tree size and species categories should be kept low[b]. For example Table 1 shows the definition of forest stand states used here, with the basal area in three tree size classes (small, medium, large), and two species categories (softwoods, hardwoods). With two levels of basal area, high or low, indicating below average or above average basal area in each species and size category, there are $2^6 = 64$ possible stand states. The example stand state #13 in Table 1, denoted (001, 100) has high basal area in large pine trees and small hardwood trees, and low basal area in the other categories (Zhou and Buongiorno 2006).

Table 1 Definition of forest stand states according to basal area level by species and tree size

	Pines ($m^2 \cdot ha^{-1}$)			Hardwoods ($m^2 \cdot ha^{-1}$)		
	Small	Medium	Large	Small	Medium	Large
Low	≤2.5	≤4.3	≤3.6	≤2.9	≤1.2	≤1.7
High	>2.5	>4.3	>3.6	>2.9	>1.2	>1.7
State #13	0	0	1	1	0	0

The example state #13 has high basal area in large pine trees and small hardwoods, and low basal area in other categories. Source: Zhou and Buongiorno (2006).

In MDPs, the evolution of stands over time is described by $\mathbf{T} = [p(s'|s)]$, a matrix of yearly transition probabilities between stand states s and s'. These probabilities cannot be established directly by direct observation due to lack of sufficient data in each stand. Simulation methods are used instead. A typical stochastic simulation model of stand growth suitable for this purpose has the form:

$$\mathbf{y}_{t+1} = \mathbf{G}_t\mathbf{y}_t + \mathbf{i}_t + \boldsymbol{\varepsilon}_t \quad (1)$$

where $\mathbf{y}_t = [y_{ijt}]$ is the vector of the number of trees per unit area of species i and size j at time t. $\mathbf{G}_t$ is a matrix of parameters describing the probability that a tree of a particular species stays alive and grows into a higher size class from t to $t+1$. The vector $\mathbf{i}_t$ refers to recruitment, the number of trees of a particular species that enters the smallest diameter class from t to $t+1$. Both $\mathbf{G}_t$ and $\mathbf{i}_t$ vary depending on the stand condition, $\mathbf{y}_t$, through the stand density that affects tree growth and recruitment. $\boldsymbol{\varepsilon}_t$ is a vector of random disturbances.

In the following applications the parameters of model (1) are estimated from observations on permanent sample plots in the mixed loblolly pine (*Pinus taeda* L.)-hardwood forests of the Southern United States (Schulte et al. 1998)[c]. The differences between the deterministic predictions of model (1), and the observations on the plots give observations on the random shocks $\boldsymbol{\varepsilon}_t$ due to ice storms, wind, insect outbreaks, abnormal weather, etc.... that have affected forest growth during the observation period (Zhou and Buongiorno 2004).

To get the transition probabilities between stand states $p(s'|s)$ model (1) is used to predict the future stand state, s', of a random initial stand in state s, by bootstrapping a random shock $\boldsymbol{\varepsilon}_t$ from the set of the observed shocks. This is repeated a sufficient number of times to obtain stable estimates of the transition probabilities $p(s'|s)$ (Zhou 2005, p. 51).

Due to the short time typically covered by plot observations, the transition probability matrix $\mathbf{T} = [p(s'|s)]$ obtained in this manner reflects mostly small scale, high frequency disturbances. In addition, catastrophic, low frequency events, such as hurricanes must be added where needed, based on weather and geographic data. Table 2 shows the non-zero elements of the 64 × 64 transition probability matrix computed in this way and used in the results of this paper. The entry in row and column 000,000 means that if a stand has now low basal area in all species and size categories, the probability is 0.797 that it will be in the same state next year. The entry in row 000,000 and column 000,001 means that the same stand has a 0.027 probability of growing into a stand with high basal area in the largest size class. Table 2 also indicates that regardless of its current state, there is a 0.025 probability that the stand will be next year in state 000,000 due to a catastrophic event, a hurricane in this example (Zhou and Buongiorno 2006).

The resulting **T** matrix is then used to predict future states:

$$\mathbf{\Pi}_{t+1} = \mathbf{\Pi}_t \mathbf{T} \quad t = 0, 1, 2... \tag{2}$$

where $\mathbf{\Pi}_t = [\pi_{st}]$ is the vector of probabilities of stand state s at time t, and the steady-state probabilities, $\mathbf{\Pi} = \mathbf{\Pi}_t = \mathbf{\Pi}_{t+1}$ are obtained by solving the system of equations:

$$\mathbf{\Pi}(\mathbf{I}-\mathbf{T}) = \mathbf{0} \tag{3}$$

where $\mathbf{\Pi} = [\pi_s]$ is the vector of long-term steady state probabilities for each stand state s, and $\mathbf{I}$ is the identity matrix. As equation (3) indicates, the long-term probabilities are independent of the initial state. Each probability is also interpreted as the fraction of a large forest landscape that is in each state s in the long run when only natural disturbances occur. As indicated below, an equation equivalent to (3) can also be used to predict the steady state probabilities when a particular management regime is followed indefinitely.

Markov price model

The financial returns from forest management depend critically on the price of wood. Perfect prediction of future prices is not possible, but probable price levels and changes can be drawn from the behavior of past prices. In parallel with the procedure for forest stands, the method describes market fluctuations by transition probabilities between price levels. For example, based on past data prices may be classified as low, medium, or high. And the transition probabilities between price levels are obtained from a price model such as[d]:

$$Q_t = \alpha + \beta Q_{t-1} + \nu_t \tag{4}$$

where Q_t is the price in year t and α, β are parameters estimated by regression analysis of past data. The matrix of transition probabilities between price levels is $\mathbf{M} = [p(m'|m)]$ where $p(m'|m)$ is the probability that next year's price level is m' given the current price level m. These probabilities are obtained by simulations and repeated sampling from the distribution of the residual ν_t. An example of price transition probabilities used later in this paper is in Table 3 (Zhou 2005). Given a currently low price level (less than $\$84 \cdot m^{-3}$), the probability that it will be low following year is 0.82. The probability that it will be medium (between $\$84 \cdot m^{-3}$ and $\$94 \cdot m^{-3}$) is 0.10, and the probability that it will be high (above $\$94 \cdot m^{-3}$) is 0.08. Table 3 illustrates the high price autocorrelation (price stickiness): there is a high probability that the price stays at the same level from year to year.

System states and transition probabilities

Given transition probability matrices between stand states $\mathbf{T} = [p(s'|s)]$ and price levels $\mathbf{M} = [p(m'|m)]$, the transition probabilities of the forest-price system are:

$$\mathbf{S} = [p(j|i)] = [p(s'|s)p(m'|m)] \tag{5}$$

where i is the current system state (stand state s and price level m), and j is next year's system state (stand state s' and price level m'). For example, with the 64 stand states in Table 2 and the three price levels in Table 3, there are 192 system states and 192 × 192 transition probabilities.

Decisions and immediate rewards

Decisions consist in moving instantly from one stand state to another by harvesting some of the trees. It is assumed that market prices are exogenous, i.e. the decisions have no effect on prices. A decision produces a stand state with less basal area, less carbon sequestered in the stand, and a different composition of tree species and size. It also generates revenues that depend on the volume of timber harvested and on the price at decision time[e]. A policy consists of a set of decisions applied systematically to each stand-price state. If the policy is applied to an entire forest, decisions alter the distribution of stand states and thus the landscape diversity, and possibly the part of the forest that stays in an old-growth (late-seral) state.

Ecological criteria include a variety of indices concerning the state of the stand and of the landscape induced by a policy. Although several diversity indices are possible (Magurran 1988), Shannon's index (Shannon 1948) was used here throughout. For example, applied to tree species diversity Shannon's index is:

$$H = -\sum_{a=1}^{n} f_a \ln(f_a) \tag{6}$$

where H is the tree-species diversity of a stand state, f_a is the fraction of trees of species a, and n is the number of species. The fraction f_a is preferably based on the basal area of trees rather than their number, to give more importance to the large trees. With this index, species

Table 2 Transition probabilities between stand states

State # at *t*	Stand compositiona[1]	State # at *t* + 1 year (transition probability)
1	000,000	1(0.797), 2(0.027), 3(0.011), 5(0.055), 9(0.062), 17(0.019), 33(0.029)
2	000,001	1(0.025), 2(0.782), 4(0.014), 6(0.064), 10(0.053), 18(0.027), 34(0.035)
3	000,010	1(0.025), 3(0.78), 4(0.029), 7(0.049), 11(0.064), 19(0.025), 35(0.028)
4	000,011	1(0.025), 4(0.817), 8(0.049), 12(0.059), 20(0.022), 36(0.027)
5	000,100	1(0.036), 5(0.777), 6(0.022), 7(0.045), 13(0.063), 21(0.028), 37(0.030)
6	000,101	1(0.025), 5(0.010), 6(0.794), 8(0.058), 14(0.053), 22(0.028), 38(0.032)
7	000,110	1(0.025), 7(0.826), 8(0.027), 15(0.067), 23(0.027), 39(0.029)
8	000,111	1(0.025), 8(0.857), 16(0.046), 24(0.031), 40(0.042)
9	001,000	1(0.025), 9(0.842), 10(0.022), 13(0.060), 25(0.025), 41(0.026)
10	001,001	1(0.025), 9(0.016), 10(0.843), 12(0.012), 14(0.052), 26(0.017), 42(0.035)
11	001,010	1(0.025), 11(0.822), 12(0.047), 15(0.050), 27(0.023), 43(0.033)
12	001,011	1(0.025), 12(0.893), 16(0.052), 44(0.030)
13	001,100	1(0.025), 13(0.875), 14(0.018), 15(0.033), 29(0.018), 45(0.031)
14	001,101	1(0.025), 10(0.013), 13(0.010), 14(0.851), 16(0.047), 30(0.023), 46(0.031)
15	001,110	1(0.025), 15(0.850), 16(0.050), 31(0.026), 47(0.050)
16	001,111	1(0.025), 12(0.013), 15(0.013), 16(0.917), 32(0.019), 48(0.013)
17	010,000	1(0.067), 17(0.745), 18(0.020), 21(0.051), 25(0.083), 49(0.034)
18	010,001	1(0.025), 2(0.033), 17(0.012), 18(0.739), 20(0.011), 22(0.054), 26(0.102), 50(0.024)
19	010,010	1(0.025), 3(0.039), 19(0.731), 20(0.026), 23(0.047), 27(0.092), 51(0.030)
20	010,011	1(0.025), 4(0.040), 20(0.750), 24(0.058), 28(0.095), 52(0.032)
21	010,100	1(0.025), 5(0.027), 21(0.763), 22(0.023), 23(0.038), 29(0.099), 53(0.025)
22	010,101	1(0.025), 6(0.036), 18(0.012), 22(0.800), 24(0.036), 30(0.066), 54(0.024)
23	010,110	1(0.025), 7(0.037), 23(0.795), 24(0.030), 31(0.089), 55(0.025)
24	010,111	1(0.025), 8(0.066), 24(0.745), 32(0.120), 56(0.013)
25	011,000	1(0.025), 9(0.035), 25(0.828), 26(0.022), 29(0.061), 57(0.030)
26	011,001	1(0.025), 10(0.039), 26(0.837), 28(0.016), 30(0.053), 58(0.030)
27	011,010	1(0.025), 11(0.036), 27(0.816), 28(0.048), 31(0.045), 59(0.030)
28	011,011	1(0.025), 12(0.040), 28(0.859), 32(0.045), 60(0.031)
29	011,100	1(0.025), 13(0.045), 25(0.012), 29(0.825), 30(0.028), 31(0.041), 61(0.023)
30	011,101	1(0.025), 14(0.031), 30(0.879), 32(0.048), 62(0.017)
31	011,110	1(0.025), 15(0.030), 31(0.863), 32(0.050), 63(0.032)
32	011,111	1(0.025), 16(0.062), 28(0.014), 32(0.872), 64(0.027)
33	100,000	1(0.066), 33(0.738), 34(0.019), 35(0.013), 37(0.050), 41(0.054), 49(0.060)
34	100,001	1(0.025), 2(0.045), 34(0.776), 38(0.050), 42(0.044), 50(0.060)
35	100,010	1(0.025), 3(0.059), 35(0.703), 36(0.056), 39(0.051), 43(0.051), 51(0.055)
36	100,011	1(0.025), 4(0.062), 36(0.702), 40(0.062), 44(0.055), 48(0.011), 52(0.084)
37	100,100	1(0.025), 5/90.038), 37(0.772), 38(0.015), 39(0.040), 45(0.054), 53(0.056)
38	100,101	1(0.025), 6(0.029), 38(0.812), 40(0.023), 46(0.052), 54(0.059)
39	100,110	1(0.025), 7(0.065), 39(0.786), 40(0.020), 47(0.040), 55(0.065)
40	100,111	1(0.025), 8(0.049), 40(0.845), 48(0.031), 56(0.049)
41	101,000	1(0.025), 9(0.047), 41(0.777), 42(0.026), 43(0.011), 45(0.041), 57(0.072)
42	101,001	1(0.025), 10(0.056), 41(0.019), 42(0.780), 46(0.052), 58(0.068)
43	101,010	1(0.025), 11(0.060), 43(0.774), 44(0.027), 47(0.058), 59(0.056)
44	101,011	1(0.025), 12(0.049), 44(0.822), 48(0.049), 60(0.038), 64(0.016)

Table 2 Transition probabilities between stand states *(Continued)*

45	101,100	1(0.025), 13(0.058), 41(0.015), 45(0.789), 46(0.023), 47(0.035), 61(0.055)
46	101,101	1(0.025), 14(0.034), 45(0.015), 46(0.810), 48(0.050), 62(0.065)
47	101,110	1(0.025), 15(0.039), 47(0.850), 48(0.043), 63(0.043)
48	101,111	1(0.025), 16(0.074), 48(0.800), 64(0.101)
49	110,000	1(0.025), 17(0.039), 33(0.014), 49(0.752), 50(0.024), 53(0.049), 57(0.098)
50	110,001	1(0.025), 18(0.047), 34(0.025), 50(0.768), 54(0.055), 58(0.079)
51	110,010	1(0.025), 19(0.039), 35(0.018), 51(0.750), 52(0.035), 55(0.046), 59(0.087)
52	110,011	1(0.025), 20(0.051), 28(0.010), 36(0.017), 51(0.014), 52(0.730), 56(0.031), 60(0.122)
53	110,100	1(0.025), 21(0.035), 37(0.021), 53(0.775), 54(0.017), 55(0.031), 61(0.096)
54	110,101	1(0.025), 22(0.032), 38(0.015), 53(0.017), 54(0.783), 56(0.042), 62(0.086)
55	110,110	1(0.025), 23(0.044), 39(0.015), 55(0.783), 56(0.031), 63(0.102)
56	110111	1(0.025), 24(0.061), 40(0.036), 52(0.012), 56(0.757), 64(0.109)
57	111,000	1(0.025), 25(0.062), 41(0.018), 57(0.821), 58(0.023), 61(0.050)
58	111,001	1(0.025), 26(0.036), 42(0.018), 57(0.016), 58(0.852), 62(0.052)
59	111,010	1(0.025), 27(0.058), 43(0.016), 59(0.832), 60(0.033), 63(0.037)
60	111,011	1(0.025), 28(0.025), 44(0.025), 60(0.860), 64(0.064)
61	111,100	1(0.025), 29(0.062), 45(0.014), 61(0.835), 62(0.022), 63(0.043)
62	111,101	1(0.025), 30(0.042), 46(0.019), 58(0.019), 61(0.019), 62(0.842), 64(0.034)
63	111,110	1(0.025), 31(0.036), 47(0.025), 59(0.017), 63(0.846), 64(0.050)
64	111,111	1(0.025), 32(0.068), 48(0.017), 64(0.890)

Source: Zhou and Buongiorno (2006). [1]Basal area in pulpwood, small sawtimber, and large saw timber of pines (first three digits), or hardwoods (last three digits) 1 = higher than current average, 0 = lower than current average.

diversity is between 0 and $\ln(n)$, and it is at the maximum when f_a is the same for all species. The tree size diversity of a stand state is calculated in a similar way.

The forest landscape diversity is also expressed similarly with f_s, the fraction of the forest landscape in state s. Another useful criterion is the fraction of the forested landscape in "old growth" state. For some forest types, old growth or late-seral stands have been defined in previous studies (Hummel and Calkin 2005). Here, old-growth is defined as the most frequent states that develop in the steady state predicted with equation (3) in the absence of harvest (Zhou and Buongiorno 2006). With the data in Table 2, the five old-growth states are 001,011; 001,111; 011,111; 101,111; and 111,111, which all have high basal area in the largest softwoods and hardwood trees.

Table 3 Annual transition probability matrix between price levels

	Price $t+1$		
	Low	Medium	High
Price t	$< \$84 \cdot m^{-3}$	$[\$84 \cdot m^{-3}, \$94 \cdot m^{-3}]$	$> \$94 \cdot m^{-3}$
Low	0.82	0.10	0.08
Medium	0.11	0.78	0.11
High	0.07	0.12	0.81

The data on the expected timber volume and ecological criteria by stand state are obtained during the simulations with model (1) that give the transition probabilities. The immediate ecological reward of a decision is the characteristic of the state induced by the decision, for example, its tree diversity. The immediate financial reward is equal to the change in volume obtained by harvesting the stand from the current state to another, multiplied by the price level at decision time. The carbon sequestered was estimated from the amount of growing stock left after harvest, assuming $1.24\ t \cdot m^{-3}$ of CO_2e for pulpwood and $1.57\ t \cdot m^{-3}$ for sawtimber (AFC 2014).

In sum, each immediate reward, monetary or ecological, is summarized by a vector $V = [v_{ik}]$ where v_{ik} is the immediate reward when the stand-price system is in state i and the decision is k (which changes the stand state or leaves it intact).

Consequences of management policies

A policy, c, is described by a particular decision matrix:

$$\mathbf{D}_c = [p_c(k|i)] \quad (7)$$

where $p_c(k|i)$ is the probability of harvesting a stand to state k, given the stand-price state i under policy c. If the harvest policy is deterministic (the same action is taken all the time in each state) the probability of a particular decision is 0 or 1. To investigate the consequences of continuing the current management, independently of the price level, $\mathbf{D}_c$ is based on the frequency at which harvests have changed the state of the sample plots during the observation period.

The transition probabilities between stand-price states given a particular policy are then obtained from:

$$\mathbf{E}_c = [p_c(j|i)] = [p_c(k|i)p(j|k)] \tag{8}$$

where $p_c(j|i)$ is the probability of ending in stand-price state j after one period from stand-price state i under policy c, and $p(j|k)$ is the probability of ending in stand-price state j after one period from stand state k. Substituting $\mathbf{E}_c$ for $\mathbf{T}$ in equation (3) gives the probability vector $\mathbf{\Pi}_c = [\pi_i^c]$ where π_i^c is the steady state probability of stand-price state i under policy c.

Predicting expected undiscounted rewards

The expected undiscounted reward, $\mathbf{R}$, such as the expected tree species diversity, or the expected annual timber revenue, that results in the long run by following a specific policy is:

$$\mathbf{R} = \mathbf{V}'\mathbf{\Pi}_c \tag{9}$$

where $\mathbf{V}' = [r_i^c]$ is a row vector of the immediate reward in stand-price state i under policy c. For financial returns r_i^c is the timber income when the stand-price state is i and the policy is c. The expected long-run diversity of tree species and size, the expected basal area and carbon sequestered, and the expected annual production under a particular policy are obtained in similar fashion. The expected long-run forest landscape diversity is based on equation (6) where p_a is replaced by the steady state probability of each stand state, π_s^c, under the policy c.

The expected interval between harvests (or cutting cycle), in years, is the inverse of the yearly probability of a harvest. Let $h_i^c = 1$ if the decision called for by policy c is a harvest when the stand-price state is i. Then, the annual probability of a harvest is:

$$\pi_h = \sum_i \pi_i^c h_i^c \tag{10}$$

And the expected cutting cycle is:

$$C = \frac{1}{\pi_h} \tag{11}$$

Predicting expected discounted rewards

Some criteria are typically discounted, future outcomes receiving less weight than current ones. This is especially the case for timber revenues, but ecological criteria may also be discounted (Boscolo et al. 1997; Howarth 2009). For a particular management policy, c, the present value of a criterion is obtained by solving the following system of equations (Hillier and Lieberman 2005, p. 785):

$$V_i^c = r_i^c + \frac{1}{1+\rho}\sum_j p_c\left(j\middle|i\right) V_j^c \quad \forall i \tag{12}$$

where V_i^c is the present value of the criterion of interest over an infinite horizon for an initial stand-price state i, when the harvest policy c is followed, with an immediate reward r_i^c and a yearly interest rate ρ. In the results presented below, the interest rate was set at 3.08% per year, the real yield on AAA corporate bonds from 2001 to 2013, using the consumer price index as deflator (U.S. Government 2014). This choice implies a long-term, low-risk investment by a conservative investor.

Then, the average expected present value per ha over an entire forest with a particular initial distribution of stands and probability of price level is:

$$PV = \sum_{i=1}^{192} V_i^c \pi_i^0 \tag{13}$$

where π_i^0 is the initial probability of stand-price state i.

Optimizing management policies

In parallel with the predictions of the effects of specific management policies, the optimization of policies differs depending on whether the rewards are discounted or undiscounted.

Optimizing expected discounted rewards

The policy that maximizes an expected discounted reward over an infinite horizon, such as the net present value of the timber income, without any other constraint, is found by solving the following linear programming program, originally due to d'Epenoux (1963).

$$\begin{aligned} &\max_{y_{ij}} \sum_i \sum_k r_{ik} y_{ik} \\ &\text{subject to:} \\ &\sum_k y_{jk} - \frac{1}{1+\rho}\sum_i \sum_k p(j|i,k) y_{ik} = \beta_j \quad \forall j \\ &y_{ik} \geq 0 \forall i,k \end{aligned} \tag{14}$$

where y_{ik} is the total discounted probability of being in state i and making decision k over an infinite horizon[f]. β_j is the probability that the system starts in stand-price state j. This probability must be strictly positive, i.e. $\beta_j > 0$ and $\sum \beta_j = 1$. After finding the best solution, y_{ik}^*, the best decisions are obtained from:

$$D_{ik} = \frac{y^*_{ik}}{\sum_k y^*_{ik}} \quad \forall i, k \tag{15}$$

where D_{ik} is the probability of making decision k when the stand-price state is i. The optimal policy is deterministic ($D_{ik} = 0$ or 1), and it is independent of the initial condition, i.e. of the β_j's (Hillier and Lieberman 2005, p. 921), although the value of the objective function, the maximum discounted value of the reward, depends very much on the initial condition[g].

Optimizing expected undiscounted rewards

Another linear program gives the policy that maximizes an unconstrained undiscounted reward (Manne 1960). Examples of undiscounted criteria are the expected species diversity observed yearly in steady state, or the expected annual wood harvest.

$$\begin{aligned} &\max_{y_{ij}} \sum_i \sum_k r_{ik} z_{ik} \\ &\text{subject to:} \\ &\sum_k z_{jk} - \sum_i \sum_k p(j|i,k) z_{ik} = 0 \qquad \forall j \end{aligned} \tag{16}$$

where z_{ik} is the steady-state probability of stand-price state i and decision k.

Given the best solution, z^*_{ik}, the corresponding best decisions are:

$$D'_{ik} = \frac{z^*_{ik}}{\sum_k z^*_{ik}} \quad \forall i, k \tag{17}$$

The best decisions obtained with equation (17) are still deterministic ($D'_{ik} = 0$ or 1) and independent of the initial stand-price state. And, the maximum value of the objective function, such as the maximum expected undiscounted species diversity over an infinite time horizon, is also independent of the initial condition.

Multiple objectives

Policies that best meet multiple objectives simultaneously are obtained by modifying the objective function of models (14) and (16), or/and by adding constraints. In the results shown below the models were kept linear by expressing both the objective function and the constraints in undiscounted or discounted terms.

For example, in a discounted framework, a policy that maximizes the net present value of timber income, while keeping a specific fraction of the forest area in old growth state, is obtained with a model consisting of equations (14)–(15) and the additional constraint:

$$\sum_i \sum_k o_{ik} y_{ik} \geq \alpha O^* \tag{18}$$

where $oik = 1$ if state k is an old-growth state, $oik = 0$ otherwise. O^* is the maximum, unconstrained, discounted value of the fraction of old growth, obtained with model (16) to (17), and α is the desired fraction of this maximum that must be maintained by the policy.

As an example of an undiscounted framework, the maximum expected annual harvest that keeps the sequestered carbon dioxide equivalent (CO_2e) at a specific level is obtained by solving model (16)–(17) with r_{ik} equal to the amount of timber harvest resulting from decision k in state i, and the additional constraint:

$$\sum_i \sum_k c_{ik} y_{ik} \geq C^* \tag{19}$$

where c_{ik} is the amount of CO_2e stored in the stand state that results from decision k in state i, and C^* is the desired expected value of the stored CO_2e.

Results

Effects of disturbances on predicted stand growth

Figure 1 illustrates the importance of recognizing random shocks in predicting the evolution of forest ecosystems. The data result from long-term simulations of a forest stand of mixed hardwoods-loblolly pines in the south of the United States, growing without human intervention (Zhou and Buongiorno 2004). The figure shows long-term predictions of total stand basal area, and basal area of pine trees. The continuous line refers to predictions obtained with the deterministic part only of model (1), ignoring the random shocks, $\boldsymbol{\varepsilon}_t$. The jagged shows instead the predictions obtained with the random shocks reflecting the high-frequency, low impact disturbances derived from the plot data, without catastrophic events.

While the total basal area is qualitatively similar for both models, the predictions of the basal area of pine trees are very different. Ignoring the random shocks, pines totally disappear in about 300 years. Instead, when the random disturbances are taken into account the basal area of the pine trees is never less than 20 $m^2 \cdot ha^{-1}$, about a third of the total basal area. Forest scientists agree that stochastic disturbances "can be a major determinant of forest structure" (Oliver and Larson 1990). In the case of loblolly pine, recurrent fires are typical in the south of the United States. Lobllolly pines are more fire resistant than hardwoods. Fires also cause openings that favor pine regeneration. The simulations with stochastic shocks are thus more likely to predict correctly the long-term evolution of loblolly pine-hardwood stands than the deterministic version.

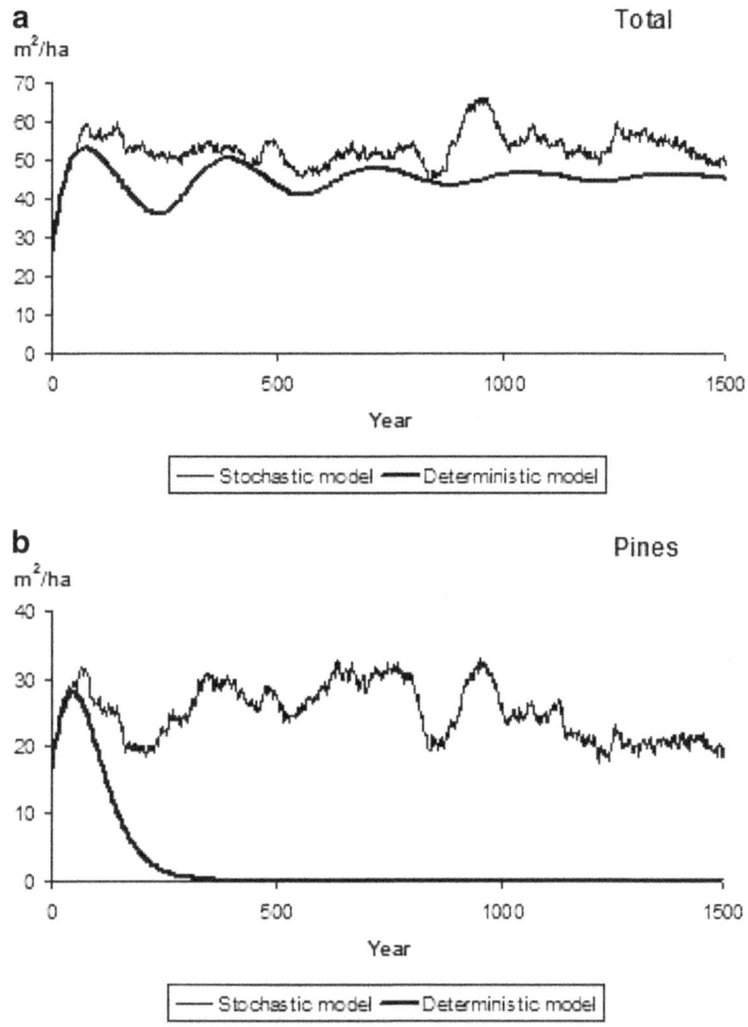

Figure 1 Simulated total basal area (a) and basal area of pine trees (b) in an uneven-aged loblolly pine stand with a deterministic model (smooth line) or a stochastic model with low-impact high-frequency shocks but no catastrophic event (jagged line). Source: Zhou and Buongiorno (2004).

Effects of catastrophes on forest landscape

In addition to the high-frequency, low-impact disturbances, the forests are also subject to low-frequency catastrophes, such as hurricanes in the region investigated here. The long-term effects of catastrophic events on the forest landscape were predicted with the Markov model in Table 2, with and without the 0.025 probability of a catastrophe appearing in the first column. The fraction of the landscape that would be present in the long run (steady state) in each of the 64 possible states was predicted with equation (3).

Figure 2a shows the prediction without catastrophic events, while Figure 2b shows the prediction with catastrophes. Both figures show the fraction of the landscape area occupied in the steady state by each of the five old-growth states, and the aggregate fraction occupied by the 59 other states. Without catastrophic disturbances, 81% of the landscape was in an old growth state, and 38% consisted of stands in state #16 (001,111) with high basal area in hardwoods in all three size categories, and high basal area in large pine trees. However, when catastrophic events were taken into account, only 29% of the landscape was in an old-growth state and the dominant old-growth state #16 occupied only 12% of the landscape area. In sum, the long-run expected diversity of the forest landscape measured with Shannon's index (6) was 80% higher due to catastrophic disturbances than it would have been without them.

Economic and ecological consequences of current management

The economic consequences of the current management, were expressed by the net present value (NPV) of timber income over an infinite horizon. The NPV was derived

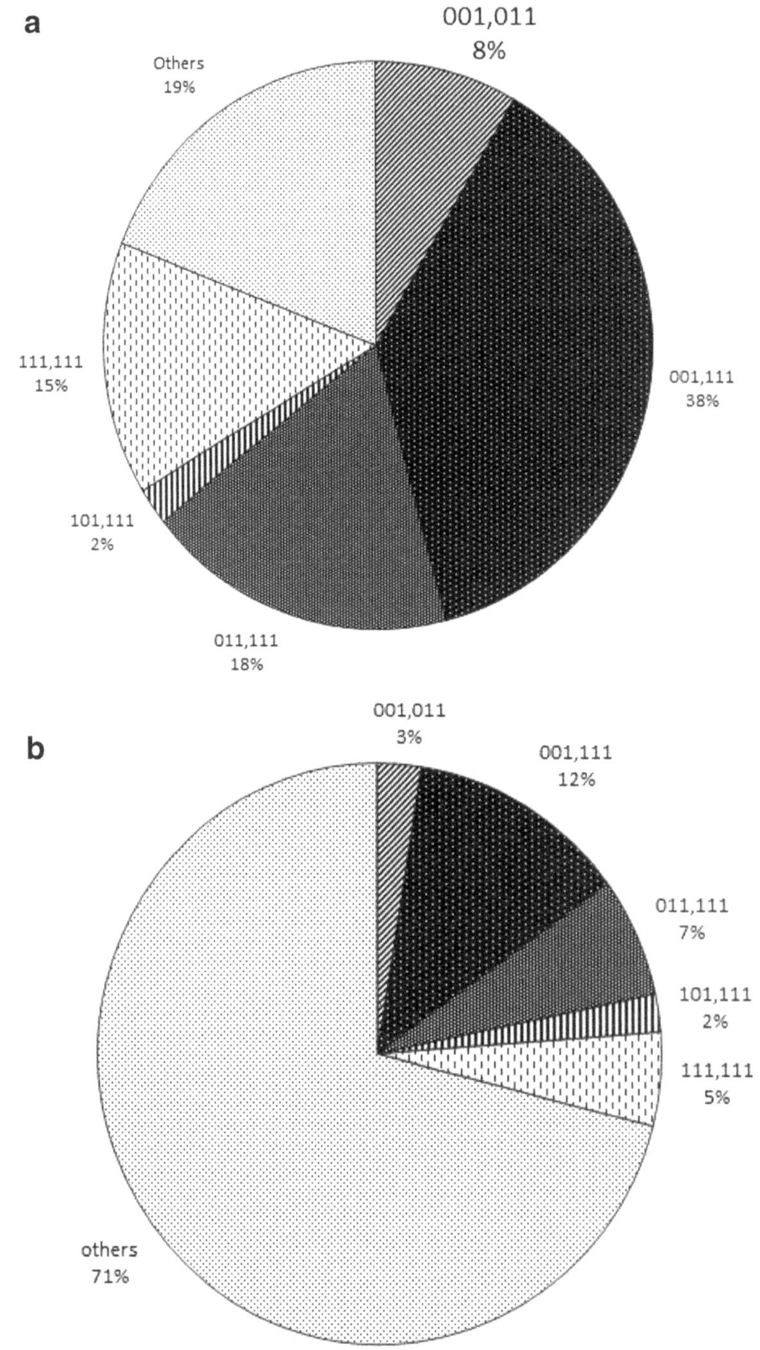

Figure 2 Expected distribution of forest area in steady state, without catastrophic disturbances (a) and with catastrophic disturbances (b).

with equation (12) with the initial distribution of stands reflected in the plot data and a low initial price level[h]. Other consequences, such as the annual wood production and the diversity indices were undiscounted expected values in steady state, obtained with equation (9), while equation (11) gave the expected length of the cutting cycle. The results in Table 4 compare the consequences of the current management with those arising from forest growth alone without management.

Continuing the current management indefinitely generated an NPV of $1,339 $\cdot$ ha^{-1} in timber revenues and an average expected production of 1.9 $m^3 \cdot ha^{-1} \cdot y^{-1}$ over the entire forest. Due to this harvest the expected average basal area was 2 $m^2 \cdot ha^{-1}$ lower with the current management than it would be without human intervention.

Using the diversity indices without management as a basis of comparison, the current management increased tree species diversity slightly (1%). Like catastrophic events,

Table 4 Long-term effects of the current management on economic return and ecological criteria, compared with the outcome without management

Criteria	Unit	With current management	Without management
NPV of harvested wood	$\$.ha^{-1}$	1339	–
Sawtimber harvest	$m^3 \cdot ha^{-1} \cdot y^{-1}$	1.5	–
Pulpwood harvest	$m^3 \cdot ha^{-1} \cdot y^{-1}$	0.4	–
Basal area	$m^2 \cdot ha^{-1}$	19	22
Species diversity[a]	%	101	100
Size diversity[a]	%	98	100
Landscape diversity[a]	%	103	100
Old growth fraction	%	18	29
CO_2e sequestered	$t.ha^{-1}$	203	244
Cutting cycle	year	11	–

[a]Relative to the level without management.

management also increased landscape diversity, though to a lesser extent (3%). However, by harvesting large trees it decreased the expected tree size diversity by 2%.

The fraction of old-growth in the forest maintained by the current management was 11% lower than it would be in a natural forest, and the value of the average carbon sequestration was 42 $t \cdot ha^{-1}$ lower. All these outcomes stemmed from harvests in individual stands that occurred at average intervals of 7 years under the current management.

Optimizing financial objectives

The policy that maximized the expected NPV was obtained by solving model (14)–(15), with a uniform distribution of initial stand-price states (i.e. $\beta_j = 1/192 \, \forall \, j$). This best policy was then used to obtain with equations (12)–(13) the expected NPV given the stand states distribution in the region of interest, as reflected by the plot data, and the current price level ("low" in 2014 according to the definitions in Table 3).

The results in Table 5 show how the best decisions adapted to the stand state and price level observed at decision time. For example, for a stand in state #64 (111,111), at low price, the best decision was to cut the stand to state #52 (110,011) by reducing the basal area of the large pine trees and of the small hardwood trees. If instead the current price was average, the best decision was to cut the stand to state #3 (000,010) which meant reducing further the basal area of the small and medium pine trees. Last, if the price was high, the best decision was to reach stand state #1 (000,000) by reducing the basal area to a low level in all species and size categories. In contrast, under current management, the plot data revealed that a stand in this same state #64 (111,111) was left intact 85% of the time and cut to state #44 (101,011) only 15% of the time.

The expected consequence of this best financial policy for timber production and the various ecological indicators is in Table 6. The expected NPV and annual timber production were more than four times higher than with the current policy[i]. However, the expected basal area, and the diversity of tree species and size were all lower. The landscape diversity was less than a third of its level with the current policy, and no old growth was left in the entire landscape. The amount of CO_2e sequestered was less than half, while the frequency of harvests was about double that with the current policy.

Optimizing ecological objectives

Two criteria were used as applications of policies that best meet ecological objectives: maximizing tree species diversity, or carbon sequestration. The results in Table 7 show that at most, tree species diversity could be 7% higher than with the current management. Maximizing tree species diversity would not hurt financial returns, as in fact the expected NPV was $\$713 \cdot ha^{-1}$ higher than with the current management as it led to substantially higher harvest of sawtimber and pulpwood. This more intensive harvest, however, reduced the fraction of the forest in old-growth state by 5 percentage points. Accordingly, the landscape diversity was reduced by approximately 30%, and the amount of CO_2e sequestered was 52 $t \cdot ha^{-1}$ less. The tree size diversity index was also reduced slightly.

Figure 3a shows the expected fraction of the total forest area in four stand states, starting in the current condition and evolving over a century with the policy that maximized the expected tree species diversity. The states in Figure 3a dominated the steady-state landscape when state #1 (000,000), with low basal area in all species and size categories occupied 20% of the forest landscape, state #7 (000,110) with high basal area in small and medium size hardwoods occupied 14.5%, state #8 (000,111) with in addition high basal area in large hardwoods occupied 14%, and state #16 (001,111) with in addition high basal area in large pine trees occupied 13%. With the tree species maximization policy, state #1 quickly over reached its steady-state level, while the other states approached it more slowly.

Figure 3b shows how the tree species diversity index of the entire forest evolved from its initial level to the steady state with the policy that maximized expected tree species diversity. The diversity index is relative to the steady-state diversity of the unmanaged forest. With this policy it took approximately 30 years for the species diversity index to reach 90% of its maximum steady-state value of 108%.

Maximizing carbon sequestration

The amount of CO_2e sequestered in the trees living biomass in steady state could be increased by 42 $t \cdot ha^{-1}$

Table 5 Management policy that maximizes the net present value of harvested wood, depending on stand state and price level

Stand state #	Stand composition[1]	Price		
		Low	Medium	High
		Best decision[2]		
1	000,000	–	–	–
2	000,001	1	1	1
3	000,010	–	–	1
4	000,011	3	3	1
5	000,100	1	1	1
6	000,101	1	1	1
7	000,110	3	3	1
8	000,111	3	3	1
9	001,000	1	1	1
10	001,001	1	1	1
11	001,010	3	3	1
12	001,011	3	3	1
13	001,100	1	1	1
14	001,101	1	1	1
15	001,110	3	3	1
16	001,111	3	3	1
17	010,000	1	1	1
18	010,001	1	1	1
19	010,010	3	3	1
20	010,011	3	3	1
21	010,100	–	1	1
22	010,101	21	1	1
23	010,110	3	3	1
24	010,111	3	3	1
25	011,000	1	1	1
26	011,001	1	1	1
27	011,010	3	3	1
28	011,011	3	3	1
29	011,100	21	1	1
30	011,101	21	1	1
31	011,110	3	3	1
32	011,111	3	3	1
33	100,000	1	1	1
34	100,001	1	1	1
35	100,010	3	3	1
36	100,011	–	3	1
37	100,100	1	1	1
38	100,101	1	1	1
39	100,110	3	3	1
40	100,111	36	3	1
41	101,000	1	1	1
42	101,001	1	1	1
43	101,010	3	3	1
44	101,011	36	3	1
45	101,100	1	1	1
46	101,101	1	1	1
47	101,110	3	3	1
48	101,111	36	3	1
49	110,000	1	1	1
50	110,001	1	1	1
51	110,010	3	3	1
52	110,011	–	3	1
53	110,100	21	1	1
54	110,101	21	1	1
55	110,110	3	3	1
56	110111	52	3	1
57	111,000	1	1	1
58	111,001	1	1	1
59	111,010	3	3	1
60	111,011	52	3	1
61	111,100	21	1	1
62	111,101	21	1	1
63	111,110	3	3	1
64	111,111	52	3	1

[1]Basal area in pulpwood, small sawtimber, and large saw timber of pines (first three digits), or hardwoods (last three digits) 1 = higher than current average, 0 = lower than current average. [2]Stand state # resulting from the best harvest decision, "–" indicates no harvest.

Table 6 Long-term effects of maximizing economic returns (NPV) from harvested wood on NPV and ecological criteria, compared with the outcomes due to the current management

Criteria	Unit	Current management	Maximizing NPV
NPV of harvested wood	$\$ \cdot ha^{-1}$	1339	5535
Sawtimber harvest	$m^3 \cdot ha^{-1} \cdot y^{-1}$	1.5	5.3
Pulpwood harvest	$m^3 \cdot ha^{-1} \cdot y^{-1}$	0.4	2.6
Basal area	$m^2 \cdot ha^{-1}$	19	12
Species diversity[a]	%	101	96
Size diversity[a]	%	98	89
Landscape diversity[a]	%	103	31
Old growth fraction	%	18	0
CO_2e sequestered	$t \cdot ha^{-1}$	203	98
Cutting cycle	year	11	5

[a]Relative to the level without management.

Table 7 Long-term effects of maximizing expected tree species diversity or CO_2e sequestered, compared with the outcome due to the current management

Criteria	Unit	Current management	Maximizing tree species diversity	Maximizing CO_2e sequestered
NPV of harvested wood	$\$ \cdot ha^{-1}$	1339	3363	119
Sawtimber harvest	$m^3 \cdot ha^{-1} \cdot y^{-1}$	1.5	4.3	0.1
Pulpwood harvest	$m^3 \cdot ha^{-1} \cdot y^{-1}$	0.4	1.1	0.1
Basal area	$m^2 \cdot ha^{-1}$	19	17	22
Species diversity[1]	%	101	108	99
Size diversity[1]	%	98	96	100
Landscape diversity[1]	%	103	71	101
Old growth fraction	%	18	13	15
CO_2e sequestered	$t \cdot ha^{-1}$	203	151	245
Cutting cycle	year	11	10	86

[1]Relative to the level without management.

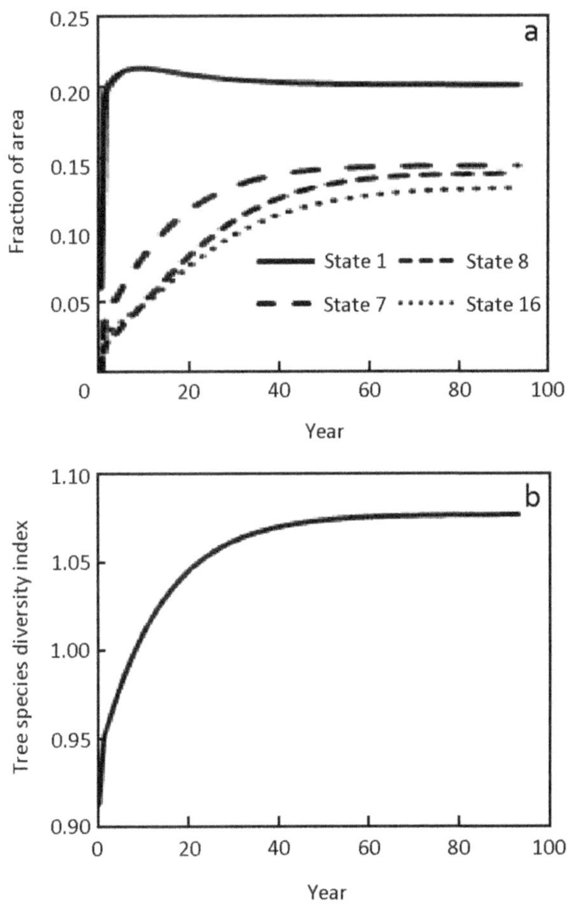

Figure 3 Predicted changes of the expected fraction of forest area in selected stand states (a), and of the expected tree species diversity index (b), given the initial distribution of stand states revealed by the forest inventory and the policy that maximized tree species diversity. State 1 = (000,000), state 7 = (000,110), state 8 = (000,111), state 16 = (001,111).

compared to the current management (last column of Table 7), but in doing so, wood production would be very low (0.2 $m^3 \cdot ha^{-1} \cdot y^{-1}$). In fact, the level of CO_2e sequestered in this way would be almost the same as that obtained by letting the forest grow naturally without any harvest (Table 4, last column). Indeed, the hands-off option maybe superior as it kept 29% of the landscape in old-growth state against only 15% when maximizing CO_2e sequestration, at the small opportunity cost of $\$119 \cdot ha^{-1}$ in foregone NPV from timber revenues[j]. The other criteria were similar to those obtained by natural forest growth.

Constrained financial and ecological objectives

Figure 4 shows the results of a series of optimizations in which the objective was to maximize the NPV of timber revenues given its initial stands distribution and current low price, subject to increasing constraints on the fraction of old growth maintained on the forest[k]. Figure 4 documents the opportunity cost, in terms of the NPV of foregone revenues, of higher levels of old growth. While approximately $\$5,300 \cdot ha^{-1}$ of NPV could be obtained without any old growth, this was reduced to about $\$1500 \cdot ha^{-1}$ when 22% of the landscape was kept in old growth. The marginal cost of old growth increased at increasing levels of old growth, from a low of $\$368 \cdot ha^{-1}$ from 0 to 5% of old growth to a high of $\$1145 \cdot ha^{-1}$ for 21% to 22% of old growth.

As an example of application with multiple ecological objectives, Figure 5 shows the results of maximizing annual wood production (sawtimber and pulpwood), subject to constraints on the amount of CO_2e sequestered in the growing stock. Both objectives were undiscounted and the constraint was as in equation (19). As indicated by Figure 5, there was no conflict between wood production and carbon sequestration up to approximately 100 $t \cdot ha^{-1}$

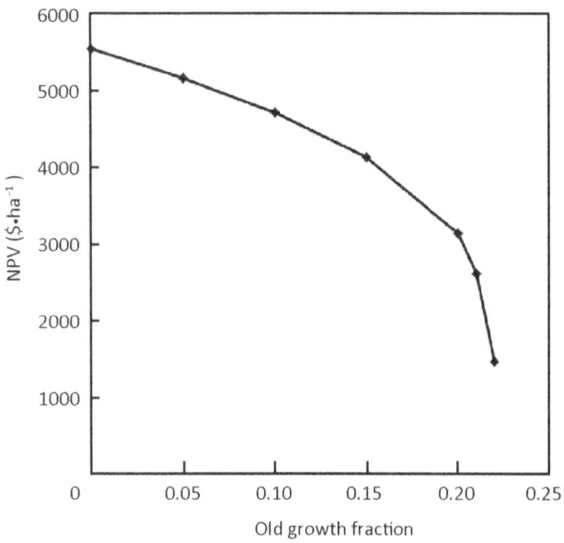

Figure 4 Maximum net present value (NPV) of harvested wood at increasing levels of the fraction of the forest landscape kept in an old-growth state. The old growth fraction is the constant annual equivalent of the present value of the old growth fraction over an infinite horizon.

of CO_2e. Beyond that point, wood harvest decreased rapidly at an increasing rate. Increasing carbon storage from 100 to 150 $t \cdot ha^{-1}$ decreased wood production by 1.2 $m^3 \cdot ha^{-1} \cdot y^{-1}$, and increasing it from 200 to 250 $t \cdot ha^{-1}$ decreased production by 3.3 $m^3 \cdot ha^{-1} \cdot y^{-1}$. As shown above, the maximum carbon storage achievable was about 245 $t \cdot ha^{-1}$, and at that point there was hardly any wood harvest.

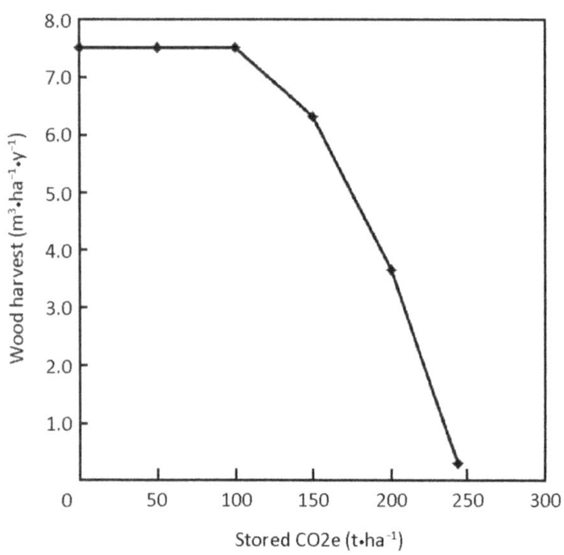

Figure 5 Maximum expected annual wood harvest at increasing levels of CO_2e stored in growing stock. Both the annual harvest and the stored CO_2e are undiscounted.

Discussion and conclusions

In a 1999 interview, G.B. Dantzig, the inventor of linear programming, remarked that "all problems that are solved under deterministic means have that fundamental weakness—they don't properly take uncertainty into account" (Dantzig 1999). Forestry planning is no different. Risk, uncertainty, and stochastic behavior are critical parts of how forest ecosystems work. As illustrated above, simulations of mixed loblolly pine-hardwood stands reveal how ignoring random disturbances due to biological or catastrophic shocks in forest growth models can lead to wrong ecological predictions, such as the disappearance of pines in this context. To these natural sources of risk must be added the high risk that stems from price fluctuations which complicates decision making with financial objectives.

The methods suggested in this paper to handle risk in forest decision making follow the modeling approach outlined by Holling et al. (1986). The first step is to "bring the world to the laboratory" with possibly complex and non-linear stochastic models, and then simplify the models to allow for efficient optimization with Markov decision process models. MDPs are attractive for their simplicity while keeping the essence of planning problems under risk[1]. They recognize that the future is unknown and that it may be described only in terms of probabilities. And, they lead to best adaptive management policies whereby decisions depend entirely on the systems state, i.e. the state of knowledge of the decision maker at decision time.

The fact that with MDP models the future state of a forest ecosystem depends only on its current state should not be viewed as a shortcoming, but recognition of fact. Predictions can rely only on current knowledge. This current knowledge may include past behavior of the system (such as past tree growth) if necessary. Thus, Markov models are not necessarily "memory less" and apparent failure of Markov models to correctly represent forest growth (Binkley 1980; Roberts and Hruska 1986; Johnson et al. 1991) may be due to incomplete description of the current ecosystem state rather than to a shortcoming of the Markov model. Similarly, Markov models of price changes are very general, embracing random walk, rational expectations, autoregressive, and "any stochastic model in which the price is conditional on previous prices" (Taylor 1984). In sum, Markov chains allow the simplification of complex, multi-dimensional stochastic processes and make their optimization easier or at all possible (Holling et al. 1986; Insley and Rollins 2005).

There exists considerable potential and flexibility to enrich the models presented in this paper as they are relatively small compared with the size of the problems that can be solved with current linear programming software. Still, to be successful in forest ecosystem management the MDPs should adhere to the general principles of parsimony and simplicity. To this end, the number of system states must be kept as small as possible, with a few state variables (tree species and size categories, prices) and a few levels of these variables.

In the same spirit of simplicity, multiple objectives such as optimizing financial returns subject to ecological constraints, of vice-versa, have been treated here with linear models. This requires expressing both the objective function and the constraints in either discounted or undiscounted criteria. While discounting financial returns from forestry has long been a standard procedure (Faustmann 1849), it is less so for ecological criteria. Nevertheless, it is plausible to give more weight to the present ecological characteristics of a forest ecosystem than to their future values. Indeed, it seems preferable to maintain desirable current states (such as old-growth stands) rather than lose them, even if they could be restored later (albeit after a long delay). It has also been argued on theoretical grounds that "the future benefits provided by a generic public good–environmental quality–should be discounted at a rate that is close to the market rate of return for risk-free financial assets" (Howarth 2009), a principle that has been applied for example in discounting future carbon sequestered in forests (Boscolo et al. 1997)[m].

The methods presented here can be expanded in many directions, in particular to deal with the new carbon markets and the highly stochastic prices for CO_2e sequestered in forests. The treatment is parallel to the stochastic prices of timber. This would increase substantially the size of the problem, but still keep it well within the capabilities of current linear programming software.

Other issues may require deeper modifications, to deal with non-stationary processes. For example, "climate change can affect forests by altering the frequency, intensity, duration, and timing of fire, drought, introduced species, insect and pathogens outbreaks, hurricanes, windstorms, ice storms, or landslides" (Dale et al. 2001). This implies a change in the transition probabilities over time. Although non-stationary problems can in principle be converted to stationary ones by a reformulation (Bertsekas 1995, p. 167), efficient numerical methods are still elusive (Ghate and Smith 2013). Nevertheless, the consequences of changes in transition probabilities on the steady-state criteria can be readily explored with the methods described in this paper. In sum, given the conceptual generality and the well-developed theory of MDPs, coupled with the powerful solution techniques available, the MDP approach is well suited to deal with risk in forest ecosystem management, and to develop practical adaptive management policies with both economic and ecological objectives.

Endnotes

[a]In this paper the terms "risk" or "uncertainty" are synonymous. Although Knight (1921) defines uncertainty as the absence of probabilities, it may be argued that probabilities, objective (data based) or subjective (opinion based) always exist to some degree.

[b]While modern linear programming software places almost no limit (computationally) on the number of states, it is best for all practical purposes (identifying states of nature, doing the computation, and laying out the recommendations) to use the minimum number of states needed for sufficient accuracy and realism of applications.

[c]For other applications, any stochastic model of forest growth can be used to compute the transition probabilities used in the MDP approach.

[d]If the price series is not stationary, the price data are first de-trended, and the trend is added to the discount rate in present value calculations.

[e]In the particular application described here, prices vary by tree size and species, and all the prices change in parallel with the probabilistic changes of the price index described in Table 3.

[f]That is: $y_{ik} = z_{0ik} + dz_{1ik} + d^2 z_{2ik} + d^3 z_{3ik} + \ldots$ where z_{tik} is the probability of state i and decision k at time t and $d = 1/(1+\rho)$ is the discount factor (Hillier and Lieberman 2005, p. 921).

[g]As applied to forestry, the MDP described by equations (10) and (11) is a generalization of (Faustmann 1849) formula recognizing that future stand states and prices are known only as probability distributions. The classical, deterministic Faustmann formula is a special case in which the transition probabilities are 0 or 1 (Buongiorno 2001).

[h]According to Timber Mart South (2014), the average price of softwood sawtimber stumpage in the South of the United States in the second quarter of 2014 was $25 per short ton or approximately $\$19 \cdot m^{-3}$, placing it in the low range according to the definitions in Table 3.

[i]Previous results also show that the optimum adaptive policy derived here is superior in terms of NPV to an optimum fixed policy that converts stands to a chosen state at fixed intervals (Zhou et al. 2008b).

[j]The reduction in old-growth fraction from 29% to 15% was due to the best decision calling for a harvest when the stand was in the old-growth state #32 (011,111). Although the harvest occurred only every 86 years on average, this was equivalent to a low frequency natural catastrophe reducing substantially the fraction of the landscape in old-growth state.

[k]In Figure 4, the old growth fraction is the right-hand side of equation (18) with O^* replaced by its annual constant perpetual equivalent, $\rho O^*/(1+\rho)$.

[l]See Buongiorno and Gilless (2003, p. 337–371) for an introduction to Markov and MDP models in forestry.

[m]Alternatively, models (10)–(11) and (12–13) can be extended to deal with discounted objective functions and undiscounted constraints, or the reverse, by introducing non-linear constraints (e.g. Rollin et al. 2005; Zhou 2005), but at the cost of the attendant numerical difficulties.

Competing interests
The authors declare that they have no competing interests.

Authors' contributions
This paper was based on research carried jointly by JB and MZ. JB wrote the first draft of the paper which was revised and approved by MZ. Both authors read and approved the final manuscript.

Acknowledgements
The preparation of this paper was supported in part by the USDA Forest Service, Southern Research Station, through a cooperative research agreement with Joseph Buongiorno, directed by Jeff Prestemon. It was also supported in part by the USDA McIntire-Stennis fund WVA00105. We are also grateful to the organizers of the workshop on Risk and Uncertainty in Ecosystems Dynamics, Beijing Forestry University, October 13–17, 2014, for motivating the study.

Author details
[1]University of Wisconsin-Madison, Madison, USA. [2]West Virginia University, Morgantown, USA.

References

AFC (Alabama Forestry Commission) (2014) How much carbon have your trees stored? Alabama Forestry Commission. http://www.forestry.state.al.us/HowMuchCarbonHaveYourTreesStored.aspx?bv=5&s=0. Accessed 26 Feb 2015

Bertsekas DP (1995) Dynamic programming and optimal control, vol 2. Athena Scientific, Belmont, p 292

Binkley CS (1980) Is succession in hardwood forests a stationary Markov process? For Sci 26:566–570

Boscolo M, Buongiorno J, Panayotou T (1997) Simulating options for carbon sequestration through improved management of a lowland tropical rainforest. Environ Dev Econ 2:241–263

Buongiorno J (2001) Generalization of Faustman's formula for stochastic forest growth and prices with Markov decision process models. For Sci 47(4):466–474

Buongiorno J, Gilless JK (2003) Decision methods for forest resource management. Academic Press/Elsevier, Amsterdam, p 439

d'Epenoux F (1963) A probalistic production and inventory problem. Manage Sci 30(1):98–108

Dale VH, Joyce LA, McNulty S, Neilson RP, Ayres MP, Flannigan MD, Hanson PJ, Irland LC, Lugo A, Peterson CJ, Simberloff D, Swanson FJ, Stocks BJ, Wotton BM (2001) Climate change and forest disturbances. BioScience 51(9):723–734

Dantzig GB (1999) Interview of Georges B. Dantzig by P. Horner. First published in the October 1999 issue of *OR/MS Today*. http://www.orms-today.org/orms-6-05/dantzig.html. Accessed 26 Feb 2015

de Finetti B (1937) La prévision, ses lois logiques, ses sources subjectives. Annales de l'Institut Henri Poincaré 7(1):1–68

Faustmann M (1849) Calculation of the value which forest land and immature stands possess for forestry. Allgemaine forst-und jagdzeitung 15:441–455

Ghate A, Smith RL (2013) A linear programming approach to nonstationary infinite horizon Markov decision processes. Oper Res 61(2):413–425

U.S. Government (2014) Economic report of the President. U.S. Government Printing Office. http://www.gpo.gov/fdsys/browse/collection.action?collectionCode=ERP. Accessed 26 Feb 2015

Hillier FS, Lieberman GJ (2005) Introduction to operations research, 8th edn. Mc Graw Hill, Boston, p 1061

Holling CS, Dantzig GB, Winkler C (1986) Determining optimal policies for ecosystems. In: Kallio M et al (eds) Systems analysis in forestry and forest industries. TIMS studies in management science 21, North Holland, Amsterdam, pp 463–473

Howarth RB (2009) Discounting, uncertainty, and revealed time preferences. Land Econ 85(1):24–40

Hummel S, Calkin DE (2005) Costs of landscape silviculture for fire and habitat management. For Ecol Manage 207(3):385–404

Insley M, Rollins K (2005) On solving the multirotational timber harvesting problem with stochastic prices: a linear complementary formulation. Amer J Agric Econ 87:735–755

Johnson SE, Ferguson IS, Li RW (1991) Evaluation of a stochastic diameter growth model for mountain ash. For Sci 37:1671–1681

Kant S, Alavalapati RR (2014) Handbook of forest resource economics. Part 4: Economics of risk, uncertainty and natural disturbances. Routledge, New York, p 559

Kaya I, Buongiorno J (1987) Economic harvesting of uneven-aged northern hardwood stands under risk: A markovian decision model. For Sci 33(4):889–907

Knight FH (1921) Risk, uncertainty, and profit. Houghton Mifflin, Boston

Lembersky MR (1976) Maximum average annual volume for managed stands. For Sci 22:69–81

Lembersky MR, Johnson KN (1975) Optimal policies for managed stands: An infinite Markov decision process approach. For Sci 21:109–122

Lin CR, Buongiorno J (1998) Tree diversity, landscape diversity, and economics of maple-birch forests: Implications of markovian models. Manage Sci 44(10):1351–1366

Magurran AE (1988) Ecological diversity and its measurement. Princeton University Press, Princeton, NJ, p 179

Manne AS (1960) Linear programming and sequential decisions. Manage Sci 6:259–267

Oliver CD, Larson BC (1990) Forest stand dynamics. McGraw-Hill, Chicago, p 467

Roberts MR, Hruska AJ (1986) Predicting diameter distributions: a test of the stationary Markov model. Can J For Res 16:130–135

Rollin F, Buongiorno J, Zhou M, Peyron JL (2005) Management of mixed-species, uneven-aged forests in the French Jura: From stochastic growth and price models to decision tables. For Sci 51(1):64–75

Schulte B, Buongiorno J, Lin CR, Skog K (1998) SouthPro: A computer program for uneven-aged management of loblolly pine stands. USDA Forest Service, Forest Products Laboratory, General technical report FPL-GRT-112, Madison, p 47

Shannon CE (1948) A mathematical theory of communication. Bell Syst Tech J 27(379–423):623–656

Taylor CR (1984) Stochastic dynamic duality: Theory and empirical applicability. Amer J Agric Econ 66:351–357

Timber Mart-South (2014) South wide pine stumpage prices. http://www.timbermart-south.com/prices.html. Accessed 26 Feb 2015

White DJ (1993) A survey of applications of Markov decision processes. J Oper Res Soc 44(11):1073–1096

Zhou M (2005) Managing resources for multiple purposes under risk: Markov models of southern mixed loblolly pine-hardwood forests. University of Wisconsin-Madison, Dissertation, p 135

Zhou M, Buongiorno J (2004) Nonlinearity and noise interaction in a model of forest growth. Ecol Model 180:291–304

Zhou M, Buongiorno J (2006) Forest landscape management in a stochastic environment, with an application to mixed loblolly pine-hardwood forests. For Ecol Manage 223:170–182

Zhou M, Buongiorno J, Liang J (2008a) Economic and ecological effects of diameter caps: A Markov decision model for Douglas-fir/Western hemlock forests. For Sci 54(4):397–407

Zhou M, Liang J, Buongiorno J (2008b) Adaptive versus fixed policies for economic and ecological objectives in forest management. For Ecol Manage 254:178–187

21

Decomposition dynamics of coarse woody debris of three important central European tree species

Steffen Herrmann[1,2*], Tiemo Kahl[1] and Jürgen Bauhus[1]

Abstract

Background: Coarse woody debris (CWD) is an important element of forest structure that needs to be considered when managing forests for biodiversity, carbon storage or bioenergy. To manage it effectively, dynamics of CWD decomposition should be known.

Methods: Using a chronosequence approach, we assessed the decomposition rates of downed CWD of *Fagus sylvatica*, *Picea abies* and *Pinus sylvestris*, which was sampled from three different years of tree fall and three different initial diameter classes (>10 – ≤ 20 cm, >20 – ≤40 cm, >40 cm). Samples originating from wind throws in 1999 were collected along a temperature and precipitation gradient. Based on the decay class and associated wood densities, log volumes were converted into CWD mass and C content. Log fragmentation was assessed over one year for log segments of intermediate diameters (>20 – 40 cm) after 8 and 18 years of decomposition.

Results: Significantly higher decomposition constants (k) were found in logs of *F. sylvatica* (0.054 year^{-1}) than in *P. abies* (0.033 year^{-1}) and *P. sylvestris* (0.032 year^{-1}). However, mass loss of *P. sylvestris* occurred mainly in sapwood and hence k for the whole wood may be overestimated. Decomposition rates generally decreased with increasing log diameter class except for smaller dimensions in *P. abies*. About 74 % of the variation in mass remaining could be explained by decomposition time (27 %), tree species (11 %), diameter (17 %), the interactive effects between tree species and diameter (4 %) as well as between decomposition time and tree species (3 %) and a random factor (site and tree; 9.5 %), whereas temperature explained only 2 %.
Wood fragmentation may play a more important role than previously thought. Here, between 14 % and 30 % of the decomposition rates (for the first 18 years) were attributable to this process. Carbon (C) density (mgC · cm^{-3}), which was initially highest for *F. sylvatica*, followed by *P. sylvestris* and *P. abies*, decreased with increasing decay stage to similar values for all species.

Conclusions: The apparent lack of climate effects on decomposition of logs in the field indicates that regional decomposition models for CWD may be developed on the basis of information on decomposition time, tree species and dimension only. These can then be used to predict C dynamics in CWD as input for C accounting models and for habitat management.

Keywords: Dead wood; Carbon; Decay rate; Beech; Spruce; Pine

* Correspondence: steffen.herrmann@wsl.ch
[1]Institute of Forest Sciences, Faculty of Environment and Natural Resources, University of Freiburg, Tennenbacherstr. 4, D-79106 Freiburg, Germany
[2]Swiss Federal Research Institute WSL, Zürcherstrasse 111, CH-8903 Birmensdorf, Switzerland

Background

Coarse woody debris (CWD) is structurally and functionally very important for forest ecosystems, in particular for biodiversity (Siitonen 2001), the energy and nutrient cycle (Müller-Using and Bartsch 2007; Kuehne et al. 2008) and carbon storage (Harmon et al. 1986; Turner et al. 1995; Pregitzer and Euskirchen 2004; Kahl et al. 2012). Whereas the amount of CWD may comprise up to 30 % or even 40 % of the total timber volume in natural beech (Commarmot et al. 2013) and spruce (Ranius et al. 2003) forests, this share is typically less than 5 % in managed European forests (Bütler and Schlaepfer 2004; MCPFE 2007). This reduction in the amount and related quality of dead wood (Müller and Bütler 2010) has significant implications for its various functions. European lists of endangered species are often dominated by species depending on dead wood (Grove et al. 2002). For Germany, 28 % of the saproxylic beetle species are listed as threatened or regionally extinct (Seibold et al. 2014). Owing to its significance for ecosystem functioning, CWD has been recognized as an indicator of ecological sustainable forest management (MCPFE 2003). Therefore increasing efforts have been undertaken to manage CWD as a habitat component and C store in forest ecosystems. However, to manage this pool, a basic understanding of patterns and rates of dead wood decomposition in different forests is crucial. Further, for the assessment of C stocks in dead wood as part of National Greenhouse Gas inventories, detailed information on C stored in dead wood of different species and their relationship with different decay stages (which are typically captured in inventories) is necessary. So far this knowledge is incomplete and mainly based on expert opinions (Meyer et al. 2003; Rock et al. 2008; Zell et al. 2009). In addition, information on residence times of CWD in different decay classes would be very helpful to forecast its dynamics and to calculate the input and output of different decay stages in order to conserve specific habitats of dead wood dependent species (Kruys et al. 2002; Ranius et al. 2003).

Dynamics of CWD are determined by the input through tree mortality and the output through the decomposition process. The decomposition process is characterised by a decrease in dimension and mass and concomitant changes in biological, physical, and chemical properties. The main process of dead wood decomposition is the loss of organic material through respiration. In addition, its fragmentation through biological activity (in particular by insects) and physical processes can also be important. With ongoing decomposition and decreasing physical stability of CWD, fragmentation increases in importance (Harmon et al. 1986). However, leaching of soluble matter (including C) from CWD logs is of minor importance for mass loss (Spears et al. 2003; Kahl et al. 2012). The specific influence of selected environmental variables such as wood moisture and temperature on CWD decomposition can be explored in detail via lab incubation (see Herrmann and Bauhus 2012). However, to capture the complex interplay of processes responsible for decomposition in forests, long-term field measurements are necessary.

The decomposition rate of CWD is mainly dependent on climatic (wood temperature, wood moisture) and substrate specific variables (tree species, decay stage, diameter), where tree species influences chemical and physical wood properties and the decomposer community (Mackensen et al. 2003; Kahl 2008). Up to now it is not clear whether climatic or substrate specific variables are more important for the decomposition process (see also Cornwell et al. 2009; Freschet et al. 2011). Recent results indicated that climatic variables are likely to be more important for (short term) CWD mineralization than substrate specific variables (Herrmann and Bauhus 2012). However, both factors also clearly interact and must therefore be considered jointly (Herrmann and Bauhus 2012). Here, we analysed the influence of these factors on the field decomposition rate of *Fagus sylvatica* L., *Picea abies* (L.) Karst. and *Pinus sylvestris* L. along a climatic/altitudinal gradient (temperature, precipitation) using a chronosequence approach based on known ages of CWD logs. In addition, the influence of log dimension was assessed by analysing CWD pieces of different initial diameters (in three diameter classes, i.e., >10 – 20 cm, >20 – 40 cm, > 40 cm).

Our specific research questions were:

1. How do decomposition rates differ between the three tree species *Fagus sylvatica* (hardwood), *Picea abies* and *Pinus sylvestris* (softwoods)?
2. To what extent can the variation in mass and carbon remaining in CWD be explained by tree species, log dimension, wood chemistry (nutrients, lignin) and microclimatic conditions?

In addition, we tested if it was possible to determine the dry density of CWD logs with the drill resistance method.

Methods

Study sites

For the chronosequence approach that relates mass (i.e., wood density × volume) to the age of CWD, logs were sampled at sites where the previous stand had been wind-thrown and where no or only little salvage logging had taken place. Thus we could be certain about the period that logs had undergone decomposition. The study sites were mainly located in southern Germany, except for the oldest site that was situated in northern Germany (Lower Saxony, near the city of Hannover). All selected sites were hit by severe winter storms in 1999, 1990 and 1972 (Table 1).

Table 1 Description of the study sites

Site	Year of tree fall	Elevation[a] (m a.s.l.)	Soil type[b]	Average temperature[c] per year (°C)	Precipitation[c] (mm per year)
Forêt de Ha-guenau (FdH)	1999	150–170	Gleyic Cambisol	10.98	645
Bienwald (Biw)	1999	100–130	Cambisol	10.0	680–700
Röttlerwald (Roe)	1999	510–590	Luvisol and Cambisol	9–10	900–1080
Lotharpfad (Lot)	1999	950	Podsolic Cambisol	7.0	1700
Hofstatt (Hof)	1990	320–340	Luvisol	8.8	800
Silbersandgrube (Sig)	1990	550	Luvisol	7.8	665
Kiekenbruch (Kib)	1972	70	Podsolic Cambisol	8.5	657

[a]according to GPS; [b]FAO (2006); [c]according to forest office information

Sampling design

Downed CWD of *F. sylvatica*, *P. abies* and *P. sylvestris* from three different years of tree fall (1972 (except for beech), 1990 and 1999) was sampled in three different initial diameter classes (>10 – ≤ 20 cm, >20 – ≤40 cm, >40 cm). Samples originating from wind throws in 1999 were collected along an altitudinal gradient from the Rhine valley up to the top of the black forest (Table 2). This represents a climatic gradient with regard to temperature and precipitation (see Table 1). Unfortunately, no pine CWD was found at the highest elevation.

In our design, site was not independent from year since tree fall. Generally we sampled at least 5 replicates for each combination of species, site, and diameter. Where clearly different diameter classes were present along the length of a single log, we selected in some cases several sample-points at a minimum distance of 2 m, where discs were extracted. Discs were approximately 10 cm thick. On average, we extracted 2.5 sample discs per log. These samples were treated as separate samples, because the variations in wood density from one log and within the same decay class were comparable to the variations within the same decay class when collected from a large population of logs in the field (see also Müller-Using and Bartsch 2009). Separate samples from the same log can also differ considerably in composition of the microflora, as has been shown by others (Shigo 1986; Schwarze et al. 1999). In addition, our preliminary analysis showed that samples from the same log were not autocorrelated according to the Durbin Watson-Test (DW = 1.7).

Table 2 Year of storm event (i.e., decomposition time) and elevation for the 7 study sites

Year of storm event	Elevation (m a.s.l.)		
	100	600	900
1999	Biw (spruce, pine), FdH (beech)	Roe (beech, spruce, pine)	Lot (beech, spruce)
1990		Hof (beech), Sig (spruce, pine)	
1972	Kib (spruce, pine)		

Biw Bienwald, *FdH* Forêt de Haguenau, *Roe* Röttlerwald, *Lot* Lotharpfad, *Hof* Hofstatt, *Sig* Silbersandgrube, *Kib* Kiekenbruch

Decay class was assessed for each log and measurement position according to a 4 stage classification system (Albrecht 1991 for *P. abies* and *P. sylvestris*), modified for *F. sylvatica* by Müller-Using and Bartsch (2009). Based on the decay class and the associated known wood densities, log volumes assessed in field inventories can be converted into CWD mass and C content (Grove et al. 2009). However, because the internal log condition may not be represented by a visual decay classification (Meyer 1999), large errors for assessed C pools can result. Therefore, we also assessed the internal density of logs with the drill resistance method (Kahl et al. 2009). At each point, where log diameter was measured, two drill resistance measurements were conducted, one horizontally and one vertically. In general, at least one stem disc per tree, representing the dominating decay class, was cut with a chainsaw at a drill resistance measurement position for further calibration and analysis. If more than one diameter or decay class was analysed per log, the number of extracted discs increased accordingly. At the site 'Lot' no discs could be removed for most beech logs, as their position was very close to a hiking trail. In this case samples were taken with a wood corer (according to the method applied above).

Density assessment

Density was assessed gravimetrically and via drill resistance measurements using the Resistograph 3450S (RINNTECH, Heidelberg, Germany) with a resolution of 0.01 mm and a maximum drilling depth of 44 cm (for further details see Kahl at al. 2009).

To determine density of wood samples gravimetrically, wooden bars were cut with a band saw along the drill resistance measurement lines for each sampled stem disc. These were further cut into small cubes (about 2 cm thick) and dried in a fan-forced oven at 105 °C until constant weight was achieved. Dry density was measured for each cube (without further sample preparation) via the water displacement method and related to the drill resistance value for calibration.

Log fragmentation assessment

Log fragmentation was assessed over one year for log segments of intermediate diameters (>20 – 40 cm) of *F. sylvatica*, *P. abies* and *P. sylvestris* after 8 and 18 years of decomposition (representing initial and advanced decay stages), with three replicates per species and year. Nylon fabric (1 mm × 1 mm mesh, ca. 180 cm length and about 120 cm width) was placed underneath each log and metal pegs were used to hold up the material. After one year, the material that had fallen onto the nylon mesh was collected and quantified.

Chemical analysis

Following density assessment, a subset of all samples with a total of 3 replicates per species, decay stage and diameter class was ground and analysed for C concentration. C concentration was determined by combustion at 950 °C with a Leco Truspec™ CN analyser (St. Joseph, MI, USA). Wood cores were taken from intact wood (1 sample per diameter class per species and site, except for the sites 'Biw', 'Lot' and 'FdH') to estimate initial nutrient concentrations. N concentration was determined according to the method described above for C concentration. Concentrations of P, S, Ca, K, Mg, Na, Al, Fe, and Mn were analysed with Inductively Coupled Plasma-Optical Emission Spectroscopy (ICP-OES) (Spectro, Kleve, Germany), following HNO_3 digestion in pressure chambers. Lignin concentration was measured according to the methods proposed by TAPPI (1976) and Effland (1977).

Data analysis

To calculate decomposition constants (k) based on mass loss, mean dry density from gravimetric density assessment was used if available. Otherwise we used mean dry density based on drill resistance. For each measurement position, the volume of the stem disc represented by this position was calculated based on formula 1:

$$V = L * \pi * \left(\frac{d}{2}\right)^2 \tag{1}$$

We determined the k values of each individual log or measurement position based on the single negative exponential decay model (Olson 1963):

$$k = \frac{-\ln\left(\frac{M(t)}{M(0)}\right)}{t} \tag{2}$$

Here $M(t)$ is the mass (g) remaining at time t (year) and $M(0)$ is the original mass (g) at time 0 and k is the decomposition rate constant (year^{-1}). The average k value for each species and size class was then calculated on the basis of these individual k values.

The time when 50 % of the original mass would have been decomposed was calculated by:

$$t_{0.5} = \frac{0.693}{k} \tag{3}$$

To model CWD decomposition, exponential decay functions, which assume a homogeneous substrate that decays at a constant rate, have been used in the majority of cases (Mackensen et al. 2003). This may be an oversimplification of the complex processes occurring during dead wood decomposition. Recent modelling approaches have also considered the different wood constituents such extractives, cellulose or lignin, which undergo, owing to their chemical resistance, different decomposition dynamics (Tuomi et al. 2011). However, we did not employ such model, because the temporal resolution of our data would not have permitted a robust fitting of the decomposition process using a model with many parameters. To facilitate comparison of results with that of other studies, we fitted the exponential decay functions for all tree species.

Original dry mass of undecayed wood of the three tree species was calculated based on original diameter and dry density values from literature studies (Trendelenburg and Mayer-Wegelin 1964). Reference values for dry density ($g \cdot cm^{-3}$) were: *F. sylvatica*: Lot: 0.65, all other sites: 0.68; *P. abies*: Lot: 0.4, Kib: 0.46, all other sites: 0.43; *P. sylvestris*: Kib > 40 cm: 0.45, all other sites and diameter classes: 0.49. Different values reflect site (i.e., climatic) differences and variation along stem height (pine), as observed by Trendelenburg and Mayer-Wegelin (1964). Additionally, wood cores (three replicates) were taken from undecomposed wood from the different sites (except for Lot), tree species and diameter (stratified as >30 cm and <30 cm, and 30 cm for pine at Roe) to confirm that density values from the literature were appropriate.

If the original diameter was different from the current diameter (i.e., bark and/or (sap) wood was already decomposed), it was reconstructed based on the thickness of intact parts on the same log or adjacent logs.

Statistical analysis

Statistical analyses were conducted using SPSS Statistics 17.0 (SPSS Inc., Chicago, IL, USA) and R (R Development Core Team). All significance testing was done with $p < 0.05$. The analysis for even distribution of the residuals was conducted graphically via normal qq-plot and scatter plot (residuals vs. predicted values; and vs. fitted values (linear regression model)), as well as via Kolmogorov-Smirnov test (Dormann and Kühn 2009).

A transformation (usually ln) was conducted, if residuals were not evenly distributed.

A two-way ANOVA was conducted to test for significant differences in wood density and in C content between the three species as well as between the different decay stages. A one-way ANOVA followed by Fisher's LSD or Dunnett-T3 post-hoc test, if variances were not homogenous, was conducted to further analyse possible differences in 1) *k* (ln-transformed) between the three species as well as between the different diameter classes at the species level, 2) mass remaining at the species level for each period, beginning at t_0, for which logs had undergone decomposition, 3) wood density between species for each decay stage and to compare the average dry density of the different decay stages for each species and 4) C concentration between the different decay stages of each species. In the case of *P. Sylvestris* (only one sample in decay stage 4, a two-sample *t*-test was applied for comparison between decay stages.

The influence of substrate specific, climatic and environmental variables on the mass of logs remaining after different periods of decomposition in the field was analysed based on a linear mixed-effects model. Possible influencing variables were identified using a Spearman rho correlation matrix. The climatic variables (temperature and precipitation) were analysed for collinearity and centered for further use in the model. The residuals were further analysed for even distribution graphically via box-cox transformation. To decide if a model is better than a previous model, we used the explained variation (r^2) and the AIC as 1st and 2nd criteria. Eta squared was used as an effect size measure.

To assess the residence time (based on Kruys et al. 2002), in total and for each decay stage, the average dry density ($g \cdot cm^{-3}$) and the average decomposition time (years) per decay stage were calculated from the whole dataset in a first step. For further calculations, each decay stage was defined as the interval between the midpoints of dry densities of decay classes (e.g., 0.5–1.49 for decay stage 1). Afterwards a dry density value was assigned for each decay stage interval based on the exponential relationship between decay stage and average dry density. In the same way, the decomposition time was assigned, which was calculated based on the exponential relationship between average dry density and average decomposition time. Finally the residence time per decay stage was defined as the decomposition time at the upper bound of each decay stage.

Results

Drill resistance measurements

Up to 77 % of the variation in wood density could be explained with drill resistance for beech (Fig. 1). For spruce and pine, only 58 % and 50 % of the variation in density were explained with drill resistance, respectively.

Decomposition rates (*k*) and mass loss

The average *k* value of *F. sylvatica* (0.054 $year^{-1}$; 95 % CI: 0.048–0.059 $year^{-1}$) was significantly different from the one of *P. abies* (0.033 $year^{-1}$; 95 % CI: 0.03–0.035 $year^{-1}$) and *P. sylvestris* (0.032 $year^{-1}$; 95 % CI: 0.028–0.035 $year^{-1}$) across all sites and diameter classes ($p < 0.001$) (Table 3). The two conifer species were not significantly different from each other. The corresponding average diameters of logs were 34.8 (SD ± 14.8) cm for *F. sylvatica*, 33.9 (SD ± 13.9) cm for *P. abies* and 32.2 (SD ± 12.1) cm for *P. sylvestris* indicating that the effects of species were not biased through differences in log dimensions between species. The time until 50 % of the initial mass was decomposed was 13 years for *F. sylvatica*, 21 years for *P. abies* and 22 years for *P. sylvestris*. The average *k* values until 18 years for *P. abies* and *P. sylvestris* were 0.033 $year^{-1}$; 95 % CI: 0.031–0.036 $year^{-1}$ and 0.034 $year^{-1}$; 95 % CI: 0.03–0.037 $year^{-1}$, respectively.

Owing to different decomposition rates, the patterns of mass remaining differed accordingly between the species (Fig. 2). Whereas the average mass remaining in *P. sylvestris* CWD was similar to that of *P. abies* until 18 years of decomposition (about 59 % each (SD ± 12, resp. 11 %; *P. abies*)), it was considerably higher at year 36 (*P. sylvestris*: 46 % (SD ± 15), *P. abies*: 39 % (SD ± 18)). At this time, nearly all sapwood of *P. sylvestris* logs was decomposed and heartwood started to decompose. Average mass remaining decreased with increasing decomposition time in all species ($p < 0.05$, Fig. 2).

With increasing diameter class, decomposition rates decreased significantly in all species, except between the diameter classes <20 cm and >20 – 40 cm for *P. abies* (Table 3).

Prediction of mass remaining

About 74 % of the variation in mass remaining could be explained by decomposition time (27 %), tree species (11 %), diameter (17 %), the interactive effects between tree species and diameter (4 %) as well as between decomposition time and tree species (3 %) and a random factor with site and tree (9.5 %), whereas temperature explained only 2 % of the variation (Table 4). If we subtract the random factor, which cannot be predicted, 64 % of the variation in mass remaining may be predicted for the tree species investigated here.

If initial nutrient concentrations were considered as additional independent variables, a slight model improvement was achieved, when the concentration of manganese was included. Based on a reduced data set ($n = 172$ instead of $n = 287$), an additional 1 % of the total variance could be explained.

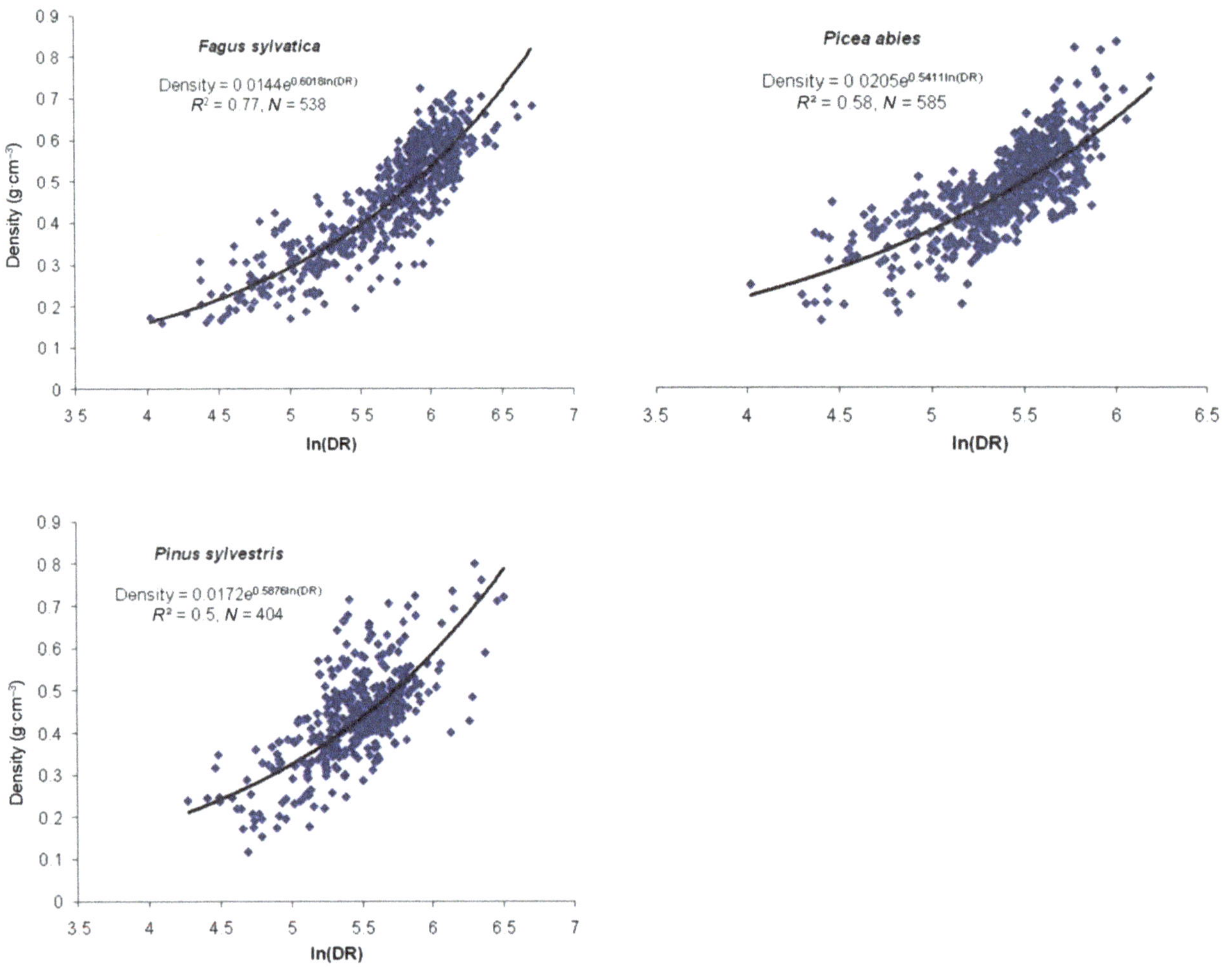

Fig. 1 Relationship between drill resistance (ln(DR)) and dry density for *F. sylvatica*, *P. abies* and *P. sylvestris*

Log fragmentation

The amount of fragmented material differed between tree species and age of logs, but in relation to total mass it was very low (Table 5). The maximum value was 1.8 % per annum for one *P. abies* log (Table 5). For *F. sylvatica* and *P. abies*, bark fragmentation dominated after the first 8 years of decomposition. In *P. sylvestris*, only a minor amount of bark fragmentation and no wood fragmentation occurred during this period of time. In older logs of *F. sylvatica* and *P. abies* (18 years), fragmented material originated mostly from wood in logs and was much higher in the former than the latter species. For *P. sylvestris*, bark fragmentation dominated in older logs. Fragmentation was very patchy between individual logs and often occurred as a single event, as indicated by the high standard deviation. However, when fragmentation is related to the decomposition rate, the actual contributions are considerable. When compared to the total *k* value for first 18 years of decomposition, fragmentation over one year accounted for 27 % for beech, 14 % for spruce and 30 % for pine, respectively.

Table 3 Annual decomposition constants *k* for different diameter classesin *Fagus sylvatica*, *Picea abies* and *Pinus sylvestris*

Species	Diameter class (cm)			Combined
	< 20	20–40	> 40	
F. sylvatica ($n = 96$)	0.078 (0.027)[a]	0.055 (0.027)[b]	0.035 (0.015)[c]	0.054 (0.028)
P. abies ($n = 107$)	0.034 (0.011)[a]	0.036 (0.012)[a]	0.027 (0.013)[b]	0.033 (0.013)
P. sylvestris ($n = 83$)	0.050 (0.015)[a]	0.030 (0.01)[b]	0.021 (0.01)[c]	0.032 (0.015)

Values are means with standard deviation in parentheses. Different letters indicate significant differences between diameter classes ($p < 0.05$)

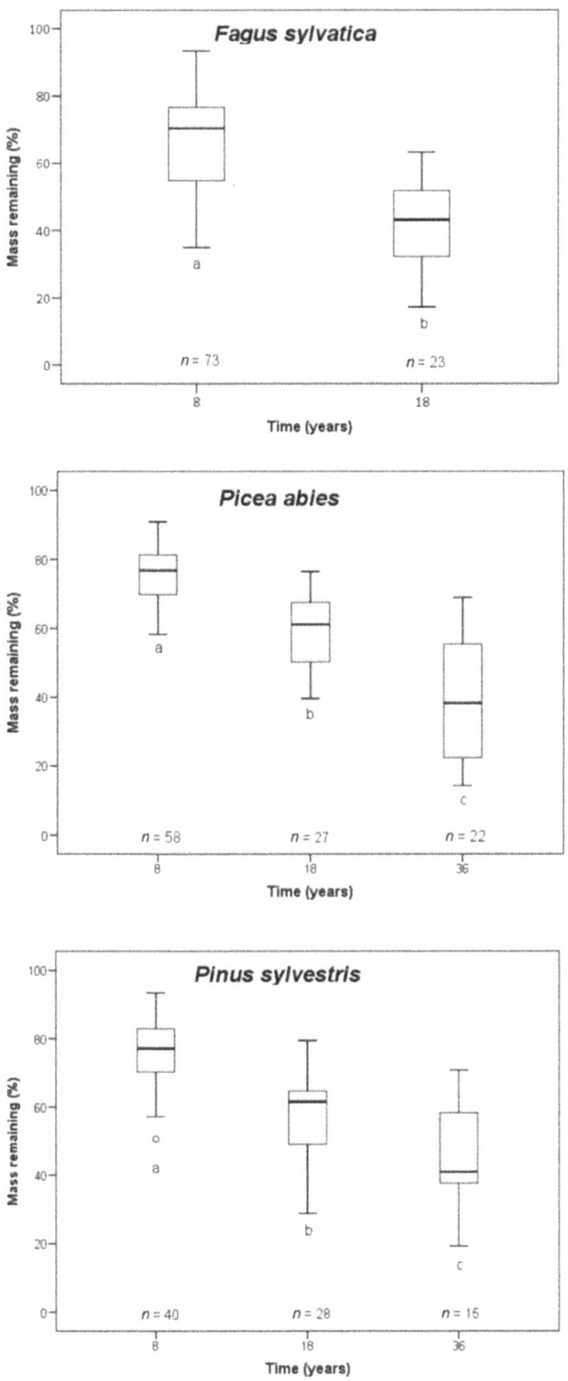

Fig. 2 Remaining mass of log segments of *Fagus sylvatica* (n = 96), *Picea abies* (n = 107) and *Pinus sylvestris* (n = 83); boxplots display median, lower and upper quartile, minimum and maximum values; points outside boxplots represent outliers; different letters indicate significant differences (p < 0.05)

Residence time per decay class

The residence time per decay class increased with decay class (i.e., higher degree of decomposition). This increase was most pronounced in pine and least pronounced in beech (Table 6). In addition, with decreasing decomposition rate from *F. sylvatica* to *P. sylvestris*, the proportion of time, relative to the entire decay process, that logs stayed in the first decay stage became shorter, whereas the time in the last decay stage increased.

Wood density, carbon content and decay stage

Average dry density values differed significantly between species and decay stages. Specifically, average dry densities were significantly different between *F. sylvatica*, *P. abies* and *P. sylvestris* for decay stage 2 ($p < 0.001$). For decay stage 3, dry density of *F. sylvatica* was significantly different from those of *P. abies* and *P. sylvestris* ($p < 0.001$ and $p < 0.01$), which were similar to each other. No significant differences between species were found for decay stage 4.

Average dry density decreased with increasing decay stage in all species ($p < 0.001$; Fig. 3).

The decrease in density was highest from decay stage 2 to 3 and the density variation was highest in decay stage 3 for all three species, but in particular for *F. sylvatica*. In dead wood of *F. sylvatica*, the decrease in density was not significant between decay stages 3 and 4. In contrast, a significant decrease between decay stages 3 and 4 was found for *P. abies* ($p < 0.05$). For *P. sylvestris* there was only one value in decay stage 4.

Similar to dry density, carbon concentration was significantly higher for *F. sylvatica*, compared to *P. abies* and *P. sylvestris*, which were not different from each other (Table 7). With advancing decomposition and decreasing dry density, an increase in C concentration was observed for all three species (except between decay stages 3 and 4 for beech and spruce). C concentration differed significantly between decay stages 1 and 2, as well as 3 for *F. sylvatica* and *P. abies*. For *P. abies*, decay stage 1 was significantly different also from decay stage 4, while decay stages 1 and 2 were significantly different from decay stage 3 for *P. sylvestris* (Table 7).

The effect of decay stage on carbon (C) density was dependent on species ($p = 0.000$). The pattern of C density and decay stage followed the pattern of wood density and decay stage closely (Table 7 and Fig. 3).

Discussion

Decomposition rates (k) and mass loss

To our knowledge, this is the first study that systematically assessed CWD decomposition rates and dynamics of *Fagus sylvatica*, *Picea abies* and *Pinus sylvestris* across different sites and diameter classes in Central Europe. Average k values of *F. sylvatica* were significantly higher than those of *P. abies* and *P. sylvestris* across all sites and diameter classes.

In other studies, decomposition rates of *F. sylvatica* varied from 0.056 year^{-1} (Kahl 2008) to 0.089 year^{-1}

Table 4 Linear mixed-effects model (ANOVA table) to predict the remaining massof *Fagus sylvatica*, *Picea abies* and *Pinus sylvestris*

Source	Sum of Squares	df	*F*	Sig.	Eta squared
Decomp. time	12162.2	1	129.449	0.000	27.06
Tree species	5058.7	2	12.768	0.000	11.26
Diameter	7655.1	1	97.979	0.000	17.03
Temperature	789.4	1	7.273	0.009	1.76
Decomp. time × Species	1405	2	6.066	0.003	3.13
Species × Diameter	1873.8	2	10.8	0.000	4.17

Conditional *R*-squared: 0.739 (Marginal R^2: 0.644)

(Müller-Using and Bartsch 2009), with intermediate values of 0.06–0.075 $year^{-1}$ (Christensen et al. 2005; based on 86 European beech forest reserves). In comparison to these findings, our *k* value for beech of 0.054 $year^{-1}$ (SD ± 0.028, 52 %) is within the range of variation and very similar to the estimate by Kahl (2008).

The difference between *k* values of beech may be attributed to the uncertainty over the cause of death, as observed by Kahl (2008). In that study, a *k* value of 0.075 $year^{-1}$ (SD ± 0.034) for logs that died naturally (infected by *Fomes fomentarius* prior to death) and a *k* value of 0.025 $year^{-1}$ (SD ± 0.012) for wind thrown logs was calculated (mean diameter > 40 cm). This indicates that decomposition had advanced already in trees affected by *F. fomentarius* before they died or fell. We calculated a *k* value of 0.035 $year^{-1}$ (SD ± 0.015) for the diameter class > 40 cm; which is similar to the estimate by Kahl (2008). Most of the trees died "naturally" due to *F. fomentarius* in the study by Müller-Using and Bartsch (2009), which may mainly explain the high *k* value in that study. It can be assumed that most logs investigated in our study originated from trees that were still vital compared to trees that died "naturally". In freshly fallen logs, colonization by (cord-forming) basidiomycetes differs from logs that have already been affected by fungi before they fell and wood can be expected to decay slower (Boddy and Heilmann-Clausen 2008). Further, tissues that have died "naturally" may be predisposed to microbial colonization, whereas artificially detached living tissues (i.e., due to wind breakage) may maintain metabolic activities against decomposers (see Yin 1999). In addition, different *k* values might result (to some extent) from different tree diameters in other studies. Colonization by (cord-forming) basidiomycetes seems to progress slower in large diameter logs (Boddy and Heilmann-Clausen 2008). Although this is not supported by the results of this study, different site, i.e., microclimatic conditions might further lead to an increase or decrease in *k* values, for example through canopy removal (Hagemann et al. 2010; Forrester et al. 2012). Managed forests may also harbor a lower diversity of wood decaying fungi, which may lead to different decay rates (Stenlid et al. 2008; Purahong et al. 2014a, b). With regard to the origin of dead wood and its diameters, our data are likely more representative of managed forests rather than strict reserves.

The decomposition rate of 0.033 $year^{-1}$ (SD ± 0.013, 39 %) for *P. abies* calculated in our study was comparable to *k* values in other studies (see Rock et al. 2008). In detail, it was in the range of variation of the estimate by Kahl (2003; 0.027 $year^{-1}$ (SD±0.023)) in Central Germany and identical to *k* values found by Naesset (1999; *k* = 0.033 $year^{-1}$; mean diameter: 13 cm) in southeastern Norway. It was also similar to values in European (Shorohova and Kapitsa 2014) and Russian boreal forests (Krankina and Harmon 1995; Krankina et al. 1999; Tarasov and Birdsey 2001; Harmon et al. 2000; Yatskov et al. 2003).

The conditions of the decomposition process in the study by Naesset (1999) were very similar to our situation. Their starting point of the decomposition process was the date of cutting. All logs were free from rot at the time of cutting and the decomposition took place in open areas.

Table 5 Annual fragmentation loss for bark and wood of *Fagus sylvatica*, *Picea abies* and *Pinus sylvestris*

Species	CWD age (yrs)	Bark (% total fragmentation loss)	Wood (% total fragmentation loss)	Fragmentation (% total mass)
F. sylvatica (*n* = 6)	8	99.6 (0.4)	0.4 (0.4)	0.27 (0.37)
	18	28.5 (27.2)	71.5 (27.2)	0.02 (0.01)
P. abies (*n* = 6)	8	98.3 (3)	1.7 (3)	0.13 (0.22)
	18	0	100	0.65 (1.02)
P. sylvestris (*n* = 6)	8	100	0	0.01 (0.01)
	18	73.8 (18.8)	26.2 (18.8)	0.18 (0.20)

Values are means with standard deviation in parentheses

Table 6 Absolute and relative residence times per decay stage for CWD logs of *Fagus sylvatica*, *Picea abies* and *Pinus sylvestris*

Decay stage	Residence time (Years)					
	Fagus sylvatica		*Picea abies*		*Pinus sylvestris*	
	yrs	%	yrs	%	yrs	%
1	6.7	12.2	8.6	9.8	6	6.0
2	10.7	19.5	15.3	17.4	14	14.0
3	15.8	28.8	25.3	28.8	28.3	28.4
4	21.6	39.4	38.5	43.9	51.5	51.6

Relative residence times were calculated as percent of the sum of residence times for each species

Compared to beech, decomposition rates k from the literature are much less variable for spruce. One explanation for this may be that the sensitivity to influencing factors (as cause of death, decomposer community or differences in diameter) is less pronounced in spruce. This may partly be attributable to a higher decay resistance of spruce and more uniform substrate conditions (see also Cornwell et al. 2009; Weedon et al. 2009) when compared to beech.

However, we observed a significant effect of diameter on k values also for spruce. Since all comparable studies in boreal regions were conducted mainly on small diameter trees, the variation between k values should also be small. In our study, the influence of climatic variables on

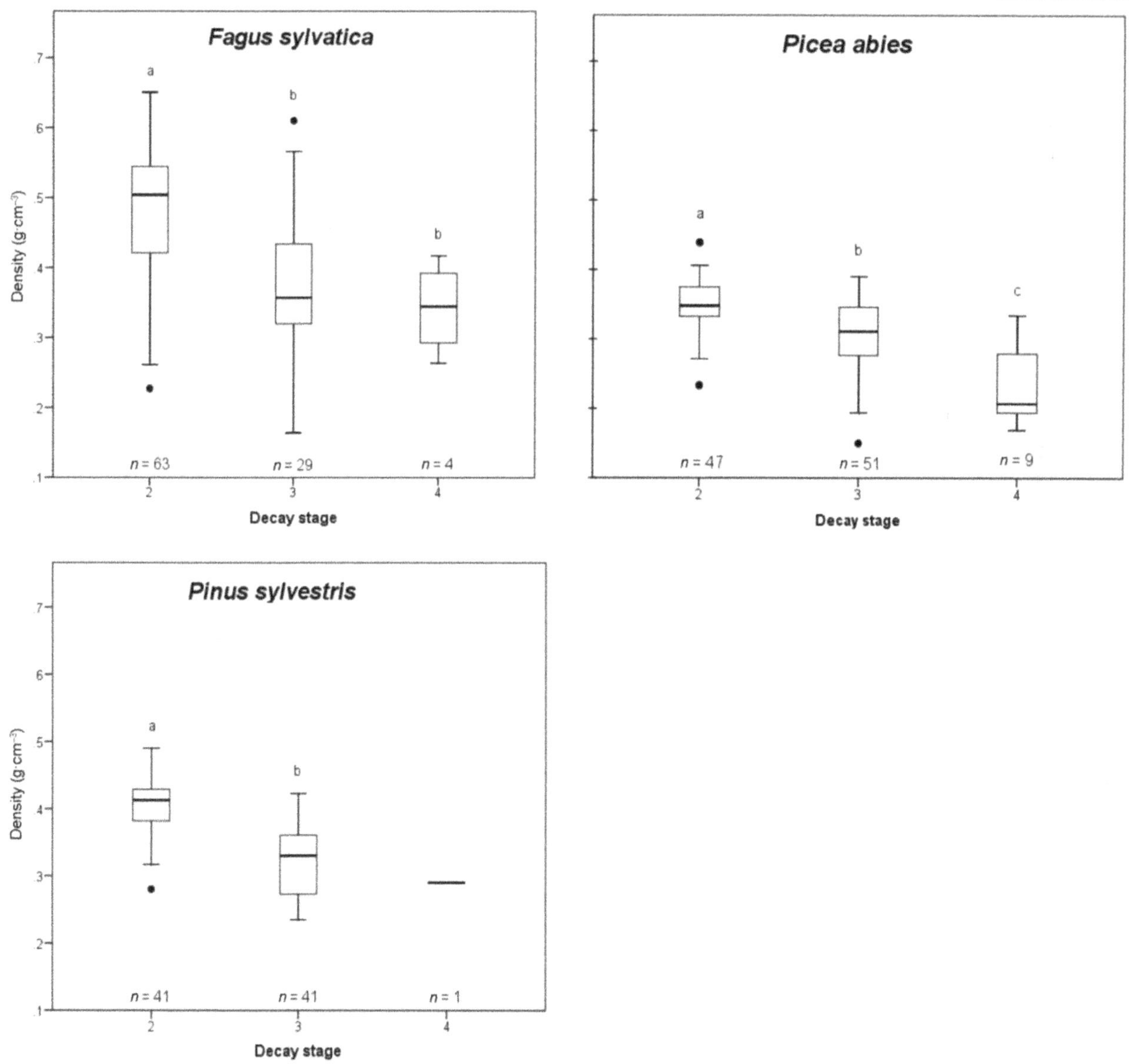

Fig. 3 Dry density (g cm^{-3}) in log segments of *Fagus sylvatica*, *Picea abies* and *Pinus sylvestris*; boxplots display median, lower and upper quartile, minimum and maximum values; points outside boxplots represent outliers; different letters indicate significant differences ($p < 0.05$)

Table 7 Carbon concentrations (%) and contents (mg cm^{-3}) in relation to decay stage (DS), mean and standard deviation (SD), as well as no. of samples, of log segments of *Fagus sylvatica*, *Picea abies* and *Pinus sylvestris*

Species		Decay stage												*p* value (Anova)	
		1			2			3			4				
		Mean	SD	*n*	Mean	SD	*n*	Mean	SD	*n*	Mean	SD	*n*	DS	Species
F. sylvatica[a]	%	46.3[a]	0.52	4	47.5[b]	0.62	39	47.5[b]	0.73	22	47.4[ab]	0.31	2	< 0.05	< 0.001
	mg cm^{-3}	315[a]	3.6		238[b]	36.8		174[c]	48.3		161[c]	52.7			
P. abies[b]	%	47.[a]	0.89	7	49.9[b]	1.51	38	50.1[b]	1.73	34	49.8[b]	2.73	9	< 0.01	
	mg cm^{-3}	203[a]	9.2		170[b]	17.4		155[c]	18.1		117[c]	34.1			
P. sylvestris[b]	%	48.8 [a]	1.13	5	49.2 [a]	1.54	34	50.2[b]	1.31	31	51.4		1	< 0.05	
	mg cm^{-3}	235[a]	11.4		198[b]	22.6		162[c]	28.5		149[d]			< 0.01	

Small letters indicate significant differences between tree species for total C concentration as well as between the different decay stages within each species

the decomposition, also of spruce logs, was low. For example, we found no difference between *k* values for the sites Kib ('cold and wet') and Sig ('warm and dry'), which may simply reflect different types of climatic limitations for wood decaying fungi when compared to the 'warm and wet' optimum, which was not represented in our design.

Our *k* value for *P. sylvestris* (0.032 year^{-1} (SD ± 0.015, 47 %)) was similar to values found in the boreal forests of Europe (Shorohova and Kapitsa 2014) and Russia (Krankina and Harmon 1995; Harmon et al. 2000; Wirth et al. 2000; Yatskov et al. 2003; see also Rock et al. 2008).

The *k* value of pine determined in our study was mainly attributable to the loss of sapwood (as heartwood was found to be still largely intact for most logs after 36 years of decomposition time). Hence, if a longer total decomposition period was analysed, the average decomposition constant might be lower than the one we reported here.

Similar to spruce, variation in the decomposition constant *k* of pine logs was also very low. This may again point to a higher decay resistance and a more uniform decomposition process. In the gymnosperm wood of spruce and pine, there was no highly variable spatial pattern of intact and highly decomposed patches next to each other, as was observed for beech. This observation is in accordance with much higher variation in log respiration rates in dead wood of *F. sylvatica* than in *P. abies* and *P. sylvestris* (Herrmann and Bauhus 2012).

At a more general level, *k* values of angiosperm wood are typically higher, on average 77 %, than in gymnosperm wood (Weedon et al. 2009). A similar difference was also found by Russell et al. (2014). For comparison, our decomposition rate for beech CWD was about 60 % higher than for spruce and pine.

An estimation of decomposition rates for the federal state of Brandenburg (Northeastern Germany) based on a literature review and expert consultation (Rock et al. 2008) produced *k* values for *F. sylvatica*, *P. abies* and *P. sylvestris* of 0.067 year^{-1}, 0.0525 year^{-1} and 0.0575 year^{-1}, respectively. In comparison to our results the estimated *k* values of *P. abies* and in particular of *P. sylvestris* appear to be too high. The deviations between expert estimates of decay rates and our data underpin the importance of actual measurements. In particular for low decay rates, small absolute errors result in proportionally very large errors when calculating mass loss. For example, using a *k* value of 0.0525 year^{-1} for *Picea abies* (Rock et al. 2008), which is 59 % higher when compared to 0.033 year^{-1} (determined in this study) would shorten the period until 50 % of CWD mass was lost by about 37 % (8 years).

In contrast, the chronosequence approach used in our study may cause an underestimation of *k*, because slow-decaying logs may have a higher probability of being included in the sampling (Kruys et al. 2002; Herrmann and Prescott 2008). However, this problem increases with the decomposition time covered and it is negligible, when the observation period is shorter than the minimum decomposition time for logs. For example, if we assume a) high decomposition rates of $k = 0.09$, and b) that logs can still be identified in the field when they contain 20 % of the original mass, no logs should be lost form the sample population before 18 years.

In our study, decomposition rates increased with decreasing diameter class, except for *P. abies*, where the *k* values between 20 and 40 cm tended to be slightly higher than in smaller logs (<20 cm) (Table 4). Lower *k* values for smaller diameters (< 10 cm compared to > 25 cm) of *P. abies* were also observed by Naesset (1999), who assumed that it was most likely caused by branches that prevent the logs from soil contact. In our study, the diameters < 20 cm for *P. abies* were most often sampled in crown sections of the logs, where also soil contact was less often encountered than for 20–40 cm diameter logs. However, *k* values commonly decrease with increasing diameter (Graham and Cromack 1982; Stone et al. 1998; Chambers et al. 2000; Mackensen et al. 2003), even if sometimes inconsistent relationships have been reported (see

Herrmann and Prescott 2008). Hence, where decomposition models for CWD with large variation in dimensions are required, it is advisable to consider log diameter.

We are aware that our assumption of a constant decay rate based on the single negative exponential decay model may be questionable and that there are more sophisticated models, e.g., as applied in the Yasso model (Tuomi et al. 2011). However, our dataset was not sufficient in terms of the variation in decomposition periods of logs to fit the temporal decay dynamics of a more sophisticated model in a robust way. Since there are only very few field experiments that followed mass loss in individual CWD pieces over the long term, the 'real' temporal mass loss pattern is not known. Even if, the mass loss rate of some wood constituents slowed down with increasing time, this may be partially compensated by fragmentation, which was found to be of increasing importance in later decomposition stages (see e.g., Lambert et al. 1980). The mass loss rate may even increase in later stages as observed in the study by Kahl (2008).

Prediction of mass remaining

A substantial proportion of the variation (64 %) in mass remaining could be predicted by decomposition time, tree species, original log diameter, and their interactive effects, whereas temperature had a very small, and initial nutrient concentrations had almost no influence. The observed influence of manganese concentrations on mass remaining may be explained by the relevance of manganese for lignin degradation by white-rot fungi (Hofrichter et al. 2009).

Similar to our findings, a literature review on CWD decomposition had indicated that tree species had a stronger influence on decomposition and nutrient dynamics than the abiotic environment (Laiho and Prescott 2004). And no effect of lignin or nitrogen concentrations was observed in an attempt to model CWD decay based on data about 300 cases of stem, branch and root woody debris decay from North America (Yin 1999). Also, most of the explained variation of mass remaining in the study by Yatskov et al. (2003) could be attributed to time since death (50 %), log position (8 %) and tree species (6 %). However, in some studies, environmental variables did explain a significant proportion of CWD decomposition over time (e.g., Russell et al. 2014). About 80 % of the variance in the decomposition rate constant was explained by a multiple regression model with the factors of tree species, mean diameter, mean temperature in July, sum of precipitation per year and a lag time (Zell et al. 2009).

In contrast to Zell et al. (2009), we observed no significant improvement in our prediction if mean temperature in July instead of mean annual temperature was used in the model.

In a global meta-analysis of wood decomposition rates of angiosperms and gymnosperms, significant relationships between wood traits and decomposition rates were only observed for angiosperms (Weedon et al. 2009). For angiosperms, positive relationships between k and the nutrients N and P, and a negative relationship between k and C:N ratios were found. We suggest that the mass loss of CWD of the species investigated in our study can be sufficiently well predicted for most management and modelling tasks by the factors tree species, time since commencement of decomposition, and log diameter, which can be obtained easily from forest inventories. Predictions will likely become less accurate, when CWD originates from different processes, e.g., when trees also die standing and decay slowly for many years before logs hit the ground, or when the wood decomposition by fungi commences in the living tree (e.g., Kahl 2008). In our field study, most trees were felled as living trees by storms.

In contrast to this field study, about 60 % of the variation in CO_2 flux of CWD of the same species was explained by climatic variables (wood moisture and wood temperature) in a lab incubation experiment (Herrmann and Bauhus 2012). In the same study, temperature explained more than 90 % of CWD respiration of individual *Fagus sylvatica* and *Picea abies* logs over one year in the field. This comparison shows that scaling up from short-term CWD respiration measurements to the long-term dynamics of CWD mass loss is difficult, since the combined effects of temperature, moisture, and interactions between substrate quality and microorganisms need to be considered (Herrmann and Bauhus 2008). In order to capture the complex interplay of processes (i.e., respiration, fragmentation) responsible for decomposition in forests, long-term field measurements are necessary.

CWD fragmentation

The limited study on wood fragmentation indicated that this process may contribute considerably to annual mass loss (max. 30 % of k for pine) although values were lower than in some other studies (e.g., 63 % found by Lambert et al. 1980). Our logs were mainly at the beginning and in the middle of the decomposition process. Fragmentation was found to be of increasing importance with advanced decomposition (Harmon et al. 1986; Müller-Using and Bartsch 2009) as well as in higher elevations (Lambert et al. 1980). However, our study also showed that fragmentation was highly variable within and between logs, and that single events, such as activities of animals searching for food in the logs, may contribute substantially to mass loss.

Measuring fragmentation over the course of only one year was certainly not sufficient to assess the significance of this process in the medium to long-term. This would deserve a separate study. However, it would be very difficult to include fragmentation in models predicting CO_2 release from CWD decomposition, since a large proportion of the material lost from CWD through fragmentation is simply transferred to a different pool, the litter layer. And we have no information on the decomposition rate of fragmented wood in the litter layer.

Drill resistance measurements

In a study that used a similar measurement device, 65 % of the variation in wood density of *P. abies* could be explained by drill resistance (Kahl et al. 2009). The potential to determine wood density in a large number of samples in a short period of time may compensate for the lack of precision for individual pieces in large inventories. In addition, determining drill resistance may be the only form to collect data on wood density in strict forest reserves, where the collection of stem discs is not permitted.

Predicting carbon density in CWD

The loss of mass and thus carbon in CWD is a continuous process. Using distinct decay stages based on visual assessment of logs can be a useful approach to capture the variation in CWD mass and C density in forest inventories. However, the usefulness of this approach depends on how distinct decay stages differ in wood density and C concentration.

Here, we observed a high variation in CWD density within decay classes and hence only few significant differences between adjacent decay classes within a given species. Density variation was particularly high for *F. sylvatica* logs in decay stages 2 and 3. This might suggest, among other things, that decay stages 2 and 3 were most difficult to distinguish by the visual classification system. In some cases, decomposition was found to be more (or less) advanced than 'suggested' by the tree surface (i.e., bark or sapwood condition). This could be the result of case hardening (drying out of the outer sapwood) as decomposition took mostly place in open areas (see also Yin 1999). In addition, decomposition of *F. sylvatica* logs was spatially highly variable. Intact and decomposed patches (with different densities) separated by demarcation lines composed of melanin, which is characteristic of white rot fungi (see also Schwarze et al. 1999; Kahl 2008), were found in direct proximity within the same tree disc. Owing to such patterns of log colonization by fungi, density variation may initially increase with decomposition. For example, an increase in density variation with progressing decomposition, with the maximum in decay stage 3 and a decline towards decay stage 5, was also observed by Yatskov et al. (2003).

The absence of distinct differences in wood density between decay classes has also been observed in other studies. Little change (respectively an overlap) in density between the two least (1 and 2) and most advanced (4 and 5) decay classes has been found also in the study by Yatskov et al. (2003). Similar to our results, no significant difference in density of decay stages 3 and 4 of *F. sylvatica* were also found by Müller-Using and Bartsch (2009). One main characteristic of log sections in decay stage 4 in our study was the close proximity of highly decomposed material to areas of relatively intact wood. Since the maximum decomposition time for woody debris of *F. sylvatica* in our study was about 16 years, decay stage 4 comprised mainly log segments with a diameter < 20 cm. Hence, our data may not be representative for larger diameter CWD of beech.

Wood densities for the different tree species converged for advanced decay stages, which is in accordance with other studies (Yatskov et al. 2003). We found no significant differences between species for CWD densities in decay stage 4.

The increase in C concentration with advancing decomposition observed in our study was also found by Müller-Using and Bartsch (2007) for beech. It suggests an increase of lignin and a decrease of cellulose, as lignin has a higher proportion of C than cellulose. The increase in C concentration cannot be explained by a decrease of minerals in relation to C. Similar to our results, C concentrations were lower for angiosperm CWD when compared to gymnosperm CWD in a study that analysed C concentrations of 60 tree species (Harmon et al. 2013). Unlike our results, a decrease in C concentration with increasing decay class was observed for angiosperm CWD in that study. In contrast to C concentration, C content was highest for beech, followed by pine and spruce. This can be explained by the higher density of sound wood.

The pattern of C density ($mgC \cdot cm^{-3}$) and decay stage was similar to that of wood density and decay stage. C density decreased parallel with density and mass loss, as has been found by others (Müller-Using 2005). Carbon density in CWD converged across species with increasing decay (similar to the relationship between density and decay stage). Hence for highly decayed logs, for which it may also be difficult to determine the species origin, one common wood and C density may be assumed.

To assess C in CWD from inventories that record decay stage, our values of dry wood and C density per species and decay stage could be used for calculation purposes. However, the actual assessment of C concentration in inventories appears to be meaningful only for beech, where the difference between the measured C concentration and the default value of 50 % was 3.4 %–4.6 %.

Conclusions

To our knowledge, this is the first study that assessed the decomposition dynamics of CWD logs of *F. sylvatica*, *P. abies* and *P. sylvestris* across different sites and diameter classes in central Europe. In comparison to studies at individual sites, sampling of CWD across different sites provides a more robust estimation of decomposition rates because a wider range in climatic conditions and decomposer communities are captured.

Our results indicated that a reasonable prediction of the decomposition of CWD should be possible on the basis of a few easily obtainable parameters, since variables such as diameter and tree species had the greatest influence on mass loss over time. Since the variation in climatic parameters such as average annual temperature and precipitation had only a minor effect on the decomposition process, these variables might be neglected for the prediction of CWD mass loss in this temperature and precipitation range.

The decomposition rates determined in our study, although they may be further refined, provide a basis for the management of CWD. The decomposition constants will allow the development of decomposition models based on information about the input of CWD according to species and diameter. Based on the derived C-densities per decay stage, C contents in CWD can be calculated to serve as input to C accounting of forest ecosystems.

Competing interests

The authors declare that they have no competing interests.

Authors' contributions

SH planned and conducted the study including field sampling and analysis and wrote the majority of the manuscript. TK conducted linear mixed effects modelling and contributed to the manuscript. JB conceived and guided the study and co-wrote the manuscript. All authors read and approved the final manuscript.

Acknowledgements

We are grateful to Renate Nitschke and Germar Csapek, and the many students, who helped in the field and in the lab. We thank Sarah Grimm from Seminar for Statistics ETH Zurich for statistical advice. Oliver Jakoby provided helpful comments to improve the manuscript. This research was funded by a German Science Foundation grant to Jürgen Bauhus (DFG - BA 2821/4-1).

References

Albrecht L (1991) Die Bedeutung des toten Holzes im Wald. Forstw Cbl 110(1991):106–113

Boddy L, Heilmann-Clausen J (2008) Basidiomycete community development in temperate Angiosperm wood. In: Boddy L, Frankland J, Van West P (eds) Ecology of Saprotrophic Basidiomycetes. Academic, London, pp 211–238

Bütler R, Schlaepfer R (2004) Wie viel Totholz braucht der Wald? Schweiz Z Forstwes 155(2):31–37

Chambers JQ, Higuchi N, Schimel JP, Ferreira LV, Lemack JM (2000) Decomposition and carbon cycling of dead trees in tropical forests of central Amazon. Oecologia 122:380–388

Christensen M, Hahn K, Mountford EP, Odor P, Standovar T, Rozenbergar D, Diaci J, Wijdeven S, Meyer P, Winter S, Vrska T (2005) Dead wood in European beech (*Fagus sylvatica*) forest reserves. For Ecol Man 210:267–282

Commarmot B, Brändli U-B, Hamort F, Lavnyy V (2013) Inventory of the largest primeval beech forest in Europe. A Swiss-Ukrainian scientific adventure. Swiss Federal Research Institute WSL, Birmensdorf; Ukrainian National Forestry University, L'viv; Carpathian Biosphere Reserve, Rakhiv. p 69

Cornwell WK, Cornelissen JHC, Allison SD, Bauhus J, Eggleton P, Preston CM, Scarff F, Weedon JT, Wirth C, Zanne AE (2009) Plant traits and wood fates across the globe: rotted, burned, or consumed? Global Change Biol 15:2431–2449

Dormann CF, Kühn I (2009) Angewandte Statistik für die biologischen Wissenschaften. 2., durchgesehene, aktualisierte, überarbeitete und erweiterte Auflage. Helmholtz Zentrum für Umweltforschung-UFZ.S 245

Effland MJ (1977) Modified procedure to determine acid-insoluble in wood and pulp. Tappi 60(10):143–144

FAO (2006) World reference base for soil resources 2006. A framework for international classification, correlation and communication. World soil resources report No. 103. 2006 edition. Food and agriculture organization of the united nations (FAO). Rome, 2006. p 128 http://www.fao.org/ag/Agl/agll/wrb/doc/wrb2006final.pdf

Forrester JA, Mladenoff DJ, Gower ST, Stoffel JL (2012) Interactions of temperature and moisture with respiration from coarse woody debris in experimental forest canopy gaps. For Ecol Man 265:124–132

Freschet GT, Weedon JT, Aerts R, van Hal JR, Cornelissen JHC (2011) Interspecific differences in wood decay rates: insights from a new short-term method to study long-term wood decomposition. J Ecol 100(1):161–170

Graham RL, Cromack KJR (1982) Mass, nutrient content, and decay rate of dead boles in rain forests of Olympic National Parc. Can J For Res 12:511–521

Grove S, Meggs J, Goodwin A (2002) A review of biodiversity conservation issues relating to coarse woody debris management in the wet eucalypt production forests of Tasmania. Forestry Tasmania, Hobart, p 72

Grove SJ, Stamm L, Barry C (2009) Log decomposition rates in Tasmanian *Eucalypt obliqua* determined using an indirect chronosequence approach. For Ecol Man 258:389–397

Hagemann U, Moroni MT, Gleißner J, Makeshin F (2010) Disturbance history influences downed woody debris and soil respiration. For Ecol Man 260:1762–1772

Harmon ME, Franklin JF, Swanson FJ, Sollins P, Gregory V, Lattin JD, Anderson NH, Cline SP, Aumen NG, Lienkaemper GW, Cromack KJ, Cummins KW (1986) Ecology of coarse woody debris in temperate ecosystems. Adv Ecol Res 15:133–302

Harmon ME, Krankina ON, Sexton J (2000) Decomposition vectors: a new approach to estimating woody detritus decomposition dynamics. Can J For Res 30:76–84

Harmon ME, Fasth B, Woodall CW, Sexton J (2013) Carbon concentration of standing and downed woody detritus: Effects of tree taxa, decay class, position, and tissue type. For Ecol Man 291:259–267

Herrmann S, Bauhus J (2008) Comparison of methods to quantify respirational carbon loss of coarse woody debris. Can J For Res 38:2738–2745

Herrmann S, Bauhus J (2012) Effects of moisture, temperature and decomposition stage on respirational carbon loss from coarse woody debris (CWD) of important European tree species. Scand J For Res 28(4):346–357

Herrmann S, Prescott CE (2008) Mass loss and nutrient dynamics of coarse woody debris in three Rocky Mountain coniferous forests: 21 year results. Can J For Res 38:125–132

Hofrichter M, Liers C, Kinne M, Ullrich R (2009) Abbau und Transformation von Lignin und Huminstoffen durch Pilze. 81-97. In: Ökologische Rolle von Pilzen. Rundgespräche der Kommission für Ökologie, Bd. 37. Hrsg. Bayerische Akademie der Wissenschaften, Verlag Dr. Friedrich Pfeil, 2009

Kahl T (2003) Abbauraten von Fichtentotholz (Picea abies (L.) Karst.) – Bohrwiderstandsmessungen als neuer Ansatz zur Bestimmung des Totholzabbaus, einer wichtigen Größe im Kohlenstoffhaushalt mitteleuropäischer Wälder. Magisterarbeit Friedrich-Schiller-Universität Jena, S 98

Kahl T (2008) Kohlenstofftransport aus dem Totholz in den Boden. Dissertation Universität Freiburg. (http://www.freidok.uni-freiburg.de/volltexte/5522/)

Kahl T, Wirth C, Mund M, Böhnisch G, Schulze ED (2009) (2009) Using drill resistance to quantify the density in coarse woody debris of Norway spruce.Eur. J For Res 128:467–473

Kahl T, Mund M, Bauhus J, Schulze ED (2012) Dissolved organic carbon from European beech logs: Patterns of input to and retention by surface soil. Écoscience 19(4):1–10

Krankina ON, Harmon ME (1995) Dynamics of the dead wood carbon pool in northwestern Russian boreal forest. Water Air Soil Polut 82:227–238
Krankina ON, Harmon ME, Griazkin AV (1999) Nutrient stores and dynamics of woody detritus in a boreal forest: modeling potential implications at the stand level. Can J For Res 29:20–32
Kruys N, Jonnsson BG, Stahl G (2002) A stage-based matrix model for decay-class dynamics of woody debris. Ecol Appl 12(3):773–781
Kuehne C, Donath C, Müller-Using SI, Bartsch N (2008) Nutrient fluxes via leaching from coarse woody debris in a *Fagus sylvatica* forest in the Solling Mountains, Germany. Can J For Res 38:2405–2413
Laiho R, Prescott CE (2004) Decay and nutrient dynamics of coarse woody debris in northern coniferous forests: a synthesis. Can J For Res 34:763–777
Lambert RL, Lang GE, Reiners WA (1980) Loss of mass and chemical change in decaying boles of a subalpine balsam fir forest. Ecology 61(6):1460–1473
Mackensen J, Bauhus J, Webber E (2003) Decomposition rates of coarse woody debris – a review with particular emphasis on Australian tree species. Austr J Bot 51:27–37
MCPFE (2003) State of Europe's Forests 2003. The MCPFE Report on Sustainable Forest Management in Europe. MCPFE Liaison Unit Vienna and UNECE/FAO. URL on 21.10.2015: www.foresteurope.org/documentos/forests_2003.pdf
MCPFE (2007) State of Europe's Forests 2007. The MCPFE Report on Sustainable Forest Management in Europe. MCPFE Liaison Unit Warsaw, UNECE and FAO. URL on 21.10.2015: http://www.foresteurope.org/documentos/state_of_europes_forests_2007.pdf
Meyer P (1999) Totholzuntersuchungen in nordwestdeutschen Naturwäldern: Methodik und erste Ergebnisse. Forstwiss Cbl 118:167–180
Meyer P, Bartsch N, Wolff B (2003) Methoden der Totholzerfassung im Wald. Forstarchiv 74:263–274
Müller J, Bütler R (2010) A review of habitat thresholds for dead wood: a baseline for management recommendations in European forests. Eur J For Res 129:981–992
Müller-Using SI (2005) Totholzdynamik eines Buchenbestandes im Solling. Berichte des Forschungszentrums Waldökosysteme der Universität Göttingen, Reihe A, Bd. 193:187S
Müller-Using SI, Bartsch N (2007) Totholz im Elementhaushalt eines Buchenbestandes. Forstarchiv 78:12–23
Müller-Using SI, Bartsch N (2009) Decay dynamic of coarse and fine woody debris of a beech (*Fagus sylvatica* L.) forest in Central Germany. Eur J For Res 128:287–296
Naesset E (1999) Decomposition rate constants of Picea abies logs in southeastern Norway. Can J For Res 29:372–381
Olson JS (1963) Energy storage and the balance of producers and decomposers in ecological systems. Ecology 44:322–331
Pregitzer KS, Euskirchen ES (2004) Carbon cycling and storage in world forests: biome patterns related to forest age. Global Change Biol 10:1–26
Purahong W, Kahl T, Schloter M, Bauhus J, Buscot F, Krüger D (2014a) Comparing fungal richness and community composition in coarse woody debris in Central European beech forests under three types of management. Mycol Progress. doi:10.1007/s11557-013-0954-y
Purahong W, Hoppe B, Kahl T, Schloter M, Schulze E-D, Bauhus J, Buscot F, Krüger D (2014b) Changes within a single land-use category alter microbial diversity and community structure: Molecular evidence from wood-inhabiting fungi in forest ecosystems. J Environ Manage 139:109–119
Ranius T, Kindvall O, Kruys N, Jonsson BG (2003) Modeling dead wood in Norway spruce stands subject to different management regimes. For Ecol Manage 182:13–29
Rock J, Badeck F-W, Harmon ME (2008) Estimating decomposition rate constants for European tree species from literature sources. Eur J For Res 127:301–313
Russell MB, Woodall CW, Fraver S, D'Amato AW, Domke GM, Skog KE (2014) Residence Times and Decay Rates of Downed Woody Debris Biomass/Carbon in Eastern US Forests. Ecosystems 17:765–777
Schwarze FW, Engels J, Mattheck C (1999) Holzzersetzende Pilze in Bäumen. Strategien der Holzzersetzung. Rombach Ökologie 5
Seibold S, Brandl R, Buse J, Hothorn T, Schmidl J, Thorn S, Müller J (2014) Association of the extinction risk of saproxylic beetles and the ecological degradation of forests in Europe. Conservation Biology. (in press)
Shigo AL (1986) A new tree biology. Shigo and Trees Associates, Durham
Shorohova E, Kapitsa E (2014) Influence of the substrate and ecosystem attributes on the decomposition rate of coarse woody debris in European boreal forests. For Ecol Manage 315:173–184
Siitonen J (2001) Forest management, coarse woody debris and saproxylic organisms: Fennoscandian boreal forests as an example. Ecol Bull 49:11–41
Spears JDH, Holub SM, Harmon ME, Lajtha K (2003) The influence of decomposing logs on soil biology and nutrient cycling in an old-growth mixed coniferous forest in Oregon, U.S.A. Can J For Res 33:2193–2201
Stenlid J, Penttilä R, Dahlberg A (2008) Wood-decay basidiomycetes in boreal forests: distribution and community development. In: Boddy L, Frankland J, Van West P (eds) Ecology of Saprotrophic Basidiomycetes. Academic, London, pp 239–262
Stone JN, MacKinnon A, Parmintor JV, Lertzman KP (1998) Coarse woody debris decomposition documented over 65 years on southern Vancouver Island. Can J For Res 28:788–793
TAPPI (1976) Acid-soluable lignin in wood and pulp UM 250. Technical association of the pulp and paper industry, Atlanta
Tarasov ME, Birdsey RA (2001) Decay rate and potential storage in the Leningrad region. Ecol Bull 49:137–147
Trendelenburg R, Mayer-Wegelin H (1964) Das Holz als Rohstoff', 2. völlig überarbeitete Auflage. Carl Hanser Verlag, München
Tuomi M, Laiho R, Repo A, Liski J (2011) Wood decomposition model for boreal forests. Ecol Model 222:709–718
Turner DP, Koerper GJ, Harmon ME, Lee JJ (1995) A Carbon Budget for Forests of the Conterminous United States. Ecol Appl 5:421–436
Weedon JT, Cornwell WK, Cornelissen JHC, Zanne AE, Wirth C, Coomes DA (2009) Global meta-analysis of wood decomposition rates: a role for trait variation among tree species? Ecol Lett 12:45–56
Wirth C, Schulze E-D, Luhker B (2000) How fire exerts control over carbon balances in Siberian Scots pine forests. In: The role of boreal forests and forestry in the global casbon budget. Abstracts of International Science Conference, 8–12 May 2002, Edmonton
Yatskov M, Harmon ME, Krankina ON (2003) A chronosequence of wood decomposition in the boreal forests of Russia. Can J For Res 33:1211–1226
Yin X (1999) The decay of forest woody debris: numerical modeling and implications based on some 300 data cases from North America. Oecologia 121(1):81–98
Zell J, Kändler G, Hanewinkel M (2009) Predicting constant decay rates of coarse woody debris – A meta-analysis approach with a mixed model. Ecol Model 220:904–912

22

Estimation of conservation value of myrtle (*Myrtus communis*) using a contingent valuation method: a case study in a Dooreh forest area, Lorestan Province, Iran

Neda Amiri[1*], Seyed F. Emadian[2], Asghar Fallah[3], Kamran Adeli[4] and Hamid Amirnejad[5]

Abstract

Background: Around 2000 plant species occur naturally in Lorestan Province of which 250 species are medicinal and myrtle is one of them. Myrtle is a shrub whose leaves and fruits have medicinal value and thus, if managed and harvested properly, could produce sustained economic benefits. In recent years, however, over half of the myrtle site areas was destroyed, due to inappropriate management and excessive harvesting practices. Thus, coming up with a practical harvesting approach along with identifying those factors damaging the sites, seems to be very crucial.

Methods: In our investigation, we calculated the conservation value per hectare of myrtle in the Dooreh forest area in Lorestan Province. Using the Contingent Valuation (CV) and Double Bounded Dichotomous Choice (DBDC) methods, we determined the willingness to pay (WTP) for myrtle conservation. The WTP was estimated with a logit model for which indices were obtained based on a maximum precision criterion.

Results: The results showed that 86.67 per cent of people were willing to pay for the conservation of these myrtle sites. Average monthly WTP per family was calculated as $0.79. The annual conservation value in terms of WTP for the preservation of the myrtle sites in Dooreh was estimated as $102,525. Among the variables of the model presented, education had a positive impact, while the amount proposed for payment and family size had a negative impact on the WTP.

Conclusions: Our estimate of the value of myrtle conservation should provide justification for policy makers and decision making bodies of natural resources to implement policies in order to conserve the natural sites of this species more effectively.

Keywords: Conservation value, Myrtle, Contingent valuation method, Double Bounded Dichotomous method, Logit model

Background

Non-timber forest products are goods of biological origin other than timber, derived from forests, other wooded areas and trees outside forests (FAO Forestry 1999). In this context non-timber forest products include all kinds of animals, resin, tannin, medicinal and industrial plants and non-food products including materials or plant parts having medicinal properties, are edible and of industrial use (Noubissie et al. 2008). In recent years, issues related to non-timber forest products, particularly the conservation of their natural habitat as well as their production, has attracted considerable interest in the world. The main reasons are the role of these products in providing livelihood and food for rural communities in order to reduce poverty along with supporting various environmental goals such as conservation of biodiversity (FAO Forestry 1999). Non-timber forest products (NTFPs) are accepted as a veritable means of achieving poverty alleviation because of their role in sustaining this livelihood, for food security and environmental objectives such as biodiversity conservation (FAO Forestry 1999).

* Correspondence: amiri.neda90@yahoo.com
[1]Department of Forestry, Sari Agricultural and Natural Resource University, Sari, Iran
Full list of author information is available at the end of the article

Valuation of NTFPs has received a great deal of attention (Lindal 1994). Many attempts have been made to value these non-marketed forest products to their full extent, especially the NTFPs. Extraction and increased trade of NTFPs has been advocated as an alternative to timber-focused utilization on the grounds that it could be pursued without causing tropical deforestation. Consequently, NTFPs are thought to be capable of playing a constructive role in forest conservation and development. This assumes that their harvest, even at a commercial level, would have little or no adverse ecological impact on a forest. This can be questioned, however, since the impact on the forest ecosystem of an increase in harvesting NTFPs is not fully known (Gunatilake et al. 2006; Peters 1994). Lescuyer (1996) argues that the establishment of a monetary value for each NTFPs leads to a change in the modes of use of the resource which, in turn, can cause or accelerate a process of restrictive appropriation of these resources. In many parts of the world, NTFPs provide off-farm employment to a large part of the rural population and accounts for a considerable share of household income. These resources are essential, especially for the rural poor and women, and may, in some cases, provide them with the only source of personal income (Rodda 1991; Falconer 1996). But economic problems may sometimes result in increasing the harvest of those NTFPs that have higher values, particularly those that can be sold on the international market. This is the kind of situation that may sometimes result in very high costs in terms of damage to forest ecosystems (Kengen 1997).

Conservation valuation seeks to create awareness about the need for decisions to optimize the economic management of resources and assists the development of a strategy for sustainable development in a region. Valuation of environmental services and functions is important for a number of reasons, such as: 1) recognizing and understanding of environmental and ecological interests by humans, 2) providing environmental issues to decision makers and country planning experts and 3) providing the link between economic policies and income derived from natural resources.

Economic valuation of natural and environmental resources includes two categories, i.e., use values and non-use values (Pearce and Pearce 2001; Molaei 2009). Use values are associated with the consumption of products, while non-use values are related to enhancement of their utility to people without direct consumption of the products. Non-use values which reflect the concept of conservation consist of existence values, bequest values and option values. Existence value refers to the intrinsic value of a resource, even if people have never seen or used the resource (Torras 2000). Bequest value or future generation value is the desirability resulting from knowledge in preserving natural resources for future generations (Pearce and Pearce 2001). Option value measures the preference of individuals to preserve natural resources for possible future use (Lee and Han 2002). Alternatively, existence value can be considered the willingness of a society to pay for the conservation of natural resources, bequest value may be understood as the willingness to pay (WTP) for conservation of resources for the benefit of future generations, while option value can be defined as the willingness of society to pay for the conservation of resources for possible future use (Lee and Han 2002).

Researchers have attempted to estimate forest conservation values using various approaches. For example, Echeverria et al. (1995) estimated the value of existence benefits in Costa Rica's forests, based on a Contingent Valuation (CV) method and came up with a value of \$238 per hectare of forest per year. Lehtonen et al. (2003) estimated the average conservation value of forests in southern Finland in terms of the willingness to pay at \$23.80 per household per year, using CV in conjunction with Double Bounded Dichotomous questionnaires and a logit regression model. In another study, Leong et al. (2005) studied the conservation value of forests in mountainous areas of Malaysia and estimated conservation values ranging from 20 to 27 Ringgits per individual respondent. Gurluk (2006) used a Contingent Valuation approach to study the value of ecosystem services in a region in Turkey and estimated it at \$76.94 per year per household. Tao et al. (2012), also using a CV method, estimated the annual household willingness to pay for conservation of Heshui forest ecosystems in China's Jiangshi Province at 238 Yuan per year. Molaei in (2009) used a Contingent Valuation method to assess the conservation value of Arasbaran forest ecosystems and showed that 81.7 % of the people from this area were willing to pay some money for conservation of the forest ecosystem. They estimated the annual WTP for the conservation of these forest ecosystems at about \$3.31 per household and \$197.32[1] per hectare of forest land. Peron and Esmaeili (2010) used the Contingent Valuation approach to estimate the conservation value of mangrove forests in Hormozgan province in Iran and found that 92 % of the respondents were willing to pay some money to conserve these mangrove forests. They estimated the average conservation value per individual at \$12.65 and at \$76.94 per hectare of forest.

Although many investigations have been carried out to estimate the conservation value of forest resources, studies on the conservation valuation of forest resources for non-timber forest products, similar to ours, are very limited. For example, Chukwuone and Okorji (2008) used CV method to estimate annual willingness to pay per person for systematic management of forest communities in

Nigeria and concluded that people were willing to pay $55.40 for conservation of their forest for non-timber forest products. Khosravi and Sabouhi (2011), used the same method to estimate the annual conservation value for all tamarisk trees in the Fahliyan riparian area in province of Fars in Iran at $4588. In the same study they estimated the monthly conservation value of each tamarisk tree at $0.163. Molaei and Kavousi Kelashmi (2011) using the same technique, but with a One-Half Bound format, estimated the conservation value of the chandelier lily at $0.192 based on a linear model and $0.285 with a logarithmic model.

Our investigation dealt with myrtle (*Myrtus communis* L.) forest sites in Dooreh, a city in Lorestan Province, Iran. Myrtle, as a non-timber forest product, is an evergreen medicinal plant. It belongs to the family *Myrtaceae* and is found naturally in many parts of the province. This valuable shrub forms dense populations at low elevations at some of its natural habitats. Given its ability to grow and develop in some unfavorable circumstances, such as warm climates, myrtle is one of those plants that plays a significant role in the conservation of its environmental habitat. This plant contains two important volatile oil substances, i.e., Myrtenol and Depantyn, which are very valuable to the pharmaceutical industry and thus, potentially myrtle can play an important role in the local economy (Makkizadeh Tafti et al. 2006). As well, this species, is of importance in maintaining environmental sustainability in semiarid regions. Because of its evergreen features it helps to make the landscape more attractive and aids the restoration of natural ecosystems (Salehnia 2008).

Unfortunately, due to excessive harvesting, dense natural stands of myrtle are nonexistent. However, reforestation and afforestation of this plant have been proposed as feasible options. These are effective strategies to increase employment, support the regional pharmaceutical industry and increase family incomes in the community (Mirabad et al. 2011). The Khorraman Pharmaceutical Company is operating as a monopoly for it is the only pharmaceutical industry in the province that uses myrtle oil extracts to produce herbal medicines such as Belfarol, Myrtol, Rectol and Aphtoplex. Each year, an average of nearly 120 t of myrtle is harvested by rural people in the region and sold to this company, resulting in a significant increase in the income of rural households, accounting for 63.38 per cent of their total monthly income during the harvest months (Amiri 2013). Frequent and excessive harvesting of myrtle shrubs by villagers, without training and knowledge of the species and its proper harvesting operation, inflict a great deal of damage on the habitat of this species (Amiri, 2013). This is consistent with the earlier findings by Karimi (2012), who reported that nearly half of the myrtle habitat was destroyed or converted into agricultural lands and the remaining half showed a dramatic decline in productivity. Overall, observations on the distribution of myrtle in Dooreh and interviews conducted with local people, have shown that the most important factors contributing to the destruction of the myrtle forest sites in order of importance are: 1) a lack of sound management practices and/or management incompetence, 2) excessive harvesting practices by the Khorraman Pharmaceutical Company to produce herbal medicines, 3) excessive harvesting practices and management by villagers for personal consumption and burning of myrtle habitats for the purpose of changing myrtle sites into agricultural land, 4) improper harvest operations by unskilled individuals, 5) lack of expertise in conservation of myrtle forests and their management, 6) weather condition in recent years and 7) excessive harvesting of myrtle branches for feeding livestock (Amiri 2013) (Fig. 1).

The purpose of our study was to estimate the conservation value of the myrtle habitat in Dooreh in order to provide the means and justification for forest policy makers and managers to support conservation of the habitat of this species and to stop its current horrifying destruction. Clearly, due to its various benefits, more attention need to be devoted and resources allocated to restore these damaged and destroyed myrtle habitats.

Methods

The study area

Our investigation was conducted in a single stand of myrtle located in Dooreh, Lorestan province. The city is located in the western part of the province, 36 km from the provincial capital, Khorramabad, at longitude 48°02′17″ and latitude 33°39′56″, at an elevation of 1118 m (Fig. 2).

Contingent valuation method

We used a Contingent Valuation method (CVM) to estimate the conservation value of myrtle forests in the Dooreh area. In this method a hypothetical market is assumed in which contingent questionnaires are distributed among respondents and completed in their presence or delivered by mail. The respondents are asked about their Willingness to Pay (WTP) for the quality and level of improvement of environmental products, or about their Willingness to Accept (WTA) a loss from inflicting damage to their resources (Bohm 1972; Dahlberg 1974; Hanemann 1991; Bateman & Willis 1999).

Double- bounded questionnaires are used to measure WTP. This method was first proposed in 1979 by Bishop and Heberlin (Venkatachalam 2003). In their approach, respondents were faced with a choice of either "Yes" or "No" with respect to a proposed price in a hypothetical market (Venkatachalam 2003). Hahnemann (1985)

Fig. 1 Two contrasting myrtle habitats (**a**) healthy (**b**) damaged

modified and adjusted the Double- bounded Choice and proposed a Double- bounded Dichotomous Choice (DDC) method. This method requires determining one higher price proposed for the respondents in addition to the initially lower price proposal. Proposing the higher price depends on the initial responses or to their feedback on the initially proposed lower price (Venkatachalam 2003).

The first section of the questionnaire contained questions regarding socio-economic aspects of individual respondents, such as age, gender, occupation, education, place of residence, family size, income and membership in environmental organizations. The main questions, relating the WTP on the part of the respondents for conservation of myrtle in Dooreh, i.e., those related to the acceptance or rejection of the bids for willingness to pay, were introduced in a Double- bounded Dichotomous Choice format. Forty questionnaires were initially completed in which open questions, without any proposed prices for willingness to pay, were asked. Thus, according to the preliminary information obtained from these questionnaires, the proposed prices were determined as 20,000 Rials for the main bid, 10,000 Rials for the lower and 30,000 Rials for the upper bid.

In order to derive the model for estimating WTP, it was assumed that the individual respondent either accepted the proposed price for conservation value to maximize his utility as described below (Eq. 1) or, alternatively, rejected it (Hanemann 1984).

$$U(1,\ Y-A; S) + \epsilon_1 \geq U(0,\ Y; S) + \epsilon_0 \tag{1}$$

Where U is respondent, Y his/her income, A the proposed price, S the socio-economic features affected

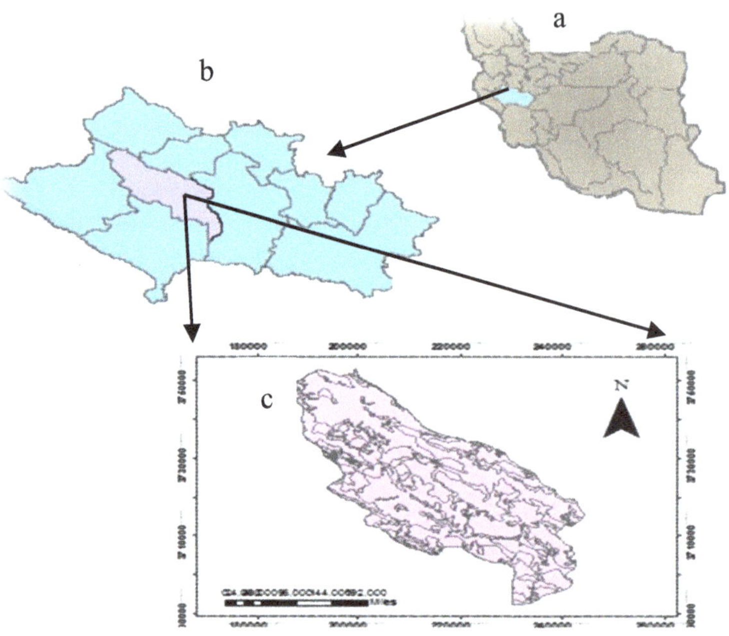

Fig. 2 The map of the study area: (**a**) Iran; (**b**) Province of Lorestan; (**c**) City of Dooreh

by his/her taste and $\in_1$, $\in_0$ are independent random variables with a zero mean. The utility difference (ΔU) due to using an environmental resource, is described as follows (Eq. 2):

$$\Delta U = U(1,\ Y-A;S)-U(0,\ Y;S)+(\in_1-\in_0) \tag{2}$$

ΔU > 0 implies the respondent maximizes his or her utility by saying "yes" and would agree to pay the amount for conservation of the myrtle (Park and Loomis 1996). The format of the double bounded questionnaire in CVM has a dependent variable with dual choice, which requires a qualitative choice model. In general, for studying regression models with a dual choice dependent variable, linear probability models, logit, probit and Tobit models are used (Nakhaei 1389). To investigate the effect of descriptive variables, such as economic and social factors, on the WTP, we used a logit regression model. The probability (P_i) that the individual accepts an offer (A) based on the logit model, can be expressed as follows (Lee and Han 2002):

$$P_i = F_n(\Delta U)\frac{1}{1+\exp(-\Delta U)} = \frac{1}{1+\exp\{-(\alpha-\beta A+\gamma Y+\theta S)\}} \tag{3}$$

Where $F_n(\Delta U)$ is the cumulative distribution function with one standard logistic difference including some of the socio-economic factors. The coefficients θ, γ, β are estimates where θ and $\gamma > 0$ and $\beta \le 0$. (Lee and Han 2002). The expected WTP was calculated using logit models by actually integer numbers between 0 and the highest proposed value (Lehtonen et al., 2003). We opted for this method because it maintained the stability and compatibility of theoretical constraints, statistical efficiency and congregate ability (White and Lovett 1999).

The logit model utilizes maximum likelihood specifications and is the most common technique for estimating expected WTP (Piri *et al.*, 2009). This was calculated by integer numbers between 0 and the highest proposed value: (Eq. 4).

$$E(WTP) = \int_0^{MaxA} F_n(\Delta U)dA = \int_0^{MaxA} \frac{1}{1+\exp[-\alpha^*+\beta A]}\, dA \tag{4}$$

Where E (WTP) is the expected amount of WTP and α^* (see Eq. 5) the interception adjustment value, added to the main interception (α) value, estimated after socio-economic variables were inserted in the logit model.

$$\alpha^* = (\alpha+\gamma Y+\theta S) \tag{5}$$

Logit models may be developed as either linear or semi-logarithmic functions in order to estimate both use and conservation values. In the former model, the probability of accepting the proposed price is merely a function of the proposed price itself. However, in the latter model, the probability is a function of the proposed relative price over income for simplicity and convenience in computation, we used a linear logit model to compute the mean WTP (Molaei and Kavousi Kelashmi 2011).

To estimate the conservation value of the myrtle habitat, selected households from both urban and rural communities of Dooreh were asked to participate as part of our sample population. The required sample size for simple random sampling was calculated (Cochran 1977) for the 95 % confidence interval with an allowed error of 4 % (Eq. 6). This resulted in a sample size of 270. However, to enhance accuracy, the number of household sampled in the area and participated in our study was increased to 300 (Eq. 6).

$$n = \frac{t^2 s^2}{d^2} = \frac{(1.96)^2(0.335)^2}{(0.04)^2} = 269.45 \simeq 270 \tag{6}$$

where n is the number of samples; t is Student's t value for 95 % confidence interval, s^2 is the sample variance and d is the allowed sampling error. Our interviews were focused on individuals who were independent in terms of income and when faced with the proposed amount, were able to make a decision easily. For the purpose of statistical analysis and mathematical calculations the parameters required for the logit model, we used SHAZAM and Excel.

Results

Socio-economic status of respondents

Descriptive statistics of the socio-economic characteristics of the respondents are presented in Table 1. Of the total number of respondents, 208 (69.33 %) were male and 92 (30.67 %) were female. As this table shows the average age was 35 years, the number of years of education was 13, the average size of the household consisted of 4 people with a monthly household income of 15,000 Rials ($.44), the proposed bid was 17,630 Rials ($.52) and the environmental opinion was around 4.

Conservation value of Myrtle

Out of 300 respondents, 260 (86.67 %) were willing to pay something for the conservation of Myrtle habitats of Dooreh, while the other 40 (13.33 %) were not interested to pay anything (Table 2).

Of the 107 persons who did not accept the first offer of 20,000 Rials, 67 people were not willing to pay more

Table 1 Statistical results of the socio-economic characteristics of the respondents (monetary values are in thousand Rials)

Variable	Mean	Standard deviation	Max	Min	Coefficient of variation
Age	34.97	11.58	70	20	0.331
Sex	1.307	0.4622	2	1	0.353
Marital status	1.657	0.456	2	1	0.275
Education	12.997	5.06	18	0	0.389
Household size	4	2.02	8	1	0.505
Monthly income (In thousand Rials)	15,000	6,030	2,5000	5,000	401.633
BID Rials	17,630	9,580	30,000	0	0.543
Environmental opinion	4.02	0.579	5	1	0.144

than 10,000 Rials ($0.29) and the remaining 40 individuals were not willing to pay anything.

Of the 193 people who had accepted the first offer of 20,000 Rials ($0.59), 120 rejected the 30,000 Rials ($0.88) bid and the rest agreed to pay 30,000 Rials ($0.88).

Out of 73 persons who had accepted the first offer of 30,000 Rials ($0.88), 19 were willing to pay up to 50,000 Rials ($1.47), 15 people agreed to pay the maximum bid of 100,000 Rials ($2.94), 14 persons were willing to pay a maximum amount of 150,000 Rials ($4.41) and only four respondents showed a willingness to pay up to 200,000 Rials ($5.88). The remaining 21 individuals did not want to exceed 30,000 Rials.

Forty of the respondents were not willing to pay for myrtle habitat conservation. When asked for their lack of interest, 60 per cent believed that it should be the responsibility of the government, not of individuals (Table 3).

The logit regression method provided the following model for the conservation value of the Dooreh myrtle forest:

$$\mathrm{WTP} = \alpha_0 + \alpha_1 x_1 + \alpha_2 x_2 + \alpha_3 x_3 + \alpha_4 x_4 + e_i \qquad (7)$$

where WTPi is the willingness of respondent i to pay, x_1 represents education (0 for illiterate, 10 for below diploma, 12 for diploma, 14 for post diploma degree, 16 for bachelor degree, 18 for master degree and above) ; x_2 is household size, x_3 household income in thousand Rials 5,000 ($0.15), 7,500 ($0.22), 12,500 ($0.37), 17,500 ($0.51), 22,500 ($0.67), 25,000 ($0.73)) and x_4 represents the bids in Rials (0, 10,000 ($0.29), 20,000 ($0.59), 30,000 ($0.88)).

Table 4 shows the estimated coefficients, levels of statistical significance and the effect of descriptive variables of the logit model on the dependent variable (based on maximum likelihood estimation). In arriving at the final logit model those variables, not statistically significant but with coefficients of the expected sign, were eliminated in order to obtain a more accurate model (Table 4). In the end, we obtained the following model:

$$\mathrm{WTP} = 1.970 + 0.049\mathrm{E} - 0.581\mathrm{HS} + 0.003\mathrm{I} - 0.002\mathrm{SD} \qquad (8)$$

In this equation E is education, HS is Household Size, I is Income and SD is Suggested Donation.

The results also show that variables such as price, income and household size were effective in accepting the bids at less than the 10% statistical level. As well, it should be noted that the variables price and family size had a negative effect, while income and education had a positive effect on the WTP on the part of the respondents.

Moreover, the variables bid and income were statistically significant at the one per cent level of significance and were the most important factors in the WTP to conserve the myrtle forest. As the amount of the bid increased, the willingness to pay of people reduced substantially. This means that their elasticity to pay is inversely related with the price of perceived services provided, since their low

Table 2 Response to three proposed bids for the purpose of myrtle conservation

Response		BID 1 (20,000 Rials)	BID 2 (10,000 Rials)	BID 3 (30,000 Rials)
Yes	Frequency	193	67	73
	Per cent	64.33	22.33	24.33
NO	Frequency	107	40	120
	Per cent	35.63	13.33	40
Total	Frequency	300	107	193
	Per cent	100	35.67	64.33

Table 3 Lack of willingness to pay

Lack of willingness to pay	Per cent	Frequency
Due to financial reason	12.5	5
Must be paid by the permanent land users	10	4
Must be paid by the government	60	24
Local people must maintain the habitat	17.5	7

income levels have a negative effect on their support to conserve natural myrtle forest sites.

In terms of ranking the estimated coefficients, the second most important factor for accepting the bid for myrtle conservation was the income of the respondents. The higher their income, the higher their willingness to pay. This represents an indication of the status of well-being of the community, important from a development point of view. In other words, as income increased, the awareness on the part of people towards the environment became more positive and responsive who did not view the environment only from an economic stand point, but rather considered other aspects of it as well. At lower income levels, all resources were primarily seen as economic and revenue generators.

Another variable significantly related with the WTP was the size of the household. As this increased, the willingness to pay decreased considerably. The probability of rejecting the bids by large households was higher than that of household of smaller size, for households with more family members were less capable financially and thus had a relatively lower willingness to pay for myrtle conservation.

Education of the respondents had also an important effect on their willingness to pay. The proportion of respondents with university education was the highest at 33.49 % of all participants. The level of education was directly related with household income, i.e., people with a high level of education had a more positive attitude toward the environment and the benefits that can be derived from it and, as a result, had a higher willingness to pay for the conservation of myrtle habitats.

In order to determine changes in the probability of the WTP in estimating the conservation value due to a one unit change in one of the independent variables, it was necessary to calculate their marginal effect. We also calculated elasticity to show the relative importance of the descriptive variables of the model in response to changes in the willingness to pay for the conservation of this myrtle forest ecosystem.

The proposed bid was the most important descriptive variable for determining the probability of willingness to pay for the conservation of myrtle forest resources in Dooreh. It showed that a one unit change in the amount of proposed bid changed the probability of accepting the bid by 1.8 %. Furthermore, due to the marginal effect of this variable, a one unit change in the amount of the proposed bid changed the probability of a positive response to the WTP by 0.0004534 in favor of the conservation of myrtle forest sites.

The estimated coefficient for the income variable was significant at the 1 % statistical level. With its positive sign, as expected, it meant that a one per cent change in income, the probability of accepting the amount of the proposed bid would be changed by 1.732 %. Also, in terms of the marginal effect of this variable, a one unit change in the income of the respondent, the probability of accepting the proposed bid to conserve the myrtle forest resources of Dooreh, changed by 0.665×10^{-3}.

The estimated coefficient of household size was also significant at the 1 % level and implied that a one unit change in the size of a household, the proposed bid for myrtle forest conservation changed by 1.26 %. In terms of marginal effect, a one unit change in household size changed the likelihood of a positive response on the part of the respondent in the WTP for myrtle forest conservation by 0.144.

The estimated coefficient of education was significant at the 5 % level. Therefore, one additional year spent on education, increased the probability of accepting the bid for conservation of myrtle forest by 0.284 % unit. The marginal effect of this variable suggests that one additional year spent on formal education, increased the probability of reacting positively to the willingness to pay for myrtle forest conservation by 0.012.

In addition, we calculated the McFadden coefficient of determination as 0.234, suggesting that the explanatory variables of the model explained the dependent variable

Table 4 Estimated regression coefficient of the most effective variables of the logit model

Variable	Estimated coefficient	Standard deviation	*t*-value	*p* value	Elasticity	Marginal effect
Constant coefficient	1.97	1.097	1.796	0.073	0.881	–
Education[a]	0.049	0.024	2.031	0.042	0.284	0.121×10^{-1}
Household size[b]	−0.581	0.081	−7.159	0	−1.26	−0.144
Monthly income[b] (in thousand Rials)	0.0027	0.0003	8.998	0	1.732	0.665×10^{-3}
BID (Rials)	−0.0018	0.0002	−9.122	0	−1.8	-0.456×10^{-3}

Likelihood ratio statistic = 157.656, probability (LR statistic) = 0, McFadden R-square = 0.234, Percentage of right predictions = 0.76, [a]Statistical significance at 5 % level. [b]Statistical significance at 1 % level very well. Another useful criterion for goodness of fit of the model was the classification of the respondents in two groups, i.e., those who accepted the bid for myrtle forest conservation and those who rejected it. The proportion of correct predictions of the model was 76 per cent, implying that 76 % of the respondents indeed correctly answered "Yes" or "No" when asked about their willingness to pay.

The expected average WTP according to the logit model, using maximum likelihood was calculated by numerical integration in the range of zero to the maximum amount of the proposed bid as follows:

$$E(WTP) = \int_{0}^{30000} \left(\frac{1}{1 + exp^{\{-(2.134299 - 0.001845 * BID)\}}} \right) dA = 26820 \quad (9)$$

Based on Eq. (9), the expected willingness to pay per household for the conservation of myrtle stands in Dooreh, was estimated to be 26,820 Rials ($0.79) per month which meant that each family was willing to pay an equivalent amount of 321,840 Rials ($9.47) per year to conserve the habitat of this species.

The overall conservation value of the myrtle forests in Dooreh (VMFD) was calculated using the following relationship:

$$\text{VMFD} = \text{AAWH} \times \text{NHD} \quad (10)$$

In this equation, AAWH is the annual average of WTP per households and NHD is the number of households in Dooreh.

Given the 10,831 households currently living in Dooreh (Office of Lorestan provincial governor, 2013), we obtained ultimately the following conservation value:

$$\begin{aligned}\text{VMFD} &= (26,820 \times 12) \times 10831 \\ &= 3,485,849,040 \text{ Rials}\end{aligned}$$

Thus we estimate that the entire town of Dooreh is willing to pay 3.5 billion Rials (about $102,525) annually for the conservation of their myrtle forest resources. Given that the remaining area of myrtle sites consist of only a little more than 8 ha, our final estimate of the conservation value of myrtle resources per hectare in Dooreh is:

$$\begin{aligned}\text{VMFD/ha} &= 321840 \times 10831 \div 8.2175 \\ &= 424,198,240 \text{ Rials}\end{aligned}$$

Hence, the annual conservation value per hectare of the myrtle forest resources in Dooreh was estimated as 424,198,240 Rials ($12,476).

Considering the exchange rate of 34,000 Rials per U.S. dollar, these calculations imply that each household is prepared to spend an equivalent amount of $9.47 out of their total annual income for the conservation of myrtle forest resources, valued at around $12,476.

Discussion and conclusions

Based on the results of our study, we come to the conclusion that the WTP per year per household and the conservation value per hectare in Dooreh were higher than those reported by Molaei and Kavousi Kelashmi (2011) and Khosravi and Sabouhi (2011). We suggest two reasons for this discrepancy: 1) traditional knowledge on the part of this rural community about value and potential of this species, as well as their awareness of the extent of the damage inflicted on the natural habitat of this species; 2) the increased benefits derived from the exploitation of myrtle in terms of additional seasonal income earned by the community from harvesting and selling this species and an appreciation for other environmental values of this rapidly vanishing plant. The local knowledge in the community about the medicinal value of myrtle is based on its traditional belief in and experience from the consumption of this species, as well as on the positive impact of employment and financial rewards in the manufacture of several medicinal products by the local pharmaceutical company. A fundamental problem in harvesting this species, however, is lack of proper skills in people involved in harvesting operation of this species, which has eventually led to the degradation of over half of its natural habitat.

In general, the local community now has a better understanding of how valuable myrtle is to them. Such an understanding has gradually been developed as the local people began to realize the extent of myrtle forest damage in their area and hence, the need for its conservation. In short, the main reason why local people are willing to pay something for conservation of their myrtle forest resources is entirely economical, since 86 % of them are willing to pay some money for its conservation. The remaining 14 % who were not willing to pay anything were apparently doing so for financial reasons, given their low level of income. Interestingly, for the same obvious reason, this latter group did not have a favorable response for the conservation of any other species either. This is the type of situation where the need to educate local people is felt the most, in order to enhance their understanding of environmental issues and the conservation of flora and fauna in their own community. Unfortunately, due to poor management, no effective policy has been proposed or implemented, despite its neglect over the years.

In this study, it was found that the annual WTP per household and the conservation value per hectare was lower from those reported by Lehtonen et al. (2003), Leong et al. (2005), Gurluk (2006) and Tao et al. (2012). Apparently this was because in our study several other economically valuable aspects of valuation, other than the medicinal value of myrtle, such as timber production, habitat, regulatory and information functions, were not taken into consideration. A relatively stronger currency, higher levels of average income and education could also have been mentioned as additional reasons for the higher WTP reported in their research. It is also worth mentioning that Leong et al. (2005) and Peron and Esmaeili (2010)) estimated the forest value based on per hectare per person in a given household, instead of household as a whole, which raises the estimated WTP per household per year.

In contrast, the estimated WTP per household per year and the conservation value per hectare estimated by us were higher than those mentioned by Echeverria et al. (1995) and Molaei (2009), apparently because these studies were more recent than those mentioned earlier. Clearly, this 20 year span together with inflation had a significant impact on lowering the nominal value in the local currency. In the present study, the respondents were primarily local people, whereas in earlier investigations, the respondents were not necessarily from the study area and as a result had a very low or even zero WTP.

In this study, the annual WTP per household and the conservation value per hectare were also lower from those reported by Chukwone and Okorji (2008). We cite as possible reasons the following issues: 1) the lower average income and education level of the respondents in the current study had both a significant and positive (impact on their willingness to pay; 2) valuation was based on only one NTFP, i.e., myrtle, instead of all the available NTFPs in the forest; each of which could have a further positive impact on the WTP on the part of the respondents due to their effect on the level of family income. This was consistent with the findings of Chukwone and Okorji (2008), except for the fact that in their study women had a higher WTP, primarily because they were more engaged with harvesting NTFPs than men and perhaps their higher WTP reflected a level of acknowledgment for the value of NTFPs on their part.

According to the explanatory variables of our model, as the respondents became better educated, their understanding of myrtle forests and the value of conservation of these resources became more profound, their WTP was higher. Therefore, it is expected that educating people about the importance and economic value of myrtle would increase their willingness to pay. In general, the higher their education, the higher their WTP and the more their tendency to preserve the environment. Similar results were found by Echeverria et al. (1995), Lehtonen et al. (2003), Leong et al. (2005), Tao et al. (2012), Chukwone and Okorji (2008), Peron and Esmeili (2010), Molaei (2009), and Khosravi and Sabouhi (2011).

Among the explanatory variables of the model, income levels had the highest impact on the elasticity of the willingness to pay. In other words, as incomes rise, the tendency to pay more for the conservation of the habitat of this species increased and as well, their opinion about environmental issues became more positive. This would probably indicate that implementation of appropriate financial incentives and policies with regard to rural economic development in the Dooreh forest areas and surrounding villages could lead to an improvement of economic conditions and employment in this rural community, which would then lead to an increase in their income and thus WTP to protect the myrtle forest area from further degradation. Education and income are two important factors with a significant impact on the opinion of rural people on environmental issues. With increasing income, the elasticity for the WTP to preserve their environment increased. In contrast, people with low incomes were rather more concerned to give priority to satisfy their basic needs. This results was consistent with what many investigators such as Echeverria et al. (1995), Lehtonen et al. (2003), Leong et al. (2005), Tao et al. (2012), Chukwone and Okorji (2008), Peron and Esmeili (2010), Molaei (2009), and Khosravi and Sabouhi (2011) had reported earlier.

Furthermore, smaller households with their relatively lower daily expenditures were willing to pay more for myrtle habitat conservation compared to larger households. All the same, if larger households had a higher level of income, they were also willing to pay more despite their relatively larger family size. Such a result is consistent with the findings by Tao et al. (2012), Molaei (2009) Peron and Esmeili (2010).

In general, economic hardship in the rural community living in or adjacent to the myrtle forest areas of Dooreh, induces people to rely on excessive harvesting of myrtle for additional income which eventually led to the degradation of nearly 50 % of the natural myrtle forest resource (Karimi, 2012).

The rural households in the study region have traditionally been harvesting myrtle in publicly-owned forests for their own consumption as well as for earning additional income. The problems of poverty, unemployment and a difficult economic situation dominating the rural forest communities in the region are considered to be the major causes for illegal harvesting of myrtle and the rapid degradation of its habitat. As well, the gradual but illegal encroachment of agriculture in this forest area by local farmers has had a significant impact on the process of myrtle habitat reduction and destruction.

The fact of the matter is that only a small portion of the myrtle habitat is privately owned by local farmers, with the Iranian government owning most of it. Rural owners harvest myrtle inappropriately and without the necessary skills. Also, these people tend to trespass on government owned sites for illegal harvest of myrtle. For both reasons, severe destruction of the myrtle habitat is the consequence. As mentioned earlier, the root of the problem is socio-economic and deserves special attention by the local as well as regional forest manager Estimating the value of conservation of this myrtle habitat could provide strong incentive and justification for the authorities in the Iranian Forest and Range Organization to allocate sufficient resource to restore and/or conserve the habitat of this species. Unfortunately, so far forest management approaches and policy making and implementation by the government authorities have not been effective in solving the problem of illegal harvesting and or helping conservation and restoration of this myrtle habitat. Thus, the outlook for this species appears to be rather dim, unless prompt, strong and effective action is taken by the authorities and sufficient resources allocated to support such an action.

The results of this study emphasize the need and importance of formulating and implementing appropriate policies for myrtle forest conservation and restoration by the Iranian executive authorities and policy makers managing the natural resources of our country.

Although estimating the conservation value of myrtle forest highlights the importance of lowering degradation and deforestation of these resources, by itself this is not a sufficient for very obvious reasons. When the basic needs of rural people are not met and they suffer from insufficient income and poverty, they tend to put more pressure on natural resources and, in this case, damage the myrtle by excessive harvesting. This could eventually lead to a permanent loss of the few remaining myrtle sites and their conversion to agricultural land. In such a situation, the rural community would, by no means, be willing to spend money to maintain myrtle habitat, but would be more engaged in damaging it by excess harvesting and ultimately by converting it to agricultural land. Therefore, we believe it would be equally important to educate the local population about proper harvesting methods and management of this species by trained and knowledgeable individuals.

Although the current status of the myrtle habitat represents a serious threat to its continued existence, in the end we believe and conclude that with proper management and policies, such as restoration of the myrtle forest resources, together with necessary financial and administrative support, the present situation can be turned around into an opportunity for the provincial economy and an increase in employment.

Endnotes

[1]The exchange rate between Rial and US dollar at the time of publication of this article was 34,000 Rials per US dollar.

Competing interests

The authors declare that they have no competing interests.

Authors' contributions

Research leading to this paper supported by the Sari Agricultural and Natural Resource University, and financed primarily by the university division of Higher Education. Also the branch Office of Lorestan Province of Natural Resources provided significant help and assistance during the entire process of gathering field data and completing the required questionnaires. Shortcomings and mistakes, if any, are the responsibility of the authors.

Acknowledgements

We would like to express our profound appreciation to the people of Dooreh for their hospitality and for being so receptive and cooperative during our entire field work. Our special thanks also go to the authorities and foresters in the Lorestan branch of the Office of Natural Resources for all the help and assistance provided during the entire process of gathering field data and completing the required questionnaires.

Author details

[1]Department of Forestry, Sari Agricultural and Natural Resource University, Sari, Iran. [2]Department of Forest Sciences, Gorgan Agricultural and Natural Resource University, Gorgan, Iran. [3]Department of Forestry, College of Natural Resources, Sari Agricultural Sciences and Natural Resources University, Sari, Iran. [4]Department of Forestry, Faculty of Agricultural, University of Lorestan, Lorestan, Iran. [5]Department of Agricultural Economics, Sari Agricultural and Natural Resource University, Sari, Iran.

References

Amiri N (1392). Economic Valuation of Myrtle (*Myrtus communis* L.) using market price, substitution cost, opportunity cost and Contingent Valuation methods (Case study: the city of Dooreh Lorestan Province). M.S. Thesis. Sari University of Agricultural Sciences and Natural Resources. Faculty of Natural Resources, p 84

Bateman I, Willis K (1999) Valuing Environmental Preferences, Theory and Practice of the Contingent valuation Method in the US, EU, and Developing Countries. Oxford University Press, New York, NY, pp 511–539

Bohm P (1972) Estimating Demand for Public Goods. An Experiment. Eur Econ Rev 3:111–130

Chukwuone NA, Okorji CE (2008). Willingness to pay for systematic management of community forests for conservation of non-timber forest products in Nigeria's rainforest region. Econ Poverty Environ Nat-Res Use, 117–137

Cochran WG (1977) Sampling techniques, 3rd edn. Whley and Sons, USA, p 428

Dahlberg A (1974) Geografisk rorlighet: Sociala och economiska effecter (Geographic mobility: social and economic effect). University of Umea Sweden, Department of Economics (in Swedish)

Echeverria J, Hanarahan M, Solorzano R (1995) Valuation of nonpriced amenities provided by the biological resources within the Monterverde Cloud Forest preserve, Costa Rica. Ecol Econ 13:43–52

Falconer J, Arnold JEM (1989) Forestry and food security. FAO Forestry Paper 90. Rome: Food and Agriculture Organization of the United Nations (FAO).

FAO Forestry (1999) Towards a harmonized definition of non-wood forest products. Unasylva 50(198):63–64, http://www.fao.org/docrep/x2450e/x2450e0d.htm

Gunatilake HM, Senaratne DMAH, Abeygunawardena P (1993) Role of non-timber forest products in the economy of peripheral communities of Knuckles national wilderness area of Sri Lanka: A farming systems approach. Economic Botany 47(3): 275-281

Gurluk S (2006) The estimation of ecosystem services value in the region of Misi Rural Development Project: Results from a contingent valuation survey. J Forest Policy Econ 9(3):209–218

Hanemann WM (1984) Welfare evaluation in contingent valuation experiments with discrete responses. Am J Agr Econ 66:332–341.

Hanemann, MW (1985) Some issues in continuous and discrete response contingent valuation studies. North–eastern journal of Agricultural and Resource Economics, 14: 5–13

Hanemann WM (1991) Willingness to pay and willingness to accept. How much can they differ? Am Econ Rev 81:635–647.
Karimi MA (2012) Investigating the most important degradation factors of myrtle (Myrtus communis L.) in Lorestan province (A case study in Chegnei forest area). Tehran Research Centre of Scientific-Applied University, Iran
Kengen S (1997) Forest valuation for decision making: Lessons of experience and proposals for improvement. Food and Agriculture Organization of the United Nations, No: 165.
Khosravi M, Sabouhi M (2011) Estimating conservation value of tamarisk in Fahliyan river bank using by contingent valuation method. J Environ Sci 4:73–82
Lee C, Han S (2002) Estimating the use and preservation values of national parks tourism resources using a contingent valuation method. Tour Manag 23:531–540
Lehtonen E, Kuuluvainen J, Pouta E, Rekola M, Li C (2003) Non-market benefits of forest conservation in southern Finland. Environ Sci Policy 6:195–204
Leong PC, Zakaria M, Ghani ANA, Mohd A (2005) Contingent valuation of a Malaysian Highland Forest: non market benefits accrued to local residents. J Appl Sci 5(5):916–919
Lescuyer G (1996) Monetary valuation of the non-timber forest products (NTFPs): Does it contribute to determine a sustainable management of those resources? Paper presented at the 6th International Conference of the Association for the Study of the Common Property (IASCP), Berkeley, 5-8 June 1996 (mimeo.)
Lindal M (1994) A comment on the valuation of non-wood forest benefits In: Proceedings of a workshop on evaluation of forest benefits through a total evaluation of production, environmental and social functions of forests. Forest value. Pp. 71-89. Prague-Jiloviste, 13-16 September. United Nations Economic Commission for Europe/Food and Agriculture Organization of the United Nations (FAO)
Makkizadeh tafti M, Farhoodi R, Naghdi badi H, Mehdizadeh A (2006) Determining the best treatment for germination enhancement of three medicinal plants seeds: *Rubia tinctourum* L., *Echinacea angustifolia* D.C. and *Myrtus communis* L. Quart J Sci Res Med Arom Plants Iran 2:105–116
Molaei M, Kavousi Kelashmi M (2011) Estimating conservation value of *Lilium ledebourii* using contingent valuation method with double one-dimensional choice. J Agr Econ Dev 25:322–329
Molaei M (2009). The economic-environmental valuation of Arasbaran forest ecosystem. PHD Thesis, Department of Agricultural Economics, Faculty of Agricultural Economics and Development, Tehran University
Noubissie ET, Chupezi J, Ndoye O (2008). Studies on the Socio-Economic Analysis of socio-economic non-timber forest products (NTFPs) in Central Africa. Synthesis of reports of studies in the Project GCP/RAF/398/GER. Fao. Yaounde, Cameroon, FAO GCP/RAF/398/GER Enhancing Food Security in Central Africa through the management and sustainable use of NWFP: p 43
Park T, Loomis J (1996) Joint estimation of contingent valuation survey responses. Environ Res Econ 7:149–162
Pearce DW, Pearce CJT (2001) The Value of Forest Ecosystem. A Report to the Secretariat Convention on Biological Diversity
Peron S, Esmaeili A (2010) The estimated of non-market value mangrove forest in the Hormozgan province. J Agr Econ Dev 2:162–168
Peters CM (1994) Sustainable harvest of non-timber plant resources in tropical moist forest: An ecological primer. Biodiversity Support Program. World Wildlife Fund, The Nature Conservancy and the World Resources Institute. Washington, D.C.: World Wildlife Fund.
Piri M, Mosanane Mozafari M, Javdan A (2009) Estimating existence value of forests using willingness to pay of people (Case study: Forest of Arasbaran). J Forest Wood Prod Dept Nat Res 4:343–357
Rodda A (1991) Women and the Environment. London: Zed Books
Salehnia A (2008) Investigating medicinal and environmental importance of myrtle
Tao Z, Yan H, Zhan J (2012) Economic valuation of forest ecosystem services in Heshui watershed using contingent valuation method. Proc Environ Sci 13:2445–2450
Torras M (2000) The total economic value of Amazonian deforestation, 1978–1993. Ecol Econ 33:283–297
Venkatachalam L (2003) The contingent valuation method: a review. Environ Impact Assess Rev 24:89–124
White PCL, Lovett JC (1999) Public preferences and willingness to pay for nature conservation in the North York Moors National Park. UK. J Environ Manage 55:1–13

23

An integrated assessment approach to optimal forest bioenergy production for young Scots pine stands

Tianjian Cao[1*], Kari Hyytiäinen[2], Henna Hurttala[3], Lauri Valsta[3] and Jerome K. Vanclay[4]

Abstract

Background: Bioenergy is re-shaping opportunities and imperatives of forest management. This study demonstrates, through a case study in Scots pine (*Pinus sylvestris* L.), how forest bioenergy policies affect stand management strategies.

Methods: Optimization studies were examined for 15 Scots pine stands of different initial stand densities, site types, and temperature sum regions in Finland. Stand development was modelled using the PipeQual stand simulator coupled with the simulation-optimization tool OptiFor Bioenergy to assess three forest bioenergy policies on energy wood harvest from early thinnings.

Results: The optimal solutions maximizing bare land value indicate that conventional forest management regimes remain optimal for sparse stands. Energy harvests occurred only when profitable, led to lower financial returns. A forest bioenergy policy which included compulsory energy wood harvesting was optimal for denser stands. At a higher interest rate (4 %), increasing energy wood price postponed energy wood harvesting. In addition, our results show that early thinning somewhat reduced wood quality for stands in fertile sites. For less fertile sites, the changes were insignificant.

Conclusions: A constraint of profitable energy wood harvest is not rational. It is optimal to carry out the first thinning with a flexible forest bioenergy policy depending on stand density.

Keywords: Energy wood; Policy design; Precommercial thinning; Scots pine; Simulation-optimization

Background

Global interest in bioenergy, defined as energy produced from biomass, has grown rapidly in recent years. In Finland, renewable energy sources account for 24 % of total energy consumption, and wood based fuels, including wood processing residues, account for 20 % of total energy consumption (Hakkila 2006). Modern bioenergy includes liquid biofuels, biogas, and solid biomass for heat and power generation, and forests remain an important source of bioenergy. The substitution of fossil fuels by forest biomass in bioenergy production is an important measure in climate mitigation. Carbon in forest biomass is sequestered through photosynthesis, and released by decomposition and combustion. Therefore, the utilization of forest biomass for energy is carbon-neutral (UN-Energy 2007), and in agreement with the Kyoto Protocol (UNFCCC 1997).

The environmental, social, and economic impacts of bioenergy development depend greatly on local circumstances (UN-Energy 2007). There are several policies and measures in the EU that support bioenergy and stipulate targets (McCormick and Kåberger 2007). For example, Finland aims to increase the annual use of forest chips to 12 million m^3 by 2020 (Ministry of Agriculture and Forestry 2008). Current Finnish government incentives and policy measures for energy wood production include: (1) an environmental tax exemption for wood-based fuels used for heat production, (2) a general tax refund for power plants using forest chips as fuel, and (3) a subsidy and financial aid for using small-sized trees from precommercial thinning and for the production and use of forest chips (Hakkila 2006).

Bioenergy provides opportunities to address environmental, social, and economic challenges. The issues raised by bioenergy development, such as energy security,

* Correspondence: cao@nwsuaf.edu.cn
[1]College of Forestry, Northwest A&F University, 712100 Yangling, China
Full list of author information is available at the end of the article

environmental sustainability, food security, and rural development, will mainly be dealt with by bioenergy policy. Nevertheless, bioenergy policy must be designed and managed carefully in order to avoid new environmental and social problems (UN-Energy 2007). Policy design and assessment in climate change are based heavily on models and simulations. At the European level, Wetterlund et al. (2013) developed a techno-economic-geographic optimization model to determine the optimal locations of bioenergy conversion plants with focus on the influence of different carbon cost and biofuel policies. Based on integrated modeling of biomass production and land use, Kraxner et al. (2013) presented an example of global scale assessments for bioenergy production under deforestation, land-use change, and other climate effective aspects to focus on policy design and its implementation. Their studies demonstrated that the integrated assessment approach offers a useful tool for evaluating bioenergy policy.

The source of forest bioenergy may include forest chips and processing by-products. Forest chips, including small diameter wood from young forests, logging residues, and stumps, are from thinnings, and clearcut. The potential utilization of forest biomass has been studied intensively in recent years (Malinen et al. 2001; Bolkesjø et al. 2006; Hakkila 2006; Alam et al. 2010), and diverse energy wood production systems have been proposed (Laitila et al. 2004; Jylhä and Laitila 2007; Nurmi 2007; Laitila 2008). Ahtikoski et al. (2008) investigated the economic viability of energy wood production from young stands based on six thinning methods. By linking energy wood production and forest management, Kärkkäinen et al. (2008) studied the recovery of energy wood in different cutting and climate scenarios in Finland. Their results were based on a linear programming approach (Lappi 1992) which selected the optimum combination of management regimes from a predefined set of alternatives. Kotamaa et al. (2010) integrated energy wood thinning and optimal bucking at stand level. However, their thinning regimes were based on silvicultural recommendations (Finnish Guidelines for Good Silviculture 2006), rather than optimized. There is little integration of energy wood production and forest management optimization for enhancing thinning and rotation regimes.

In contrast to energy wood from commercial thinning and clearcut, the utilization of small diameter wood from precommercial thinning may have important impacts on stand development, and consequently affect thinning and rotation regimes in the long run. Precommercial thinning is a frequently applied silvicultural treatment in conifer stands (Piene 1978; Pothier 2002; Varmola and Salminen 2004; Fahlvik et al. 2005; Huuskonen and Hynynen 2006). Because precommercial thinning increases the spacing between trees, it helps to increase the temperature of surface soil layers and offers more growing space, thus increasing future growth and yield. Precommercial thinning is usually seen as a costly silvicultural treatment. The increasing use of energy wood coupled with forest bioenergy incentives provides additional income that may offset the cost of precommercial thinning (Hakkila 2006; Ministry of Agriculture and Forestry 2006). Therefore, there is a need to investigate the optimal thinning regimes when the energy wood harvest from precommercial thinning is taken into account.

Forest management optimization has been well studied since the 1960s (Amidon and Akin 1968; Haight and Monserud 1990; Valsta 1990; Hyytiäinen et al. 2004; Pukkala 2011). Until recently, most numerical optimization studies in forestry focused on timber production. This study attempts to integrate energy wood harvesting into forest management optimization, and analyses the energy wood harvest from precommercial thinning using a simulation-optimization system for simultaneously optimizing timber and energy wood production.

The objectives of this study were: (1) to integrate bioenergy production into forest management optimization, (2) to develop a simulation-optimization tool OptiFor Bioenergy for optimizing timber and energy wood production, and (3) to demonstrate, through a case study in young Scots pine stands, how various forest bioenergy policies affect optimal thinning regimes, financial returns, as well as wood quality.

Methods

The optimization computations were conducted for 15 simulated pure Scots pine (*Pinus sylvestris* L.) stands. The initial states (Table 1) are the same as those applied earlier in background calculations made for Silvicultural Recommendations in Finland (Hyytiäinen et al. 2006). The diameter distributions were derived artificially by site type, temperature sum, and initial stand density. The investigated site types are CT (*Calluna*), VT (*Vaccinium*), and MT (*Myrtillus*) (Cajander 1909). The two temperature sum regions are 1100 (T1100) and 1300 (T1300) day degrees (d.d.), and the initial stand densities after completion of the silvicultural operations (tending and slashing) 1500 and 3000 trees·ha^{-1}, respectively. According to the 10th national forestry inventory in Finland, Scots pine dominant stands are 65 % of total forest land, in which pure Scots pine stands are 59 %. The proportion of forest land on mineral soils by site type are 2, 19 and 47 % for CT, VT, MT sites in the south, and 5, 36 and 54 % for CT, VT, MT sites in the north, respectively (Peltola and Ihalainen 2013). Therefore, the initial states used in the present study are well representative for Scots pine in Finnish conditions.

Table 1 Initial stand states of 15 simulated Scots pine stands

Plot	ST	H_{100} (m)	Age (yrs)	#tree	BA ($m^2 \cdot ha^{-1}$)	H_{dom} (m)
1	CT1300	17.3	29	1500	6.4	7.9
2	CT1300	17.6	29	2000	7.3	7.7
3	CT1300	17.5	29	3000	9.1	7.4
4	VT1100	18.2	27	1500	7.0	7.9
5	VT1100	18.4	27	2000	8.2	7.7
6	VT1100	18.4	27	3000	10.0	7.4
7	MT1100	22.1	23	1500	7.6	7.9
8	MT1100	21.9	23	2000	8.8	7.8
9	MT1100	21.5	23	3000	11.2	7.4
10	VT1300	25.7	23	1500	7.0	7.9
11	VT1300	25.4	23	2000	8.1	7.8
12	VT1300	24.8	23	3000	10.0	7.5
13	MT1300	28.3	20	1500	7.4	8.0
14	MT1300	27.9	20	2000	8.6	7.8
15	MT1300	27.3	20	3000	10.8	7.5

ST site type with temperature sum (d.d.), H_{100} dominant height at age 100 (m), *Age* initial age (yrs), *#tree* number of trees per hectare, *BA* basal area ($m^2 \cdot ha^{-1}$), H_{dom} dominant height (m)

In this study we employed the process-based growth model, PipeQual (Mäkelä 1997, 2002; Mäkelä and Mäkinen 2003), as the stand growth simulator, and models for predicting sawn wood grading of Scots pine (Lyhykäinen et al. 2009), as the sawing simulator. The process-based model was first linked with an economic optimization model in Hyytiäinen et al. (2004). They demonstrated that the standard Hooke and Jeeves' direct search algorithm (Hooke and Jeeves 1961) could be successfully applied for the stand level problem formulation that employs a detailed and computationally demanding process-based stand simulator.

The process-based model PipeQual is based on the pipe model theory (Shinozaki et al. 1964) and the theory of crown allometry (Mäkelä and Sievänen 1992). Tree growth simulation by the PipeQual simulator consists of branch, whorl, and tree levels. Tree structure is composed of foliage, fine roots, stem, branches, and transport roots. The whorl module describes the vertical structure of stem and branches. The branch module provides the annual dynamics of individual branches and their properties in each whorl. Annual growth is calculated from tree photosynthesis and respiration. PipeQual is an individual-tree model that describes the stand structure using a list of tree records. For this application the stands were represented by 10 size classes with one representative tree and the number of trees in the class. The number of trees in each sample tree class depends on the initial stand density and other stand characteristics, and evolves over time due to mortality and harvesting. No ingrowth is assumed to occur. The PipeQual is calibrated for different site qualities and temperature sum regions by adjusting the allometric ratio between fine root mass and needle mass (Hyytiäinen et al. 2004) and by comparing the model outcomes for each initial state with those from the empirical Motti simulator (Hynynen et al. 2005). More explicit description of the PipeQual simulator as a dynamic system for stand management optimization was presented in Hyytiäinen et al. (2004).

In this study, multinomial logistic regression models for predicting yields of lumber grades and by-products of individual Scots pine trees developed by Lyhykäinen et al. (2009) were linked with the PipeQual simulator. The sawn wood grading models use stem and crown dimensions as explanatory variables. The proportion of sawn wood grades is a function of the height of the crown, the height of the lowest dry branch, tree diameter at breast height, and stem volume. The end products of Scots pine include centre (grades A, B, C and D) and side (grades A, B, C and D) sawn timber, chips, bark, sawdust, as well as pulpwood.

In order to integrate forest bioenergy production into conventional stand management optimization, we developed a simulation optimization tool OptiFor Bioenergy from an earlier version OptiFor Carbon (Cao et al. 2010). The Hooke and Jeeves' direct search algorithm used in the OptiFor program has been widely employed in stand management optimization because of the advantage of its robustness (e.g. Haight and Monserud 1990; Valsta 1990). The OptiFor simulation-optimization system was tested by comparing the modified Osyczka's direct and random search (DRS) (Osyczka 1984) with a random search (RS). The DRS algorithm is clearly more efficient than the random search (Cao 2010). The result of the comparison is consistent with Valsta (1990).

The objective function of the optimization problem is based on the Faustmann (1849) formula. We formulated the optimal thinning regime for energy wood and timber production as a bound-constrained optimization problem (Eqs. 1, 2 and 3). The dynamic problem of stand management is solved as static optimization using control variables. In this formulation, the state variables $x(t)$ (tree diameter, height, etc.) are not used in the optimization algorithm. Instead, the state variable values are computed inside the stand simulator, given initial state $x(t_0)$. The bare land value of a stand is maximized by changing the set of control variables $\mathbf{u}$ (time between the i-1th and the ith thinnings, yrs) and $\mathbf{H}$ (thinning rate h_{ij} in the jth tree size class at the ith thinning),

$$\max \quad f(\mathbf{u}, \mathbf{H} : \mathbf{x}(t_0)), \tag{1}$$

$$\text{s.t.} \quad \mathbf{u} = (u_1, \ldots, u_{n+1}), u_i \in [1, 25] \tag{2}$$

$$\mathbf{H} = (h_{ij})_{n\times 3}, h_{ij}\in[0, 1] \quad (3)$$

where $f: E_n \times E_3 \rightarrow E_1$ is the objective function for the bare land value (BLV) of a thinning regime, E_n denotes n-dimensional real Euclidean space, n is the number of thinnings, $n + 1$ means final felling.

The BLV of a thinning regime for timber production can be expressed as follows (Eq. 4):

$$\mathrm{BLV} = \frac{\sum_{i=1}^{n+1}\left[\sum_{j=1}^{m}\sum_{k=1}^{l} p_k v_{ijk} - c_i\right] e^{-rt_i} - c_0}{1-e^{-rt_{n+1}}}, \quad (4)$$

where p denotes timber price, v refers to harvested timber volume, l denotes timber categories, c_i is the logging cost at the ith harvest, c_0 is the discounted stand establishment cost, r denotes interest rate, and t denotes stand age. The definition of thinning type is defined by three thinning variables, which means thinning rates in the first, the middle and the last tree size classes are selected. The decision variables of thinning type are then defined by linearly interpolating thinning rate h_{ij} in the jth tree size class to other tree sizes.

The logging cost is a function of harvested volume, productivity of felling, and on-site transport. Logging costs are composed of felling and haulage (transportation) costs and a fixed cost. The logging cost model by Kuitto et al. (1994) for timber production from commercial thinning and final felling is based on the time consumption equations for mechanized cutting and forest haulage specification. More detailed description of Kuitto's logging model was presented in Cao et al. (2006). The productivity of mechanised cutting for energy wood from early thinnings (stem wood) is based on a time consumption model for a thinning harvester equipped with an accumulating felling head (Laitila et al. 2004). Laitila (2008) reported that such a two-machine system (harvester-forwarder) is currently the most cost competitive logging system for energy wood from early thinnings in Finnish conditions. The logging cost parameters were assumed as follows: terrain class one, average travel distance 200 m, felling cost 75.67 €·h^{-1}, transportation cost 53.35 €·h^{-1}, and fixed cost 100.00 €·ha^{-1} (Cao et al. 2010).

When the first thinning is an energy wood harvest, conventional logging will be applied from the second thinning, and the energy wood logging cost (c_E) and volume v_{1j} (harvested volume in the jth tree size class at the 1st thinning) will be computed for the first thinning (Eq. 5). The objective function of BLV (Eq. 4) will be modified (III) by adding the net present value of energy wood (NPV_E), where p_E denotes the energy wood price.

$$\mathrm{NPV_E} = \left[\sum_{j=1}^{m} p_E v_{1j} - c_E\right] e^{-rt_1}. \quad (5)$$

In this study, the interest rate, costs and roadside prices are expressed in real terms. The roadside prices assumed were 52.98, 26.24 €·m^{-3} (Cao et al. 2010), for sawlog, and pulpwood, respectively. The financial returns (bare land value) were computed with a 3 % interest rate. In addition, we assumed the energy wood price of 15 €·m^{-3}, and conducted sensitivity analyses with energy wood price 15, 20, 25 €·m^{-3}, and at 2, 3 and 4 % interest rates for stands 10 and 12, respectively.

The unit silvicultural costs assumed were 142, 600, and 276 €·ha^{-1} (Cao et al. 2010), for soil preparation, sowing, and other silvicultural operations (tending and slashing), respectively. It was assumed that soil preparation had been done in the first year for all stands. Sowing was carried out in the second year for MT and VT sites. The least fertile CT site was assumed to be regenerated naturally from seeds dispersed from adjacent forest stands. It was also assumed that the tending of the seedling stands had been carried out at ages 20, 18, 15, 16, and 13 for stands 1–3, 4–6, 7–9, 10–12, and 13–15, respectively.

Conventional forest management (Policy 0) was used as a benchmark (the optimal policy for timber production only) to compare with alternative policies for energy wood and timber production. Policy I was designed for profitable energy wood harvesting, which means that the first thinning will be treated as precommercial thinning for energy wood production (small size trees from precommercial thinning) if net income from energy wood harvest is positive. In Policy II it was assumed that the first thinning for energy wood production is compulsory. In other words, no matter how unprofitable energy wood harvesting is, the first thinning is always conducted as precommercial thinning.

Results

The optimal solutions indicate that a conventional forest management regime (Policy 0) was still optimal for sparse stands (Fig. 1a and b). However, Policy II (compulsory energy harvest) turned out to be optimal for denser stands with an initial density of 3000 trees·ha^{-1} (Fig. 1c). With Policy II, the number of thinnings was reduced, and the rotation length was shortened slightly. The most significant rotation change was for stand 12 (VT1300, 3000 trees·ha^{-1}) in which the rotation was shortened by 20 years.

It is somewhat surprising that Policy I (economic harvest), which favours profitable energy wood harvesting, was neither optimal for sparse nor optimal for dense stands (Table 2). For denser stands (3000 trees·ha^{-1}),

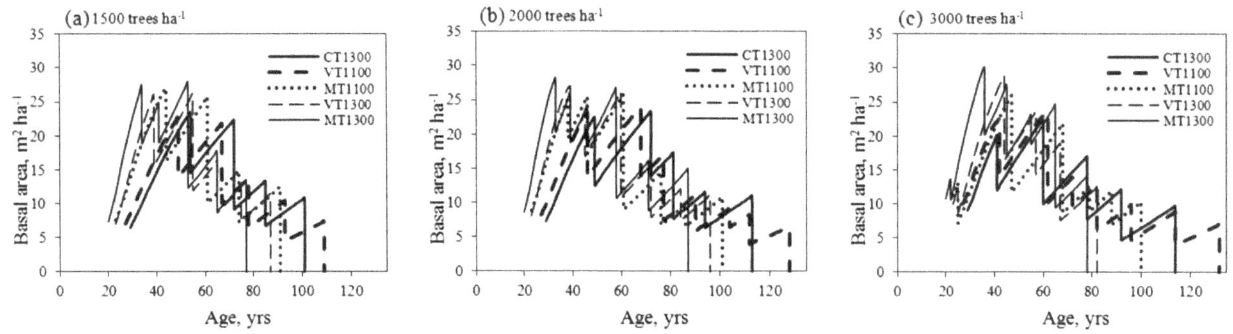

Fig. 1 Basal area development based on optimal solutions for 15 Scots pine plots, with an initial density of 1500 (**a**), 2000 (**b**), and 3000 trees·ha^{-1} (**c**), energy wood price 15 €·m^{-3}, interest rate 3 %

Policy I increased the intensity of the first thinning. For sparser stands (1500 trees·ha^{-1}), Policy I significantly reduced thinning intensity to an unrealistically low level, and brought the first thinning earlier (Table 3).

Thinning from below was optimal at the first thinning for sparser stands (initial density 1500 trees·ha^{-1}) with Policy 0. With Policy II, thinning from above was optimal at the first thinning for dense stands with an initial density of 3000 trees·ha^{-1}. The dominant height at the first thinning for the energy wood harvest was lower than eight meters in fertile sites (stands 9, 12 and 15), but higher (9.7 m and 9.2 m for stands 3 and 6, respectively) in less fertile sites. The first thinning with the energy wood harvest was clearly earlier than conventional first thinning. With optimal energy wood harvesting, Policy I resulted in a later first thinning compared to Policy II. The first thinning was postponed by 2–7 years (2, 6, 6 and 7 years for stands 3, 6, 9 and 12, respectively) except stand 15 (Table 3).

Table 2 Bare land value (BLV, €·ha^{-1}) based on Policy 0 (conventional harvest), Policy I (economic harvest) and Policy II (compulsory bioenergy harvest) for 15 Scots pine plots, energy wood price 15 €·m^{-3}, interest rate 3 %

Plot	P0	PI	PII
1	876	853	839
2	801	779	783
3	717	714	730
4	481	473	443
5	413	403	407
6	302	343	350
7	1548	1519	1471
8	1440	1416	1412
9	1274	1297	1338
10	2400	2373	2341
11	2288	2241	2279
12	2088	2062	2188
13	3714	3623	3606
14	3638	3546	3553
15	3336	3336	3495

P0 Policy 0, *PI* Policy I, *PII* Policy II

The maximized bare land values were higher systematically for lower initial densities from all combinations of site type and temperature region (Table 2). Following Policy 0 (conventional harvest), the higher bare land values were obtained from fertile sites with a greater temperature sum (Tables 1 and 2). Stand 13 from the MT site with a temperature sum of 1300 (d.d.) resulted in the highest bare land value (3714 €·ha^{-1}). Stand 6 from the VT site with a temperature sum of 1100 (d.d.) led to the lowest bare land value (302 €·ha^{-1}). On the same sites (the same site type and temperature sum), sparse stands (stands 1, 4, 7, 10 and 13 with an initial density of 1500 trees·ha^{-1}) resulted in greater financial returns than denser stands.

Increasing energy wood price from 15 to 25 €·m^{-3} resulted in greater financial returns for the denser stand 12. However, for the sparse stand 10, the financial returns were insensitive to a higher energy wood price (Fig. 2). At a 2 % interest rate, the timing of energy wood harvesting was also insensitive to energy wood price (Fig. 3a). Nevertheless, at a higher interest rate, increasing energy wood price postponed energy wood harvesting (Fig. 3).

Our results reveal that a negative net income from energy wood harvesting might increase financial returns for the whole rotation realized in later thinnings and the final harvest. The net incomes of energy wood harvesting were all negative for optimal solutions with Policy II, even though a relatively high energy wood price of 15 €·m^{-3} was applied. The net incomes from energy wood harvesting were – 68.4,–141.1, 224.9,–293.5, and – 287.1 €·ha^{1} for stands 3, 6, 9, 12, and 15, respectively. However, the total financial returns increased by 2–16 % compared to Policy 0. With Policy II, financial returns from fertile sites (stands 9, 12, 15) increased by 5 %, but

Table 3 Timing (Age, yrs), dominant height (H_{dom}, m) and intensity ($Thin_r$, proportion of removed trees, %) of the first thinning based on different energy wood policies for 15 Scots pine stands, interest rate 3 %

Stand	P0			PI			PII		
	Age	H_{dom}	$Thin_r$	Age	H_{dom}	$Thin_r$	Age	H_{dom}	$Thin_r$
1	54	12.5	27	53	12.3	7	46	10.8	6
2	49	11.5	31	46	11.0	24	43	10.5	29
3	43	10.0	42	41	9.7	50	41	9.7	47
4	49	12.3	27	49	12.3	2	43	11.2	7
5	46	11.6	28	46	11.6	25	33	9.7	24
6	42	10.4	45	40	10.0	48	36	9.2	51
7	44	13.5	27	40	12.6	9	30	9.7	4
8	38	11.8	16	38	11.8	15	25	8.0	25
9	31	9.5	47	32	9.8	55	25	7.7	51
10	39	13.7	27	36	12.7	6	34	11.9	3
11	39	13.4	23	26	8.4	16	25	8.1	19
12	32	10.5	34	31	10.2	44	25	7.8	49
13	34	13.8	21	22	8.4	1	24	9.2	3
14	33	13.0	19	22	8.2	27	23	8.5	29
15	22	7.8	39	22	7.8	39	22	7.8	40

Note: Grey area means the 1st thinning is a precommercial thinning for an energy wood harvest
P0 Policy 0, *PI* Policy I, *PII* Policy II

the variation on less fertile sites was large (financial returns increased by 2 and 16 % for stands 3 and 6, respectively).

According to our results, energy wood harvesting somewhat reduced the proportions of sawn wood grades for the most fertile site (stand 15, MT, T1300). However, for the less fertile sites (stands 3, 6, 9, 12), the changes were insignificant (Table 4). The highest reduction (from 63.55 to 51.83 $m^3 \cdot ha^{-1}$), 1.17 % in the proportion of sawn wood grades, was for the centre sawn wood of

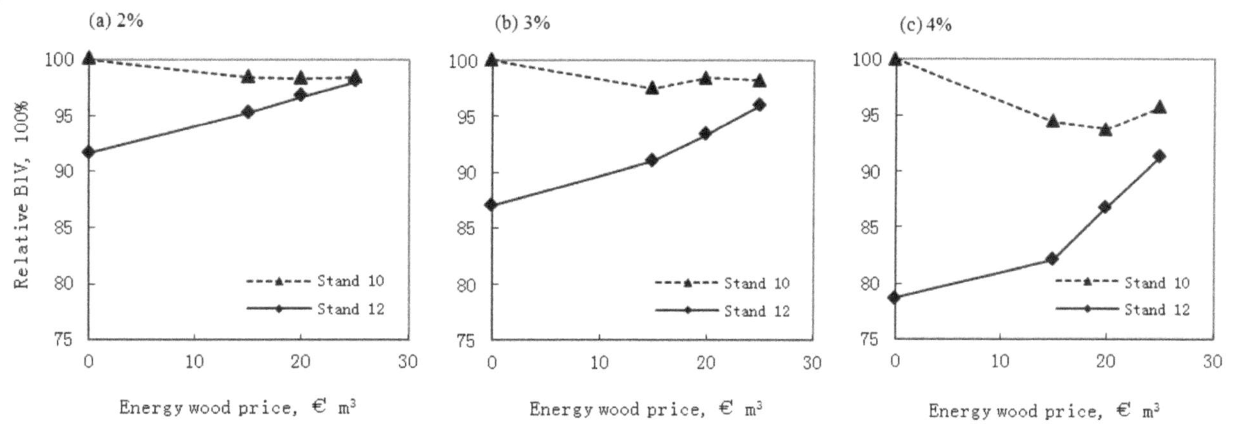

Fig. 2 Relative BLV (bare land value) based on P0 (energy wood price 0 €·m^{-3}) and PII (energy wood prices 15, 20, 25 €·m^{-3}) for stands 10 (1500 trees·ha^{-1}) and 12 (3000 trees·ha^{-1}), interest rates 2 (**a**) 3 (**b**) and 4 % (**c**)

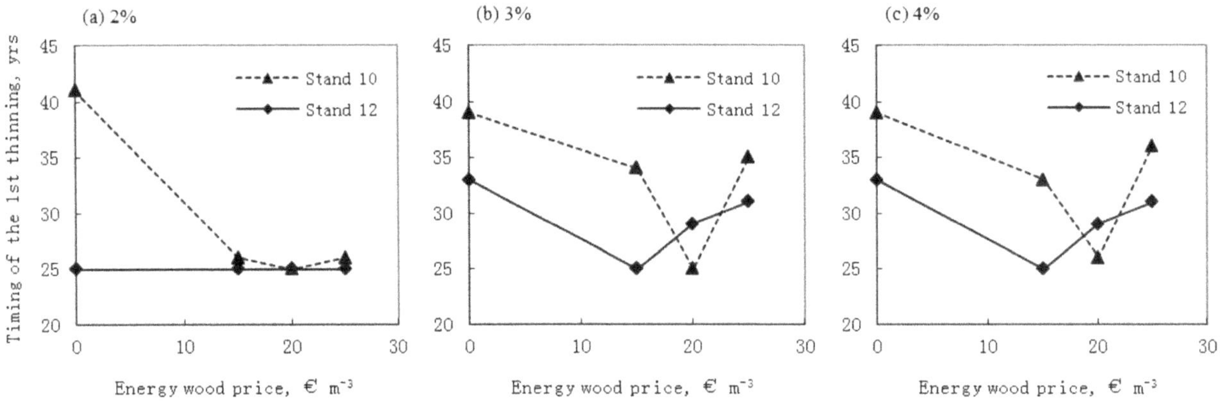

Fig. 3 Timing of the 1st thinning based on Policy 0 (energy wood price 0 €·m^{-3}) and Policy II (energy wood prices 15, 20, 25 €·m^{-3}) for stands 10 (1500 trees·ha^{-1}) and 12 (3000 trees·ha^{-1}), interest rates 2 (**a**) 3 (**b**) and 4 % (**c**)

grade A (the most valuable grade) for stand 15. The side sawn wood of grade A was slightly decreased by 0.06, 0.25, 0.26 and 0.45 % for stands 3, 9, 12, and 15, respectively (Table 4).

Precommercial thinning clearly increased branch diameter growth. Compared to Policy 0, Policy II resulted in thicker branches, a higher crown base, and a lower crown ratio at the second thinning (Fig. 4). Stands 9, 12, and 15 with earlier precommercial thinning (at age 25, 25, and 22, respectively) had more significant differences in branch thickness, crown base, and ratio than stands 3 and 6 with later precommercial thinning (at age 41 and 36, respectively). The most significant changes in branch thickness and crown base appeared in the most fertile site. The thickest dry branch increased by 7 % (0.13 cm), and the height of the crown base increased by 9 % (0.6 m) for stand 15 (H_{100} = 27.3 m).

Discussion

Pukkala (2011) developed a simulation-optimization system, in which carbon balances of harvested wood, end use of products, harvesting and processing releases, and substitution effects were incorporated. In his study energy wood was collected only in clearcut, precommercial thinning as energy harvesting was not taken into account. Our results agree that the utilization of forest bioenergy shortened the rotation length slightly as reported in Pukkala (2011), but revealed a more significant change (shortened 20 year) for stand 12 from the VT site in the present study. This was caused by the optimal thinning regimes in which two low thinnings were limited in Pukkala (2011), while the present study allows more frequent thinnings and more flexible thinning types that led to longer rotation.

The requirement of positive harvest profit for the first thinning is common in practical stand management. Ahtikoski et al. (2008) also made this requirement when evaluating the feasibility of energy wood harvests from young stands. However, our optimization results reveal that an energy wood harvest plays a more important role as an investment in early thinnings. The profitability of the first thinning depends on the unit costs of energy wood harvesting and conventional logging technology,

Table 4 Proportion of sawn wood grades (%) of total harvested volume over the rotation with Policy 0 (conventional harvest), and percentage changes (+/–%) of Policy II (compulsory energy harvest) compared to Policy 0 for five Scots pine stands (initial density 3000 trees·ha^{-1}), energy wood price 15 €·m^{-3}, interest rate 3 %

Sawn wood grades	Stand 3		Stand 6		Stand 9		Stand 12		Stand 15	
	P0	PII	P0	PII	P0	PII	P0	PII	P0	PII
A-side	1.20	−0.06	2.24	+0.14	1.89	−0.25	2.06	−0.26	1.54	−0.45
B-side	0.74	+0.01	1.30	+0.09	1.35	+0.37	1.89	−0.02	1.90	−0.22
C&D-side	1.43	+0.02	1.56	+0.05	1.65	+0.14	1.69	+0.18	1,96	+0.03
A-centre	7.50	+0.02	10.35	+0.51	9.98	−0.07	10.34	+0.10	10.12	−1.17
B-centre	0.88	+0.04	1.47	+0.13	1.66	+0.65	2.26	+0.22	2.27	−0.07
C&D-centre	0.01	+0.00	0.04	−0.01	0.03	−0.01	0.05	0.02	0.02	0.01

A-D denote sawn wood grades, e.g. A-side means grade A (side sawn wood), C&D-centre means grades C and D (centre sawn wood)
P0 Policy 0, *PII* Policy II

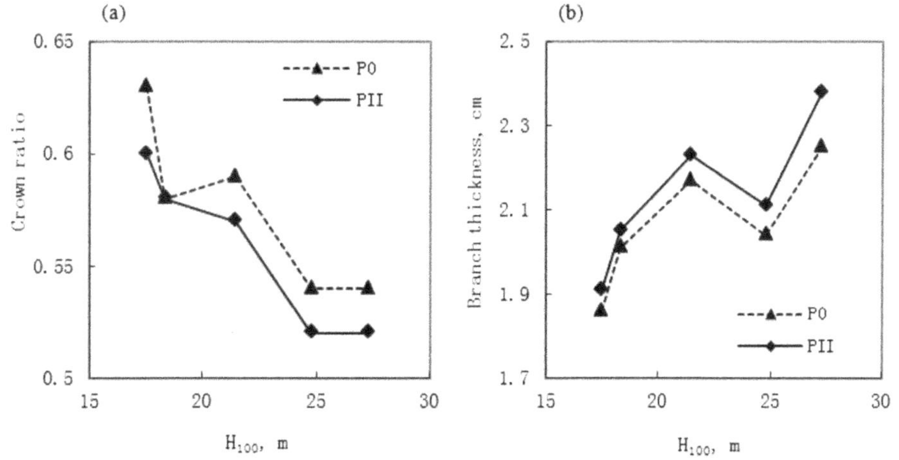

Fig. 4 Crown ratio of dominant trees (**a**) and branch thickness (**b**) at the second thinning with Policy 0 and Policy II for five Scots pine stands (initial density 3000 trees·ha^{-1}) by H_{100} (dominant height at 100 year, m), energy wood price 15 €·m^{-3}, interest rate 3 %

as well as the roadside prices of energy wood, pulpwood, and sawlog. As illustrated in Fig. 5, the unit costs of energy wood harvesting and conventional logging decrease with a lower number of stems and higher volume. However, the unit cost of energy wood harvesting is clearly less sensitive to variations in stand density and volume. In other words, energy wood logging is more competitive when more small size stems are removed.

The effects of precommercial thinning on growth and yield have been investigated based on thinning experiments for conifer cultures, such as Scots pine (Ruha and Varmola 1997; Varmola and Salminen 2004; Huuskonen and Hynynen 2006), balsam fir (Pothier 2002), and lodgepole pine (Johnstone 2005). In general, our results on the effects of early thinnings on growth and yield were in line with earlier empirical studies for Scots pine (Mäkinen and Isomäki 2004; Varmola and Salminen 2004; Huuskonen and Hynynen 2006). We contribute to this literature by extending the analysis over the whole rotation period and optimizing the timing and intensity of the thinnings with the option of an energy wood harvest for the first thinning.

Many forest managers expect greater sawlog volume increment and shorter rotations after a stand has been precommercially thinned (Pothier 2002). It has been reported that the strategy of precommercial thinning depends on the aims of stand management, such as even spacing, high yield, or both quality and yield (Vuokila 1972; Ruha and Varmola 1997; Varmola and Salminen 2004). Ruha and Varmola (1997) stated that the intensity of precommercial thinning depends on the aim of the first commercial thinning. Our results confirmed that optimal thinning regimes vary with different forest bioenergy policies. In contrast to Ruha and Varmola (1997), in the present study the whole rotation was taken into account, not only the first commercial thinning. The positive effects of precommercial thinning on diameter and volume growth at the first commercial thinning were reported for balsam fir by Pothier (2002), and for lodgepole pine by Johnstone (2005). However, they pointed

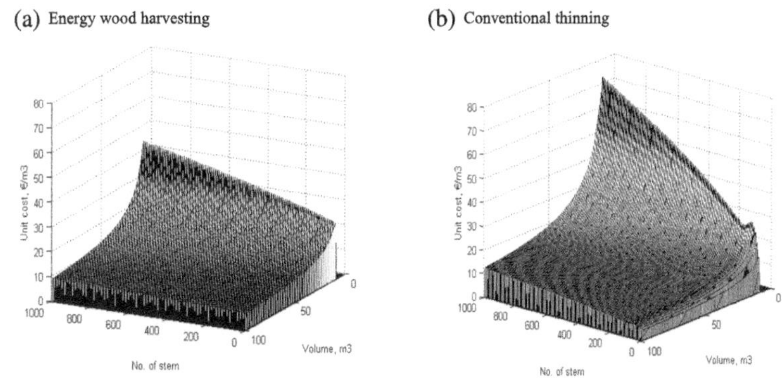

Fig. 5 Unit logging costs of energy wood harvesting (**a**) and conventional thinning (**b**)

out that there were no significant differences in merchantable volume at maturity (Pothier 2002) or, even after 20 years (Johnstone 2005). Meanwhile, Pothier (2002) found that the effect of precommercial thinning on rotation length depends on the management goals. This is supported by the results of the present study. With Policy II, the rotation length was somewhat shortened for dense stands.

Huuskonen and Hynynen (2006) reported that early and light precommercial thinning (H_{dom} 3 m, to a density of 3000 trees·ha^{-1}) significantly increased the first commercial thinning removal by 40 %. Postponing the first commercial thinning by 10 years, increased the present value of the stumpage revenues by 65 % at a 4 % interest rate (Huuskonen 2008). According to Varmola and Salminen (2004), precommercial thinning to 1600 and 2200 trees·ha^{-1} was moderate. Our results confirmed that early energy wood harvesting increased financial returns for dense stands (with an initial density of 3000 trees·ha^{-1}) by 2–16 % at a 3 % interest rate (Table 2). The intensity of energy wood harvesting was in the range of 40–51 % (Table 3), and the density after energy wood harvesting was close to the target density recommended by Varmola and Salminen (2004).

Precommercial thinning has positive effects on growth and yield. On the other hand, it might also result in negative effects on wood quality. Precommercial thinnings create more growing space and accelerate tree growth and, consequently, branch growth. This lowers timber grades for timber production from commercial thinnings and final felling. Economic analysis of thinning regimes and timber quality has been conducted for Scots pine (Hyytiäinen et al. 2004). Precommercial thinning was excluded in their studies. Positive effects of precommercial thinning on diameter growth and branch development have been reported for jack pine (Tong and Zhang 2005) and Scots pine (Varmola and Salminen 2004; Fahlvik et al. 2005). Their studies lacked financial information. Compared to earlier studies, our optimization results show that the differences in the effects of precommercial thinning on the end products' distribution were not significant. However, in this study, price differences between sawn wood grades were not considered and an average sawlog price was used.

There are several limitations of the data and models used in this study. In energy wood harvesting practice, young stands from which energy wood is harvested, stand densities clearly exceed 3000 trees·ha^{-1}. In addition, a significant proportion of removed stem number may comprise broadleaved tree species, especially on fertile sites, but the dynamics of these mixed stands is beyond the scope of the present study of even-aged stand management. The constraint of thinning rate formulated in our optimization problem allows extremely heavy thinning. Such heavy thinning led to a low stand density, and may increase the risks of wind and snow damages. On the other hand, natural regeneration is likely to occur when such low stand density is kept for a long time. Nevertheless, the ingrowth model was excluded in the PipeQual simulator. Future studies which link natural hazard models with the OptiFor system may prevent heavy thinning and improve the applicability of optimization results.

Energy wood production may lead to biomass losses. Interactions between energy wood production and carbon sequestration have been reported recently (Alam et al. 2010; Melin et al. 2010; Mangoyana 2011; Pukkala 2011). These effects are beyond the scope of the present study. Future numerical studies including forest bioenergy production and carbon benefits, and wood quality would be interesting topics as well.

Conclusions

According to our optimization results, it is optimal to carry out the first thinning as an early energy wood harvest (Policy II) for initially dense stands. For sparse stands, it is optimal to follow conventional forest management regimes (Policy 0). In addition, we found that a constraint of profitable energy wood harvest (Policy I) is not rational and may lead to a very early timing and extremely low intensity of the first thinning. This was caused by a thinning strategy that aimed at minimizing the negative outcome of a conventional first thinning.

Competing interests
The authors declare that they have no competing interests.

Authors' contributions
TC developed the simulation-optimization tool OptiFor Bioenergy, conducted the calculations and drafted the manuscript. All authors jointly revised the manuscript. All authors read and approved the final manuscript.

Acknowledgments
This study was partly supported by GSForest in Finland and National Natural Science Foundation of China (NSFC 31170586). We thank Juha Laitila for the opportunity to link their energy wood harvesting model to the OptiFor simulation-optimization system. We also thank Anssi Ahtikoski, Marc Hanewinkel, and Niels Strange for their valuable comments.

Author details
[1]College of Forestry, Northwest A&F University, 712100 Yangling, China. [2]Department of Economics and Management, University of Helsinki, P.O. Box 27, FIN-00014 Helsinki, Finland. [3]Department of Forest Sciences, P.O. Box 27, FIN-00014 Helsinki, Finland. [4]Forest Research Centre, Southern Cross University, P.O. Box 157, Lismore NSW 2480, Australia.

References
Ahtikoski A, Heikkilä J, Alenius V, Siren M (2008) Economic viability of utilizing biomass energy from young stands–the case of Finland. Biomass Bioenergy 32:988–996
Alam A, Kilpeläinen A, Kellomäki S (2010) Potential energy wood production with implications to timber recovery and carbon stocks under varying thinning and climate scenarios in Finland. Bioenerg Res 3:362–372

Amidon EL, Akin GS (1968) Dynamic programming to determine optimal levels of growing stock. For Sci 14:287291

Bolkesjø TF, Trømborg E, Solberg B (2006) Bioenergy from the forest sector: economic potential and interactions with timber and forest products markets in Norway. Scand J For Res 21:175–185

Cajander AK (1909) Über Waldtypen. Acta For Fenn 1:1–175

Cao T (2010) Silvicultural decisions based on simulation-optimization systems. Dissertationes Forestales 103. pp 48 + 53. ISBN 978-951-651-296-2, ISSN 1795-7389

Cao T, Hyytiäinen K, Tahvonen O, Valsta L (2006) Effects of initial stand states on optimal thinning regime and rotation of *Picea abies* stands. Scand J For Res 21:388–398

Cao T, Valsta L, Mäkelä A (2010) A comparison of carbon assessment methods for optimizing timber production and carbon sequestration in Scots pine stands. For Ecol Manage 260:1726–1734

Fahlvik N, Ekö PM, Pettersson N (2005) Influence of precommercial thinning grade on branch diameter and crown ratio in *Pinus sylvestris* in southern Sweden. Scand J For Res 20:243–251

Faustmann M (1849) Berechnung des wertes welchen Waldboden sowie noch nicht haubare Holzbestände für die Waldwirtschaft besitzen. Allgemeine Forst- und Jagd-Zeitung 15:441–455

Finnish Guidelines for Good Silviculture (2006) Metsätalouden kehittämiskeskus Tapio, Helsinki, pp 100 (in Finnish)

Haight RG, Monserud RA (1990) Optimizing any-aged management of mixed-species stands: II. Effects of decision criteria. For Sci 36:125–144

Hakkila P (2006) Factors driving the development of forest energy in Finland. Biomass Bioenergy 30:281–288

Hooke R, Jeeves TA (1961) "Direct search" solution of numerical and statistical problems. J Assoc Comput Mach 8:212–229

Huuskonen S (2008) The development of young Scots pine stands-precommercial and first commercial thinning. Dissertations Forestales pp 61:62

Huuskonen S, Hynynen J (2006) Timing and intensity of precommercial thinning and their effects on the first commercial thinning in Scots pine stands. Silva Fenn 40(4):645–662

Hynynen J, Ahtikoski A, Siitonen J, Sievänen R, Liski J (2005) Applying the MOTTI simulator to analyse the effects of alternative management schedules on timber and non-timber production. For Ecol Manage 207:5–18

Hyytiäinen K, Hari P, Kokkila T, Mäkelä A, Tahvonen O, Taipale J (2004) Connecting a process-based forest growth model to stand-level economic optimization. Can J For Res 34:20602073

Hyytiäinen K, Ilomäki S, Mäkelä A, Kinnunen K (2006) Economic analysis of stand establishment for Scots pine. Can J For Res 36:1179–1189

Johnstone WD (2005) The effects of juvenile spacing on 7-year-old lodgepole pine in central British Columbia. West J Appl For 20(3):160–166

Jylhä P, Laitila J (2007) Energy wood and pulpwood harvesting from young stands using a prototype whole-tree bundler. Silva Fenn 41(4):763–779

Kärkkäinen L, Matala J, Härkönen K, Kellomäki S, Nuutinen T (2008) Potential recovery of industrial wood and energy wood raw material in different cutting and climate scenarios for Finland. Biomass Bioenergy 32:934–943

Kotamaa E, Tokola T, Maltamo M, Packlén P, Kurttila M, Mäkinen A (2010) Integration of remote sensing-based bioenergy inventory data and optimal bucking for stand-level decision making. Eur J Forest Res 129:875–886

Kraxner F, Nordström E, Havlik P, Gusti M, Mosnier A, Frank S, Valin H, Fritz S, Fuss S, Kindermann G, McCallum I, Khabarov N, Böttcher H, See L, Aoki K, Schmid E, Máthé L, Obersteiner M (2013) Global bioenergy scenarios – Future forest development, land-use implications, and trade-offs. Biomass Bioenergy 57:86–96

Kuitto JP, Keskinen S, Lindroos J, Ojala T, Rajamäki J, Räsänen T, Teräväinen J (1994) Mechanized cutting and forest haulage. Metsäteho Report 410. Painovalmiste KY, Helsinki. (In Finnish)

Laitila J (2008) Harvesting technology and the cost of fuel chips from early thinnings. Silva Fenn 42(2):267–283

Laitila J, Asikainen A, Korhonen KT, Nuutienn Y (2004) Cost factors of the production of small-sized wood chips and supply logistics. (Pienpuuhakkeen tuotannon kustannustekijät ja toimituslogistiikka). Finnish Forest Research Institute, Working Papers, 3. pp 57

Lappi J (1992) A linear programming package for management planning. Research Papers 414. The Finnish Forest Research Institute, 134

Lyhykäinen H, Mäkinen H, Mäkelä A, Usenius A (2009) Predicting lumber grade and by-product yields for Scots pine. For Ecol Manage 258:146–158

Mäkelä A (1997) A carbon balance model of growth and self-pruning in trees based on structural relationships. For Sci 43(1):7–23

Mäkelä A (2002) Derivation of stem taper from the pipe theory in a carbon balance framework. Tree Physiol 22:891–905

Mäkelä A, Mäkinen H (2003) Generating 3D sawlog with a process-based growth model. For Ecol Manage 184:337–354

Mäkelä A, Sievänen R (1992) Height growth in open-grown trees. J Theor Biol 159:443–467

Mäkinen H, Isomäki A (2004) Thinning intensity and growth of Scots pine stands in Finland. For Ecol Manage 201:311–325

Malinen J, Pesonen M, Määttä T, Kajanus M (2001) Potential harvest for wood fuels (energy wood) from logging residues and first thinnings in Southern Finland. Biomass Bioenergy 20:189–196

Mangoyana RB (2011) Bioenergy from forest thinning: carbon emissions, energy balances and cost analyses. Renew Energy 36:2368–2373

McCormick K, Kåberger T (2007) Key barriers for bioenergy in Europe: economic conditions, know-how and institutional capacity, and supply chain co-ordination. Biomass Bioenergy 31:443–452

Melin Y, Petersson H, Egnell G (2010) Assessing carbon balance trade-offs between bioenergy and carbon sequestration of stumps at varying time scales and harvest intensities. For Ecol Manage 260:536–542

Ministry of Agriculture and Forestry (2006) Future review for the forest sector — outline of the forest council concerning focuses and aims for the forest sector. 11b/2006. ISBN 978-952-453-309-6. ISSN 1238-2531. pp 36

Ministry of Agriculture and Forestry (2008) Kansallinen mestäohjelma 2015. Lisää hyvinvointia monimuotoisista metsistä - Valtioneuvoston periaatepäätös. Maa- a metsätalousministeriön julkaisuja 3/2008 (Ministry of Agriculture and Forestry, Report 3/2008). Vammalan kirjapaino Oy. pp 45 (In Finnish)

Nurmi J (2007) Recovery of logging residues for energy from spruce (*Pices abies*) dominated stands. Biomass Bioenergy 31:375–380

Osyczka A (1984) Multicriterion optimization in engineering with FORTRAN programs. Ellis Horwood, Chichester, p 178

Peltola A, Ihalainen A (2013) Forest Resources. In: Ylitalo E (ed) Finnish Statistical Yearbook of Forestry. Finnish Forest Research Institute, Helsinki, p 440

Piene H (1978) Effects of increased spacing on carbon mineralization rates and temperature in a stand of young balsam fir. Can J For Res 8:398–406

Pothier D (2002) Twenty-year results of precommercial thinning in a balsam fir stand. For Ecol Manage 168:177–186

Pukkala T (2011) Optimizing forest management in Finland with carbon subsidies and taxes. For Policy Econ 13:425–434

Ruha T, Varmola M (1997) Precommercial thinning in naturally regenerated Scots pine stands in northern Finland. Silva Fenn 31(4):401–415

Shinozaki K, Yoda K, Hozumi K, Kira T (1964) A quantitative analysis of plant form: the pipe model theory. I. Basic analyses. Jpn J Ecol 14:97–105

Tong QJ, Zhang SY (2005) Impact of initial spacing and precommercial thinning on jack pine tree growth and stem quality. For Chron 81(3):418–428

UN-Energy (2007) Sustainable bioenergy: a framework for decision makers. United Nations. pp 61. http://www.fao.org/docrep/010/a1094e/a1094e00.htm (May 1, 2009)

UNFCCC (1997) The Kyoto protocol to the convention on climate change. United Nations Framework Convention on Climate Change. pp 24. http://unfccc.int/resource/docs/cop3/l07a01.pdf (June 12, 2015)

Valsta L (1990) A comparison of numerical methods for optimizing even aged stand management. Can J For Res 20:961–969

Varmola M, Salminen H (2004) Timing and intensity of precommercial thinning in *Pinus sylvestris* stands. Scand J For Res 19:142–151

Vuokila Y (1972) Treatment of seedling stands from the viewpoint of production. Folia For 141:36 (In Finnish with English summary)

Wetterlund E, Ledue S, Dotzauer E, Kindermann G (2013) Optimal use of forest residues in Europe under different policies—second generation biofuels versus combined heat and power. Biomass Conv Bioref 3:316

24

Sampling strategies for estimating forest cover from remote sensing-based two-stage inventories

Piermaria Corona[1*], Lorenzo Fattorini[2] and Maria Chiara Pagliarella[3]

Abstract

Background: Remote sensing-based inventories are essential in estimating forest cover in tropical and subtropical countries, where ground inventories cannot be performed periodically at a large scale owing to high costs and forest inaccessibility (e.g. REDD projects) and are mandatory for constructing historical records that can be used as forest cover baselines. Given the conditions of such inventories, the survey area is partitioned into a grid of imagery segments of pre-fixed size where the proportion of forest cover can be measured within segments using a combination of unsupervised (automated or semi-automated) classification of satellite imagery and manual (i.e. visual on-screen) enhancements. Because visual on-screen operations are time expensive procedures, manual classification can be performed only for a sample of imagery segments selected at a first stage, while forest cover within each selected segment is estimated at a second stage from a sample of pixels selected within the segment. Because forest cover data arising from unsupervised satellite imagery classification may be freely available (e.g. Landsat imagery) over the entire survey area (wall-to-wall data) and are likely to be good proxies of manually classified cover data (sample data), they can be adopted as suitable auxiliary information.

Methods: The question is how to choose the sample areas where manual classification is carried out. We have investigated the efficiency of one-per-stratum stratified sampling for selecting segments and pixels, where to carry out manual classification and to determine the efficiency of the difference estimator for exploiting auxiliary information at the estimation level. The performance of this strategy is compared with simple random sampling without replacement.

Results: Our results were obtained theoretically from three artificial populations constructed from the Landsat classification (forest/non forest) available at pixel level for a study area located in central Italy, assuming three levels of error rates of the unsupervised classification of satellite imagery. The exploitation of map data as auxiliary information in the difference estimator proves to be highly effective with respect to the Horvitz-Thompson estimator, in which no auxiliary information is exploited. The use of one-per-stratum stratified sampling provides relevant improvement with respect to the use of simple random sampling without replacement.

Conclusions: The use of one-per-stratum stratified sampling with many imagery segments selected at the first stage and few pixels within at the second stage - jointly with a difference estimator - proves to be a suitable strategy to estimate forest cover by remote sensing-based inventories.

Keywords: Spatially balanced sampling; Auxiliary information; Horvitz-Thompson estimator; Difference estimator; Variance estimator; Forest monitoring

* Correspondence: piermaria.corona@entecra.it
[1]Consiglio per la ricerca in agricoltura e l'analisi dell'economia agraria - Forestry Research Centre, Viale Santa Margherita, 80 - 52100 Arezzo, Italy
Full list of author information is available at the end of the article

Background

Deforestation and forest degradation account for nearly 20 % of global greenhouse gas emissions, more than the entire global transportation sector and second only to the energy sector. Reducing Emissions from Deforestation and Forest Degradation (REDD) is a United Nations (UN) effort to create financial value for the carbon stored in forests, offering incentives for developing countries to reduce emissions from forested lands. Monitoring systems that allow for credible measurement, reporting and verification of REDD efforts are among the most critical elements for the successful implementation of any REDD mechanism (UN-REDD 2013).

Among other issues, required data refer to updated and reliable estimates of the extent of forest cover: periodic forest cover assessments are crucial for providing benchmarks for monitoring the performance of various policies by revealing what the area of existing woodland is and whether it is increasing or declining (Corona et al. 2011; Marchetti et al. 2014). Since ground inventories cannot often be periodically performed under many large-scale conditions in tropical and subtropical countries, owing to high costs and/or lack of forest accessibility (e.g. remote and/or mountainous areas), inventories based on remote sensing imagery become compulsory. Such inventories are also mandatory for constructing historical records that can be used as a forest cover baseline.

Since quite recently, large-scale wall-to-wall remote-sensing systems are available for this purpose (see e.g. the CLASlite tool by Asner et al. 2009). However, the issue of a rigorous statistical assessment of the accuracy of the sampling strategy adopted to produce estimates is, in essence, still undervalued. As a corollary, in the REDD context, estimation must be assessed to have pre-fixed requisites of statistical accuracy (UN-REDD 2011). On this issue, it is worth noting that an objective estimation of accuracy is possible only in a design-based approach, where no assumptions are made about the population under study, in such a way that accuracy stems from the sampling strategy actually adopted to carry out estimates. Thus, accuracy is real, not assumed or modelled as in model-based approaches, where accuracy crucially depends on the model which is presumed to generate the population under study.

Large-scale remote sensing-based inventories of forest cover are usually carried out by a combination of unsupervised classification of satellite imagery and subsequent manual (visual on-screen) enhancements with the highest classification accuracy taken as ground truth (e.g. Hansen et al. 2013). Because visual on-screen operations are time expensive procedures, manual classification may, as a rule, be performed only for a sample of imagery segments at a first stage selection, while forest cover within each selected segment is estimated at a second stage from a sample of pixels selected within the segment. The forest cover data arising from unsupervised classification of the satellite imagery available over the whole survey area (wall-to-wall data) are likely to be good proxies of the manually classified cover data (sample data), so that they can be adopted as suitable auxiliary information (e.g. Sannier et al. 2014).

The question is how to choose the sample areas where manual classification is to be carried out. A wide variety of strategies is available to this end. The determination of the minimum -variance strategy to estimate population totals and averages is a challenging issue when the strategies are evaluated from a design-based point of view. Indeed, under design-based inference there is a lack of optimal results, in the sense that it is not possible to determine the minimum-variance strategy, as is customary in model-based approaches (e.g. Thompson 2002, Chapter 9). Accordingly, we claim no general validity about the results achieved; these should be ascribed to the conditions under which they have been obtained or extended, at most, to similar, related cases.

When sampling spatial units, the achievement of a so called *spatially balanced sample* (SBS), i.e., a sample in which units are well spread throughout the survey area, has been the main target for a long time. In most situations, nearby units are more similar than units far apart, thus giving a poor contribution to sample information. In these cases, the presence of spatial autocorrelation should be handled by avoiding the selection of neighboring units. From the results of a recent wide investigation (Fattorini et al. 2015), the so-called one-per stratum stratified sampling (OPSS) seems to be a suitable spatial scheme, accomplishing simplicity and efficiency, compared to other schemes investigated.

The purpose of our investigation was to compare the statistical efficiency of OPSS for selecting satellite imagery segments at a first sampling stage and pixels within the selected segments at a second sampling stage with respect to the benchmark use of simple random sampling without replacement (SRSWOR) at both stages. The comparison was performed on the basis of the actual relative root mean squared errors obtained from three artificial situations which mimic forest cover estimations.

We used Landsat imagery as a reference; since it is, by and large, the most commonly applied method to support multi-temporal delineation between forest and non-forest cover types by both manual (e.g. Townshend et al. 1995) and unsupervised automated or semi-automated (e.g. Achard et al. 2002) classification procedures and their combinations (e.g. McRoberts et al. 2014; Sannier et al. 2014).

Our presentation is organized as follows. We provide the statement of the problem. Then, we delineate the two-stage estimation of forest cover from a general point of view. Subsequently, the estimation strategies achieved using SRSWOR and OPSS, respectively, are detailed, together with problems related to variance estimation and a study is performed to compare SRSWOR and OPSS. In order to be realistic, a real map reporting the satellite classification (forest/non forest) is considered, from which the manual classification is artificially generated using assumed classification error rates. Concluding remarks are given.

It is worth noting that, even if not pursued by us, alternative criteria could be considered for selecting a sampling strategy, besides statistical efficiency. For example, cost issues could be addressed, such as when, in a multi-scene situation, image acquisition and processing costs are reduced by using multi-stage designs selecting a sample of scenes at the first stage.

Methods

Statement of the problem

Consider a survey area covered by a grid of N rectangles of pre-fixed size, referred to as *segments* or *primary sampling units* (PSUs) by Sannier et al. (2014). Denote by U the population of the N segments. In turn each segment is constituted by M satellite pixels referred to as *secondary sampling units* (SSUs). Denote by P_j the population of the M pixels within the j-th segment. As a whole, there is a target population of $N \times M$ pixels to be sampled by a two-stage scheme.

The pixel-level information consists of two dichotomous variables $x_{i,j}$ and $y_{i,j}$. The variable $x_{i,j}$ arises from forest/non-forest satellite classification and is equal to 1 if the i-th pixel of the j-th segment is classified as forest and 0 otherwise. Similarly, the variable $y_{i,j}$ arises from forest/non forest on-screen interpretation and has the value 1 if the i-th pixel of the j-th segment is interpreted as forest and 0 otherwise. While the $x_{i,j}$ variables arise from satellite spectral classes regrouped into forest and non-forest thematic classes and readily available from satellite maps for all the pixels in the population, the $y_{i,j}$ variables arise from time consuming efforts by forest experts, based on satellite imagery in combination with available very high resolution imagery and Google Earth, as well as map archives such as Bing Maps, Google Map, national maps and local maps (e.g. Sannier et al. 2014). For these reasons, the $x_{i,j}$ s are referred to as *map data* and will be used as auxiliary information in forest cover estimation, while the $y_{i,j}$s are higher quality data (e.g. more accurate data and/or with higher spatial resolution), referred to as *reference data*. These data will be taken as ground truth, given that field data cannot be collected due to the difficulty of access in dense forests. Owing to the high cost of reference data collection, these data cannot be known for all the pixels in the population but only for a sample.

For each segment $j \in \mathsf{U}$, denote x_j as the fraction of pixels of the j-th segment classified as forest from satellite information, i.e.,

$$x_j = \frac{1}{M}\sum_{i \in \mathsf{P}_j} x_{i,j}$$

Because all the $x_{i,j}$s are known, the x_js are known for each $j \in \mathsf{U}$. Accordingly, their population mean

$$\bar{X} = \frac{1}{N}\sum_{j \in \mathsf{U}} x_j$$

is also known and represents the fraction of the grid area classified as forest from the satellite map. Similarly, for each segment $j \in \mathsf{U}$, denote by y_j the fraction of pixels of the j-th segment interpreted as forest by experts, i.e.,

$$y_j = \frac{1}{M}\sum_{i \in \mathsf{P}_j} y_{i,j}$$

As stated earlier, the $y_{i,j}$s, and subsequently the y_js, cannot be completely known owing to their recording costs. Accordingly their population mean

$$\bar{Y} = \frac{1}{N}\sum_{j \in \mathsf{U}} y_j$$

is unknown. It represents the fraction of grid area interpreted as forest by experts and is taken as the ground truth. Usually $\bar{Y}$ is referred to as the *forest cover*. It constitutes the target parameter to be estimated on the basis of a sampling strategy.

On the basis of empirical investigations, Sannier et al (2014) provide evidence that satellite classifications come close to resemble expert interpretation. The close matching between map and reference data suggests the use of x_js as accurate and effective proxies for the y_js.

Two-stage estimation

At the first stage, consider a sampling scheme without replacement to select a sample of n segments $\mathsf{S} \subset \mathsf{U}$, where $n < N$. Denote by π_j and π_{jh} $(h > j \in \mathsf{U})$ the first- and second-order segment inclusion probabilities induced by the first-stage scheme (see e.g. Hedayat and Sinha 1991, section 1.3). Similarly, at the second stage, consider a sampling scheme without replacement to select a sample of m pixels $\mathsf{Q}_j \subset \mathsf{P}_j$, where $m < M$ for each segment $j \in \mathsf{S}$ selected at the first stage. Denote by $\tau_{i,j}$ and $\tau_{ih,j}$ $\left(h > i \in \mathsf{P}_j\right)$ the first- and second-order pixel inclusion probabilities induced at the second-stage. All these probabilities are known and previously established before sampling.

If the second stage were not performed, i.e., if all the pixels of the segments selected in the first stage were interpreted, the y_js would be recorded without error and

$$\hat{\hat{Y}}_1 = \sum_{j\in S} \frac{y_j}{\pi_j} \quad (1)$$

would constitute the (virtual) one-stage Horvitz-Thompson (HT) estimator of $\bar{Y}$. From the theory of HT estimation, (1) is design-unbiased with design-based variance

$$V_S\left(\hat{\bar{Y}}_1\right) = \frac{1}{N^2}\sum_{h>j\in U}\left(\pi_j\pi_h - \pi_{jh}\right)\left(\frac{y_j}{\pi_j} - \frac{y_h}{\pi_h}\right)^2 \quad (2)$$

which represents the variance due to the first stage, i.e. the uncertainty only due to the selection of segments.

Actually, at the second stage, a sample of pixels is selected within each selected segment. From the theory of two-stage estimation (Särndal et al. 1992, Chapter 4), the two-stage HT estimator of $\bar{Y}$ is given by

$$\hat{\bar{Y}}_2 = \frac{1}{N}\sum_{j\in S}\frac{\hat{y}_j}{\pi_j} \quad (3)$$

where

$$\hat{y}_j = \frac{1}{M}\sum_{i\in Q_j}\frac{y_{i,j}}{\tau_{i,j}},\ j\in S \quad (4)$$

is the HT estimator of y_j obtained from the sample of pixels Q_j selected at the second stage for each segment j selected at the first stage.

According to Särndal et al. (1992), the two-stage HT estimator (3) is design-unbiased with design-based variance

$$V\left(\hat{\bar{Y}}_2\right) = V_S\left(\hat{\bar{Y}}_1\right) + \frac{1}{N^2}\sum_{j\in U}\frac{V_Q\left(\hat{y}_j\right)}{\pi_j} \quad (5)$$

where the second term represents the increase in variance due to the second stage, i.e., the uncertainty due to the estimation of the y_js in the selected segments and

$$V_Q\left(\hat{y}_j\right) = \frac{1}{M^2}\sum_{h>i\in P_j}\left(\tau_{i,j}\tau_{h,j} - \tau_{ih,j}\right)\left(\frac{y_{i,j}}{\tau_{i,j}} - \frac{y_{h,j}}{\tau_{h,j}}\right)^2 \quad (6)$$

is the variance of the HT estimator (4) for any $j\in U$. Henceforth the subscript S will denote expectation and variance with respect to the first sampling stage, the subscript Q will denote expectation and variance with respect to the second sampling stage, conditional to the sample S selected in the first stage, while expectation and variance with respect to both stages are presented without a subscript.

Because the x_js are known for each segment, they may be used as auxiliary information exploited at the design or estimation level. Because nothing ensures that the x_js are invariably positive, they cannot be used at the design level for constructing a so-called probability-proportional-to-size scheme (Hedayat and Sinha 1991), in which the π_js are proportional to the x_js. Alternative ways to exploit the x_js at the design level are the adoption of an x-based stratification or balancing schemes (e.g. Deville and Tillé 2004) ensuring that the sample estimate of the auxiliary mean agrees with the known population mean $\bar{X}$.

If the x_js constitute good proxies for the y_js, an alternative estimation strategy is to adopt a first stage scheme where the π_js are not affected by the x_js and to use the difference (D) estimator, in order to exploit auxiliary information at the estimation level by predicting the y_js by means of the x_js (Särndal et al. 1992, Section 6.3). Once again, if all the pixels of the segments selected at the first stage were interpreted, the y_js would be recorded without error and

$$\tilde{\bar{Y}}_1 = \bar{X} + \frac{1}{N}\sum_{j\in S}\frac{e_j}{\pi_j} \quad (7)$$

would constitute the (virtual) one-stage D estimator of $\bar{Y}$, where $e_j = y_j - x_j$ denotes the error obtained from predicting y_j by means of x_j. From the theory of D estimation, (7) is design-unbiased with a design-based variance

$$V_S\left(\tilde{\bar{Y}}_1\right) = \frac{1}{N^2}\sum_{h>j\in U}\left(\pi_j\pi_h - \pi_{jh}\right)\left(\frac{e_j}{\pi_j} - \frac{e_h}{\pi_h}\right)^2 \quad (8)$$

which represents the variance due to the first stage, i.e., the uncertainty only due to the selection of segments. On the other hand, the two-stage D estimator turns out to be

$$\tilde{\bar{Y}}_2 = \bar{X} + \frac{1}{N}\sum_{j\in S}\frac{\hat{e}_j}{\pi_j} \quad (9)$$

where $\hat{e}_j = \hat{y}_j - x_j$ is the second-stage estimate of e_j. From Särndal et al. (1992), the two-stage estimator (9) is design-unbiased with a design-based variance

$$V\left(\tilde{\bar{Y}}_2\right) = V_S\left(\tilde{\bar{Y}}_1\right) + \frac{1}{N^2}\sum_{j\in U}\frac{V_Q\left(\hat{y}_j\right)}{\pi_j} \quad (10)$$

where the second term is the same as that of equation (5) and represents the increase in the variance due to the second stage.

Regarding the criteria for exploiting information from remote sensing at the estimation level in forest inventories, many recent papers make use of the generalized regression (GREG) estimator (e.g. Opsomer et al. 2007; Mandallaz et al. 2013; McRoberts et al. 2014). While the D estimator is used when an auxiliary variable (proxy) strictly resembles the survey variable, the GREG estimator is used when one or more auxiliary variables are strongly correlated with the survey variable. Given a close match between map and reference data, the use of the D estimator is suitable in remote sensing-based inventories. It should be noted that the D estimator is theoretically simpler than the GREG estimator, based on the fact that it is unbiased with an exact variance expression. On the other hand, the GREG estimator is not unbiased, but unbiased up to the first order of approximation, its variance is unknown and must be approximated up to the first order approximation (Särndal et al. 1993, Result 6.6.1).

Finally, for the variance estimation of both HT and D estimators, suitable solutions depend on the characteristics of the schemes adopted at the first and second stages. On this topic, it should be noted that at both stages we are dealing with the selection of spatial units (segments or pixels) from a regular grid of units, i.e., we are moving in the framework of spatial sampling. In order to reduce the selection of neighboring units (which tend to be more similar than units far apart), it is a common feature of spatial schemes that the second-order inclusion probabilities are zero or very close to zero for units that are close in distance. Moreover, explicit expressions for the second-order inclusion probabilities might be lacking for the most common spatial schemes. In these cases it is not possible to perform standard design-unbiased variance estimation and *ad-hoc* solutions should be pursued.

Two-stage simple random sampling

The simplest way to select spatial units from a grid is to adopt SRSWOR. Accordingly, the use of SRSWOR for selecting segments at the first stage as well as pixels within segments at the second stage is considered as a benchmark. In this case, the first-stage inclusion probabilities are $\pi_j = n/N$ and $\pi_{jh} = \{n(n-1)\}/\{N(N-1)\}$ for each $h > j \in \mathrm{U}$, and $\tau_{i,j} = m/M$ and $\tau_{ih,j} = \{m(m-1)\}/\{M(M-1)\}$ for each $h > i \in \mathrm{P}_j$. From these inclusion probabilities the two-stage HT estimator (3) reduces to (Cochran 1977, Chapter 10)

$$\hat{\hat{Y}}_2 = \frac{1}{n}\sum_{j\in \mathrm{S}} \hat{y}_j \tag{11}$$

where the second-stage HT estimator (4) reduces to

$$\hat{y}_j = \frac{1}{m}\sum_{i\in \mathrm{Q}_j} y_{i,j}$$

which represents the fraction of pixels interpreted as forest out of the m pixels selected within segment j. Moreover, equation (2) reduces to

$$\mathrm{V_S}\left(\hat{\hat{Y}}_1\right) = \frac{N-n}{N}\frac{S_y^2}{n} \tag{12}$$

where

$$S_y^2 = \frac{1}{N-1}\sum_{j\in \mathrm{U}}\left(y_j - \bar{Y}\right)^2$$

is the population variance of the y_js, while equation (6) reduces to

$$\mathrm{V_Q}\left(\hat{y}_j\right) = \frac{M-m}{M}\frac{S_j^2}{m}$$

where S_j^2 is the variance of the $y_{i,j}$s within segment j. Because the $y_{i,j}$s are 0-1 variables, S_j^2 can be rewritten as

$$S_j^2 = \frac{M}{M-1} y_j\left(1-y_j\right)$$

from which equation (6) ultimately reduces to

$$\mathrm{V_Q}\left(\hat{y}_j\right) = \frac{M-m}{M}\frac{S_j^2}{m} = \frac{M-m}{M-1}\frac{y_j\left(1-y_j\right)}{m} \tag{13}$$

Replacing equations (12) and (13) into equation (5), the variance of $\hat{\hat{Y}}_2$ reduces to

$$\mathrm{V}\left(\hat{\hat{Y}}_2\right) = \frac{N-n}{N}\frac{S_y^2}{n} + \frac{1}{Nn}\frac{M-m}{M-1}\frac{1}{m}\sum_{j\in \mathrm{U}} y_j\left(1-y_j\right) \tag{14}$$

(see also Cochran 1977, equation 10.8).

With SRSWOR a design-unbiased estimator of variance is given by

$$\hat{V}_2 = \frac{N-n}{N}\frac{s_{\hat{y}}^2}{n} + \frac{1}{Nn}\frac{M-m}{M}\frac{1}{m-1}\sum_{j\in \mathrm{S}} \hat{y}_j\left(1-\hat{y}_j\right) \tag{15}$$

where

$$s_{\hat{y}}^2 = \frac{1}{n-1}\sum_{j\in \mathrm{S}}\left(\hat{y}_j - \hat{\hat{Y}}_2\right)^2$$

is the sample variance of the $\hat{y}_j$s (see Cochran 1977, equation 10.15 and the subsequent proof).

If the x_js are used as proxies for the y_js, from equation (9) the D estimator with SRSWOR reduces to

$$\tilde{\hat{Y}}_2 = \bar{X} + \frac{1}{n}\sum_{j\in \mathrm{S}} \hat{e}_j \tag{16}$$

Since under SRSWOR equation (8) reduces to

$$V_S\left(\tilde{\bar{Y}}_1\right) = \frac{N-n}{N}\frac{S_e^2}{n}$$

where

$$S_e^2 = \frac{1}{N-1}\sum_{j\in U}\left(e_j-\bar{E}\right)^2$$

is the population variance of the e_js and $\bar{E} = \bar{Y}-\bar{X}$ their population mean, from equation (10) the variance of $\tilde{\bar{Y}}_2$ reduces to

$$V\left(\tilde{\bar{Y}}_2\right) = \frac{N-n}{N}\frac{S_e^2}{n} + \frac{1}{Nn}\frac{M-m}{M-1}\frac{1}{m}\sum_{j\in U} y_j\left(1-y_j\right) \quad (17)$$

Under SRSWOR, an unbiased estimator of variance is given by

$$\tilde{V}_2 = \frac{N-n}{N}\frac{s_{\hat{e}}^2}{n} + \frac{1}{Nn}\frac{M-m}{M}\frac{1}{m-1}\sum_{j\in S}\hat{y}_j\left(1-\hat{y}_j\right) \quad (18)$$

where

$$s_{\hat{e}}^2 = \frac{1}{n-1}\sum_{j\in S}\left(\hat{e}_j-\hat{\bar{e}}\right)^2$$

is the sample variance of the $\hat{e}_j$s and $\hat{\bar{e}} = \tilde{\bar{Y}}_2-\bar{X}$ their sample mean. The proof for the unbiasedness of (18) is similar, *mutatis mutandis*, to the proof adopted by Cochran (1977) to demonstrate the unbiasedness of (15).

Two-stage one-per-stratum stratified sampling

When sampling units from a grid, a wide variety of spatial sampling schemes is available besides SRSWOR. To obtain a SBS sample, in which units are well spread throughout the survey area, has been the main target for a long time. SBSs can be obtained using spatial versions of traditional sampling schemes such as stratified or systematic sampling (e.g. Thompson 2002, Chapters 11, 12) or by schemes explicitly constructed to avoid or reduce the selection of contiguous units such as the generalized random-tessellation stratified sampling method by Stevens and Olsen (2004), the drawn-by-drawn sampling scheme excluding the selection of contiguous units by Fattorini (2006), the local pivotal method of first type by Grafström et al. (2012), the spatially correlated Poisson sampling by Grafström (2012) and the doubly balanced spatial sampling by Grafström and Tillé (2013).

In this setting, the choice of effective strategies to perform forest cover estimation is a challenging issue. Our decision of adopting OPSS at both stages is based on a study recently carried out by Fattorini et al. (2015). In the presence of effective auxiliary information, as usually occurs in forest cover estimation, all these schemes are likely to provide good and similar performance of the D estimator. Accordingly, the use of the OPSS seems suitable. Its performance is similar to the more complex explicitly-constructed spatial schemes but, in contrast to these schemes, it straightforwardly provides SBS samples and can be well understood and readily planned even by non-statisticians. On the other hand, when using these spatial schemes, the sample selection is computationally intense and becomes practically impossible to apply in large populations of pixels, as those occurring in forest cover estimation. In this case, suboptimal implementations of the schemes are necessary (Grafström et al. 2014 and their references).

Under OPSS, the grid U of N segments is partitioned into n blocks of contiguous segments $U_1, \ldots, U_n$, each consisting of N/n segments, where one segment is randomly selected from each block. Then, at the second stage, the selected segment $j\in S$ is partitioned into m blocks of contiguous pixels $P_{j(1)}, \ldots, P_{j(m)}$, each consisting of M/m pixels, where again one pixel is randomly selected from each block. The scheme probably constitutes the first and simplest way to weaken the selection of contiguous polygons (Thompson 2002, Chapters 11, 12) and has a long standing in statistical literature (Breidt 1995). It should be pointed out that in forest cover estimation, Sannier et al. (2014) use OPSS at the first stage to select segments, while they use simple random sampling with replacement to select pixels at the second stage within the selected segments. We have considered the use of OPSS at both stages. In this case, the first order inclusion probabilities of segments and pixels are n/N and m/M, respectively, as is the case in SRSWOR, while those of the second order are 0 if two segments or pixels belong to the same block and are n^2/N^2 or m^2/M^2 otherwise.

From these inclusion probabilities the two-stage HT estimator coincides with the estimator (11) obtained with SRSWOR. On the other hand, with OPSS, equation (2) turns out to be

$$V_S\left(\hat{\bar{Y}}_1\right) = \frac{N-n}{Nn^2}\sum_{l=1}^{n} S_{y(l)}^2 \quad (19)$$

where

$$S_{y(l)}^2 = \frac{1}{N/n-1}\sum_{j\in U_l}\left(y_j-\bar{Y}_l\right)^2$$

is the variance of the y_js within the l-th block of segments U_l and $\bar{Y}_l$ their mean, while equation (6) turns into

$$V_Q\left(\hat{y}_j\right) = \frac{M-m}{Mm^2}\sum_{k=1}^{m} S_{j(k)}^2$$

where $S_{j(k)}^2$ is the variance of the $y_{i,j}$s within block k of segment j. Because the $y_{i,j}$s are 0-1 variables, $S_{j(k)}^2$ can be rewritten as

$$S_{j(k)}^2 = \frac{M/m}{M/m-1} y_{j(k)}\left(1-y_{j(k)}\right)$$

where

$$y_{j(k)} = \frac{1}{M/m}\sum_{i \in P_{j(k)}} y_{i,j}$$

is the fraction of pixels interpreted as forest within block k of segment j. Thus, equation (6) ultimately reduces to

$$V_Q\left(\hat{y}_j\right) = \frac{1}{m^2}\sum_{k=1}^{m} S_{j(k)}^2 \tag{20}$$

Replacing equations (19) and (20) in equation (5), the variance of $\hat{\hat{Y}}_2$ reduces to

$$V\left(\hat{\hat{Y}}_2\right) = \frac{N-n}{Nn^2}\sum_{l=1}^{n} S_{y(l)}^2 + \frac{1}{Nn}\frac{1}{m^2}\sum_{j \in U}\sum_{k=1}^{m} y_{j(k)}\left(1-y_{j(k)}\right) \tag{21}$$

If the x_js are used as proxies for the y_js, the two-stage D estimator coincides once again with the estimator (16) obtained with SRSWOR. On the other hand, under OPSS, equation (8) reduces to

$$V_S\left(\tilde{\tilde{Y}}_1\right) = \frac{N-n}{Nn^2}\sum_{l=1}^{n} S_{e(l)}^2 \tag{22}$$

where

$$S_{e(l)}^2 = \frac{1}{N/n-1}\sum_{j \in U_l}\left(e_j-\bar{E}_l\right)^2$$

is the variance of the e_js within the l-th block of segments U_l and $\bar{E}_l$ their mean. Thus, from equation (10) the variance of $\tilde{\tilde{Y}}_2$ reduces to

$$V\left(\tilde{\tilde{Y}}_2\right) = \frac{N-n}{Nn^2}\sum_{l=1}^{n} S_{e(l)}^2 + \frac{1}{Nn}\frac{1}{m^2}\sum_{j \in U}\sum_{k=1}^{m} y_{j(k)}\left(1-y_{j(k)}\right) \tag{23}$$

From equations (21) and (23) it is apparent that the variances of the two-stage estimators $\hat{\hat{Y}}_2$ and $\tilde{\tilde{Y}}_2$ depend on the variances of the y_js or e_js within the n blocks partitioning the grid of segments, as well as on the variances of the $y_{i,j}$s within the m blocks partitioning each segment. Because a single segment and a single pixel is selected within their corresponding blocks, it is not possible to estimate block variances from the sample information. Thus, there is no possibility to obtain design-unbiased estimators for these variances.

Conservative estimation for (21) and (23) can be attempted by using equations (15) and (18), respectively, as if SRSWOR were adopted at both stages. Because SRSWOR tends to provide greater variances than OPSS, with OPSS equations (15) and (18) tend to overestimate the actual variances (21) and (23). However, in one stage sampling, the bias induced by presuming SRSWOR when the actual scheme is OPSS, has been theoretically investigated by Mihályffy (2001); and nothing ensures that it is invariably positive. Thus, nothing ensures that (15) and (18) are invariably conservative. From tedious but conceptually simple algebra, the expectations of (15) and (18) with OPSS turn out to be

$$E\left(\hat{V}_2\right) = \frac{N-n}{N(n-1)}\left\{\frac{N-1}{N}S_y^2 - V_S\left(\hat{\hat{Y}}_1\right)\right\} + \frac{1}{N^2}\left\{\frac{N-n}{n} - \frac{M-m}{M(m-1)}\right\}\sum_{j \in U} V_Q\left(\hat{y}_j\right) + \frac{1}{N^2}\frac{M-m}{M(m-1)}\sum_{j \in U} y_j\left(1-y_j\right) \tag{24}$$

and

$$E\left(\tilde{V}_2\right) = \frac{N-n}{N(n-1)}\left\{\frac{N-1}{N}S_e^2 - V_S\left(\tilde{\tilde{Y}}_1\right)\right\} + \frac{1}{N^2}\left\{\frac{N-n}{n} - \frac{M-m}{M(m-1)}\right\}\sum_{j \in U} V_Q\left(\hat{y}_j\right) + \frac{1}{N^2}\frac{M-m}{M(m-1)}\sum_{j \in U} y_j\left(1-y_j\right) \tag{25}$$

where $V_S\left(\hat{\hat{Y}}_1\right)$, $V_Q(\hat{y}_j)$ and $V_S\left(\tilde{\tilde{Y}}_1\right)$ are provided by equations (19), (20) and (22), respectively.

Analytical tests for artificial populations

The performance of the four possible strategies obtained by combining SRSWOR and OPSS with the HT and D estimators, were tested for three artificial populations.

Populations

For generating the populations of y_js we started from a real survey area located in central Italy, consisting of a 20 km by 20 km square area. For this area, the Landsat classifications (forest/non forest) were available for each 1 ha pixel (Fig. 1). The area was partitioned into a grid of $N = 400$ segments of 100 ha. Each segment consisted of a square grid of $M = 100$ pixels. For each segment j, the fraction of forest cover x_j was readily calculated as the average of the $x_{i,j}$s available from the satellite map. The fraction for the whole area turned out to be $\bar{X} = 0.57$. The x_js were subsequently used to generate the forest cover values y_js for the three populations. We assumed that α was the probability that a pixel classified as forest from satellite information was interpreted as forest and β,

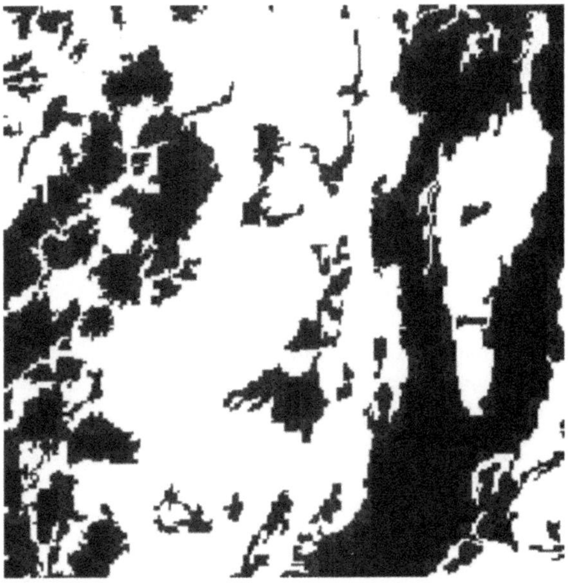

$\bar{X}=0.57$

Fig. 1 Satellite map consisting of 40,000 square pixels with a 100 m side length classified as forest (white) and non forest (black) for an area of 20 km by 20 km located in central Italy

the probability that a pixel classified as non-forest from satellite information, was interpreted as non-forest. Thus, the interpreted forest cover in the segment j was generated as

$$y_j = \frac{1}{M}\left\{\sum_{i \in \mathrm{P}_{j(F)}} u_{i,j} + \sum_{i \in \mathrm{P}_{j(NF)}} v_{i,j}\right\}$$

where $\mathrm{P}_{j(F)}$ and $\mathrm{P}_{j(NF)}$ were the sets of pixels of the j-th segment classified as forest and non-forest from satellite information and the $u_{i,j}$s and $v_{i,j}$s were independent Bernoulli random variables with parameter α and $1-\beta$, respectively. Clearly, for each j, the generated set of the M Bernoulli random variables gives rise to the reference values $y_{i,j}$s.

In order to obtain three different populations, the y_js were generated from the x_js presuming $\alpha = 0.80$ and $\beta = 0.70$ for population 1 (referred to as P1), $\alpha = 0.80$ and $\beta = 0.75$ for population 2 (P2) and $\alpha = 0.85$ and $\beta = 0.85$ for population 3 (P3). To avoid excessive, unrealistic fragmentation of the map, when a cluster of ten or fewer contiguous pixels of one class was completely surrounded by pixels of the other class, the cluster was assigned to the other class.

The resulting populations P1, P2, and P3 (Fig. 2a, b, c) had forest cover of 0.62, 0.59, and 0.57, respectively. For each population, the graph of y_js vs x_js gave rise to point scatters clumped around the line $x = y$ (Fig. 3a, b, c) with R^2 values of 0.80, 0.92 and 0.99.

Sampling

The closest matching between satellite and real classifications, also proved by Sannier et al (2014) via empirical investigations, rendered the x_js accurate and effective proxies for the y_js to be suitably used at the estimation level by the D estimator.

We considered SRSWOR and OPSS schemes. In first instance, these were used without exploiting auxiliary information by means of the HT estimator, followed by the exploitation of auxiliary information at the estimation level by means of the D estimator. We presumed the forest cover estimation to be as follows: for each of the four strategies, final samples of $n \times m$ = 100, 400 and 2000 pixels were assumed, corresponding to 0.25 %, 1 % and 5 % sampling fractions. Samples of 100 pixels were obtained by selecting n = 4, 10 or 25 segments at the first stage and m = 25, 10 or 4 pixels at the second stage, in such a way that the total number of selected pixels turned out to be 100. Similarly, samples of 200 pixels were obtained by opting for n = 16, 20 or 25 and m = 25, 20 or 16, respectively and samples of sizes 2000 were obtained by n = 40, 50 or 100 and m = 50, 40 or 20.

Performance indicators

Owing to the simplicity of the sampling schemes adopted, there was no need for simulation to determine the performance of the four sampling strategies. Each strategy gave rise to design-unbiased estimators, so that their accuracy could be determined exactly from their variance expressions, rather than approximated by Monte Carlo distributions, as is customary in more complex cases. More precisely, the variances of HT and D estimators obtained with SRSWOR were determined using equations (14) and (17), respectively, while those from OPSS were determined by equations (21) and (23). From these quantities, the values of relative standard errors (RSE) were determined as the ratio $\mathrm{SE}/\bar{Y}$ of the square root of the variance (SE) to the value under estimation, i.e., $\bar{Y}$.

For the estimation of variances, no investigations were necessary with SRSWOR, because in this case unbiased estimators were possible for the HT estimator from equation (15) and for the D estimator from equations (18). On the other hand, with OPSS the expectations of the estimators (15) and (18) were analytically determined, respectively, from equations (24) and (25). Finally the values of the relative bias (RB) were obtained as the ratio of the bias, i.e., the expectation minus the actual variance, to the actual variance. Because variances are squared quantities and as such are difficult to interpret, it is customary to estimate the relative standard error as the ratio of the square root of the estimated variance to the estimate of the population mean. As the ratio of two estimates, it is not possible to derive the expectation of the

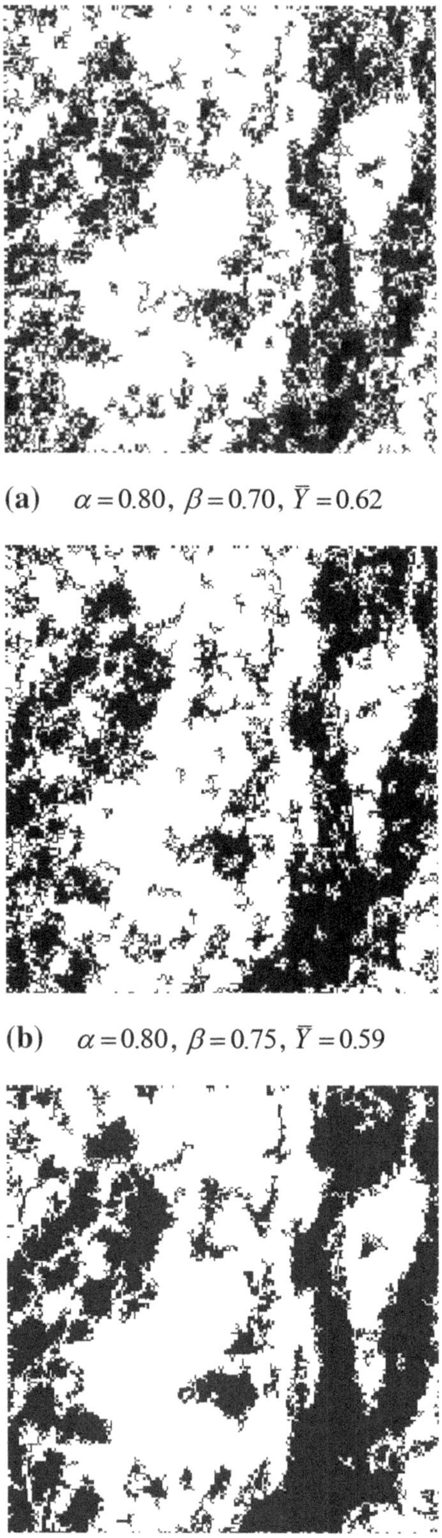

(a) $\alpha = 0.80,\ \beta = 0.70,\ \bar{Y} = 0.62$

(b) $\alpha = 0.80,\ \beta = 0.75,\ \bar{Y} = 0.59$

(c) $\alpha = 0.85,\ \beta = 0.85,\ \bar{Y} = 0.57$

Fig. 2 Reference maps artificially generated from the satellite map of Fig. 1, presuming three probability levels $\alpha = 0.80, 0.80, 0.85$ that a pixel classified as forest from satellite information was interpreted as forest and three probability levels $\beta = 0.70, 0.75, 0.85$ that a pixel classified as non-forest from satellite information was interpreted as non-forest

relative standard error estimator (ERSEE) analytically and therefore, we considered the first order approximation of this expected value. After some mathematical manipulation, the approximate expectation of the relative standard error estimator (AERSEE) turns out to be

$$\text{AERSEE} = \text{RSE}(1 + \text{RB})^{1/2}$$

Results and discussion

Table 1 presents the RSE values for each of the populations, strategies and combinations of n and m. It also shows, in parentheses, the AERSEE values from OPSS. The results of Table 1 motivate the following inferences.

Owing to the strong correlation between map and reference data, the exploitation of map data as auxiliary information in the D estimator proves to be highly effective with respect to the HT estimator, in which no auxiliary information is exploited. With SRSWOR the decrease in RSE varies from about 20 % to 75 % and are more marked for the P3 population when the y_js are maximally correlated with the x_js. With OPSS, the improvement involved by using the D estimator are even more pronounced. In this case the decrease in RSE varies from about 30 % to 80 %.

The use of OPSS provides considerable improvement with respect to the use of SRSWOR. When the HT estimator is used, i.e., when no auxiliary information is exploited, the improvements are similar in each of the three populations. Decreases in RSE vary from about 10 % to 35 % and are more marked with the greatest sampling effort, i.e., of 2000 pixels. When the D estimator is used, the improvements involved when OPSS is used are even more pronounced, with decreases in RSE varying from about 20 % to 55 % and were more marked for the P3 population, when the y_js were maximally correlated with the x_js.

Given the repartitioning of the sampling effort between the two sampling stages, for a specific number of selected pixels $n \times m$, the performance tends to increase for the three populations and four strategies when the number of selected segments n increases, with fewer pixels m selected within them.

When no auxiliary information is exploited (HT), the performance tends to improve as the forest cover increases. When auxiliary information is exploited, the performance tends to improve as the correlation between auxiliary and survey variable increases.

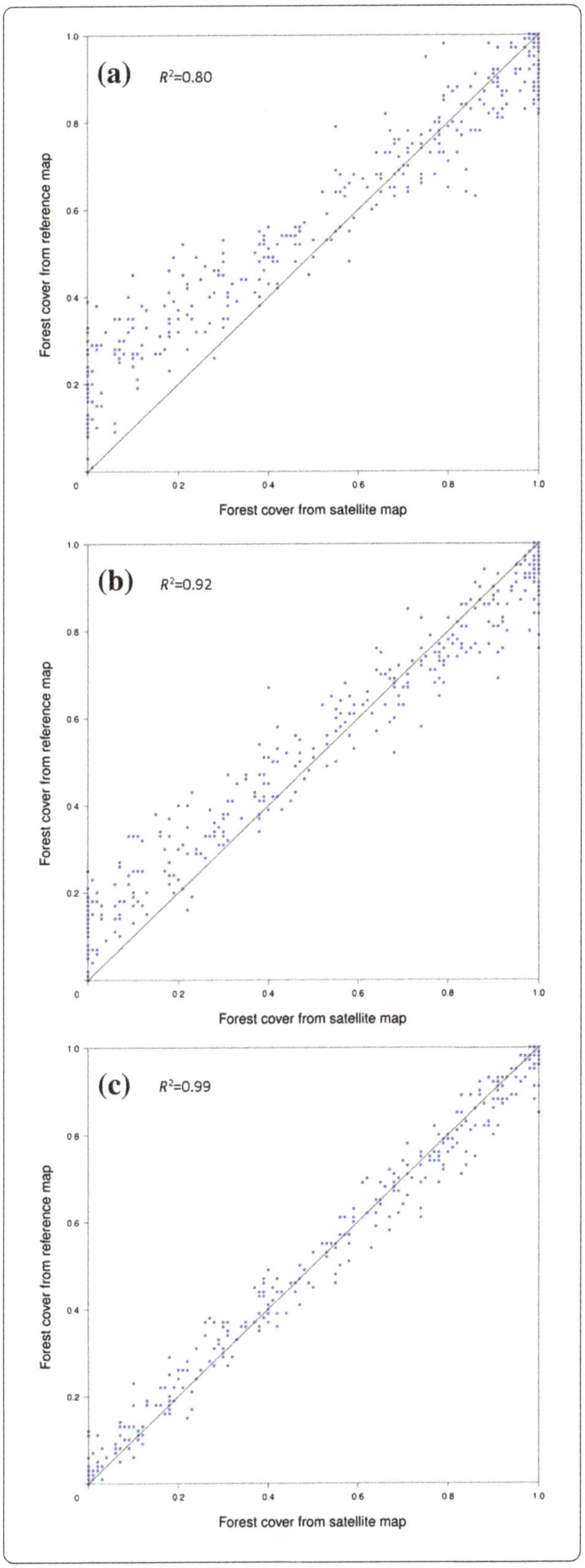

Fig. 3 Scatter plots of forest fractions from the map of Fig. 1 vs. forest fractions from the three reference maps of Fig. 2 for the 400 segments partitioning the squared area with a side length of 20 km located in central Italy

Given the estimation of RSE with OPSS, true values are overestimated up to a maximum of 3 percentage points. In relative terms, overestimation is moderate when accuracy is poor and tends to increase as accuracy increases. Estimates of the small RSEs tend to be about twice the size of the actual values.

Conclusions

Remote sensing from satellites is important in data collection on forest cover on a large scale and is conceived as an essential tool for REDD monitoring (UN-REDD 2013). Given this perspective, the remote sensing-based forest cover inventory procedure proposed by us proves to be suitable for any type of imagery (including that from supervised classifications or from active sensors, such as SAR interferometry or radargrammetry that can overcome eventual classification problems due to cloudiness).

From the results of our study, the use of OPSS with many image segments selected at the first stage and few pixels at the second stage - jointly with the D estimator - proves to be a suitable strategy to estimate forest cover by remote sensing-based inventories. OPSS straightforwardly provided SBSs with segments well dispersed within the survey area and pixels well spread within the selected segments. As well, the use of many segments selected at the first stage and few pixels at the second stage further increases the distribution of selected pixels onto the survey area, avoiding clumping of sample pixels within the selected segments. Finally, the D estimator exploits map data information as effective proxies of reference data. At least for the populations we investigated, the performance of this strategy was appealing. For large n and small m values, the RSEs are invariably less than 7 % even with the low sampling fraction of 0.25 % of pixels, decreasing to 1 %-1.5 % with a sampling fraction of 5 %. Because the performance of this strategy improves as the forest cover increases, better results can be expected with forest cover greater than 60 %, as frequently happens in tropical areas where coverage of about 80 % is customary (Sannier et al. 2014).

A less satisfactory issue concerns the estimation of RSE with OPSS. The proposed strategy tends to overestimate the actual values and is more marked when RSE is small. If a moderate overestimation is appealing, because it avoids the dangerous incidence of concluding that a strategy is accurate when it is not, some relevant overestimations obtained in our study are unsuitable, because they likely mask the accuracy gained by the use of OPSS jointly with the D estimator. As pointed out by Grafström (2012),

Table 1 Values (in %) of relative standard errors obtained from the four strategies adopted in estimating remote sensing-based forest cover for the three artificial populations P1, P2 and P3 and each combination of *n* and *m*. Values in brackets represent the approximate expectations of the standard error estimators adopted with OPSS

Population	*nxm*	*N*	*m*	SRSWOR + HT	OPSS + HT	SRSWOR + D	OPSS + D
P1	100	4	25	25.0	22.1 (23.2)	14.2	10.3 (10.3)
$\bar{Y} = 0.62$		10	10	18.2	14.4 (15.3)	12.9	7.8 (8.1)
$R^2 = 0.80$		25	4	14.2	9.2 (10.8)	11.7	6.6 (7.3)
	400	16	25	12.3	9.8 (11.5)	7.1	4.7 (5.3)
		20	20	11.1	8.9 (10.7)	6.5	4.8 (5.7)
		25	16	10.0	8.0 (9.9)	5.9	4.7 (5.7)
	2000	40	50	7.1	5.2 (7.0)	3.4	2.2 (3.2)
		50	40	6.3	4.4 (6.6)	3.2	2.4 (3.5)
		100	20	4.5	2.9 (5.0)	2.8	1.9 (3.5)
P2	100	4	25	28.7	25.9 (27.1)	13.4	8.7 (8.7)
$\bar{Y} = 0.59$		10	10	20.4	16.7 (17.7)	12.8	7.1 (7.4)
$R^2 = 0.92$		25	4	15.4	10.2 (12.2)	11.8	6.4 (7.0)
	400	16	25	14.2	11.4 (13.5)	11.8	4.2 (4.7)
		20	20	12.7	10.3 (12.4)	6.2	4.5 (5.3)
		25	16	11.4	9.1 (11.4)	5.7	4.5 (5.5)
	2000	40	50	8.3	5.8 (8.2)	3.0	1.9 (2.7)
		50	40	7.3	5.1 (7.6)	2.8	2.2 (3.3)
		100	20	5.2	3.4 (5.7)	2.7	1.8 (3.4)
P3	100	4	25	33.0	30.1 (31.6)	11.8	5.2 (5.1)
$\bar{Y} = 0.57$		10	10	22.9	19.1 (20.2)	12.2	5.3 (5.6)
$R^2 = 0.99$		25	4	16.7	11.0 (13.4)	11.6	5.3 (5.8)
	400	16	25	16.3	13.2 (15.6)	5.9	2.6 (3.0)
		20	20	14.6	11.9 (14.3)	5.6	3.5 (4.3)
		25	16	13.1	10.4 (13.0)	5.2	3.8 (4.8)
	2000	40	50	9.6	6.5 (9.5)	2.3	1.1 (1.6)
		50	40	8.5	5.8 (8.6)	2.3	1.6 (2.7)
		100	20	5.9	3.8 (6.3)	2.5	1.5 (3.1)

variance estimation is a bit tricky for spatial sampling in general and the construction of less biased variance estimators is a necessary step which calls for additional work in future investigations.

Competing interests
The authors declare that they have no competing interests.

Authors' contributions
The methodological development of the manuscript stems from the joint contribution of all the authors. Computations have been performed by MCP. All authors read and approved the final manuscript.

Author details
[1]Consiglio per la ricerca in agricoltura e l'analisi dell'economia agraria - Forestry Research Centre, Viale Santa Margherita, 80 - 52100 Arezzo, Italy. [2]University of Siena - Department of Economic and Statistics, Piazza San Francesco, 7 - 53110 Siena, Italy. [3]University of Molise - Department of Biosciences and Territory, Contrada Fonte Lappone snc, 86090 Pesche (Isernia), Italy.

References
Achard F, Eva HD, Stibig HJ, Mayaux P, Gallego J, Richards T, Malingreau JP (2002) Determination of deforestation rates of the world's humid tropical forests. Science 297:999–1002
Asner GP, Knapp DE, Balaji A, Paez-Acosta G (2009) Automated mapping of tropical deforestation and forest degradation: CLASlite. J Appl Remot Sens 3:1–24
Breidt FJ (1995) Markov chain designs for one-per-stratum sampling. Surv Methodol 21:63–70
Cochran WG (1977) Sampling Techniques. Wiley, New York
Corona P, Chirici G, McRoberts RE, Winter S, Barbati A (2011) Contribution of large-scale forest inventories to biodiversity assessment and monitoring. Forest Ecol Manag 262:2061–2069
Deville JC, Tillé Y (2004) Efficient balanced sampling: the cube method. Biometrika 91:893–912
Fattorini L (2006) Applying the Horvitz-Thompson criterion in complex designs: a computer-intensive perspective for estimating inclusion probabilities. Biometrika 93:269–278
Fattorini L, Corona P, Chirici G, Pagliarella MC (2015) Design-based strategies for sampling spatial units from regular grids with applications to forest surveys, land use and land cover estimation. Environmetrics 26:216–248
Grafström A (2012) Spatial correlated Poisson sampling. J Stat Plan Infer 142:139–147
Grafström A, Tillé Y (2013) Doubly balanced spatial sampling with spreading and restitution of auxiliary totals. Environmetrics 24:120–131
Grafström A, Lundström NLP, Schelin L (2012) Spatially Balanced Sampling through the Pivotal Method. Biometrics 68:514–520
Grafström A, Saarela S, Ene LT (2014) Efficient sampling strategies for forest inventories by spreading the sample in auxiliary space. Can J Forest Res 44:1156–1164
Hansen MC, Potapov PV, Moore R, Hancher M, Turubanova SA, Tyukavina A, Thau D, Stehman SV, Goetz SJ, Loveland TR, Kommareddy A, Egorov A, Chini L, Justice CO, Townshend JRG (2013) High-Resolution Global Maps of 21st-Century Forest Cover Change. Science 342:850–853
Hedayat AS, Sinha BK (1991) Design and Inference in Finite Population Sampling. Wiley, New York
Mandallaz D, Breschan J, Hill A (2013) New regression estimators in forest inventories with two-phase sampling and partially exhaustive information: a design-based Monte Carlo approach with applications to small-area estimation. Can J Forest Res 43:1023–1031
Marchetti M, Vizzarri M, Lasserre B, Sallustio L, Tavone A (2014) Natural capital and bioeconomy: challenges and opportunities for forestry. Ann Silvicult Res 38:62–73
McRoberts RE, Liknes GC, Domke GM (2014) Using a remote sensing-based, percent tree cover map to enhance forest inventory estimation. Forest Ecol Manag 331:12–18
Mihályffy L (2001) Variance estimation for stratified samples with one unit per stratum. Hungarian Statist Rev (Special Number) 6:123–133
Opsomer JD, Breidt FG, Moisen GG, Kauermann G (2007) Model-assisted estimation of forest resources with generalized additive models. J Am Statist Assoc 102:400–416
Sannier C, Mc Roberts RE, Fichet LV, Makaga EMK (2014) Using regression estimator with Landsat data to estimate proportion forest cover and net proportion deforestation in Gabon. Remote Sens Environ 151:138–148
Särndal CE, Swensson B, Wretman J (1992) Model Assisted Survey Sampling. Springer-Verlag, New York
Stevens DJ, Olsen AR (2004) Spatially Balanced Sampling of Natural Resources. J Am Statist Assoc 99:262–278
Thompson SK (2002) Sampling (2nd edn). Wiley, New York
Townshend JRG, Bell V, Desch A, Havlicek C, Justice WL, Lawrence D, Skole W, Chomentowski B, Moore III, Salas W, Tucker CJ (1995) The NASA Landsat Pathfinder Humid Tropical Deforestation Project. In: Land Satellite Information in the Next Decade, ASPRS Conference Proceedings, IV-76–IV-87. ASPRS, Virginia
UN-REDD (2011) Expert meeting on assessment of forest inventory approaches for REDD+ Meeting Report. FAO, Rome
UN-REDD (2013) National Forest Monitoring Systems: Monitoring and Measurement, Reporting and Verification (M & MRV) in the context of REDD+ Activities. FAO, Rome

Permissions

All chapters in this book were first published in FE, by Springer; hereby published with permission under the Creative Commons Attribution License or equivalent. Every chapter published in this book has been scrutinized by our experts. Their significance has been extensively debated. The topics covered herein carry significant findings which will fuel the growth of the discipline. They may even be implemented as practical applications or may be referred to as a beginning point for another development.

The contributors of this book come from diverse backgrounds, making this book a truly international effort. This book will bring forth new frontiers with its revolutionizing research information and detailed analysis of the nascent developments around the world.

We would like to thank all the contributing authors for lending their expertise to make the book truly unique. They have played a crucial role in the development of this book. Without their invaluable contributions this book wouldn't have been possible. They have made vital efforts to compile up to date information on the varied aspects of this subject to make this book a valuable addition to the collection of many professionals and students.

This book was conceptualized with the vision of imparting up-to-date information and advanced data in this field. To ensure the same, a matchless editorial board was set up. Every individual on the board went through rigorous rounds of assessment to prove their worth. After which they invested a large part of their time researching and compiling the most relevant data for our readers.

The editorial board has been involved in producing this book since its inception. They have spent rigorous hours researching and exploring the diverse topics which have resulted in the successful publishing of this book. They have passed on their knowledge of decades through this book. To expedite this challenging task, the publisher supported the team at every step. A small team of assistant editors was also appointed to further simplify the editing procedure and attain best results for the readers.

Apart from the editorial board, the designing team has also invested a significant amount of their time in understanding the subject and creating the most relevant covers. They scrutinized every image to scout for the most suitable representation of the subject and create an appropriate cover for the book.

The publishing team has been an ardent support to the editorial, designing and production team. Their endless efforts to recruit the best for this project, has resulted in the accomplishment of this book. They are a veteran in the field of academics and their pool of knowledge is as vast as their experience in printing. Their expertise and guidance has proved useful at every step. Their uncompromising quality standards have made this book an exceptional effort. Their encouragement from time to time has been an inspiration for everyone.

The publisher and the editorial board hope that this book will prove to be a valuable piece of knowledge for researchers, students, practitioners and scholars across the globe.

List of Contributors

Krishna P Poudel and Hailemariam Temesgen
Department of Forest Engineering, Resources, and Management, College of Forestry, Oregon State University, 280 Peavy Hall, Corvallis, OR 97331, USA

Andrew N Gray
USDA Forest Service, PNW Research Station, 3200 SW Jefferson Way, Corvallis, OR 97331, USA

Timo Pukkala
University of Eastern Finland, PO Box 11180101 Joensuu, Finland

Antonio Montagnoli, Sara Fusco, Mattia Terzaghi and Donato Chiatante
Department of Biotechnology and Life Science, University of Insubria, Via Dunant, 3, 21100 Varese, Italy

Alan Kirschbaum
National Park Service, Great Lakes Inventory and Monitoring Network, 2800 Lake Shore Drive East, Ashland, WI 54806, USA

Dirk Pflugmacher
Department of Forest Ecosystems and Society, Oregon State University, 321 Richardson Hall, Corvallis, OR 97331, USA

Warren B Cohen
USDA Forest Service, Pacific Northwest Research Station, Forestry Sciences Laboratory, 3200 SW Jefferson Way, Corvallis, OR 97331, USA

Gabriella S Scippa
Department of Biosciences and Territory, University of Molise, 86090 Pesche, Italy

Benoit Lafleur
Institut de recherche sur les forêts, Université du Québec en Abitibi-Témiscamingue, 445 boul. de l'Université, Rouyn-Noranda, QC J9X 5E4, Canada
Centre d'étude de la forêt, Université du Québec à Montréal, 141 Avenue du Président-Kennedy, Montréal, QC H2X 1Y4, Canada

Nicole J Fenton and Yves Bergeron
Institut de recherche sur les forêts, Université du Québec en Abitibi-Témiscamingue, 445 boul. de l'Université, Rouyn-Noranda, QC J9X 5E4, Canada

Heather H. Halbritter and Louis C. Bender
Extension Animal Sciences and Natural Resources, New Mexico State University, Las Cruces, NM, USA

WeiSheng Zeng
Academy of Forest Inventory and Planning, State Forestry Administration, Beijing 100714, China

Erkki Tomppo
Natural Resources Institute Finland, Helsinki, Finland

Sean P. Healey
Forest Service, United States Department of Agriculture, Ogden, UT, USA

Klaus V. Gadow
University of Göttingen, Göttingen, Germany
Beijing Forestry University, Beijing, China
Department of Forestry and Wood Technology, University of Stellenbosch, Stellenbosch, South Africa

Colin J Ferster, Nicholas C Coops and Baozhang Chen
Department of Forest Resources Management, University of British Columbia, 2424 Main Mall, Vancouver, BC V6T 1Z4, Canada

JA (Tony) Trofymow
Canadian Forest Service (Pacific Forestry Centre), Natural Resources Canada, 506 Burnside Road West, Victoria, BC V8Z 1M5, Canada
Department of Biology, University of Victoria, PO Box 1700, Station CSC, Victoria, BC V8W 2Y2, Canada

Thomas Andrew Black
Faculty of Land and Food Systems, University of British Columbia, 2357 Main Mall, Vancouver, BC V6T 1Z4, Canada

Matti Maltamo, Lauri Korhonen and Jari Kouki
University of Eastern Finland, School of Forest Science, P.O. Box 111, FI-80101 Joensuu, Finland

Annukka Pesonen
Blom Kartta Oy, Kauppakatu 15, 80100 Joensuu, Finland

Mikko Vehmas
City of Joensuu, 80100 Joensuu, Finland

Kalle Eerikäinen
Natural Resources Institute Finland, Joensuu Unit, P.O. Box 68, FI-80101 Joensuu, Finland

Shipra Shah
Department of Forestry, College of Agriculture, Fisheries and Forestry, Fiji National University, asinuP.O. Box 7222, Fiji Islands

DP Sharma
Department of Silviculture and Agroforestry, Dr. Y.S. Parmar University of Horiculture and Forestry, Solan, Himachal Pradesh, India

Mohammad Redowan
Department of Forestry and Environmental Science, Shahjalal University of Science and Technology, Sylhet 3114, Bangladesh

David E Calkin and Matthew P Thompson
Forestry Sciences Laboratory, US Forest Service Rocky Mountain Research Station, 800 East Beckwith, Missoula, MT 59801, USA

Mark A Finney
Fire Sciences Laboratory, US Forest Service Rocky Mountain Research Station, 5775 Highway 10 West, Missoula, MT 59808, USA

Jesse L. Morris
Department of Forest, Rangeland, and Fire Sciences, College of Natural Resources, University of Idaho, Moscow, ID 83844, USA

R. Justin DeRose
USDA Forest Service, Forest Inventory Analysis, Rocky Mountain Research Station, Ogden, UT 84401, USA

Andrea R. Brunelle
Department of Geography, University of Utah, Salt Lake City, UT 84112, USA

Timo Pukkala
University of Eastern Finland, PO Box 111, 80101 Joensuu, Finland

Erkki Lähde and Olavi Laiho
Joen Forest Program Consulting, Rauhankatu 41, 80100 Joensuu, Finland

Helena M. Henttonen
Natural Resources Institute Finland (Luke), Economics and Society unit, Jokiniemenkuja 1, 01370 Vantaa, Finland

Annika Kangas
Natural Resources Institute Finland (Luke), Economics and Society unit, Yliopistokatu 6, 80100 Joensuu, Finland

Hayato Iijima and Takuo Nagaike
Yamanashi Forest Research Institute, Saisyoji 2290-1, Fujikawa, Yamanashi 400-0502, Japan

Mike Mentis
Business & the environment, Postnet Suite 10102, Private Bag X7005, 3650 Hillcrest, South Africa

Arne Pommerening and Anders Muszta
Department of Forest Resource Management, Faculty of Forest Sciences, Swedish University of Agricultural Sciences SLU, Skogsmarksgränd, SE-901 83 Umeå, Sweden

Klaus J Puettmann
Department of Forest Ecosystems and Society, Oregon State University, Corvallis, OR 97330, USA

Scott McG Wilson
Consultant Forester and Forest Ecologist, 3Thorngrove Crescent, Aberdeen, Scotland AB15 7FH, UK

Susan C Baker
School of Biological Sciences, University of Tasmania, Private Bag 55, Hobart, TAS 7001, Australia

Pablo J Donoso
Department of Forests and Society, Faculty of Forest Sciences and Natural Resources, Universidad Austral de Chile, Isla Teja s/n, Valdivia, Chil

Lars Drössler
Southern Swedish Forest Research Centre, Swedish Agricultural University, PO Box 49, Alnarp SE-23053, Sweden

Girma Amente
Haramaya University, P. Box 138, Diredawa, Ethiopia

Brian D Harvey
Institut de recherche sur les forêts, Université du Québec en Abitibi-Témiscamingue, 445, Boulevard de l'Université, Rouyn-- Noranda, Québec J9X 5E4, Canada

Thomas Knoke
Institute of Forest Management, TUM School of Life Sciences Weihenstephan, Technische Universität München, Hans-Carl-von-Carlowitz-Platz 2, Freising 85354, Germany

Yuanchang Lu
Department of Forest Management and Statistics IFRIT, Chinese Academy of Forestry, Dongxiaofu 2, Xiangshan Road, Beijing, Haidian 100091, China

Susanna Nocentini
Department of Agricultural, Food and Forestry Systems, University of Florence, v. S. Bonaventura 13, Florence I-50145, Italy

Francis E Putz
Department of Biology, University of Florida, P.O. Box 118526, Gainesville, FL 32611-8526, USA

Toshiya Yoshida
Uryu Experimental Forest, Field Science Center for Northern Biosphere, Hokkaido University, Moshiri, Horokanai 074-0741, Japan

Jürgen Bauhus
Chair of Silviculture, Faculty of Environment and Natural Resources, University of Freiburg, Tennenbacherstr. 4, Freiburg 79085, Germany

Ahmed El-Guellab, Hugo Asselin and Yves Bergeron
NSERC/UQAT/UQÀM Industrial Chair in Sustainable Forest Management, Université du Québec en Abitibi-Témiscamingue, 445, boulevard de l'Université, Rouyn-Noranda, Québec J9X 5E4, Canada

Sylvie Gauthier
Natural Resources Canada, Canadian Forest Service - Laurentian Forestry Centre, 1055 du P.E.P.S., P.O. Box 3800, Sainte-Foy, Québec G1V 4C7, Canada

Adam A. Ali
Centre for Bio-Archeology and Ecology (UMR5059 CNRS), Université Montpellier 2, 163 rue Auguste Broussonet, Montpellier F-34090, France

Joseph Buongiorno
University of Wisconsin-Madison, Madison, USA

Mo Zhou
West Virginia University, Morgantown, USA

Steffen Herrmann
Institute of Forest Sciences, Faculty of Environment and Natural Resources, University of Freiburg, Tennenbacherstr. 4, D-79106 Freiburg, Germany
Swiss Federal Research Institute WSL, Zürcherstrasse 111, CH-8903 Birmensdorf, Switzerland

Tiemo Kahl and Jürgen Bauhus
Institute of Forest Sciences, Faculty of Environment and Natural Resources, University of Freiburg, Tennenbacherstr. 4, D-79106 Freiburg, Germany

Neda Amiri
Department of Forestry, Sari Agricultural and Natural Resource University, Sari, Iran

Seyed F. Emadian
Department of Forest Sciences, Gorgan Agricultural and Natural Resource University, Gorgan, Iran

Asghar Fallah
Department of Forestry, College of Natural Resources, Sari Agricultural Sciences and Natural Resources University, Sari, Iran

Kamran Adeli
Department of Forestry, Faculty of Agricultural, University of Lorestan, Lorestan, Iran

Hamid Amirnejad
Department of Agricultural Economics, Sari Agricultural and Natural Resource University, Sari, Iran

Tianjian Cao
College of Forestry, Northwest A&F University, 712100 Yangling, China

Kari Hyytiäinen
Department of Economics and Management, University of Helsinki, P.O. Box 27, FIN-00014 Helsinki, Finland

Henna Hurttala and Lauri Valsta
Department of Forest Sciences, P.O. Box 27, FIN-00014 Helsinki, Finland

Jerome K. Vanclay
Forest Research Centre, Southern Cross University, P.O. Box 157, Lismore NSW 2480, Australia

Piermaria Corona
Consiglio per la ricerca in agricoltura e l'analisi dell'economia agraria - Forestry Research Centre, Viale Santa Margherita, 80 - 52100 Arezzo, Italy

Lorenzo Fattorini
University of Siena - Department of Economic and Statistics, Piazza San Francesco, 7 - 53110 Siena, Italy

Maria Chiara Pagliarella
University of Molise - Department of Biosciences and Territory, Contrada Fonte Lappone snc, 86090 Pesche (Isernia), Italy